MIMO Processing for 4G and Beyond

Fundamentals and Evolution

Edited by
Mário Marques da Silva
Francisco A. Monteiro

CRC Press
Taylor & Francis Group
Boca Raton   London   New York

CRC Press is an imprint of the
Taylor & Francis Group, an **informa** business

CRC Press
Taylor & Francis Group
6000 Broken Sound Parkway NW, Suite 300
Boca Raton, FL 33487-2742

First issued in paperback 2016

© 2014 by Taylor & Francis Group, LLC
CRC Press is an imprint of Taylor & Francis Group, an Informa business

No claim to original U.S. Government works

Version Date: 20140411

ISBN 13: 978-1-138-03397-9 (pbk)
ISBN 13: 978-1-4665-9807-2 (hbk)

Library of Congress Cataloging-in-Publication Data

MIMO processing for 4G and beyond : fundamentals and evolution / editors, Mário Marques da Silva, Francisco A. Monteiro.
    pages cm
    Summary: "Multiple-input multiple-output was from the beginning a disruptive idea which opened horizons to up-until-then unimaginable data rates and system capacities in wireless systems. This book offers a fresh look at MIMO signal processing, namely its detection (including in the frequency domain) and precoding. It also looks at its combination with OFDM, UWB and CDMA and the impact at the system-level. MIMO remains a pillar of high-speed systems beyond 4G which incorporate massive MIMO and network coding at the physical layer and these topics are also addressed. The book brings together some highly cited authors from first-class institutions. "-- Provided by publisher.
    Includes bibliographical references and index.
    ISBN 978-1-4665-9807-2 (hardback)
    1. MIMO systems. I. Silva, Mário Marques da, editor of compilation. II. Monteiro, Francisco A., editor of compilation.

TK5103.4836.M56 2014
621.3845'6--dc23   2014002246

Visit the Taylor & Francis Web site at
http://www.taylorandfrancis.com

and the CRC Press Web site at
http://www.crcpress.com

Contents

Foreword

In the transmission of information over a wireless channel, the channel is modeled classically as a linear system black box with an input and output, that is, a single input and a single output (SISO). The input is the connection point from the power amplifier of the transmitter to the transmitting antenna terminal and the output is the connection point from the receiving antenna terminal to the radio frequency (RF) front-end filter of the receiver. The antennas are modeled as a structure that radiates EM waves that propagate through space. The simplest such antenna structure is a radiating electric dipole element. With the presence of multipath propagation in the channel, it becomes evident that the electric field at the receiver location undergoes variations in amplitude over distances in space of the order of a wavelength. As a result, variations of the classical wireless channel were employed, where multiple receiving antenna elements were introduced, or in other words, antenna structures with multiple interconnection points to the receiver. These antennas were designed to achieve so-called receiver diversity. The channel could then be modeled as having a single input and multiple outputs, or in the current terminology SIMO (single input multiple output). Classical receiver techniques to process the multiple outputs were referred to as combining techniques. The three classical combining techniques are selection, equal gain, and maximal ratio combining.

Alternative processing approaches of the signals at the multiple outputs taking the phase of the received signal into account (or taking the timing of the received signal into account) result in the receiver antenna structure behaving as a directional antenna, where the antenna gain is dependent on the direction of arrival—the technique is referred to as beamforming, or antenna array, or phased array, and works well when the timing of the received signals at the antenna elements is on the order of a carrier period; that is, elements are spaced on the order of a wavelength. Such a technique can also be employed with multiple antenna elements at the transmitter and a single element at the receiver. This results in the system being modeled as a SISO system with directivity. In either the case of transmitter or receiver beamforming, the system may be modeled as a SISO system with directivity.

Over the past 30 years of wireless research, another variation was introduced, where multiple transmitting antenna elements are introduced, and where an information signal is fed into the different elements with various delays. The relative delays between the signals are much larger than the carrier period and on the order of the transmission symbol period. This technique became known as delay transmission diversity and in effect simulated a channel with multipath propagation. Such a channel exhibits what is referred to as frequency diversity in that the frequency response has a variation over the frequency band of the transmitted signal. The variation with frequency is measured by the coherence bandwidth parameter and is inversely related to the relative delay of the signals being fed into the transmitting antenna elements. A SISO channel with the channel multipath having a small delay spread exhibits what we call frequency flat fading, with its effect being uncorrectable with any signal processing techniques at the receiver. The introduction of relative delays at the different transmitting antenna elements basically transforms the channel into a frequency-selective fading channel, where appropriate signal processing techniques can then be introduced at the receiver to correct the effect of fading. Such a receiver utilizes an equalizer and the resulting signal is insensitive to the precise location of the receiving antenna over several wavelengths. One form of this scheme, for wideband signals, is a spread spectrum system (e.g., CDMA*) with a Rake receiver.

The next development in antenna schemes came with the introduction of a multiplicity of antenna elements at the transmitter but with different information signals being fed into the different antenna elements. This work was first reported by DaSilva and Sousa in the mid-1980s as a way to solve the flat fading problem with narrowband systems, such as those with 30 kHz RF channels (1G in North America). The flat fading problem of narrowband cellular systems is a result of narrow RF bandwidth together with a small portable receiver terminal which is not conducive to receiver antenna diversity. The idea was to find a transmitter diversity technique that requires a relatively simple receiver. To motivate the idea we imagine such a SISO system with a multipath where, at the receiver, there are a series of nulls and peaks in the signal strength over distances of the order of half of a wavelength. If the single antenna at the receiver is located at a null, then we have a very weak signal. To attempt to mitigate the flat fading, conceptually we imagine the transmitting antenna being placed in a motor and move it so that the nulls at the receiver also move and we get diversity. However, we wish to implement such a system without moving parts, hence we can propose what is called switching diversity, where the signal at the transmitter is periodically switched between two antenna elements. Such switching may have undesirable effects in broadening the spectrum. We therefore considered that switching between two antennas is a form of transmitting two orthogonal signals in the two antenna elements and, in this way, we were led to a scheme where a signaling constellation of two dimensions is created where each dimension is associated with a transmitting antenna element. In the basic scheme we have the constellation points: (−1, −1), (−1, 1), (1, −1), and

---

* Code division multiple access.

(1,1). We then realized that if we introduce a rotation of this constellation in the plane we obtain in effect a modulation scheme that, in a sense, has fading resistant properties.

A variation of the above technique of transmitter diversity was introduced afterward by Alamouti where, instead of associating the signals transmitted on the different antennas with a 4-point constellation, a scheme is devised where two symbols s_1 and s_2 are transmitted over two successive time slots as follows: At time slot 1 we send s_1 on antenna 1 and s_2 on antenna 2, and in time slot 2 we send $-s_2^*$ on antenna 1 and s_1^* on antenna 2. This scheme was later coined as space–time coding, although some coding theorists would argue that this is a misnomer as there is no coding theory involved in the classical sense of coding theory.

The next major development, by Foschini and others, occurred with consideration of enhanced receiver antenna structures where it is possible to install a multiplicity of antenna elements and achieve some degree of decorrelation between the signals in the different receiver antenna elements. We consider a system with m receiving antenna elements and n transmitting antenna elements; that is, a system with multiple inputs and multiple outputs (MIMO). The channel can then be modeled by a channel gain matrix $\mathbf{H} = (h_{i,j})$, where $h_{i,j}$ is the path gain between the ith transmitting antenna element and the jth receiving antenna element. We model the inputs to the different antenna elements by the vector $\mathbf{x}$, where the component x_i is fed into the ith antenna element and where the vector $\mathbf{y}$ models the outputs at the receiver antenna. Thus, we have $\mathbf{y} = \mathbf{Hx}$. We observe that if the channel has no multipath components, then we have $h_{i,j} = 1$ for all i,j, and use of the multiple antennas has no benefit in the sense that we cannot separate the signals from the different transmitter elements to, in effect, recover all the data. In fact, the signals at each receiver antenna element are all equal to the sum of the inputs to the different transmitter elements. However, in the case that $m = n$, and if the matrix $\mathbf{H}$ is invertible, we can easily compute $\mathbf{x} = \mathbf{H}^{-1}\mathbf{y}$. Hence, we can recover x from y; that is, we can decouple the channel and make it behave as if there are n parallel noninteracting paths. The next observation is that the behavior of this scheme is very much dependent on the eigenvalues of the matrix $\mathbf{H}$, because in a sense they determine the signal strength in each of the decoupled paths and eventually the performance in the presence of noise in each path. The scheme works well if the environment contains many multipath components.

The analysis of this scheme is usually presented by diagonalizing the matrix $\mathbf{H}$ as follows: $\mathbf{H} = \mathbf{M\Lambda M}^*$, where Λ is the diagonal matrix of eigenvalues of $\mathbf{H}$ and $\mathbf{M}$ is the matrix whose column vectors are the associated eigenvectors. In this form we can view the effect of $\mathbf{H}$ on the channel input signal as three transformations. First, the multiplication by $\mathbf{M}^*$ is a form of pre-coding of the channel input vector $\mathbf{x}$, or changing of coordinate system to make it simple to apply the channel transformation. Then, the multiplication by Λ, in effect, models the different independent paths of the channel and, finally, the effect of multiplication by $\mathbf{M}$ converts back into a coordinate system that corresponds to the system under which the real signal observations are made. In the case that the matrix $\mathbf{H}$ is not a square, then a similar decomposition can be achieved using the well-known singular-value decomposition (SVD): $\mathbf{H} = \mathbf{U\Lambda V^*}$, where $\mathbf{U}$ is an $m \times m$ real or complex unitary matrix; $\mathbf{\Lambda}$ is an $m \times n$ rectangular diagonal matrix; and $\mathbf{V^*}$ is the conjugate transpose of $\mathbf{V}$, which is an $n \times n$ real or complex unitary matrix. We may view the matrix $\mathbf{H}$ as providing the transformation between the input signal $\mathbf{x}$ and the output signal $\mathbf{y}$, hence we have $\mathbf{y} = \mathbf{Hx} = \mathbf{U\Lambda V^* x}$.

Considering the input to the channel as a vector of information symbols, the "signals" in this linear system are finite dimensional vectors. These transformations become very familiar if we draw a comparison with the area of linear time invariant (LTI) systems with which we are familiar in an introduction to linear systems. In those systems, the familiar concepts are time domain, frequency domain, convolution, and transfer function. The direct method of computation of the output is based on a convolution and is done in one step. The indirect method based on the frequency domain uses three steps. In a sense performing the multiplication $\mathbf{y} = \mathbf{Hx}$ is analogous to convolution. Now, using the SVD decomposition of $\mathbf{H}$, we first compute $\mathbf{V^* x}$ (analogous to converting the input into the "frequency domain"). Then, multiplication by the diagonal matrix $\mathbf{\Lambda}$ (simple) is like multiplying by the transfer function. Finally, multiplication by the matrix $\mathbf{U}$ corresponds to converting back into the analogue of the time domain. In the design of linear systems, we prefer the frequency domain because it simplifies our understanding of the system and, in a sense, the above SVD form is preferable to understanding the behavior of the multipath channel.

Now, MIMO is a technology that has been widely adopted in state-of-the-art cellular systems such as LTE and systems such as WiFi and WiMax, where configurations up to 4×4 (4 transmitting elements and 4 receiving elements) have been adopted. For instance, bit-rate capacities per link are often quoted in bps/Hz where the ITU* specifies the target of 15 bps/Hz in the downlink for the LTE[†] advanced system. Nevertheless, to achieve these targets, a radio environment with rich multipath propagation is required. However, when a network assessment is made, the benefits of MIMO are reduced. To achieve a high degree of multipath, we need to consider multipath over a wide range of angles of arrival, but in a networking context we also have the option of beamforming at the transmitter in order to reduce interference. Using a narrow beam at the transmitter greatly reduces the interference on neighboring terminals. However, this beamforming then reduces the multipath that is required by a MIMO link and the effectiveness of MIMO then decreases. Ultimately, a MIMO system utilizes propagation modes (or paths), which have a wide range of angles of departure from the transmitter and angles of arrival at the receiver. However, these propagation modes, in a sense, cause the

---

* International telecommunications union.
[†] Long-term evolution.

occupation of more space by extra signal components, which degrade the network where a number of links are designed to operate simultaneously.

The standard MIMO technique assumes a number of antenna elements that are relatively close together. The goal is to achieve uncorrelated signals at the different antenna elements, where the correlation decreases with increased distance between elements, but at the same time have all the elements as close to one another so that the overall antenna occupies less space. This is an important issue, especially for portable terminals and the result is that the form factor of the terminal puts a limit on the number of elements that we may have at the terminal. Nevertheless, there is another approach where we may place the antenna elements at significant distances from each other, especially on the infrastructure side of the link where each element is connected to a controller using various approaches such as optical links—often referred to as front-haul. This approach is referred to as virtual MIMO and it is one of the approaches currently being considered in cellular systems employing the so-called remote RF in highly dense cell environments. In other words, a base station feeds a number of remote RF units where the signals from/to these units are jointly processed at the base station.

As we look into the future, there are a number of possibilities to generalize MIMO techniques. Currently, most of the systems consider that each input and output corresponds to an antenna radiating element as in an antenna array. The set of elements may be replaced by a generalized radiating antenna structure that has a number of ports to input and output signals. Such a structure may be quite different from a set of discrete elements as in the current configurations. One could also imagine applications and scenarios where the wireless channels and the propagation modes or paths are under the control of the system designer.

We have discussed MIMO from the standpoint of the transfer of signals from a transmitter to a receiver, where the essence of the scheme has been the increase in the dimensionality of the signal space available for the transmission of information signals. There is also the important aspect of mapping or coding of generalized information signals to the input ports of a MIMO link, and the associated processing of the signals at the output ports in order to extract the information at the receiver. A vast amount of research has been performed in the areas of channel modeling, channel sounding, detection, modulation, coded modulation, and adaptation to the three major wireless standards of today. A great deal of this work is documented in the following chapters of this book. Looking into the future, we can expect this research to continue, especially targeting new antenna structures where the system permits a larger form factor (more antenna elements), and also the targeting of various cooperative approaches including hybrids of standard MIMO and cooperative MIMO. A key issue will be the design of schemes with improved channel sounding techniques over the current standards, where the overhead for channel sounding quickly grows with the rank of the MIMO scheme (number of antennas). The network of the future may very well be a two-tier network, where each transmission from network infrastructure to mobile hand-held terminal occurs over two hops. The first hop would heavily employ MIMO techniques between the base stations and a relay station (i.e., not hand held), and can utilize MIMO with a high rank. The second hop would be between the relay station and the terminal. Many different scenarios in building the base station to relay station infrastructure, depending on the nature of traffic demand, are possible, and different MIMO structures will be required to be employed in each scenario.

Elvino Sousa
University of Toronto

Preface

This book presents a detailed and comprehensive summary of the most important enhancements in multiple-input multiple-output (MIMO) systems for 4G, including their evolution and future trends. Moreover, this book includes descriptions of the fundamentals of MIMO techniques, associated signal processing, and receiver design.

The previous books in this series, titled *Transmission Techniques for Emergent Multicast and Broadcast Systems* (CRC Press) and *Transmission Techniques for 4G Systems* (CRC Press), focused on the transition from 3G to LTE and on the transmission techniques to be employed in future 4G cellular systems, respectively. Therefore, these books covered wide areas.

The purpose of this book is to concentrate in a single place several important ongoing research and development (R&D) activities in the field of MIMO systems and their associated signal processing that are expected to be employed in 4G and 5G systems. This is a hot and important topic which allows achieving the increased throughputs demanded by emergent services. Moreover, this book also aims to provide a comprehensive description of MIMO fundamentals and theory, with special interest for those needing to improve their skills in the subject, such as corporate/industrial employees or graduate students. Therefore, this book aims to serve a wide range of potential readers. It can be used by an engineer with a BSc to learn about the latest R&D on MIMO systems, for the purpose of an MSc or a PhD program, or for business activities; it can also be used by academic, institutional, or industrial researchers in order to support the planning, design and development of prototypes and systems. It is worth noting that the contributing authors have been working on many international R&D projects related to the subject of the book, and are highly cited in the MIMO field, which represents an added value.

MIMO systems were initially introduced in IEEE 802.11n (commonly referred to as Wi-Fi) and in Release 7 of the universal mobile telecommunication system (UMTS) by the Third Generation Partnership Project (3GPP). These systems were intended to efficiently use network and radio resources by transmitting multiple data streams over a common radio channel. This is achieved by using multiple antennas at the transmitter and at the receiver sides.

This book is divided into chapters written by different invited authors, leading researchers in MIMO, covering the various topics associated with MIMO systems and MIMO processing, starting from the fundamental concepts and conventional receiver design, up to the most advanced and recently proposed processing techniques.

The use of multiple antennas at both the transmitter and receiver aims to improve performance or increase the symbol rate of the systems, without an increase of the spectrum bandwidth, but it usually requires higher implementation complexity. In the case of a frequency selective fading channel, the different transmitted symbols will be affected by interference coming from other symbols, and this effect is usually known as intersymbol interference (ISI). This effect tends to increase with the used bandwidth. By exploiting diversity, multi-antenna systems can be employed to mitigate the negative effects of both fading and ISI. Space–time coding, such as the pioneering scheme proposed by Alamouti, is an example of a scheme that can be applied to improve the performance by exploiting diversity.

MIMO systems may also target high bit-rate services over a common channel, as proposed by Foschini. One of the most important properties of MIMO systems is resource sharing among many user equipment (UE). MIMO based on spatial multiplexing (multilayer transmission) aims to achieve higher data rates in a given bandwidth. This rate increase corresponds to the number of transmitter antennas. In the case of spatial multiplexing using the conventional postprocessing approach, the number of receiver antennas must be equal to or higher than the number of transmitter antennas (although there is active research for underdetermined systems where that restriction is removed).

MIMO processing can be split into two different categories: postprocessing and preprocessing. Single-user MIMO (SU-MIMO) considers data being transmitted from a single user to another individual user, and corresponds to having postprocessing only or a combination of both, if channel state information (CSI) is available at the transmitter side. With a sufficient number of receiver antennas, it is possible to resolve all data streams, as long as the antennas are sufficiently spaced to minimize the correlation between them. Another type of preprocessing for MIMO is commonly referred to as multiuser MIMO (MU-MIMO).[*] The approach behind MU-MIMO is similar to spatial multiplexing typically employed in the uplink.[†] Nevertheless, MU-MIMO is normally implemented in the downlink. This allows sending different data streams to different UEs. Using the preprocessing approach, the traditional constraint on spatial multiplexing (where the number of receiver antennas must be equal to or higher than the number of transmitter antennas) can be reversed. Therefore, the spatial multiplexing concept can be employed in the downlink, where the transmitter (BS) accommodates a high number of transmitter antennas and the receiver (UEs) can only accommodate a single or a reduced

---

[*] Also referred to as MIMO broadcast.

[†] Because the base station (BS) typically has the ability to accommodate a higher number of receiver antennas to perform the nulling algorithm.

number of antennas. With this approach, multiple data streams may be sent to multiple users at the same time, all in the same frequency bands. Alternatively, instead of implementing the spatial multiplexing principle described above, the MU-MIMO can also be performed using a beamforming algorithm. One should emphasize that most of the MU-MIMO systems require accurate downlink CSI at the transmitter side.

Chapter 1 of this book starts by exposing the different wireless communication standards that make use of MIMO systems, followed by a description of the various MIMO techniques, whether SU-MIMO or MU-MIMO, including space–time coding, spatial multiplexing, and beamforming. Finally, Chapter 1 ends with a description of advanced MIMO applications, such as base station cooperation, multihop relay, and multiresolution transmission schemes.

Chapter 2 focuses on receiver processing associated with MIMO signal detection, and clarifies how spatial multiplexing is achieved via an insightful geometric interpretation of the MIMO detection problem based on lattice theory. The chapter covers the most important detection algorithms and lays the bases for the reader to better understand subsequent chapters on precoding (Chapter 3), on OFDM detection (Chapter 4), and detection in systems with large antenna arrays (Chapter 10).

Chapter 3 tells us that, when CSI is available at the transmitter side, the traditional singular value decomposition combined with water-filling power allocation among the channel's singular values, applied to spatial multiplexing MIMO, is not optimal. Also capitalizing on a lattice interpretation of the precoding problem, the optimal precoders are derived. The preprocessing associated to MU-MIMO is also described under the same framework in Chapter 3, as the only difference is that the receiver antennas are not co-located. This latter concept comprises multiple streams of data simultaneously allocated to different users, using the same frequency bands.

Optimized MIMO schemes and processing for block transmission techniques are described in Chapter 4 for the orthogonal frequency division multiplexing (OFDM) transmission technique, and in Chapter 5 for single carrier–frequency domain equalization (SC–FDE), while Chapter 6 presents MIMO processing and optimization for wideband code division multiple access (WCDMA). Moreover, the description of the ultra-wideband (UWB) transmission technique* and corresponding MIMO processing and optimizations are provided in Chapters 7 and 8. The MIMO optimizations for different transmission techniques include enhanced state-of-the-art receivers, in order to improve the overall system performance, capacity, and coverage.

Although all of these subjects are nonspecific to any particular system, Chapter 9 focuses on the performance analysis of possible schemes for 4G systems using different combinations of MIMO schemes (SU-MIMO and MU-MIMO) and transmission techniques, cellular environments, relay techniques, and services.

---

* UWB transmission technique is envisaged for 5G systems.

To allow the deployment of the emergent services, such as video broadcast or video-on-demand, the combination of these enhancements is accomplished by adaptive transmission techniques. In Chapter 9, the authors combine several of these techniques for the purpose of implementing 4G services. A number of simulation results obtained using link level and system level simulations are presented, allowing for comparison among the several different techniques and schemes, and enabling the reader to identify the key factors required by emergent wireless systems.

In the final three chapters, the book presents very recent and exciting extensions to MIMO. After more than a decade of research on MIMO detection methods, an efficient way to put in practice a receiver for the 8×8 antenna configuration, as required by the LTE-Advanced standard, is still an issue that with which manufacturers have to deal. Chapter 10 shows that recent developments in detection algorithms (combined with the new nature of the problem when the number of dimensions becomes very high) actually open doors to the use of much larger antenna arrays. Then, in Chapters 11 and 12, the very recent concept of "adding bits in the air" (now known as physical layer network coding) is explored and combined in different ways with multiple-antenna terminals. This brings together coding and signal processing in a way that will redefine the design of the physical layer. It seems that many more interesting discoveries in the history of radio communications are still to come.

Mário Marques da Silva
Francisco A. Monteiro

MATLAB® and Simulink® are registered trademarks of The MathWorks, Inc. For product information, please contact:

The MathWorks, Inc.
3 Apple Hill Drive
Natick, MA 01760-2098 USA
Tel: 508 647 7000
Fax: 508-647-7001
E-mail: info@mathworks.com
Web: www.mathworks.com

Editors

Mário Marques da Silva is an associate professor at the Universidade Autónoma de Lisboa, and a researcher at Instituto de Telecomunicações in Lisbon, Portugal. He earned his BSc in electrical engineering in 1992, and MSc and PhD in telecommunications/electrical engineering in 1999 and 2005, respectively, both from Instituto Superior Técnico, University of Lisbon.

Between 2005 and 2008, he was with the NATO Air Command Control & Management Agency (NACMA) in Brussels (Belgium), where he managed the deployable communications of the new Air Command and Control System Program. He has been involved in several telecommunications projects. His research interests include networking and mobile communications, namely, block transmission techniques (OFDM, SC-FDE), interference cancellation, space–time coding, MIMO systems, smart and adaptive antennas, channel estimation, software defined radio, IP technologies, and network security. He is also a Cisco certified CCNA instructor.

Dr. Marques da Silva has authored books such as *Multimedia Communications and Networking* (CRC Press), *Transmission Techniques for Emergent Multicast and Broadcast Systems* (CRC Press), and *Transmission Techniques for 4G Systems* (CRC Press), as well as authored several dozens of journal and conference papers. He is a member of the Institute of Electrical and Electronics Engineers (IEEE) and a member of the Armed Forces Communications and Electronics Association (AFCEA), as well as a reviewer for a number of international scientific IEEE journals and conferences. Finally, he has chaired many conference sessions and has been serving in the organizing committee of relevant EURASIP and IEEE conferences.

Francisco A. Monteiro is an assistant professor at ISCTE—University Institute of Lisbon and a researcher at Instituto de Telecomunicações in Lisbon, Portugal. In the past, he has been a teaching assistant at the Department of Electrical and Computer Engineering at Instituto Superior Técnico, University of Lisbon.

He obtained his PhD in engineering from the University of Cambridge, United Kingdom, in 2012, and had previously received both the Licenciatura degree and an MSc in electrical and computer engineering from Instituto Superior Técnico, University of Lisbon, in 1999 and 2003, respectively. His research has always been focused on signal processing for wireless communications. He acts as a frequent reviewer for a number of IEEE journals and conferences, and has been serving in the organizing committee of relevant EURASIP and IEEE conferences.

In 2008, Dr. Monteiro was a visiting researcher at the Department of Electrical and Computer Engineering of the University of Toronto for four months, with scholarships from the Royal Academy of Engineering, the Calouste Gulbenkian Foundation and the Cambridge Philosophical Society. At Cambridge, he was a member of Fitzwilliam College (where he received a College Senior Scholarship in 2007/2008), and carried out his research in the Digital Technology Group at the Computer Laboratory, while being affiliated with the Department of Engineering.

He received the Best Paper Award (Conference Prize) at the European Conference on Wireless Technology 2007 in Munich, Germany. He was also awarded the Young Engineer Best Paper Award at the European Conference on Wireless Technology 2004, in Amsterdam, the Netherlands. His MSc thesis was awarded 3rd place at the Innovation Young Engineer Competition presented by the Portuguese Engineers Institution in 2002.

Dr. Monteiro is a member of several societies of the IEEE, a life member of Fitzwilliam College, a Fellow of the Cambridge Philosophical Society, a junior member of the Isaac Newton Institute for Mathematical Sciences, and a life member of student societies for the promotion of science and technology (namely, the Trinity Mathematical Society, and the Cambridge University Scientific Society).

Contributors

Ben Allen
University of Bedfordshire
Luton, United Kingdom

Américo Correia
ISCTE—Instituto Universitário
   de Lisboa
and
Instituto de Telecomunicações
Lisbon, Portugal

Mário Marques da Silva
Universidade Autónoma de Lisboa
and
Instituto de Telecomunicações
Lisbon, Portugal

Armin Dekorsy
University of Bremen
Bremen, Germany

Rui Dinis
FCT—Universidade Nova de Lisboa
and
Instituto de Telecomunicações
Caparica, Portugal

David J. Edwards
University of Oxford
Oxford, United Kingdom

Dzevdan Kapetanović
Interdisciplinary Centre for
   Security Reliability and
   Trust (SnT)
University of Luxembourg
Luxembourg

Ioannis Krikidis
University of Cyprus
Nicosia, Cyprus

Xiaoli Ma
Georgia Institute of
   Technology
Atlanta, Georgia

Wasim Q. Malik
Massachusetts Institute
   of Technology
and
Harvard Medical School
Cambridge, Massachusetts

Francisco A. Monteiro
ISCTE—Instituto Universitário de
   Lisboa
and
Instituto de Telecomunicações
Lisbon, Portugal

Paulo Montezuma
FCT—Universidade Nova de
    Lisboa
and
UNINOVA
Caparica, Portugal

Fredrik Rusek
Lund University
Lund, Sweden

João Carlos Silva
ISCTE—Instituto Universitário de
    Lisboa
and
Instituto de Telecomunicações
Lisbon, Portugal

Vit Sipal
Dublin Institute of Technology
Dublin, Ireland

Elvino Sousa
University of Toronto
Toronto, Ontario, Canada

Nuno Souto
ISCTE—Instituto Universitário de
    Lisboa
and
Instituto de Telecomunicações
Lisbon, Portugal

John S. Thompson
The University of Edinburgh
Edinburgh, United Kingdom

Ian J. Wassell
University of Cambridge
Cambridge, United Kingdom

Meng Wu
University of Bremen
Bremen, Germany

Dirk Wübben
University of Bremen
Bremen, Germany

Qi Zhou
Ratrix Technologies
Atlanta, Georgia

Chapter 1

MIMO Techniques and Applications

Mário Marques da Silva and Américo Correia

Contents

1.1 Evolution of Cellular Systems and the New Paradigm of 4G

The first generation of cellular networks (1G) were analog, deployed between 1980 and 1992. 1G included a myriad of cellular systems, namely, the total access communication system (TACS), the advanced mobile phone system (AMPS), and the Nordic mobile telephony (NMT), among others. These systems were of low reliability, low capacity, low performance, and without roaming capability between different networks and countries. The multiple access technique adopted was frequency division multiple access (FDMA), where signals of different users are transmitted in different (orthogonal) frequency bands.

The second generation of cellular networks (2G), like the global system for mobile communications (GSM), were widely used between 1992 and 2003. This introduced the digital technology in the cellular environment, with a much better performance, better reliability, higher capacity, and even with the roaming capability between operators, due to its high level of standardization and technological advancements. The multiple access technique used by GSM was time division multiple access (TDMA), where signals generated by different users were transmitted in different (orthogonal) time slots.

Narrowband code division multiple access (CDMA) system was adopted in the 1990s by IS-95 standard, in the United States. IS-95 was also a 2G system. Afterward, the Universal Mobile Telecommunications System (UMTS), standardized in 1999 by the Third Generation Partnership Project (3GPP)* release 99 (see Table 1.1), proceeded with its utilization, in this particular case using the wideband CDMA (WCDMA). The UMTS consists of a third-generation cellular system (3G).

The CDMA concept relies on different spread spectrum transmissions, each one associated with a different user's transmission, using a different (ideally orthogonal) spreading sequence [Marques da Silva et al. 2010].

The long-term evolution (LTE) can be viewed as the natural evolution of 3G,† using a completely new air interface, as specified by 3GPP release 8, and enhanced in its release 9. Its initial deployment took place in 2010. The LTE comprises an

---

* 3GPP is responsible for specifying and defining the architecture of the European 3G and 4G evolution.
† In fact, LTE is sometimes referred to as 3.9G.

Table 1.1 Comparison of Several Different 3GPP Releases

FDD TDD	WCDMA TD-SCDMA	HSPA TD-HSDPA	HSPA+ TD-HSUPA	LTE TD-LTE	LTE/IMT Advanced
Deployment	2003	2006/8	2008/9	2010	2014
3GPP release	99	5/6	7	8/9	10/11/12
Downlink data rate	384 kbps	14.4 Mbps[a]	28 Mbps[a]	>160 Mbps[b]	1 Gbps nomadic; 100 Mbps mobile
Uplink data rate	128 kbps	5.76 Mbps[a]	11 Mbps[a]	>60 Mbps[b]	500 Mbps nomadic; 50 Mbps mobile
Switching	Circuit + packet switching	Circuit + packet switching	Circuit + packet switching	IP based (packet switching)	IP based (packet switching)
Transmission technique	WCDMA/TD-SCDMA	WCDMA/TD-SCDMA	WCDMA/TD-SCDMA	Downlink: OFDMA Uplink: SC-FDMA	Downlink: OFDMA Uplink: SC-FDMA
MIMO	No	No	Yes	Yes	Yes[c]
Multihop relay	No	No	No	No	Yes
Adaptive modulation and coding	No	Yes	Yes	Yes	Yes
Cooperative systems	No	No	No	No	Yes
Carrier aggregation	No	No	No	No	Yes

a Peak data rates.
b Assuming 20 MHz bandwidth and 2 × 2 MIMO.
c 8 × 8 MIMO.

air interface based on orthogonal frequency division multiple access (OFDMA)* in the downlink and single carrier–frequency division multiple access (SC-FDMA) in the uplink. This allows a spectral efficiency improvement by a factor of 2–4, as compared to the high-speed packet access[†] (HSPA), making use of new spectrum, different transmission bandwidths from 1.4 MHz up to 20 MHz, along with multiple-input–multiple-output (MIMO) systems and the all-over IP[‡] architecture [Marques da Silva et al. 2010; Marques da Silva 2012].

With the aim of fully implementing the concept of "anywhere" and "anytime," as well as to support new and emergent services, users are demanding more and more from the cellular communication systems. New requirements include increasing throughputs and bandwidths, enhanced spectrum efficiency, lower delays, and network capacity, made available by the air interface.[§] These are the key requirements necessary to deliver the new and emergent broadband data services. In order to meet these requirements, the LTE-Advanced[¶] was initially specified in release 10 of 3GPP, and improved in its release 11 and 12. The LTE-Advanced consists of a fourth-generation cellular system (4G), being expected to be fully implemented in 2014. It aims to support peak data rates in the range of 100 Mbps for vehicular mobility to 1 Gbps for nomadic access (in both indoor and outdoor environments). 4G aims to support current and emergent multimedia services, such as social networks and gaming, mobile TV, high-definition television (HDTV), digital video broadcast (DVB), multimedia messaging service (MMS), or video chat, using the all-over IP concept and with improved quality of service** (QoS) [Marques da Silva 2012].

The specifications for International Mobile Telecommunications–Advanced (IMT-Advanced[††]) were agreed upon at the International Telecommunications Union–Radio communications (ITU-R) in [ITU-R 2008]. ITU has determined that "LTE-Advanced" should be accorded the official designation of IMT-Advanced. IMT-Advanced is meant to be the international standard of the next-generation cellular systems.[‡‡]

New topological approaches such as cooperative systems, carrier aggregation, multihop relay, advanced MIMO systems, as well as block transmission techniques allow an improved coverage of high-rate transmission, and improved system performance and capabilities, necessary to fit the advanced requirements of IMT-A [ITU-R 2008].

---

* In opposition to WCDMA utilized in UMTS.
† Standardized in 3GPP Releases 5, 6, and 7 (see Table 1.1).
‡ Internet protocol.
§ Where the bottleneck is typically located.
¶ LTE-Advanced is commonly referred to as LTE-A.
** Important QoS parameters include the definition of the required throughput, bit error rate, end-to-end packet loss, delay, and jitter.
†† IMT-Advanced is commonly referred to as IMT-A.
‡‡ Similarly, IMT2000 corresponds to a set of third-generation cellular system standards, namely, IEEE 802.16e, CDMA2000, WCDMA, and so on.

1.1.1 Evolution from 3G Systems into Long-Term Evolution

The third generation of cellular system comprises different evolutions. The initial version, specified by 3GPP release 99, marked a sudden change in the multiple access technique (see Table 1.1). While the GSM was based on TDMA, the 3G makes use of WCDMA to achieve an improved spectrum efficiency and cell capacity. This evolution allowed a rate improvement from a few dozens of kbps up to 384 kbps for the downlink and 128 kbps for the uplink. These rates were improved in the following updates, achieving 28 Mbps in the downlink of HSPA+ (3GPP release 7). In order to respond to the increased speed demands of the emergent services, higher speeds became possible with the already deployed LTE, supporting 160 Mbps in the downlink (as defined by 3GPP release 8), and even higher speeds with some additional improvements to the LTE baseline introduced in 3GPP release 9 (e.g., advanced MIMO systems). The LTE air interface was the result of a study item launched by 3GPP named Evolved UTRAN (E-UTRAN). The goal was to face the latest demands for voice, data, and multimedia services, improving spectral efficiency by a factor 2–4, as compared to HSPA release 7. The LTE can be viewed as a cellular standard for 3.9G (3.9 Generation).

The LTE air interface relies on a completely new concept that introduced several technological evolutions as a means to support the performance requirements of this new standard. This includes block transmission technique using multicarriers, multiantenna systems (MIMO), base station (BS) cooperation, multihop relaying, as well as the all-over IP concept.

The air interface of LTE considers the OFDMA transmission technique in the downlink and SC-FDMA in the uplink. Depending on the purpose, different types of MIMO systems are considered in 3GPP release 8. The modulation employed in LTE comprises quadrature phase shift keying (QPSK), 16-QAM, or 64-QAM (quadrature amplitude modulation), using adaptive modulation and coding (AMC). In the presence of noisy channels, the modulation order is reduced and the code rate is increased. The opposite occurs when the channel presents better conditions.

The LTE comprises high spectrum flexibility, with different spectrum allocations of 1.4, 3, 5, 10, 15, and 20 MHz. This allows a more efficient spectrum usage and a dynamic spectrum allocation based on the bandwidths/data rates required by the users [Astely et al. 2009].

Intracell interference is avoided in LTE by allocating the proper orthogonal time slots and carrier frequencies between users in both uplink and downlink. However, intercell interference is a problem higher than in the case of UMTS,[*] especially for users at the cell edge. Intercell interference can be mitigated by imple-

---

[*] Owing to the lower-power spectral density of WCDMA signals, the level of interferences generated in UMTS tends to be lower.

menting mechanisms such as interference cancelation schemes, reuse partitioning, and advanced BS cooperation.

Another important modification of the LTE, as compared to UMTS, is the all-IP architecture (i.e., all services are carried out on top of IP), instead of the circuit* plus packet† switching network adopted by UMTS.

An important improvement of the LTE, compared to the UMTS, relies on its improved capability to support multimedia services.

The multimedia broadcast and multicast service (MBMS), already introduced in 3GPP release 6 (HSPA), aims to use spectrum-efficient multimedia services, by transmitting data over a common radio channel. MBMS is a system that allows multiple mobile network users to efficiently receive data from a single content provider source by sharing radio and transport network resources. While conventional mobile communications are performed in unicast‡ mode, multimedia services are normally delivered in either broadcast or multicast mode. In broadcast mode, data are transmitted in a specific area (MBMS service area) and all users in the specific MBMS service area are able to receive the transmitted MBMS data. Very often, broadcast communications are established in a single direction (i.e., there is no feedback from the receiver into the transmitter). In multicast mode, data are transmitted in a specific area but only registered users in the specific MBMS service area are able to receive the transmitted MBMS data.

The LTE introduced a new generation of MBMS, titled evolved MBMS (eMBMS). This is implemented in LTE in two types of transmission scenarios [Astely et al. 2009]:

■ Multicell transmission: multimedia broadcast over a single-frequency network (MBSFN)§ on a dedicated frequency layer or on a shared frequency layer. The group of cells that receive the same MBSFN multicast data service is referred to as MBSFN area.¶

■ Single-cell transmission: single cell–point to multipoint (SC-PTM) on a shared frequency layer.

Multicell transmission in single-frequency network (SFN) area is a way to improve the overall network spectral efficiency. In MBSFN, when different cells transmit the same eMBMS multimedia data service, the signals are combined, in order to provide diversity for user equipment (UE) located at a cell boundary. This results in an improved performance and better service quality.

---

* Circuit switching is employed in UMTS for voice service.
† Packet switching is employed in UMTS for data service.
‡ Unicast stands for a communication whose data destination is a single station.
§ The MBSFN allows delivering services such as mobile television.
¶ Within an MBSFN area, if one or more cells are not required to broadcast the multimedia data service, the transmission can be switched off and the corresponding resources can be released to regular unicast or other services.

1.1.2 WiMAX—IEEE802.16

WiMAX stands for Worldwide Interoperability for Microwave Access and allows fixed and mobile access. WiMAX, standardized by the Institute of Electrical and Electronics Engineers (IEEE) as IEEE 802.16, was initially created in 2001 and updated by several newer versions. It consists of a technology that implements a wireless metropolitan area network (WMAN) [Eklund et al. 2002; Andrews et al. 2007; Peters and Heath 2009]. WiMAX provides wireless Internet access to the last mile, with a range of up to 50 km. Therefore, it can be viewed as a complement or competitor of the existing asynchronous digital subscriber line (ADSL) or cable modem, providing service with the minimum effort in terms of required infrastructures. However, fixed WiMAX can also be viewed as a backhaul for Wi-Fi (IEEE 802.11), cellular BS, or mobile WiMAX. Since the standard only defines the physical layer and medium access control (MAC) sublayer, it can be associated with either IP version 4 (IPv4) or IP version 6 (IPv6).

In order to allow the operation of WiMAX in different regulatory spectrum constraints faced by operators in different geographies, this standard specifies channel sizes ranging from 1.75 MHz up to 20 MHz, using either time division duplexing (TDD) or frequency division duplexing (FDD), with many options in between [Yarali and Rahman 2008].

The initial version of WiMAX was updated by several newer versions, namely

- IEEE 802.16-2004, also referred to as IEEE 802.16d. This version only specified the fixed interface of WiMAX, without providing any support for mobility (IEEE 802.16-2004). This version of the standard was adopted by European Telecommunications Standards Institute (ETSI) as a base for the HiperMAN.*
- IEEE 802.16-2005, also referred to as IEEE 802.16e. It consists of an amendment to the previous version. It introduced support for mobility, handover, and roaming, among other new capabilities (IEEE 802.16e). In addition, to achieve better performances, MIMO schemes were introduced.
- Relay specifications are included in IEEE 802.16j amendment. The incorporation of multihop relay capability in the foundation of mobile IEEE 802.16-2005 is a way to increase both the available throughput by a factor of 3–5 and/or coverage (and higher channel reuse factor), or even to fill the "coverage hole" of indoor coverage [Oyman et al. 2007; IEEE 802.16; Peters and Heath 2009]. Multihop relay capability was also included in IMT-Advanced [Astely et al. 2009].

In addition to these versions, requirements for the next-version Mobile WiMAX titled IEEE 802.16 m (IEEE 802.16 m) were completed. The goal of

---

* High-performance metropolitan area network.

IEEE 802.16 m version is to reach all the IMT-Advanced requirements as proposed by ITU-R in ITU-R 2008, making this standard a candidate for the IMT-A. Advances in IEEE 802.16 m include wider bandwidths (up to 100 MHz, shared between uplink and downlink), adaptive and advanced TDMA/OFDMA access schemes, advanced relaying techniques (already incorporated in IEEE 802.16j), advanced multiple-antenna systems, adaptive modulation schemes such as hierarchical constellations and AMC, and frequency adaptive scheduling, among other advanced techniques.

The original version of the standard specified a physical layer operating in the range of 10–66 GHz, based on OFDM and TDMA technology. IEEE 802.16-2004 added specifications for the 2–11 GHz range (licensed and unlicensed), whereas IEEE 802.16-2005 introduced the scalable OFDMA (SOFDMA) with MIMO (either space–time coding based, spatial multiplexing based, or beamforming) or advanced antenna systems (AAS) (IEEE 802.16e), instead of the simple OFDM with 256 subcarriers considered by the previous version.

In terms of throughputs and coverage, these two parameters are subject to a trade-off (IEEE 802.16e): typically, mobile WiMAX provides up to 10 Mbps per channel (symmetric), over a range of 10 km in rural areas (line-of-sight [LOS] environment) or over a range of 2 km in urban areas (non-LOS environment) [Ohrtman 2008]. With the fixed WiMAX, this range can normally be extended. Mobile version considers an omnidirectional antenna, whereas fixed WiMAX uses a high-gain antenna (directional). Throughput and ranges may always change. Nevertheless, by enlarging one parameter, the other has to reduce; otherwise, the bit error rate (BER) would suffer degradation. In the limit, WiMAX can deliver up to 70 Mbps per channel (in LOS, short distance and fixed access), and may cover up to 50 km (in LOS for fixed access), with a high-gain antenna [Ohrtman 2008], but not both parameters simultaneously. Contrary to UMTS where handover is specified in detail, mobile WiMAX has three possibilities but only the first one is mandatory: hard handover (HHO), fast base station switching (FBSS), and macro-diversity handover (MDHO). FBSS and MDHO are optional, as it is up to the manufacturers to decide on their implementation specifications. Therefore, there is the risk that handover is not possible for these advanced handover schemes between two BS from different manufacturers. Another drawback on the use of WiMAX is the maximum speed allowed in mobility, which is limited to 60 km/h. For higher speeds, the user may experience a high degradation in performance.

The WiMAX version currently available (IEEE 802.16-2005) incorporates most of the techniques also adopted by LTE (from 3GPP). Such examples of techniques are OFDMA, MIMO, advanced turbo coding, all-over IP architecture, and so on. In addition, the inclusion of multihop relay capabilities (IEEE 802.16j) aims to improve the speed of service delivery and coverage by a factor of 3–5. Moreover, IEEE 802.16 m integrates and incorporates several advancements in transmission techniques that meet the IMT-Advanced requirements, including 100 Mbps mobile and 1 Gbps nomadic access, as defined by ITU-R [ITU-R 2008].

1.1.3 LTE-Advanced and IMT-Advanced

The 4G aims to support the emergent multimedia and collaborative services, with the concept of "anywhere" and "anytime," facing the latest bandwidth demands. The LTE-Advanced (standardized by 3GPP) consists of a 4G system. Based on LTE, the LTE-Advanced presents an architecture using the all-over IP concept [Bhat et al. 2012]. The support for 100 Mbps in vehicular and 1 Gbps for nomadic access* is achieved with the following mechanisms:

■ Carrier aggregation composed of multiple bandwidth components (up to 20 MHz) in order to support transmission bandwidths of up to 100 MHz.
■ AAS increasing the number of downlink transmission layers to eight and uplink transmission layers to four. Moreover, LTE-Advanced introduced the concept of multiuser MIMO, in addition to the single-user MIMO previously considered by the LTE.
■ Multihop relay (adaptive relay, fixed relay stations, configurable cell sizes, hierarchical cell structures, etc.) in order to achieve a coverage improvement and/or an increased data rate.
■ Advanced intercell interference cancelation (ICIC) schemes.
■ Advanced BS cooperation, including macro-diversity.
■ Multiresolution techniques (hierarchical constellations, MIMO systems, OFDMA multiple access technique, etc.).

Standardization of LTE-Advanced is part of 3GPP release 10 (completed in June 2011), and enhanced in its release 11 (December 2012) and in its release 12 (March 2013).

The IMT-Advanced refers to the international 4G system, as defined by the ITU-R [ITU-R 2008]. Moreover, the LTE-Advanced was ratified by the ITU as an IMT-Advanced technology in October 2010 [ITU 2010].

Within 4G, voice, data, and streamed multimedia are delivered to the user based on an all-over IP packet-switched platform, using IPv6. The goal is to reach the necessary QoS and data rates in order to accommodate the emergent services.

Owing to the improvements in address spacing with the 128 bits made available by IPv6, multicast and broadcast applications will be easily improved, as well as the additional security, reliability, intersystem mobility, and interoperability capabilities. Moreover, since the 4G system relies on a pool of wireless standards, this can be efficiently implemented using the software defined radio (SDR) platform, being currently an interesting R&D† area in many industries worldwide.

---

* 1 Gbps as a peak data rate in the downlink, whereas 500 Mbps is required for the uplink.
† Research and development.

1.2 MIMO Techniques

The basic concept behind MIMO techniques relies on exploiting the multiple propagation paths of signals between multiple transmit and multiple receive antennas. The use of multiple antennas at both the transmitter and receiver aims to improve performance or symbol rate of systems, without an increase of the spectrum bandwidth, but it usually requires higher implementation complexity [Rooyen et al. 2000; Marques da Silva and Correia 2001, 2002a,b, 2003; Hottinen et al. 2003; Marques da Silva et al. 2012].

In the case of frequency selective fading channel, different symbols suffer interference from each other, whose effect is usually known as intersymbol interference (ISI). This effect tends to increase with the increase of the symbol rate.

By exploiting diversity, multiantenna systems can be employed to mitigate the effects of ISI. The antenna spacing must be larger than the coherence distance to ensure independent fading across different antennas [Foschini 1996, 1998; Rooyen et al. 2000]. Alternatively, different antennas should use orthogonal polarizations to ensure independent fading across different antennas.

The various configurations, shown in Figure 1.1, are referred to as single input single output (SISO), multiple input single output (MISO), single input multiple output (SIMO), or multiple input multiple output (MIMO). SIMO and MISO architectures are a form of receive and transmit diversity schemes, respectively. Moreover, MIMO architectures can be used for combined transmit and receive diversity, as well as for the parallel transmission of data or spatial multiplexing. When used for spatial multiplexing, MIMO technology promises high bit rates in a narrow bandwidth; therefore, it is of high significance to spectrum users. MIMO systems transmit different signals from each transmit element so that the receive antenna array receives a superposition of all the transmitted signals. Figure 1.2 presents a generic diagram of a MIMO system.

For M transmit and N receive antennas, the spectral efficiency, expressed in bits per seconds per hertz (bit/s/Hz), is defined by [Telatar 1995; Foschini 1998]

$$C_{EP} = \log_2\left(\det\left(I_N + \frac{\beta}{M}HH'\right)\right) \tag{1.1}$$

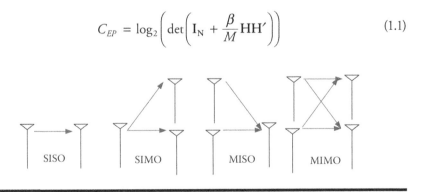

Figure 1.1 Multiple antenna configurations.

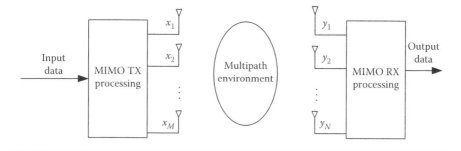

Figure 1.2   Generic diagram of a MIMO scheme.

where $\mathbf{I_N}$ is the identity matrix of dimension $N \times N$, $\mathbf{H}$ is the channel matrix, $\mathbf{H'}$ is the transpose-conjugate of $\mathbf{H}$, and β is the signal-to-noise ratio (SNR) at any receive antenna. Foschini and Telatar demonstrated that the capacity grows linearly with $m = \min(M, N)$, for uncorrelated channels [Foschini 1996; Telatar 1999].

MIMO schemes are implemented based on multiple-antenna techniques. These multiple-antenna techniques can be of different forms:

- Space–time block coding (STBC)
- Multilayer transmission
- Space division multiple access (SDMA)
- Beamforming

STBC is essentially a MISO system. Nevertheless, the use of receive diversity makes it a MIMO, which corresponds to the most common configuration for this type of diversity. STBC-based schemes focus on achieving a performance improvement through the exploitation of additional diversity, while keeping the symbol rate unchanged. Open-loop transmit diversity (TD) schemes achieve diversity without previous knowledge of the channel state at the transmitter side. On the contrary, closed-loop TD requires knowledge about the channel state information (CSI) at the transmitter side. STBC, also known as Alamouti scheme, is the most known open-loop technique [Alamouti 1998] (see Figure 1.3). 3GPP specifications define several TD schemes, such as the standardized closed-loop modes 1 and 2 [3GPP 2003a], or the open-loop STBC [3GPP 2003b] for two transmit antennas. The selective transmit diversity (STD) is a closed-loop TD. This is not exactly an STBC scheme. Nevertheless, since this is also a TD scheme, this subject is dealt along with STBC.

However, multilayer transmission and SDMA belong to another group, titled spatial multiplexing (SM), whose principles are similar but whose purposes are quite different. The goal of the MIMO based on multilayer transmission[*] scheme

---

[*] The same principle is applicable to SDMA, but where different transmit antennas correspond to different users that share the spectrum.

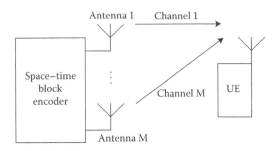

Figure 1.3   Generic block diagram of an open-loop transmitter scheme.

relies on achieving higher symbol rates in a given bandwidth. This rate increase corresponds to the number of transmit antennas. Using multiple transmit and receive antennas, together with additional postprocessing, multilayer transmission allows exploiting multiple and different flows of data, increasing the throughput and the spectral efficiency. In this case, the number of receive antennas must be equal to or higher than the number of transmit antennas (although there is active research for underdetermined systems where that restriction is removed). The increase of symbol rate is achieved by "steering" the receive antennas to each one (separately) of the transmit antennas, to receive the corresponding symbol stream. This can be achieved through the use of the nulling algorithm.

Finally, beamforming is implemented by antenna array with certain array elements at the transmitter or receiver being closely located to form a beam array (typically separated half wavelength). This scheme is an effective solution to maximize the SNR, as it steers the transmit (or receive) beam toward the receive (or transmit) antenna. As a result, an improved performance or coverage can be achieved with beamforming.

1.2.1 Space–Time Coding

Although space–time coding is essentially a MISO system, the use of receiver diversity makes it a MIMO, which corresponds to the most common configuration for this type of diversity. The STBC is a TD technique, being particularly interesting for fading channels, where it is difficult to have multiple receive antennas. A possible scenario for its application is the downlink transmission of a cellular environment, where the BS uses several transmit antennas and the mobile terminal typically has a single antenna [Alamouti 1998].

STBC-based schemes focus on achieving a performance improvement through the exploitation of additional diversity, while keeping the symbol rate unchanged [Alamouti 1998; Tarokh et al. 1999]. Symbols are transmitted using an orthogonal block structure, which enables a simple decoding algorithm at the receiver [Alamouti 1998; Marques da Silva et al. 2004, 2009].

The Alamouti's TD scheme requires some processing from the transmitter, and can be implemented either in the time domain or in the frequency domain. In this latter case, it is referred to as space–frequency block coding.* The current description focuses on the time-domain coding. The extension to frequency domain coding is straightforward.

Let us consider the *l*th transmitted symbol defined by

$$s_l(t) = a_l\, h_T(t - nT_S) \tag{1.2}$$

where a_l refers to the symbol selected from a given constellation, T_S denotes the symbol duration, and $h_T(t)$ denotes the adopted pulse-shaping filter. The signal $s_l(t)$ is transmitted over a time-dispersive channel. The received signal is sampled, and the resulting time-domain block is y_l, which is then subject to frequency domain equalization.

In the following, we will adopt the following notation: lower- and upper-case signal variables correspond to time and frequency domain variables, respectively. The mapping between one and the other is achieved through discrete Fourier transform (DFT) and inverse DFT (IDFT) operations (i.e., $X_l = \mathrm{DFT}\{x_l\}$ and $x_l = \mathrm{IDFT}\{X_l\}$).

The *l*th frequency-domain block before the receiver's equalization process is $Y_l = DFT(y_l)$, with

$$Y_l = \sum_{m=1}^{M} S_l^{(m)} H_l^{(m)} + N_l \tag{1.3}$$

In the above expression, $H_l^{(m)}$ denotes the channel frequency response for the *l*th time domain block (the channel is assumed invariant in the frame) and the *m*th transmit antenna. N_l is the frequency-domain block channel noise for the *l*th block.

1.2.1.1 Space–Time Block Coding for Two Antennas

Considering the STBC with two transmit antennas (STBC2), the *l*th time-domain block to be transmitted by the *m*th antenna ($m = 1$ or 2) is $s_l^{(m)}$, with [Alamouti 1998]

$$
\begin{aligned}
s_{2l-1}^{(1)} &= a_{2l-1} \\
s_{2l-1}^{(2)} &= -a_{2l}^{*} \\
s_{2l}^{(1)} &= a_{2l} \\
s_{2l}^{(2)} &= a_{2l-1}^{*}
\end{aligned}
\tag{1.4}
$$

---

* The Alamouti transmit diversity scheme can also be applied simultaneously to space, time, and frequency. In this case, it is referred to as space–time–frequency block coding (STFBC).

where a_l refers to the symbol selected from a given constellation (e.g., a QPSK constellation) under an appropriate mapping rule, to be transmitted in the lth time domain block. Considering the matrix–vector representation, we define $\mathbf{s}_l^{[1,2]}$ as

$$\mathbf{s}_l^{[1,2]} = \begin{bmatrix} a_{2l-1} & a_{2l} \\ -a_{2l}^* & a_{2l-1}^* \end{bmatrix} \tag{1.5}$$

where different rows of the matrix refer to transmit antenna order and different columns refer to symbol period orders.

The Alamouti postprocessing for two antennas becomes [Alamouti 1998; Marques da Silva et al. 2009]

$$\tilde{A}_{2l-1} = \left[Y_{2l-1} H_l^{(1)*} + Y_{2l}^* H_l^{(2)} \right] \beta$$
$$\tilde{A}_{2l} = \left[Y_{2l} H_l^{(1)*} - Y_{2l-1}^* H_l^{(2)} \right] \beta \tag{1.6}$$

where $\beta = \left[\sum_{m=1}^M |H_l^{(m)}|^2 \right]^{-1}$.

Defining $\mathbf{Y}_l^{[1,2]} = \begin{bmatrix} Y_{2l-1} & Y_{2l}^* \\ Y_{2l} & -Y_{2l-1}^* \end{bmatrix}$ and $\mathbf{H}_l^{[1,2]} = \begin{bmatrix} H_l^{(1)*} & H_l^{(2)} \end{bmatrix}^T$, Equation 1.6 can be expressed in the matrix–vector representation as $\tilde{\mathbf{A}}_l^{[1,2]} = \left[\mathbf{Y}_l^{[1,2]} \times \mathbf{H}_l^{[1,2]} \right] \times \beta$, where $\tilde{\mathbf{A}}_l^{[1,2]} = \begin{bmatrix} \tilde{A}_{2l-1} & \tilde{A}_{2l} \end{bmatrix}^T$.

Finally, the decoded symbols become

$$\tilde{A}_{2l-j} = A_{2l-j} \overbrace{\sum_{m=1}^M \left| H_l^{(m)} \right|^2 \beta}^{\text{Desired Symbol}} + \overbrace{N_{2l-j}^{eq}}^{\text{Noise Component}} \qquad j = 0,1 \tag{1.7}$$

where $N_{k,l}^{eq}$ denotes the equivalent noise for detection purposes.

1.2.1.2 Space–Time Block Coding for Four Antennas

Orthogonal codes of rate one, using more than two antennas, do not exist for modulation orders higher than two. Schemes with 4 and 8 antennas with code rate one only exist in the case of binary transmission. If orthogonality is essential (fully loaded systems with significant interference levels), a code with $R < 1$ should be employed.

The current description focuses on a nonorthogonal scheme with $M = 4$ transmit antennas (STBC4) [Marques da Silva et al. 2009], presenting a code rate one.

The symbol construction for the STBC with four antennas can be generally written as [Marques da Silva et al. 2009]

$$\mathbf{s}_l^{[1,4]} = \begin{bmatrix} \mathbf{s}_l^{[3,4]} & \mathbf{s}_l^{[1,2]} \\ \mathbf{s}_l^{[1,2]^*} & -\mathbf{s}_l^{[3,4]^*} \end{bmatrix} \tag{1.8}$$

where $\mathbf{s}_l^{[3,4]}$ is the same as $\mathbf{s}_l^{[1,2]}$ (see Equation 1.5), by replacing the subscripts 1 by 3 and 2 by 4, as well as by replacing $2l$ by $4l$ (e.g., $a_{2l-1} \rightarrow a_{4l-3}$).

Similar to Equations 1.4 and 1.5, and considering the STBC with four transmit antennas as in Equation 1.8, the time-domain blocks to be transmitted by the mth antenna ($m = 1, 2, 3,$ or 4) are $s_l^{(m)}$, which can be expressed in the matrix–vector representation as

$$\mathbf{s}_l^{[1,4]} = \begin{bmatrix} a_{4l-3} & a_{4l-2} & a_{4l-1} & a_{4l} \\ -a_{4l-2}^* & a_{4l-3}^* & -a_{4l}^* & a_{4l-1}^* \\ a_{4l-1}^* & a_{4l}^* & -a_{4l-3}^* & -a_{4l-2}^* \\ -a_{4l} & a_{4l-1} & a_{4l-2} & -a_{4l-3} \end{bmatrix} \tag{1.9}$$

Note that in Equation 1.9, different rows of the matrix refer to different transmit antennas and different columns refer to different symbol periods.

The lth frequency-domain block, before the receiver's equalization process, is $y_l = \text{IDFT}\{Y_l\}$, with Y_l as defined by Equation 1.3.

Let us define

$$\tilde{\mathbf{A}}_l^{[1,4]} = \begin{bmatrix} \tilde{A}_{4l-3} & \tilde{A}_{4l-2} & \tilde{A}_{4l-1} & \tilde{A}_{4l} \end{bmatrix}^T \tag{1.10}$$

$$\mathbf{Y}_l^{[1,4]} = \begin{bmatrix} \mathbf{Y}_l^{[3,4]} & -\mathbf{Y}_l^{[1,2]^*} \\ \mathbf{Y}_l^{[1,2]} & \mathbf{Y}_l^{[3,4]^*} \end{bmatrix} \tag{1.11}$$

$$\mathbf{H}_l^{[1,4]} = \begin{bmatrix} H_l^{(1)^*} & H_l^{(2)} & H_l^{(3)} & H_l^{(4)^*} \end{bmatrix}^T \tag{1.12}$$

with $\mathbf{Y}_l^{[3,4]}$ as defined for $\mathbf{Y}_l^{[1,2]}$ by replacing the subscripts 1 by 3 and 2 by 4, as well as by replacing $2l$ by $4l$ (e.g., $Y_{2l-1} \rightarrow Y_{4l-3}$).

The postprocessing STBC for four antennas ($M = 4$) becomes

$$\tilde{\mathbf{A}}_l^{[1,4]} = \begin{bmatrix} \mathbf{Y}_l^{[1,4]} \times \mathbf{H}_l^{[1,4]} \end{bmatrix} \times \beta \tag{1.13}$$

Finally, the decoded symbols become [Marques da Silva and Dinis 2011]

$$
\tilde{A}_{4l-j} = \overbrace{A_{4l-j} \sum_{m=1}^{M} \left| H_l^{(m)} \right|^2}^{\text{Desired Symbol}} - \overbrace{C \cdot A_{4l-p}}^{\text{Residual Interference}} + \overbrace{N_{4l-j}^{eq}}^{\text{Noise Component}} \tag{1.14}
$$

with $j = 0,1,2,3$, and $p = 3 - j$. We also define

$$
C = \frac{2 \operatorname{Re}\left\{ H_l^{(1)*} H_l^{(4)} - H_l^{(2)} H_l^{(3)*} \right\}}{\sum_{m=1}^{M} \left| H_l^{(m)} \right|^2} \tag{1.15}
$$

which stands for the residual interference coefficient generated in the STBC decoding process of order four.

The signal processing for nonorthogonal scheme with eight transmit antennas is defined in [Marques da Silva et al. 2009] and in the references therein.

It is worth noting that although the described STBC scheme is a MISO, by adopting receive diversity, this can be viewed as a MIMO system.

1.2.2 Selective Transmit Diversity

The STD is a closed-loop TD, where the transmitter selects one out of multiple transmit antennas to send data, based on the one that presents a more beneficial propagation environment.

Assuming the downlink direction, the STD scheme comprises a low-rate feedback link from the receiver (MS) telling the transmitter (BS), which antenna should be used in transmission. It is worth noting that by adopting receive diversity this can also be viewed as a type of MIMO technique.

As depicted in Figure 1.4, the STD comprises a common or dedicated pilot sequence, which is also transmitted. Different antennas with specific pilot patterns/codes enable antenna selection. Then, the transmitter has a link quality information about the M (number of transmit antennas) links. Based on link information, it transmits a single symbol stream over the best antenna. Then, another decision process is made. The receiver is supposed to reacquire the carrier phase $\theta_k(t)$ after every switch between antennas. The antenna switch (AS) has the capability to switch every slot duration.

The decision for the antenna selection of the STD using WCDMA signals combined with the Rake receiver should select the signal from the transmit antenna whose multipath diversity order is higher. In other words, it should select the mth transmit antenna, experiencing a channel with L multipaths, which maximizes the

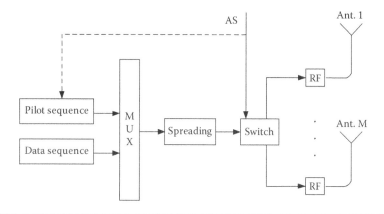

Figure 1.4   Scheme of a selective transmit diversity with feedback indication.

quantity $\max[\sum_{l=1}^{L} |h_{m,l}|^2]$. This process is similar to the selective combiner that can be employed in receive diversity.

1.2.3 Multilayer Transmission

The goal of the MIMO based on multilayer transmission scheme relies on achieving higher data rates in a given bandwidth, whose increase rate corresponds to the number of transmit antennas [Foschini 1996, 1998; Nam and Lee 2002].

Currently, MIMO systems are typically employed to improve the spectral efficiency. This is achieved by increasing the throughput sent in a certain bandwidth. A widely employed technique for this purpose relies on the use of multilayer transmission, as opposed to space–time coding (which aims to provide a performance improvement by exploiting diversity).

In multilayer transmission, the number of receive antennas must be equal to or higher than the number of transmit antennas. The increase of symbol rate is achieved by "steering" the receive antennas to each one (separately) of the transmit antennas, in order to receive the corresponding data stream. This can be achieved through the use of the nulling algorithm. With a sufficient number of receive antennas, it is possible to resolve all data streams, as long as the antennas are sufficiently spaced so as to minimize the correlation [Marques da Silva et al. 2005].

Let us consider the generic scheme of the multilayer MIMO depicted in Figure 1.5. The $M \times N$ MIMO scheme is spectral efficient and resistant to fading, where the BS uses M transmit antennas and the MS uses N receive antenna [Marques da Silva et al. 2004, 2012]. As long as the antennas are located sufficiently far apart, the transmitted and received signals from each antenna undergo independent fading.

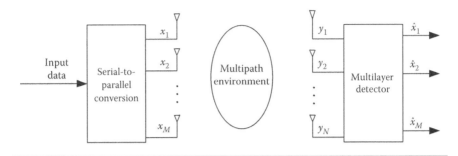

Figure 1.5   Generic diagram of the *M* × *N* multilayer MIMO scheme.

As depicted in Figure 1.6, two different multilayer MIMO schemes can be considered: scheme 1 and scheme 2 [Marques da Silva et al. 2004]. Scheme 1[*] directly allows an increase of symbol rate whose increase rate corresponds to the number of transmit antennas. Scheme 2 allows the exploitation of diversity, without increasing the data rate. The diversity combining is performed using any combining algorithm, preferably the MSE-based (mean square error) combining algorithm. In case of the scheme 2 (depicted in Figure 1.6b), antenna switching is performed at a symbol rate, where gray dashed lines represent the signal path at even symbol periods, in case of two transmit antennas. Output signals are then properly delayed and combined to provide diversity using a combining algorithm.

Let us focus on the MIMO based on multilayer transmission, as depicted in Figure 1.5, where the system is equipped with *M* transmit and *N* receive antennas (with $N \geq M$). At the transmitter side, the data stream is demultiplexed into *M* independent substreams, and then each substream is encoded (modulated) and transmitted by *M* antennas. Each spread data symbol x_m is sent to the

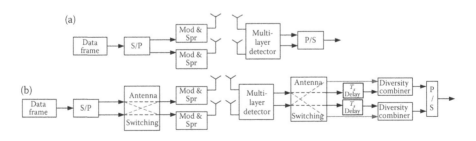

Figure 1.6   Diagram of the 2 × 2 multilayer MIMO alternatives: (a) scheme 1 and (b) scheme 2.

---

[*] Note that, in Figure 1.6, "Mod & Spr" stands for modulator and spreader.

*m*th transmit antenna. It is considered that the radiated signals are propagated through multipath frequency selective fading channels. At the receiver, the detector estimates the transmitted symbols from the received signals at the *N* receive antennas.

The low pass equivalent transmitted signals at the *M* antennas are given by

$$
\begin{cases}
x_1(t) = \sqrt{E_S} \cdot s_l^{(1)} \cdot h_T(t - nT_S) \\
\vdots \\
\\
\vdots \\
\\
\vdots \\
x_M(t) = \sqrt{E_S} \cdot s_l^{(M)} \cdot h_T(t - nT_S)
\end{cases}
\tag{1.16}
$$

where s_l refers to the symbol selected from a given constellation (e.g., a QPSK constellation) under an appropriate mapping rule, to be transmitted in the *l*th time domain block from the *m*th transmit antenna (with $2 \le m \le M$). Moreover, as in Equation 1.2, T_S refers to the symbol interval and $h_T(t)$ is the pulse waveform. $s_l^{(1)}$ and $s_l^{(M)}$ are the data symbols fed to the first and to the *M*th transmit antenna. The baseband equivalent of the *N*-dimensional received vector $\mathbf{y} = [y_1\ y_2\ \cdots\ y_N]^T$ at sampling instants is expressed by

$$
\mathbf{y} = \sum_{l=1}^{L} \mathbf{H}_l \mathbf{x} + \mathbf{z}
\tag{1.17}
$$

where $\mathbf{x} = [x_1\ x_2\ \cdots\ x_M]^T$ denotes the transmitted symbol vector, with each element having the unit average power, and $\mathbf{H}_l : \mathbf{H}_1\ \mathbf{H}_2\ \cdots\ \mathbf{H}_L$ denotes the $N \times M$ channel matrix, whose elements $h_{m,n}$ at the *n*th row and *m*th column are the channel gain from the *m*th transmit antenna to the *n*th receive antenna, that is, $(h_{m,n})_l = (\alpha_{m,n}e^{j\theta_{m,n}})_l \delta(t - (\tau_{m,n})_l)$. *L* refers to the number of multipaths of the channel and the index *l* stands for the multipath order. It is considered a discrete tap-delay-line channel model where the channel from the *m*th transmit antenna to the *n*th receive antenna comprises discrete resolvable paths, expressed through the channel coefficients. The $N \times M$ sets of temporal multipaths corresponding to paths between the multiple transmit antennas and the receive antennas experience independent but identical distributed (i.i.d.) Rayleigh fading. The elements of the *N*-dimensional noise vector $\mathbf{z} = [z_1, z_2, \ldots, z_N]^T$ are also assumed to be i.i.d. complex Gaussian random variables with zero mean and variance σ_n^2.

Different types of detectors can be employed in multilayer MIMO systems. Linear detectors include the decorrelator and the minimum mean square error (MMSE). Nevertheless, the former presents the disadvantage of introducing noise enhancement. Among the suboptimal detectors, the successive interference cancelation (SIC) or the vertical Bell laboratories layered space–time (V-BLAST) detector are two examples. A detector widely employed in multilayer transmission scheme is the V-BLAST detector, which typically comprises an SIC, alongside a decorrelator or an MMSE detector. The maximum likelihood sequence estimator (MLSE) detector is also an option. Nevertheless, since the complexity and processing increases exponentially with the increase of the number of antennas, its real implementation tends to be limited to very specific scenarios.*

Lately, two types of decoders have been investigated: the lattice and the sphere detector. These types of detectors tend to achieve good performances with a reduced complexity. Different detectors for MIMO systems are dealt with in detail in Chapter 2.

1.2.4 Space Division Multiple Access

The goal of an SDMA scheme relies on improving the capacity (more users per cell), while keeping the spectrum allocation unchanged. In other words, the SDMA is a technique that allows multiple users exploiting spatial diversity as a multiple access technique, while sharing a common spectrum.

As referred to previously, both SDMA and multilayer transmission belong to the same group, titled spatial multiplexing. Therefore, the basic concept is common. Nevertheless, while in multilayer transmission an increase of the symbol rate is achieved by considering multiple antennas at the transmitter side, using the SDMA, it is assumed that each transmitter has a single antenna, and the multiple number of receive antennas allow steering the different flows of data corresponding to different users. Consequently, it is commonly employed in the uplink, where the transmitter (UE) has a single antenna, whereas the receiver (BS) has several antennas. Figure 1.7 depicts an SDMA configuration applied to the uplink. SDMA assumes that the number of antennas at the receiver is equal to or higher than the number of users who share the same spectrum. With such an approach, the receiver can decode the signals from each transmitter, while avoiding the signals from the other transmitters (interfering signals). Similar to the decoding performed in multilayer transmission, this can be achieved through the use of the nulling algorithm.

In the V-BLAST detector, the symbols from the transmit antenna with the highest SNR are first detected using a linear nulling algorithm such as zero forcing (ZF) or MMSE detector. The detected symbol is regenerated, and the corresponding signal portion is subtracted from the received signal vector using an SIC detector. This cancelation process results in a modified received signal vector with fewer interfering signal components left. This process is repeated until all symbols from all transmit antennas are detected. According to the detection-ordering scheme

---

* Owing to this reason, the MLSE detector is commonly referred to as brute force detector.

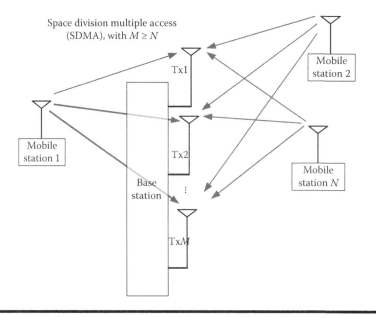

Figure 1.7   Example of SDMA scheme applied to the uplink.

in [Foschini 1996], the detection process is organized so that the symbols from the transmit antenna with the highest SNR are detected at each detection stage. Clearly, the processing of SDMA is almost the same as that of the MIMO based on multilayer transmission. Therefore, the reader should refer to Chapter 2 for a detailed description of spatial multiplexing detectors.

1.2.5 Beamforming

In STBC and spatial multiplexing MIMO schemes, the antenna elements that form an array are usually widely separated in order to form a TD array with low correlation among them. On the other hand, the beamforming is implemented by antenna array with array elements at the transmitter or receiver being closely located to form a beam. The beam is generated with the uniform linear antenna array (ULA). The ULA antenna elements spacing is typically half wavelength. Beamforming is an effective solution to maximize the SNR, as it steers the transmit (or receive) beam toward the receive (or transmit) antenna [Marques da Silva et al. 2009]. As a result, an improved performance or coverage is achieved with beamforming. In the cellular environment, this translates in a reduced number of required sites to cover a given area.

Figure 1.8 depicts a transmitting station sending signals using the beamforming, generated with the ULA. As can be seen, the beamforming allows transmitting a higher-power signal directed toward the desired station, while minimizing

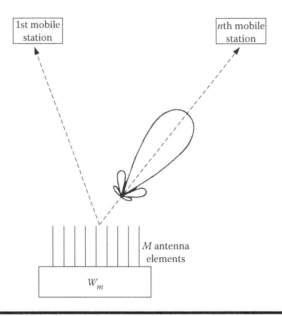

Figure 1.8 Simplified diagram of a beamforming transmitter.

the transmitted power toward the other stations. This allows a reduction of the interference.

The beamforming consists of M identical antenna elements with 120° half-power beam width (HPBW) [Schacht et al. 2003]. Each antenna element is connected to a complex weight $w_m, 1 \leq m \leq M$. An analogy can be made with a receiving array antenna.

The weighted elementary signals are summed together making an output signal as follows [Schacht et al. 2003]:

$$y(t,\theta) = \sum_{m=1}^{M} w_m x\left[t - (m-1)\frac{d}{c}\sin\theta \right] \tag{1.18}$$

where $x(t)$ is the signal sent by the first antenna element, d is the distance between elements, c is the propagation speed, and θ is the direction of arrival (or departure) for the main sector ($-60° < \theta < 60°$). In the frequency domain, the output signal can be written as

$$Y(f,\theta) = \sum_{m=1}^{M} w_m X(f) e^{-j2\pi f(m-1)\frac{d}{c}\sin\theta} \tag{1.19}$$

which makes the relationship between the output signal and the input signal to be

$$H(f,\theta) = \frac{Y(f,\theta)}{X(f)} = \sum_{m=1}^{M} w_m e^{-j2\pi f(m-1)\frac{d}{c}\sin\theta} \tag{1.20}$$

For narrowband beamforming, f is constant and θ is variable. For the beam to be directed toward the desired direction θ_1, we have

$$w_m = e^{j2\pi f(m-1)\frac{d}{c}\sin\theta_1} \tag{1.21}$$

in which, for the case of $d = \lambda/2$, Equation 1.21 results in [Schacht et al. 2003]

$$w_m = e^{j\pi(m-1)\sin\theta_1} \tag{1.22}$$

In other words, for $\theta = \theta_1$, $H(f,\theta_1)$ reduces to

$$H(f,\theta_1) = M \tag{1.23}$$

which is the maximum attainable amplitude by beamforming.

1.2.6 Multiuser MIMO

The MIMO techniques previously exposed are employed in the concept of single-user MIMO (SU-MIMO). This considers data being transmitted from a single user to another individual user. An alternative concept is the multiuser MIMO (MU-MIMO), where multiple streams of data are simultaneously allocated to different users, using the same frequency bands.

When the aim relies on achieving a performance improvement, an SU-MIMO is normally employed using an algorithm such as STBC. However, when the aim relies on achieving higher throughputs using a constrained spectrum, we have two options: in the downlink, the MU-MIMO is typically the solution; in the uplink, spatial multiplexing is normally employed (in this case, multilayer transmission).

The approach behind MU-MIMO is similar to SDMA. Nevertheless, while SDMA is typically employed in the uplink,* the MU-MIMO is widely implemented in the downlink. This allows sending different data streams into different UEs. In this case, instead of performing the nulling algorithm at the receiver side, the nulling algorithm needs to be performed using a preprocessing[†] approach at the transmitter side (BS). This is possible because the BS can accommodate a

---

* Because the nulling algorithm requires a high number of receive antennas.
[†] This is commonly referred to as precoding.

high number of transmit antennas and the UE can only accommodate a single or reduced number (lower) of receive antennas. In the downlink of an MU-MIMO configuration, the number of transmit antennas must be higher than the number of multiple data streams that are sent to multiple users, at the same time, and occupying the same frequency bands (the opposite of the SDMA approach). In this configuration, the nulling algorithm is implemented, at the transmitter side, using a preprocessing algorithm such as the ZF, MMSE, dirty paper coding, and so on. Alternatively, instead of implementing the above-described spatial multiplexing principle, the MU-MIMO can be performed using the beamforming algorithm. In any case, MU-MIMO requires accurate downlink CSI at the transmitter side. Obtaining CSI is trivial using TDD mode, being more difficult to be obtained when FDD is employed. In FDD mode, CSI is normally obtained using a feedback link.

It is worth noting that users located at the cell edge, served by MU-MIMO, may experience SNR degradation due to intercell interference, interuser interference,[*] additional path loss or due to limited BS transmit power (which results from the use of precoding). A mechanism that can be implemented to mitigate such limitation relies on employing a dynamic MIMO system, where MU-MIMO is employed everywhere (in the case of the downlink), except at the cell edge. In this area, the BS switches into SU-MIMO (using, e.g., space–time coding), which translates in an improvement of performance. Alternatively, coordinated multipoint (CoMP)[†] transmission is known as an effective mechanism that improves the performance at the cell edge, resulting in a more homogeneous service quality, regardless of the users' positions.

1.3 Advanced MIMO Applications

The challenge facing the mobile telecommunications industry today is how to continually improve the end-user experience, to offer appealing services through a delivery mechanism that offers improved speed, service attractiveness, and service interaction. In order to deliver the required services to the users at minimum cost, the technology should enable better performances, higher throughputs, improved capacities, and higher spectral efficiencies.

The basic concept of MIMO relies on the transmission of signals through multiple paths, between multiple transmit and multiple receive antennas. Instead of representing an additional interference level, these multiple paths can be used as an advantage. While, in MIMO systems, the multiple transmit or the multiple receive antennas are colocated, advanced cellular network architectures may also achieve

---

[*] Interference among users that share the spectrum and that are separated by the MU-MIMO spatial multiplexing.
[†] Sometimes also referred to as downlink base station cooperation.

the same level of diversity, but using antennas belonging to different BSs or relay nodes. Nevertheless, a certain level of synchronization or coordination is normally required between those stations. These techniques are described in the following sections.

1.3.1 Base Station Cooperation

BS cooperation aims at improving the performance of UEs located at the cells edge, or to increase the throughput of UEs in an area covered by multiple BSs.

An important requirement of 4G systems is the ability to deliver a homogeneous service, regardless of the users' positions [Lee et al. 2012]. Users located at the cell edge may experience a degradation of the SNR due to intercell interference, additional path loss, or limited eNode-B* transmit power. In case MU-MIMO is employed, the power constraint becomes more important[†] and the SNR degradation of UEs located at the cell edge can be even deeper than in SISO environments. In these scenarios, BS cooperation plays an important role, as it allows the exploitation of additional diversity or the delivery of a high and constant throughput, regardless of the users' positions.

BS cooperation stands for the ability to send or receive data from/to multiple adjacent BSs to/from UEs. Instead of employing the conventional HHO, with BS cooperation, independent antenna elements of different BSs are grouped together, forming a cluster, and the UEs may experience a throughput increase or performance improvement (i.e., SNR increase).[‡] Figure 1.9 shows the downlink BS cooperation using a cluster of three BSs. The dashed area corresponds to the area where the UE is able to simultaneously receive signals from three BSs. Using BS cooperation, adjacent BSs can simultaneously transmit the same data. In this case, a combining algorithm may be employed at the UE side, such as the maximum ratio combining (MRC), the MSE-based, or the selective combining. This technique is employed in macro-diversity, as described in Section 1.3.1.2. Alternatively, some type of preprocessing may be employed at the BSs side, such that the signals that reach the UE do not require any type of postprocessing. In this case, this cooperation is commonly referred to as coordinated multipoint transmission, as described in Section 1.3.1.1.

BS cooperation can also be employed in the uplink, being also referred to as CoMP reception. In this case, traditional receive diversity is employed, using

---

* Evolved Node-B (eNode-B) refers to the base station of LTE and 4G systems. This corresponds to the evolution of the Node-B, considered by UMTS.

[†] In MU-MIMO, the power at the transmitter side (i.e., at the BS, in case of downlink) becomes very demanding because the precoding may, instantly, require a high level of power. Consequently, using precoding, the dynamic range of signals tends to increase.

[‡] A performance improvement can be achieved through the exploitation of additional diversity using, for example, space–time coding.

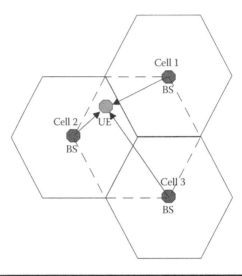

Figure 1.9   Downlink base station cooperation using a cluster of three BSs.

combining algorithms such as the MRC, or the MSE. Alternatively, independent parallel streams of data can be sent out by different UE transmit antennas (typically, a different data stream is sent out by each transmit antenna element), while different BSs decode these different streams using the above-described SDMA algorithm.

1.3.1.1 Coordinated Multipoint Transmission

Using CoMP transmission, independent antenna elements of different BSs are grouped together, forming a cluster, and the UEs can experience a throughput increase or performance improvement. A preprocessing is typically employed at the BS side such that the signals that reach the UE do not require any type of postprocessing.

In case each BS uses the MIMO scheme, the resulting MIMO can be viewed as a "giant MIMO," consisting of the combination of the independent antenna elements from different BSs (see Figure 1.10).

Coordinated multipoint transmission comprises the coordinated transmission of signals from adjacent BSs, and the corresponding reception from a UE. The signal received at the UE side consists of the sum of independent signals sent by different BSs.

CoMP transmission is an important technique that can mitigate intercell interference, improve the throughput, exploit diversity and, therefore, improve the spectrum efficiency. Note that CoMP transmission allows a spectrum efficiency improvement, even at the cell edge. CoMP transmission can be viewed as a special type of MU-MIMO.

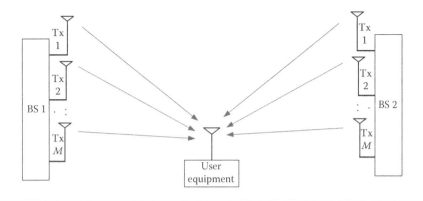

Figure 1.10   Base station cooperation (CoMP transmission) combined with downlink MIMO.

Similar to MU-MIMO, a preprocessing such as beamforming, ZF, MMSE, or dirty-paper coding [Peel 2003] is typically employed in order to assure that the UE receives a preprocessed signal coming from multiple BSs. In this case, the UE commonly employs a low-complexity and regular detector. Alternatively, noncoherent joint transmission and coordinated scheduling is employed. In this case, CoMP transmission can be associated with carrier aggregation, activating or deactivating some carriers in order to optimize the performance. This technique is employed in LTE-Advanced (3GPP Release 10).

Similar to MU-MIMO, accurate downlink CSI is required at BSs side, which consists of an implementation difficulty. In case of TDD, obtaining CSI is trivial, as the uplink and downlink channels are almost the same. Nevertheless, in case of FDD, obtaining CSI at the transmitter side is a complex task.

Depending on the way the coordination between different BSs is performed, and the way CSI is obtained (in FDD mode), two different architectures can be implemented:

▪ Centralized architecture: In this architecture, there is a central control unit (CU) that decides the transmission scheme and resources allocation to be used by different BSs. In this case, the CU is connected to different BSs of the cluster. Each UE estimates the downlink CSI of the signals received from each BS. Then, the CSI is sent back to the corresponding BS. At a third stage, CSI is sent from different BSs of the cluster to the CU through backhaul links. Based on CSI, the CU decides the transmission scheme and resources allocation to be used by each BS, and sends this information to different BSs of the cluster. A major limitation of this architecture relies on the latency, whose factor may result in performance degradation due to fast CSI variations [Diehm et al. 2010].

■ Distributed architecture: In this architecture, each BS is associated with a different CU, and the decision about the transmission scheme and resources allocation is performed independently at the BS level. This information is then exchanged among the cluster's BSs. In this case, each UE estimates the downlink CSI to different BSs and sends this joint data back not only to the BS of reference, but to all BSs. This way the CU associated with each BS has information about different downlink CSI and makes the decision accordingly. An advantage of this architecture relies on the fact that latency is much reduced, and there is no need to use backhaul links for the purpose of exchanging CSI. Nevertheless, this architecture is more subject to errors caused by the uplink transmission [Papadogiannis et al. 2009].

It is finally worth noting that the use of CoMP transmission requires performing a cost–benefit analysis, as CoMP allows a performance/network improvement at the cost of a larger amount of overhead on the over-the-air feedback links and in the backhaul links used to interconnect different BSs.

1.3.1.2 Macro-Diversity

Macro-diversity refers to the transmission of the same information by different BSs to the UEs in the downlink. Therefore, this can be viewed as a particular type of downlink BS cooperation. Macro-diversity aims at supplying additional diversity* in situations where the terminal is far from the BSs. This allows compensating the path loss affecting the transmission to a UE located at the edge of the cell. Consequently, this allows a reduction of the amount of transmit power required to reach a distant receiver, thus increasing network capacity. Therefore, macro-diversity can be viewed as a special type of BS cooperation.

There are two types of networks to be considered: the multifrequency networks (MFNs) and the SFNs. The BSs to which a terminal is linked to are referred to as the active sets.

In broadcast, the global channel impulse response (CIR) tends to be longer due to the longer distances between the transmitter (BS) and the farer of the different receivers (UEs). If the cyclic extension is long enough, the global CIR will be the sum of the independent CIRs. This can be seen from Figure 1.11. This enables the SFN concept exploiting the macro-diversity effect.

Macro-diversity is used during soft handover to ensure smooth transitions between two cells or two sectors of the same cell, increasing the multipath diversity and reducing the risk of call drop.

Using the same transmitting frequencies (from different BSs), diversity is exploited and deep fading tends to be avoided. In case the channel profile between each of the two BSs and the UE is single path, the resulting signal is viewed as

---

* Since it provides transmit diversity, macro-diversity can be viewed as a MISO system.

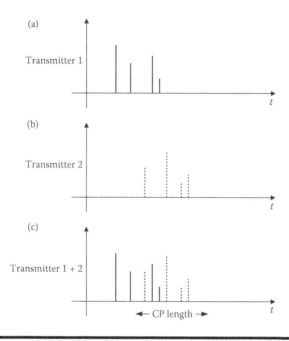

(a)

Transmitter 1

(b)

Transmitter 2

(c)

Transmitter 1 + 2

← CP length →

Figure 1.11 Global CIR (c) is composed of the sum of the CIR of the several transmitters (a + b).

a two-path channel profile, and diversity is exploited. Nevertheless, since a single receiver is employed, although diversity is exploited, the frequency selectivity increases. In fact, this effect can be viewed as one signal received from one BS interfered by the signal received from another BS. Nevertheless, this additional frequency selectivity is mitigated because OFDM signals make use of equalizers at subcarrier level, as long as the appropriate cyclic prefix is employed. Consequently, the macro-diversity presents overall benefits in terms of diversity.

The performance gain brought by macro-diversity depends on the diversity order of the channel. A two-path channel benefits more from macro-diversity than a six-path channel, because the latter already exhibits a high multipath diversity order. Since MBMS in SFN mode is simultaneously broadcasted in several cells, macro-diversity for MBMS in SFN mode does not consume network resources.

In the case of MFN, the UE is required to estimate the carrier of each BS it is linked to. This increases its power consumption. In this case, the signals received from different BSs (especially far ones) may be significantly delayed with regard to those received from near BSs. This requires extramemory at the terminal in order to store the received signals for further combining. Alternatively, an additional synchronization procedure between the BS transmitters is required. In the downlink, the combining takes place at the UE, which has to demodulate and then combine the signals received from the different BSs. The extracomplexity added by macro-diversity then depends on the receiver type. In the case of an equalizer, the UE has to equalize the channel for each BS the UE is linked to. Moreover, the UE must estimate one transmission channel per BS.

In the special case of OFDM, two main cases for macro-diversity can be distinguished:

1. BSs are synchronized, at least to allow UE's receiving signals from two or more BSs with a time difference smaller than the cyclic prefix.
2. BSs are not synchronized.

In the first case, the BSs can transmit identical signals to the UEs on the same time–frequency resource. This is possible because the signals will superimpose within the cyclic prefix. In this situation, no ISI occurs as long as the sum of the time differences plus the maximum delay of the CIR is shorter than the cyclic prefix. Therefore, the terminal can employ a single receiver to demodulate the superimposed signals. This means that it will perform a unique DFT. In this scenario, macro-diversity behaves just like TD (from a unique transmitter with multiple spaced antennas). When different BSs transmit the same data over the same subcarriers, the resulting propagation channel is equivalent to the allegorical sum of all propagation channels, which increases the diversity gain (deep fading is avoided).

When BSs send the same data over different subcarriers (MFN), the maximum diversity can be achieved since each data symbol benefits from the summation of the propagation channel powers (i.e., frequency diversity).

If the BSs are not synchronized, the terminal will need separated receiver chains to demodulate the signals from the distinct BSs. Moreover, to avoid interference, orthogonal time–frequency resources have to be allocated to different BSs. This is still very complex to fulfill; thus, in the general case, interference will occur and spectral efficiency will decrease.

Fast cell selection is one option for macro-diversity in unicast mode (selective diversity). Intra-BS selection should be able to operate on a subframe basis. An alternative to intra-BS macro-diversity scheme for unicast consists of a simultaneous multicell transmission with softcombining. The basic idea of multicell transmission is that, instead of avoiding interference at the cell border by means of intercell interference coordination, both cells are used for the transmission of the same information to a UE, thus, reducing intercell interference as well as improving the overall available transmit power. Another possibility of intra-BS multicell transmission relies on the exploitation of diversity between the cells with space–time processing (i.e., by employing STBC through two cells). Assuming BS-controlled scheduling and that fast/tight coordination between different BS is not feasible, multicell transmission should be limited to cells belonging to the same BS. For multicell broadcast, soft combining of radio links should be supported, assuming a sufficient degree of inter-BS synchronization, at least between a subset of BSs.

1.3.2 Multihop Relay

One important key to improve the coverage and capacity for high-quality multimedia broadcast and multicast transmissions in mobile networks is to provide a homogeneous service, regardless of the users' positions, that is, to allow high data rates for UEs even at the cell edge. Moreover, the distribution of UEs inside the cell can be heterogeneous. Multihop relay is a technique that can be employed to overcome these impairments.

UEs at the cell edge suffer from high propagation loss and high intercell interference from neighbor cells. Other UEs reside in areas that suffer from strong shadowing effects or require indoor coverage from outdoor BS. These impairments originate a degradation of the SNR, which translates in a reduced service quality. Thus, the overall goal of multihop relay is to bring more power to the cell edge and into shadowed areas, while inducing minimal additional interference for neighbor cells. Moreover, increased cell capacity tends also to be achieved by using multihop relay. In addition, multihop relay may also represent an increased life cycle of UEs' batteries.

There are several mechanisms that can be used to implement multihop relays [Sydir and Taori 2009]. As defined in the following, one technique employed in multihop relay consists of using TDD [Stencel et al. 2010], where the communication between the BS and the relay node (RN) is performed in a time slot different from the one utilized for the communication between the RN and the UE. On the contrary, [Schoene et al. 2008] proposed the use of multihop relay associated with FDD.

The next sections describe the most important methods utilized in multihop relay.

1.3.2.1 Adaptive Relaying

As previously described, UEs at the cell edge suffer from high propagation loss and high intercell interference from the neighbor cells. Other UEs reside in areas that suffer from strong shadowing effects. The obvious solution for this would be to decrease cell sizes by installing additional BSs, which represents an increase of network infrastructure costs. Contrarily, adaptive relay nodes can provide temporary network deployment, avoiding outage of service in the area, while keeping the investment cost at an acceptable level. The positioning of relay nodes is selected in a manner that an improvement of service quality or cell coverage is achieved.

In opposition to conventional repeaters working with the amplify-and-forward strategy, adaptive relays are understood to work in a decode-and-forward style. By performing this, relays amplify and retransmit only the wanted component of the signal they receive and suppress the unwanted portions (i.e., regeneration is performed). The disadvantage of an adaptive relay, compared to a conventional repeater, relies on the additional delay introduced into the transmission path between BS and UE. Moreover, depending on the algorithm, a signaling overhead

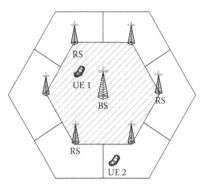

Figure 1.12   Two-hop relaying architecture.

may also be required. The gain is based on the fact that the transmission path is split up into smaller areas representing a reduction of the propagation loss.

Fixed relay stations positioned at a specified distance from the BS (see Figure 1.12) tend to increase the probability that a UE receives enough power from the BS. This deployment concept sectorizes the cell in an inner region where the UEs (e.g., UE 1 in the figure) receive signals from the BS plus some relay stations, and an outer region where only signals from relay stations are strong enough to be received by the UE (e.g., UE 2 in the figure).

1.3.2.2 Configurable Virtual Cell Sizes

Configurable virtual cell sizes were proposed in [Kudoh and Adachi 2003], consisting of a wireless multihop virtual cellular network. This configuration includes the following components:

- Central port: this corresponds to a BS acting as a gateway to the core network.
- Wireless ports: these correspond to relay stations that may or may not communicate with the UEs, and relay the signals from and to the central port.

The wireless ports that are communicating directly with the UEs are called end wireless ports. The wireless ports are stationary and can act together with the central port as one virtual BS. The central port and the end wireless ports introduce additional diversity into the cell, such that the transmit power can be reduced. This also results in a reduced interference level for other virtual cells. The differences between conventional cellular networks and virtual cellular networks can be seen from Figures 1.13 and 1.14.

From the perspective of MBMS, configurable virtual cell sizes could be employed to adapt the cell size to the user distribution and service demands. Service needs may present spatial and temporal variations.

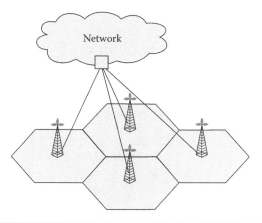

Figure 1.13   Conventional cellular network.

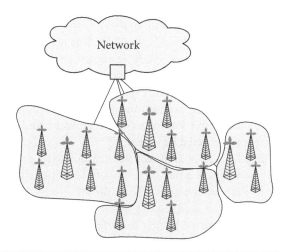

Figure 1.14   Virtual cellular network.

1.3.2.3 Multihop Relay in 3GPP

Relay nodes were introduced in 3GPP release 9 [3GPP 2010a] as a special type of eNB[*] that is not directly connected to the core network. An RN receives data that was forwarded by an eNB[†] that is connected to the evolved packet core[‡] (EPC).

---

[*] eNB stands for evolved Node-B (eNode-B), that is, a Node-B employed in LTE. Note that a Node-B is a BS in the UMTS environment.

[†] This eNB is acting as Donor eNB (DeNB).

[‡] Evolved packet core corresponds to the core of the LTE (all-IP) network.

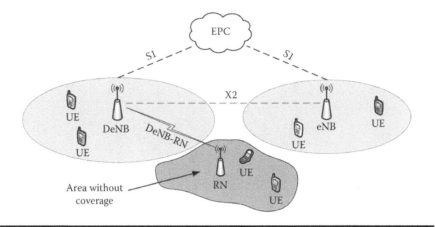

Figure 1.15 Example of relay node in E-UTRAN.

Upon receiving such data, the RN sends it to the UEs that are under its area of coverage. This technique can be seen from Figure 1.15. This is a very interesting option for operators because, compared to eNB, RNs present structures less expensive to deploy and to maintain. They can provide temporary network deployment and new services in an area (e.g., when a sporadic event that concentrates a lot of people in the same geographical area, such as a summer festival). The use of an RN allows a fast-deploying and inexpensive way to solve the problem, and can also provide coverage in small areas not covered by eNBs.

In [3GPP 2010b], four relay architectures are proposed and studied. Those architectures differ from each other in terms of the expected behavior of the RN/DeNB and how the data are sent within the EPC until it reaches the UE. That study concludes that an architecture where an RN acts as a proxy for S1/X2 has the most overall benefits, having been incorporated in 3GPP release 10 (LTE-Advanced).

The detailed RN architecture chosen for LTE-Advanced can be seen from Figure 1.16. In this architecture, we can see that the RN plays two roles simultaneously:

■ From the network point of view (particularly for the DeNB), the RN acts as an ordinary UE (denoted as Relay-UE).
■ From the UE point of view, the RN acts as a normal eNB.

This way the network can abstract itself from establishing a point-to-point connection with each and every UE. In fact, it only has to establish a connection with the RN as it would normally perform with an ordinary UE, and then forward all data that were destined to UEs to the RN (see Figure 1.15). Then, the RN, on its own, will forward all the data received from the network to the respective UEs. The process of relaying is completely transparent, that is, a UE always assumes to

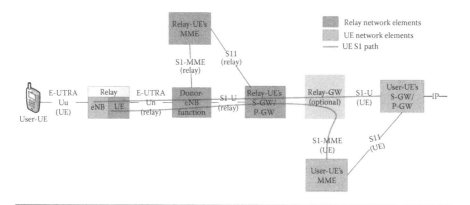

Figure 1.16 Relay architecture chosen for LTE-Advanced. (Adapted from 3GPP, Evolved Universal Terrestrial Radio Access (E-UTRA); Relay architectures for E-UTRA (LTE-Advanced), TR 36.806 V9.0.0, May 2010b.)

be connected to an eNB because RNs are viewed as eNB from the UE side, and as a consequence, no changes to the communication protocol or interfaces used by UEs have to be made.

In the context of multiresolution for Evolved-MBMS (E-MBMS),* the use of relaying fits perfectly. With the introduction of RN in E-MBMS, we can have different zones with different grades of service. Let us consider again the example depicted in Figure 1.15. In areas covered by eNBs, UEs are expected to have high signal-to-interference plus noise ratio (SINR) when they are closer to the center of the cell, reducing as the UEs approach the cell edges. If a hierarchical 64-QAM modulation with coding rate 3/4 is being employed, UEs at the cell edge will experience low levels of QoS or even service outage due to low SINR. If an RN is strategically positioned at the cell edge of existing eNBs, we can expect that area to have an improved coverage and the RN can adapt its transmitting conditions to provide a different resolution of the service in the area.† With such approach, the use of RNs means that, in E-UTRAN, when providing E-MBMS, UEs will experience higher coverage and different service resolutions for the same E-MBMS contents depending on their location and their receiving conditions.

---

* Evolved-MBMS (E-MBMS) framework constitutes the evolutionary successor of MBMS, and is envisaged to play an essential role for the LTE-A proliferation in mobile environments. E-MBMS was initially introduced in 3GPP release 8 (LTE), with further improvements to be implemented in LTE-A. The basic role of these techniques/mechanisms relies on the power and resources optimization during MBMS transmissions. The objective of E-MBMS is to provide services with different QoS requirements depending on the channel conditions experienced by different users.

† Instead of transmitting video with high frame rate and low resolution, we can half the frame rate and transmit with higher resolution.

According to [3GPP 2010a], RNs can operate in two different modes:

■ Inband mode: the link between DeNB and the RN uses the same carrier frequency as the link between RN-UE and eNB-UE.
■ Outband mode: the link between DeNB and the RN uses a different carrier frequency than that of the RN-UE link.

It is worth noting that the transmit power of RNs is lower than that of eNBs. The objective is to diminish intercell interference caused by the introduction of RNs operating in the same band and carrier frequency as eNBs (inband mode), and to limit the area of coverage of a given RN (since RNs are used to cover small areas that cannot be properly covered by existing eNBs).

In order to reduce intercell interference, RNs can also be employed in E-MBMS subframes to create "transmission gaps." These transmissions gaps are illustrated in Figure 1.17. As can be seen, in the first instance where *Time* = *t*, the DeNB transmits a segment of data labeled "Data 1" to all the UE inside its area of coverage and also to the RN. At this instance, the RN stores "Data 1" in memory and is not transmitting any data to the users inside its area of coverage, thus not creating any intercell interference. At the next instance *Time* = *t* + 1, a second set of data labeled "Data 2" is transmitted from DeNB to all the UEs within its area of coverage. On its turn, the RN is not receiving "Data 2" but, instead, is transmitting to the UEs within

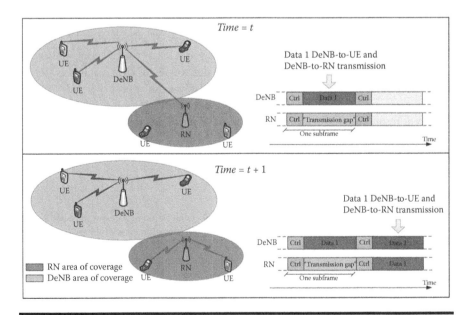

Figure 1.17   Example of transmission gaps in E-MBMS.

its area of coverage the "Data 1" that was stored in memory at *Time* = *t*, generating intercell interference.

As a result of this intercell interference pattern produced by RNs, the total intercell interference is greatly reduced (RNs are only causing interference half of the total transmission times) with a trade-off between throughput and intercell interference.

1.3.3 Multiresolution Transmission Schemes

For MBMS services, it makes sense to have two or more classes of bits with different error protection, to which different streams of information can be mapped. Depending on the propagation conditions, a given user can attempt to demodulate only the more protected bits or also the other bits that carry the additional information. By using nonuniformly spaced signal points in hierarchical modulations, it is possible to modify the different error protection levels [Vitthaladevuni and Alouini 2001, 2004].

Multiresolution techniques are interesting for applications where the data being transmitted are scalable, that is, it can be split into classes of different importance. For example, in the case of video transmission, the data from the video source encoders may not be equally important. The same happens in the transmission of coded voice. The introduction of multiresolution in a broadcast cellular system deals with source coding and transmission of output data streams. In a broadcast cellular system, there is a heterogeneous network with different terminal capabilities and connection speeds. For the particular case of video, scalable video consists of a common strategy presented in the literature that adapts its content within a heterogeneous communications environment [Li 2001; Liu et al. 2003; Vetro et al. 2003; Dogan et al. 2004; Holma 2007].

Scalable video, depicted in Figure 1.18, provides a base layer for minimum requirements, and one or more enhancement layers to offer improved quality at increasing bit/frame rates and resolutions [Li 2001]. Therefore, this method significantly decreases the storage costs of the content provider.

Besides being a potential solution for content adaptation, scalable video schemes may also allow an efficient usage of power resources in MBMS, as suggested in [Cover 1972]. This is depicted in Figure 1.18, where three separate physical resource blocks (PRBs) are provided for one MBMS service (e.g., @384 kbps): one for the base layer, at 1/3 rate of the total bit rate (128 kbps), and with a power allocation that can cover the whole cell range (UE3); another PRB for the first enhanced layer, also at 1/3 rate of the total bit rate (but aggregate rate of 256 kbps), with less power allocation than that of the base layer (UE2); and the third PRB for the second enhanced layer transmitted with small power to be received near the BS (UE1 with aggregate rate of 384 kbps).

The system illustrated in Figure 1.18 consists of three QoS regions, where the first region receives all the information, whereas the second and third regions receive the most important data. The QoS regions are associated with the geometry factor that reflects the distance of the UE from the BS antenna.

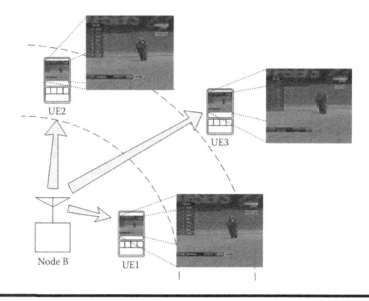

Figure 1.18   Scalable video transmission.

Scalable video transmission can also be implemented using different techniques. One possibility relies on the use of hierarchical constellations. The system illustrated in Figure 1.19 consists of 64-QAM hierarchical constellation. Strong bits blocks are QPSK demodulated, medium bits blocks are 16-QAM demodulated, and weak bits blocks are 64-QAM demodulated. Furthermore, hierarchical

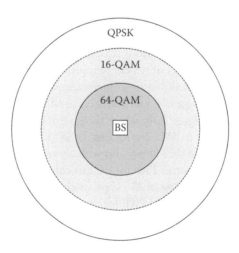

Figure 1.19   Example of the type of demodulation used inside a cell for a transmission of a 64-QAM hierarchical constellation.

constellations may also be combined with different channel coding rates. This corresponds to the concept of AMC [Souto et al. 2007].

It is worth noting that the nonuniform QAM constellation concept has already been incorporated in the DVB Terrestrial (DVB-T) standard.

Another possibility to implement scalable video transmission relies on the use of spatial multiplexing MIMO technique, where each transmit antenna sends a different data stream. The first data stream (most powerful) may include the base layer, whereas the enhancement layer may be sent by a second antenna (less powerful data stream). Depending on the power and channel conditions, a certain UE may successfully receive either the two streams or only the base layer.

1.3.4 Energy Efficiency in Wireless Communications

The tremendous expansion of mobile network terminals has contributed to the increase of the environmental footprint. Currently, more than 4 billion subscribers are using mobile phones around the world. The operation of both mobile phones and network infrastructure requires huge quantities of electrical energy, which currently represents up to 50% of the operational costs.

Energy-efficient wireless transmission techniques along with energy-efficient network architecture and protocols, hardware optimization, and renewable energy sources contribute to the implementation of the green cell networks concept, as they allow a reduction in the carbon emission footprint. Summarizing, the carbon footprint reduction can be achieved through advancements in three main different dimensions:

- Architecture and protocols: improvements in the transmission techniques and network protocols and architectures.
- Components: more efficient hardware implementation in terms of energy dependency.
- Energy provisioning: including a new strategy based on the generalized use of renewable energy sources.

The architecture and protocols dimension can be implemented in two different fronts: the network level and the link level.

The network level comprises the dimensioning and adjustment of the network as a whole such that a reduction of the energy consumption is possible without radically compromising the network performance and the spectral efficiency [Correia et al. 2010; Ferling et al. 2010]. This is achieved through the implementation of network protocols such as BS cooperation, multihop relay (see Figure 1.20), MIMO systems, hierarchical cellular structure,* or interference mitigation

---

* Using macrocells, microcells, picocells, femtocells, and so on.

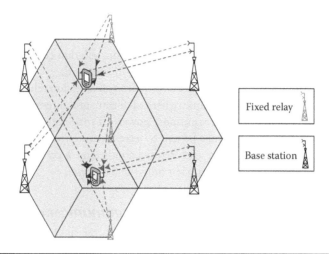

Figure 1.20   Cooperative MIMO system with fixed relay station.

schemes (including precoding). These techniques may support a reduction in the transmit power by both BSs and UE [Ericsson 2007].

Mobile adhoc networking (MANET) is a concept that allows the implementation of networks without making use of BSs, and therefore, achieving the same capability with less energy and with less visual impact and investment. Energy efficiency is also achieved by dynamically allowing an adaption of the cellular network architecture to traffic load fluctuations. In this case, the network architecture should be sufficiently dynamic such that it adapts to variations of the number of UEs, variations of the throughputs required by different UEs, or to various UEs' geographic densities. Cognitive radio brings flexibility to the network as it constitutes a mechanism that allows a dynamic and more efficient use of the spectrum. This translates in higher throughput per user or higher number of users per cell by exploiting spectrum opportunities. Cognitive radio constitutes an efficient alternative to the traditional approach where the networks were dimensioned to peak traffic scenarios.

The link level front is associated with the individual energy utilization in the interface between a UE and a BS. This includes synchronization and channel estimation techniques. Owing to multiplexing of pilot/training and data symbols, some of the available bandwidth and energy has to be consumed for accomplishing the transmission of the pilot symbols. Since the CIR can be very long, especially for block transmission schemes, the required channel estimation overheads can also be high, namely, for fast-varying scenarios. This translates in a reduction of the useful bit rate, decreasing the spectral efficiency and increasing the energy needs. A promising method for overcoming this problem relies on the employment of implicit pilots, also known as pilot embedding or superimposed pilots, which are added to the data block, instead of being multiplexed with it [Marques da Silva and Dinis 2011]. This means that we can significantly increase the pilots' density, while keeping the system capacity. Note that the MIMO implementation constitutes a mechanism to improve the spectral efficiency, throughput, or number of users per cell. Nevertheless, it requires additional energy spent in pilots, as the channel estimation is independently performed for each antenna pair, using a different pilot stream per transmit antenna. This results in an energy efficiency degradation of the MIMO relating to the SISO. The implementation of embedded pilots allows the MIMO system becoming an energy-efficient technique, as the energy spent in pilot symbols is very much reduced, and thus the MIMO can improve the SNR, coverage, or throughput, without additional transmit power (or even with a power reduction).

From the components' perspective, the basic concept relies on the development of signal processing techniques for smart components in order to obtain a reduction of the energy consumption. Moreover, the relationship between the output power and the consumed power of an equipment or component (e.g., a transmitter) should also be maximized. This can be viewed as power efficiency, and can also be improved by implementing a careful design and advanced signal processing. It is known that the energy consumed in a power amplifier (PA) represents between 50% and 80% of the electrical energy consumed in a BS [Correia et al. 2010]. OFDM transmission technique is characterized by high peak-to-average power ratio (PAPR)[*] levels, whose implementation requires a PA operating well below the saturation point. This results in poor power efficiency. A signal processing technique commonly employed to improve the energy efficiency of a PA relies on decreasing the PAPR through the use of a precoding technique, which reduces the dynamic range of a block transmission signal. This technique allows a PA operation closer to the saturation threshold, which translates in an energetic efficiency gain. Another signal processing technique currently in research relies on the implementation of smart cooling.[†] It is also worth noting that cooling systems tend to be more demanding in high-power transmitters. Reducing the required transmit power allows a simpler and less powerful cooling system.

Finally, a new energy provisioning strategy based on the use of renewable energy sources is another dimension that can be exploited in order to achieve a reduction of the carbon emission footprint. This includes the implementation of photovoltaic cells and/or wind generators in BSs. Moreover, photovoltaic cells can also be employed in mobile phones or in human clothes to charge it.

---

[*] Such PAPR level increases with the increase of the number of OFDM subcarriers.
[†] Cooling systems represent up to 30% of the consumed BS energy.

Acknowledgment

This work was supported by the FCT (Fundação para a Ciência e Tecnologia) via project PEst-OE/EEI/LA0008/2013.

References

3GPP, TR 25.214-v5.5.0, Physical layer procedures (FDD), 2003a.

3GPP, TR 25.211-v5.2.0, Physical channels and mapping of transport channels onto physical channels (FDD), 2003b.

3GPP, Feasibility study for further advancements for E-UTRA (LTE-Advanced), TR 36.912 v9.2.0, 2010a.

3GPP, Evolved Universal Terrestrial Radio Access (E-UTRA); Relay architectures for E-UTRA (LTE-Advanced), TR 36.806 V9.0.0, May 2010b.

Alamouti, S. M. A simple transmitter diversity scheme for wireless communications, *IEEE Journal on Selected Areas in Communications*, 16(8), 1451–1458, October 1998.

Andrews, J. G., A. Gosh, R. Muhamed, *Fundamentals of WiMAX: Understanding Broadband Wireless Networking*, Prentice-Hall, New Jersey, 2007.

Astely, D., E. Dahlman, A. Furuskar, Y. Jading, M. Lindstrom, S. Parkvall, LTE: The evolution of mobile broadband, *IEEE Communications Magazine*, 47(4), 44–51, April 2009.

Bhat, P., S. Nagata, L. Campoy, I. Berberana, T. Derham, G. Liu, X. Shen, P. Zong, J. Yang, LTE-advanced: An operator perspective, *IEEE Communications Magazine*, 50(2), 104–114, February 2012.

Correia, L., D. Zeller, O. Blume, D. Ferling, Y. Jading, G. István, G. Auer, L. Perre, Challenges and enabling techniques for energy aware mobile radio networks, *IEEE Communications Magazine*, 48, 66–72, November 2010.

Cover, T. Broadcast channels, *IEEE Transactions on Information Theory*, IT-18, 2–14, January 1972.

Diehm, F., P. Marsch, G. Fetweeis, The FUTON prototype: Proof of concept of coordinated multi-point in conjunction with a novel integrated wireless/optical architecture, *Proceedings of IEEE WCNCW'10*, Sydney, NSW, pp. 1–4, April 2010.

Dogan, S. et al., Video content adaptation using transcoding for enabling UMA over UMTS, *Proceedings of WIAMIS 2004*, Lisbon, Portugal, April 2004.

Eklund, C. et al., IEEE Standard 802.16: A technical overview of the Wireless MAN air interface for broadband wireless access, *IEEE Communications Magazine*, 40(6), 98–107, June 2002.

Ericsson, Sustainable energy use in mobile communications, White paper, August 2007, http://www.ericsson.com, accessed December 23, 2011.

Ferling, D., T. Bohn, D. Zeller, P. Frenger, I. Gódor, Y. Jading, W. Tomaselli, Energy efficiency approaches for radio nodes, *Proceedings of Future Networks Mobile Summit 2010*, Florence, Italy, June 2010.

Foschini, G. J. Layered space-time architecture for wireless communication in a fading environment when using multiple antennas, *Bell Laboratories Technical Journal*, 1(2), 41–59, Autumn 1996, .

Foschini, G. J., M. J. Gans, On limits of wireless communications in fading environments when using multiple antennas, *Wireless Personal Communications*, 6, 315–335, March 1998.

Holma, H., A.Toskala, *WCDMA for UMTS: HSPA Evolution and LTE*, 4th ed., John Wiley & Sons, New York, 2007.

Hottinen, A., O. Tirkkonen, R. Wichman, *Multi-Antenna Transceiver Techniques for 3G and Beyond*, John Wiley & Sons, Chichester, UK, 2003.

IEEE802.16RelayTaskGroup, 2008;http://www.ieee802.org/16/relay,accessedJanuary3, 2012.

IEEE Standard 802.16-2004, Air interface for fixed broadband wireless access systems, 2004.

IEEE Standard 802.16e-2005, Amendment to air interface for fixed and mobile broadband wireless access systems—Physical and medium access control layers for combined fixed and mobile operations in licensed bands, 2005.

IEEE Standard 802.16 m-07/002r8, IEEE802.16m System Requirements, January 2009; http://ieee802.org/16/tgm/index.html, accessed January 5, 2012.

ITU Paves Way for Next-Generation 4G Mobile Technologies, ITU press release, October 21, 2010.

ITU-R Recommendation M. 2133—Requirements, evaluation criteria and submission templates for the development of IMT-Advanced, 2008.

Kudoh, E., F. Adachi, Transmit power efficiency of a multi-hop virtual cellular system, *Proceedings of IEEE Vehicular Technology Conference, 2003 (VTC2003-Fall)*, Orlando, USA, 5, 2910–2914, October 6–9, 2003.

Lee, D., H. Seo, B. Clerckx, E. Hardouin, D. Mazzarese, S. Nagata, K. Sayana, Coordinated multipoint transmission and reception in LTE-Advanced deployment: Scenarios and operational challenges, *IEEE Wireless Communications Magazine*, 50(2), 148–155, February 2012.

Li, W. Overview of fine granularity scalability in MPEG-4 video standard, *IEEE Transactions CSVT*, 11(3), 301–317, March 2001.

Liu, J., B. Li, Y.-Q. Zhang, Adaptive video multicast over the Internet, *IEEE Multimedia*, 10(1), 22–33, January–March 2003.

Marques da Silva, M. *Multimedia Communications and Networking*, 1st ed., CRC Press Auerbach Publications, Boca Raton, USA, March 2012, ISBN: 9781439874844.

Marques da Silva, M., A. Correia, Space time diversity for the downlink of WCDMA, *IEEE— Wireless Personal and Mobile Communications—WPMC'01*, Aalborg, Denmark, September 9–12, 2001.

Marques da Silva, M., A. Correia, Space time block coding for 4 antennas with coding rate 1, *IEEE—International Symposium on Spread Spectrum Techniques and Application— ISSSTA*, Prague, Czech Republic, September 2–5, 2002a.

Marques da Silva, M., A. Correia, Space time coding schemes for 4 or more antennas, *IEEE— International Symposium on Personal Indoor and Mobile Radio Communications— PIMRC'02*, Lisbon, Portugal, September 16–18, 2002b.

Marques da Silva, M., A. Correia, Combined transmitdiversity and beamforming for WCDMA, *IEEE EPMCC'03*, Glasgow, Scotland, April 2003.

Marques da Silva, M., A. Correia, J. C. Silva, N. Souto, Joint MIMO and parallel interference cancellation for the HSDPA, *Proceedings of IEEE International Symposium on Spread Spectrum Techniques and Applications 2004 (ISSSTA'04)*, Sydney, Australia, September 2004.

Marques da Silva, M., A. Correia, R. Dinis, On transmission techniques for multi-antenna W-CDMA systems, *European Transactions on Telecommunications*, 20(1), 107–121, January 2009.

Marques da Silva, M., A. Correia, R. Dinis, N. Souto, J. Silva, *Transmission Techniques for Emergent Multicast and Broadcast Systems*, 1st ed., CRC Press Auerbach Publications, Boca Raton, USA, June 2010, ISBN: 9781439815939.

Marques da Silva, M., A. Correia, R. Dinis, N. Souto, J. Silva, *Transmission Techniques for 4G Systems*, 1st ed., CRC Press Auerbach Publications, Boca Raton, USA, November 2012, ISBN: 9781466512337.

Marques da Silva, M., R. Dinis, Iterative frequency-domain detection and channel estimation for space-time block codes, *European Transactions on Telecommunications*, 22(7), 339–351, November 2011, doi: 10.1002/ett.1484, http://dx.doi.org/10.1002/ett.1484.

Marques da Silva, M., R. Dinis, A. Correia, A V-BLAST detector approach for W-CDMA signals with frequency-selective fading, *Proceeding of the 16th IEEE Personal Indoor and Mobile Radio Communications 2005 (PIMRC'05)*, Berlin, Germany, September 11–14, 2005.

Nam, S. H., K. B. Lee, *Transmit Power Allocation for an Extended V-BLAST System*, *PIMRC'2002*, Lisbon, Portugal, September 2002.

Ohrtman, F. *WiMAX Handbook*, McGraw-Hill Communications, New York, 2008.

Oyman, O., J. N. Laneman, S. Sandhu, Multihop relaying for broadband wireless mesh networks: From theory to practice, *IEEE Communications Magazine*, 45(11), 116–122, November 2007.

Papadogiannis, A., E. Hardouin, D. Gesbert, Decentralising multicell cooperative processing: A novel robust framework, *EURASIP Journal of Wireless Communications and Networking*, 2009, 1–10, August 2009.

Peel, C. B. On "Dirty-Paper coding", *IEEE Signal Processing Magazine*, 20(3), 112–113, May 2003.

Peters, S. W., R. W. Heath, The future of WiMAX: Multihop relaying with IEEE 802.16j, *IEEE Communications Magazine*, 47(1), 104–111, January 2009.

Rooyen, P. V., M. Lötter, D. Wyk, *Space-Time Processing for CDMA Mobile Communications*, Kluwer Academic Publishers, Boston, 2000.

Schacht, M., A. Dekorsy, P. Jung, *Downlink Beamforming Concepts in UTRA FDD*, Kleinheubacher Tagung 2002, Kleinheubacher Berichte 2003.

Schoene, R., R. Halfmann, B. H. Walke, An FDD multihop cellular network for 3GPP-LTE, *VTC 2008*, pp.1990–1994, May 2008.

Souto, N., R. Dinis, J. C. Silva, Iterative decoding and channel estimation of MIMO-OFDM transmissions with hierarchical constellations and implicit pilots, *International Conference on Signal Processing and Communications (ICSPC'07)*, Dubai, United Arab Emirates, November 24–27, 2007.

Stencel, V., A. Muller, P. Frank, LTE advanced—A further evolutionary step for next generation mobile networks, *20th International Conference on Radioelektronika*, Brno, Czech Republic, pp. 19–21, April 2010.

Sydir, J., R.Taori, An evolved cellular system architecture incorporating relay stations, *IEEE Communications Magazine*, 47(6), 150–155, June 2009.

Tarokh, V., H. Jafarkhani, A. R. Calderbank, Space-time block codes from orthogonal designs, *IEEE Transactions on Information Theory*, 45(5), 1456–1467, July 1999.

Telatar, I. E. Capacity of multiantenna Gaussian channels, AT&T Bell Laboratories, Tech. Memo., June 1995.

Telatar, I. E., Capacity of multiantenna Gaussian channels, *European Transactions on Communications*, 10(6), 585–595, 1999.

Vetro, A., C. Christopoulos, H. Sun, Video transcoding architectures and techniques: An overview, *IEEE Signal Processing Magazine*, 20(2), 18–29, March 2003.

Vitthaladevuni, P. K., M.-S.Alouini, BER computation of 4/M-QAM hierarchical constellations, *IEEE Transactions on Broadcasting*, 47(3), 228–239, September 2001.

Vitthaladevuni, P. K., M.-S.Alouini, A closed-form expression for the exact BER of generalized PAM and QAM constellations, *IEEE Transactions on Communications*, 52(5), 698–700, May 2004.

Yarali, A., S. Rahman, WiMAX broadband wireless access technology: Services, architecture and deployment models, *IEEE CCECE2008*, Niagara Falls, Ontario, pp. 77–82, May 4–7, 2008.

Chapter 2

MIMO Detection Methods

Francisco A. Monteiro, Ian J. Wassell,
and Nuno Souto

Contents

2.1 Terminology: Old and New Signal Processing

Digital transmission has progressed during the last two decades of the twentieth century aiming at higher data rates, less bandwidth for a fixed data rate (i.e., higher spectral efficiency), while spending the least possible amount of energy per bit, given a certain additive noise. There is, however, a fourth dimension to the problem: the complexity involved in the "construction" of the transmitted signal (i.e., the complexity associated with the modulation and the coding scheme), and the complexity involved in the detection and decoding of the signal at the receiver. Multiple-input multiple-output (MIMO) spatial multiplexing (SM) allows unprecedented spectral efficiencies at the expense of high detection complexity due to the fact that the underlying detection problem is equivalent to the closest vector problem (CVP) in a lattice. Finding better *detection algorithms* (also referred in the literature as *detection methods* or *detection techniques*) to deal with the problem has been a central topic since the last decade of research in MIMO SM and yet an efficient technique for the 8×8 configuration in LTE-A is not a closed problem [Bai et al. 2012].

This chapter introduces the reader to the most prominent detection techniques for MIMO, namely, linear filtering-based receivers and nonlinear approaches, such as ordered successive interference cancellation (OSIC), lattice-reduction-aided, sphere decoding (SD) concept, among other more recent developments. We believe that the easiest way to cut through the MIMO detection methods is to understand the underlying lattice problem and the type of manipulations that each of these methods executes in the lattice. In addition to a unifying geometric interpretation, the lattice interpretation often offers additional insights. For example, the geometric relation between the primal and the dual lattice is clarified, leading to a simple interpretation of the optimal detection order of the layers (i.e., symbols at the receiver antennas) in OSIC.

Typically, progress toward Shannon capacity was achieved by means of channel coding concatenated with higher-order modulation, eventually at the expense of higher complexity, mostly at the receiver. Turbo-codes, low-density parity-check codes (LDPCs), and lattice-coded modulation are examples of that path of increasing complexity (mostly in detection and decoding). As it will be seen in subsequent chapters, coding still plays a crucial role in any system.

The rapid rise in the computing power available at the handset now permits rather complex baseband processing. Over a decade, we have witnessed the core problems in communication engineering being much less electronics-centric and much more algorithmic-centric. Modern communication theory is now largely entwined with problems traditionally in the domain of theoretical computer science (e.g., machine learning, data structures, algorithms, and their complexity), or, more generally, in applied mathematics (related with matrix algebra, discrete mathematics, integer programming, or combinatorics). Sometimes the separation is now only barely possible by looking at the application in mind and not by the nature of the problem itself. More generally, this can also be said of many of the aspects of information theory and coding theory. The fields of image communication or source coding (compression) always incorporated a wide variety of fundamental research. What is new is that this fusion propagated down to the physical layer, once clearly within electrical engineering.

The problem of detection in MIMO SM constitutes a clear example of an algorithmic problem at the physical layer. The problem is analogous to the CVP in lattices (sometimes also called the *nearest vector problem*, e.g., [Lenstra 2008]). The study of the problem began in the realms of theoretic computer science, cryptography, complexity theory, algorithmic number theory, and in some domains of applied mathematics. The curious reader may want to delve into details of the field. For example, a recent detection technique looks at lattices from a group theory perspective [Monteiro and Kschischang 2011]. It is shown that it is possible to approximate the typical lattices encountered in MIMO by a lattice having a trellis representation.

2.1.1 Diversity Order and Multiplexing Gain

From Shannon we know that, in the additive white Gaussian noise (AWGN) channel, a symbol error rate (SER) curve $P_s(\rho)$ is a function of the signal-to-noise ratio (SNR), ρ, and can be as steep as one wants. In the limit, the curve $P_s(\rho)$ can have an infinite negative slope. For the Rayleigh fading[*] channel, it is well known that $P_s(\rho)$ exhibits a −1 slope in the uncoded single-input single-output (SISO) case, that is,

---

[*] This limiting behavior is only characteristic for the Rayleigh fading assumption. Other fading statistics lead to different diversity orders.

one finds that $P_s(\rho) \propto \rho^{-1}$. One door that MIMO opens is the possibility of increasing (in modulo) that slope, that is, obtaining a faster reduction of the error rate as SNR increases. Formally, one defines the *diversity order*, corresponding to the slope

$$d = \lim_{\text{SNR} \to \infty} -\frac{\log(P_s(\rho))}{\log(\rho)}. \tag{2.1}$$

This *diversity order* measures how many statistically independent copies of the same symbol the receiver is able to separate. In brief, this amounts to the number of independent fading coefficients that the receiver can average in order to produce a reliable estimate of a transmitted symbol. Not surprisingly, the maximum available diversity that can be attained is $d_{\max} = N_T N_R$, in the case of uncorrelated fading, as in the model that will be introduced in Section 2.2.1.

The benefits of vector communication with spatial diversity are not limited to this increased slope. Think of a SISO setup where one switches from a 4-PAM constellation (2 bps/Hz) to an 8-PAM constellation, in this case the same error rate can be obtained by increasing the SNR by 6 dB while 1 more bit/s can be transmitted (now 3 bps/Hz) using the same bandwidth. If one changes from a 16-QAM (4 bps/Hz) to 64-QAM constellation, the same additional 6 dB are required to achieve the same SER, though the spectral efficiency is increased by a factor of 2. It is said that the *multiplexing gain* of the latter QAM constellation is higher than the one of PAM. In the MIMO general case, this gain is defined as

$$g = \lim_{\text{SER} \to \infty} \frac{\log(R)}{\log(\rho)}. \tag{2.2}$$

When plotting the symbol error rate (SER) versus the SNR, the existing diversity d in the communication link is simply the slope (in the asymptotic regime) of the SER curve. However, the interpretation of the multiplexing gain in a typical SER plot is not so straightforward. The metric g indicates how the capacity increases with the SNR, which is a common representation in information theory since Shannon but is less useful in practice. In terms of the SER, the multiplexing gain g measures how fast spectral efficiency can increase with the increase of SNR while keeping the same error rate and corresponds to the maximum number of independent *layers* or *parallel channels*, being limited by

$$g_{\max} = \min(N_T, N_R). \tag{2.3}$$

It was showed in [Zheng and Tse 2003] that there is a trade-off between d and g, that is, the famous diversity-multiplexing trade-off (DMT): increasing one leads to a decrease in the other. The only pairs (d, g) that are allowed lie on the following piecewise-linear function constructed by connecting the points defined by

$$d(k) = (N_R - k)(N_T - k), \quad \text{for } 0 \leq k \leq g_{\max}. \tag{2.4}$$

One interpretation of this trade-off is that some subset of the antennas provide for the existence of several layers, while the remaining ones assure the diversity, but they cannot all be contributing to both objectives. The operational information conveyed by the DMT curve is very often confusing and misleading in the literature. The operational meaning of the DMT curve lies in the extremities connected by each of the piecewise-linear segments. Each segment defines an operation "mode" with a pair (d, g) defined by the extremities of the segment, which then define the *maximum* value that each of the parameters can assume.

The practical relevance of the DMT has been much criticized though; some argue that it is only valid for the ideal (uncorrelated) Rayleigh channel and only at an impractically high SNR regime, while authors in the LTE literature rebuke the usefulness of DMT for ignoring other aspects of the transmission chain, which bring other sources of diversity, for example, from the frequency domain or from coding. Indeed, providing a physical interpretation of g in the finite SNR regions is impossible.

Despite the connection between the two gains, there is a division in system design between either (i) aiming at full diversity or (ii) aiming at maximizing the multiplexing gain, which in practice has been measured by the spectral efficiency gain provided by the transmission over multiple layers. The development of *space–time codes* (STC) addresses the first objective, while the SM concept aims at the latter. Physical layer designers have to opt between STC and SM. Another attempt to match them under a unified system design was presented in [El-Hajjar and Hanzo 2010]. A pragmatic approach for switching between them according to the channel conditions was considered in [Heath and Paulraj 2005]. Which one is preferable is a question that was thoroughly investigated in [Lozano and Jindal 2010]. The authors have looked into the problem of considering MIMO in systems that combine MIMO with interleaving and coding, wideband channelizations with OFDM, link adaptation (adaptive modulation and rate control), and automatic repeat request (ARQ). Their work addresses precisely some of the criticism aimed at the somewhat idealized conditions in which DMT was defined [Tse and Viswanath 2005]. The conclusion was that with all the diversity available in time and frequency (including carrier aggregation), spatial diversity becomes redundant and SM should be the *only* objective when designing the MIMO aspects of the physical layer, especially when the channel is wider than 10 MHz. Note that LTE release 8 defines channel bandwidths up to 20 MHz, LTE-Advanced (i.e., release 10) already defines a bandwidth up to 100 MHz [Parkvall et al. 2011], and 100 MHz is possible in WiMAX by carrier aggregation [Ahmadi 2009], and 802.11n is defined up to 40 MHz [Perahia and Stacey 2008].

2.1.2 Detection for Space–Time Codes

The celebrated Alamouti 2×1 space–time code (STC) was presented in 1998 [Alamouti 1998]. The scheme achieves the same diversity as the traditional maximum ratio combining (MRC) with two antennas at the transmitter while having only one antenna at the receiver, and transposing the burden of two antennas to the transmitter.* The structure of the Alamouti code permits a very simple decoding method. This advantage was extended to larger dimensions by Tarokh, Jafarkhani, and Calderbank [Tarokh et al. 1999], who discovered other orthogonal space–time block codes (OSTBC), which are also easy to decode owing to their structure. Interestingly, the Alamouti scheme realizes the optimal trade-off for the 2×1 configuration; the other OSTBC do not achieve that optimal diversity-multiplexing trade-off (DMT) [Tse and Viswanath 2005], nor does the Alamouti 2×2 system.

The computational complexity for their decoding was recently analyzed in depth [Ayanoglu et al. 2011]. Sadly, these constructions for STBCs were shown not to exist for configurations larger than 4×4. They are however instances of linear dispersion codes which, albeit not necessarily orthogonal, also spread the symbols in both space and time. The design and study of space–time codes constitutes a research field of its own [Larsson and Stoica 2003, Jafarkhani 2005, Giannakis et al. 2007]. Another family of codes, proposed earlier than STBC, is the space–time trellis codes (STTC) [Tarokh et al. 1998], where the symbols emanating from the antennas not only depend on the new data but also on the state of an encoder (as in convolutional codes).

Lattices have been essential in coding for MIMO just as they had been for the AWGN channel and afterwards for the SISO flat Rayleigh channel. A fruitful line of research, mostly lead by Oggier, Viterbo, and Belfiori, uses algebraic number theory for finding good lattice codes for MIMO [Viterbo and Oggier 2004, Oggier 2005, Oggier et al. 2006]. The 2×2 so-called Golden Code, which simultaneously achieves full diversity 4 and full multiplexing gain equal to 2, was created under that framework. One other example of STC with lattices was the discovery by [Gamal et al. 2004] of the so-called lattice space–time (LAST) codes, which realize the DMT. These codes are grounded in the Erez and Zamir constructions [Erez and Zamir 2004]. The optimum detection algorithm for LAST codes was proposed in [Su et al. 2009].

2.1.3 Detection for Spatial Multiplexing

This technique focuses *only* on the objective of increasing the data rate, leaving behind rather than obtaining any spatial diversity in an $N_T \times N_R$ configuration. If the receiver is able to correctly estimate the channel matrix (although that information is oblivious to the transmitter), signal processing at the receiver can extract the mutual interference and decouple $\min(N_T, N_R)$ streams of independent data.

---

* There is a power penalty though, given that only half of the power is collected with one antenna at the receiver rather than two. This results in an error probability curve translated by 3 dB with respect to the MRC performance curve. The extrapolation to a 2×2 configuration of the Alamouti space–time code is straightforward and that loss is recovered.

The first spatial technique, proposed by Foschini, as mentioned in Chapter 1, was D-BLAST (diagonal Bell Labs layered space–time). This technique uses error correcting codes and "rotates" over time the distribution of the different code streams (the layers) across the antennas. Considering two-dimensional space–time frames, different layers end up associated with distinct *diagonals* of a space–time grid (or frame). This creates spatial diversity besides SM. However, owing to the detection complexity it incurred it was dropped very early in favor of a much simpler approach known as vertical-BLAST (V-BLAST). One other downside with D-BLAST is that it wastes some space–time resources at the time extremities of a frame, although one can minimize that effect by appropriately dimensioning the frames [Goldsmith 2005, Section 10.6]. It should be noted that, despite its impracticability, when the number of antennas approaches infinity, D-BLAST is able to approach the capacity of the MIMO channel [Tse and Viswanath 2005, Section 8.5]. Interestingly, the V-BLAST architecture is able to approach capacity by rate adaptation across the different layers, when there is channel state information at the transmitter (CSIT) [Bölcskei et al. 2006, Chapter 5] (for a practical example, see [Magarini 2007]). Near capacity MIMO open-loop SM has been achieved using soft SD (see Section 2.5.7) concatenated with linear codes by means of an interleaver [Hochwald and ten Brink 2003].

V-BLAST does not make use of all the spatial diversity that exists in the MIMO channel as D-BLAST is able to. In the vertical version, there exists a fixed association between the parallel sub-channels and the antennas at both extremities of the link. Note that these sub-channels are also frequently called layers, even in uncoded systems, where each layer simply corresponds to the symbols from a particular transmit antenna. Consequently, the maximum spatial diversity V-BLAST offers is, at most, N_R, and even so, only if maximum likelihood detection (MLD) is used. In the case of the OSIC detector, the diversity is only $N_T - N_R + 1$, that is, only one in the common case of symmetric configurations. This latter fact was conjectured about from very early on [e.g., Paulraj et al. 2003, p. 158] but has only very recently been proven [Jiang et al. 2011] (this will be further commented on in Section 2.5.4).

Unlike the Alamouti (and other OSTBCs), both V- and D-BLAST could be trivially extended to any number of antennas, as there is no structural constraint in their design. However, to optimally detect all layers (with an error performance curve that exhibits all the diversity that these schemes still provide and with no gain penalty), involves a computational complexity that grows exponentially with the number of antennas at the receiver.

The predominance of the V-BLAST architecture made it almost synonymous with SM. Furthermore, as Golden et al. presented a simple receiver based in interference cancellation for V-BLAST [Golden et al. 1999], the name became much associated with that particular detection technique. The name successive interference cancellation (SIC) will subsequently be used in this work for the concept that underlies the detection method first proposed for V-BLAST in [Golden et al. 1999]. The V-BLAST architecture will be simply referred to throughout this work as SM.

SM does not require CSIT (since it is an open-loop architecture) and the capacity grows linearly with min(N_T, N_R). In practice, the number of antennas in some equipment is not only limited by the signal processing complexity at the receiver, but also by the physical dimension the antenna array may have; one should bear in mind that as the spacing between antennas diminishes, they become increasingly correlated and the capacity of the system diminishes [e.g., Paulraj et al. 2003, Section 4.6.1].

The major limitation in SM is the large algorithmic complexity involved in the optimum detection applying the MLD principle to achieve optimum detection. MLD captures the spatial diversity of the architecture while removing all the mutual interference between layers.

In the last decade, there was a burst of research on this problem: how to detect the received vector with a performance as close as possible to the optimal yet having a reduced complexity compared with MLD? The most abstract and general description of this problem is the CVP in a lattice, the applications of which go far beyond the MIMO detection problem in SM. The detection of some STC is also a CVP in a lattice after vectorizing the space–time matrix code words [Viterbo and Oggier 2004].

Other communication problems such as inter-symbol interference (ISI) channels [Mow 1991] and multi-user detection (MUD) [Verdú 1998] are formally the same, and may be encapsulated as a general equalization problem (as proposed in [Zhang et al. 2009]) that can be mapped as a CVP. Table 2.1 lists some terminology that is used in these different frameworks.

2.2 The Geometry of Lattices

The regularity of a lattice lends itself for the representation of problems where signals are interpreted as a point in a multidimensional space defined in some basis. One of the most important lattice problems is the CVP [Agrell et al. 2002], which consists in finding the point that is the one at the shortest distance from a given off-lattice target point.

The study of lattices began in the 1890s with Minkowski who created the then new field of *geometry of numbers* [Nguyen and Micciancio 2005, Cassels 1971]. Lattices are related with problems in the integer domain, such as continued fractions, simultaneous Diophantine equations (systems of questions where one is solely interested in integer solutions) [Schrijver 1986]; simultaneous Diophantine approximation [Lovász 1986, Lenstra 2008], (finding the closest rational numbers to a set of real numbers with the restriction that they all have the same denominator), and several other fundamental problems in number theory [Lenstra 2008, Kannan 1987], and in integer programming [Schrijver 1986, Bertsimas and Weismantel 2005]. These problems can usually be reduced to the CVP or to the shortest vector problem (SVP) in a lattice [Micciancio and Goldwasser 2002].

In the last three decades, one could have found applications of lattices in vector quantization and image coding [Dubois 1985, Zamir 1996, Agrell and Eriksson

Table 2.1   Different Names of Analogous Techniques

	MIMO	*Equalization for ISI Channels*	*Multi-User Communication*
Inversion (linear)	• Zero-forcing (ZF) • Channel inversion • Decorrelation	Zero-forcing (ZF) equalization	Decorrelation
Minimum mean squared error (MMSE)	MMSE	MMSE filtering	MMSE detection
Interference cancellation	• Nulling and cancelling • Successive interference cancellation (SIC) • V-BLAST detection	Decision feedback equalization (DFE)	• Iterative multi-user detection (MUD) • Successive interference cancellation (SIC)
Optimum detection	• Maximum likelihood detection (MLD) • Exhaustive search	Maximum likelihood sequence detection (MLSD)	• ML detection • Brute force • SD (near optimum)
Precoding	• Multiuser-MIMO • Broadcast channel (BC)	• ISI Precoding • Costas precoding • Tomlinson–Harashima precoding (THP)	Dirty paper coding (DPC)
Parallel sub-channels	• Closed-loop SU-MIMO • SVD and water filling • Communication over eigen-modes • Eigen-beam spatial division multiplexing • Precoding • Beamforming	• OFDM • Multi-tone modulation • Filter bank multicarrier	Not defined

1998], and the application of lattices in SISO communications has a long history in coding for the bandwidth-limited AWGN channel [Forney 1988] and in SISO fading channels [Boutros et al. 1996]. Despite that, it was only during the boom of research in MIMO in the last decade that lattices started to be thoroughly investigated in relation to communication problems as they are the mathematical object underlying problems such as the broadcast channel [Taherzadeh et al. 2007], the design of STCs [Viterbo and Oggier 2004], and, of course, the CVP in SM detection [Larsson 2009]. Interestingly, these MIMO communication problems triggered a series of rediscoveries and novel uses of ideas previously studied in algorithmic number theory. Examples of this are (i) the V-BLAST detection as proposed in [Wolniansky et al. 1998], which turns out to be the Babai nearest plane algorithm [Mow 2003, Galbraith 2013]; (ii) SD was already used in SISO [Viterbo and Boutros 1999] but improved ways for traversing a tree were rediscovered in [Agrell et al. 2002] making use of the much earlier findings of [Fincke and Pohst 1985] and by [Schnorr and Euchner 1994] in number theory; (iii) the use of lattice reduction techniques such as the Lenstra–Lenstra–Lovász (LLL or L^3) algorithm or Seysen's reduction remained unknown to the communications community until 2002 [Yao and Wornell 2002] and 2007, respectively (the advantages of Seysen's reduction for MIMO were simultaneously indicated by [Seethaler et al. 2007] and [Niu and Lu 2007]). However, in 2007 the 25th anniversary of the LLL algorithm was celebrated by the algorithmic number theory community with a special event and the publication of a book listing its profound implications in many problems [Nguyen and Vallée 2010].

Now that communication theory is evolving from point-to-point transmission problems to network coding ideas, lattices remain an essential tool [Feng et al. 2010, Nazer and Gastpar 2011].

Although lattices are simple to define mathematically, and have an apparent geometrical simplicity, they are, as already mentioned, closely related to many of the most difficult algorithmic problems with NP-hard complexity. As a natural consequence, lattice problems also assumed a central role in cryptography in the last decade [Micciancio and Goldwasser 2002, Micciancio and Regev 2009, Galbraith 2013], given the complexity of the algorithms and the difficulty they pose to an attacker who does not possess a trapdoor to solve the problem (such as a "good" basis for the lattice).

2.2.1 Definitions

2.2.1.1 Lattice

There are several ways of specifying a lattice* Λ. The most common method involves a set of linearly independent *generator vectors* $\mathbf{h}_i$ [Nguyen and Micciancio 2005],

---

* The term lattice has two meanings in mathematics. The same name appears in order theory (in discrete mathematics and abstract algebra) [Anderson 2003, Lidl and Pilz 2010], a subject totally unrelated to the lattices in number theory or geometry of numbers.

which constitute a *basis* for the lattice. A (real) lattice is then defined as the infinite set defined by

$$\Lambda = \left\{ \mathbf{y} \in \mathbb{R}^n : \mathbf{y} = \sum_{i=1}^{n} \mathbf{h}_i x_i = \mathbf{H} \cdot \mathbf{x}, \quad x_i \in \mathbb{Z}, \mathbf{h}_i \in \mathbb{R}^n \right\}. \tag{2.5}$$

The definition can be extended to complex lattices but, because it is possible to transform any complex lattice into a real lattice (as will be seen in Section 2.3.2), one can settle for limiting the description to real lattices.

The integer combination of real or complex n-dimensional vectors generates a discrete set of points with the properties of a *group*, namely: *closure, associativity, identity,* and *inversion* [Biggs 2002, Chapter 20]. Indeed, the shortest possible definition of a (real) lattice is the following: a lattice is a *discrete* Abelian (i.e., additive or commutative) subgroup of $\mathbb{R}^n$. The two definitions are equivalent [Siegel 1989, p. 44].

A consequence of the last definition given for Λ is that for any two elements $\mathbf{x}, \mathbf{y} \in \Lambda$, then the difference $\mathbf{x} - \mathbf{y} \in \Lambda$ (i.e., a lattice is closed under subtraction). Notice that for a structure to be a lattice, the group property by itself does not suffice; the structure also needs to be *discrete* (i.e., for each lattice point there exists a hyper-ball with radius $\varepsilon > 0$, which is centered at the lattice point and not containing any other lattice point inside; that is, the distance between lattice points is larger than ε). This caveat is sometimes forgotten by some authors. For example, the group property is preserved as the linear projection operator. However, that projection is not necessarily constituted by discrete structure [Banihashemi 1997, p. 20].

According to Equation 2.5, a lattice is an infinite set of points resulting from integer combinations of the columns of the generator matrix $\mathbf{H}$. It should be noted that some authors prefer to span the row space of a matrix, which then reflects algebraically in some of the definitions that follow.

There are other ways of specifying a lattice that do not have a set of generator vectors; however, Equation 2.5 is the most prevalent one while these other techniques remain largely unmentioned in the literature on lattices. In MIMO literature, Equation 2.5 is the only way used for specifying a lattice, perhaps because it follows directly from the natural vector description of SM. In [Lenstra 2008, Section 4], Hendrik Lenstra describes several alternative ways of specifying a lattice but comments that some are recognizably difficult to convert into Equation 2.5 or even to convert between themselves. None of the unconventional techniques seem to have played any role so far in the study of MIMO.

One of the alternative techniques to define a lattice is only applicable to the so-called *cyclic lattices*. These lattices are endowed with a specific structure that allows them to be defined by means of one modular equation (lattices with one cycle) [Paz and Schnorr 1987] or by d modular equations (said to have d-cycles) [Trolin 2004]. Interestingly, there is a connection to the field of numerical integration of multidimensional functions where cyclic lattices are closely related with the so-called *lattice rules* [Sloan and Joe 1994].

2.2.1.2 Fundamental Region

Given a certain basis of a lattice, the *fundamental region* that is associated to that basis is defined as

$$\mathcal{R}(\mathbf{H}) = \{\mathbf{Hx} : 0 < x_i < 1\}. \tag{2.6}$$

The fundamental region cannot contain any lattice point inside it. If there was at least one point inside, it could not be represented by an *integer* combination of generator vectors, which are precisely the sides defining that fundamental region (c.f. Figure 2.1). If that happens, then the set of vectors is *not* a basis of the lattice but a basis of one of its *sublattices*. A sublattice Λ' is also a lattice and the volume is $\mathrm{vol}(\Lambda') > \mathrm{vol}(\Lambda)$ (the technical definition of the volume of a lattice will shortly be given).

Note that different sets of vectors may generate the same lattice. Indeed, the number of admissible bases for a lattice is infinite; it is easy to infer from Figure 2.1 that it is always possible to select some point further distant from the origin to replace a generator and still have a fundamental region without including any lattice point in its interior. Moreover, all these different bases are related by unimodular transformations, as will be described below.

2.2.1.3 Voronoi Region

The region of the space where the lattice is embedded (i.e., in the continuous Euclidian space where the lattice exists) that contains all the points in the span of

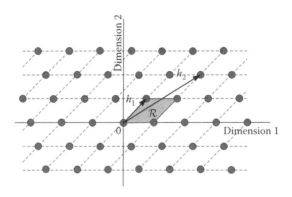

Figure 2.1 **A lattice in $\mathbb{R}^2$ and the fundamental region associated with a particular basis.**

the lattice which are closer to a given lattice point $\mathbf{x}$ than to any other point in the lattice is called the *Voronoi region* and is defined by

$$\mathcal{V}(\Lambda) = \left\{ \mathbf{z} \in \operatorname{span}(\Lambda) : \forall \mathbf{y} \in \Lambda \, \|\mathbf{x} - \mathbf{z}\| < \|\mathbf{y} - \mathbf{z}\| \right\}. \tag{2.7}$$

This (open) region is a characteristic of the lattice and independent of any particular generating matrix, and is the most interesting fundamental region amid the infinite number of other possible fundamental regions one can define to tile the entire space as it constitutes the optimal decision region for the CVP in a lattice.

2.2.1.4 Gram Matrix

The *Gram matrix* of a lattice generated by the columns of $\mathbf{H}$, as in Equation 2.5, is defined by (in the real case transposition replaces the Hermitian operator)

$$\mathbf{G} = \mathbf{H}^H \mathbf{H}. \tag{2.8}$$

By construction, the Gram matrix contains all the possible inner products between all the generator vectors: $g_{ij} = \langle \mathbf{h}_i, \mathbf{h}_j \rangle$; in particular, the diagonal elements are the squared norms $\|\mathbf{h}_i\|^2$. This fact implies that $\mathbf{G}$ is Hermitian and positive definite. Moreover, it defines a positive-definite quadratic form [e.g., Meyer 2000, Section 7.6] because

$$\|\mathbf{y}\|^2 = \|\mathbf{H}\mathbf{x}\|^2 = (\mathbf{H}\mathbf{x})^H (\mathbf{H}\mathbf{x}) = \mathbf{x}^H \mathbf{H}^H \mathbf{H}\mathbf{x}$$

$$= \mathbf{x}^H \mathbf{G}\mathbf{x} = \sum_{i=1}^{N_R} \sum_{j=1}^{N_T} g_{ij} x_i \bar{x}_j \geq 0, \quad \text{for all } \mathbf{x} \neq \mathbf{0}, \tag{2.9}$$

where $\bar{x}_i$ denotes the conjugate of x_i and $\mathbf{0}$ is the zero vector. Like the Voronoi region, the Gram matrix is another invariant of a lattice but only in respect to a particular basis.

2.2.1.5 Volume

When $\mathbf{H}$ is nonsingular, the lattice is full rank. In that case, the *volume* of the lattice (the volume of $\mathcal{R}$) is

$$\operatorname{vol}(\Lambda) = |\det(\mathbf{H})|, \tag{2.10}$$

however, for rectangular $\mathbf{H}$, the following more general definition is required:

$$\operatorname{vol}(\Lambda) = \sqrt{\det(\mathbf{H}^H \mathbf{H})} = \sqrt{\det(\mathbf{G})}. \tag{2.11}$$

The volume of the lattice is also an invariant of the lattice, that is, is independent of the choice of basis.

2.2.1.6 Unitary, Orthogonal, and Unimodular Matrices

An $n \times n$ *unitary* matrix $\mathbf{U}$ has complex entries and $\mathbf{U}^H\mathbf{U} = \mathbf{I}$ (i.e., the identity matrix) and its determinant is $\det(\mathbf{U}) = \pm 1$ (positive values account for rotations and negative values account for the existence of reflections). An *orthogonal* matrix $\mathbf{Q}$ has real entries and $\mathbf{Q}^T\mathbf{Q} = \mathbf{I}$. Both unitary and orthogonal matrices form a group [Edelman et al. 1998]. A *unimodular* matrix $\mathbf{M}$ is a square matrix with *integer entries* and with determinant $\det(\mathbf{M}) = \pm 1$ [Schrijver 1986, Section 4.3, Banerjee 1993]. The inverse of a unimodular matrix is also unimodular (this is because these matrices also form a group [Siegel 1989, Section XV, p. 148]). Unimodular matrices can always be generated by starting from an identity matrix and successively applying any of the following *elementary column operations* (or row operations according to the convention):

1. Change the signs of all the elements in a column;
2. Swap the two columns;
3. Add and integer the multiple a of one column to another columns.

Examples for these three cases are, respectively:

$$\text{i) } \mathbf{M} = \begin{pmatrix} 1 & 0 & 0 \\ 0 & -1 & 0 \\ 0 & 0 & 1 \end{pmatrix} \quad \text{ii) } \mathbf{M} = \begin{pmatrix} 0 & 1 & 0 \\ 1 & 0 & 0 \\ 0 & 0 & 1 \end{pmatrix} \quad \text{iii) } \mathbf{M} = \begin{pmatrix} 1 & 2 & 0 \\ 0 & 1 & 0 \\ 0 & 0 & 1 \end{pmatrix} \quad \text{where } a = 2.$$

All these three elementary operations which can generate any unimodular matrix have clear geometric interpretations, but for that purpose the notion of lattice equivalence up to scaling, orthogonal, and unimodular transformations will be first introduced.

2.2.1.7 Equivalent Lattices

It has already been mentioned that a basis is not unique. Furthermore, one can observe that a scaled or rotated version of a lattice is isomorphic to it, and therefore, in a geometric sense, is equivalent to it. One defines then the notion of lattice equivalence. A complex lattice generated by a basis $\mathbf{H}$ is equivalent to a lattice defined by a matrix $\hat{\mathbf{H}}$ if and only if

$$\hat{\mathbf{H}} = c \cdot \mathbf{U} \cdot \mathbf{H} \cdot \mathbf{M}, \tag{2.12}$$

where **U** is a unitary matrix, **M** is a unimodular matrix, and $c \in \mathbb{R}$. By applying a real equivalent (to be defined in Section 2.3.2), one can henceforth deal instead with $n \times n$ orthogonal matrices **Q** instead of unitary matrices **U**.

As pointed out in [Agrell and Eriksson 1998], if **U** is known and **M** is not known, it is easy to show that the lattices are the same. One option is to compute the (unique) Hermite Normal Form (HNF) for both bases and verify whether both HNF are the same [Schrijver 1986]. One alternative to this method would be to write each vector in one of the bases as an integer combination of the vectors in the other basis—as hinted by Micciancio in [Micciancio and Goldwasser 2002, p. 19]. Knowing **M**, it is also possible to find which orthogonal (unitary, in the complex model) matrix **Q** would transform one basis into the other by framing the problem as the Procrustes orthogonal problem [Higham 1989, Gower et al. 2004, Gower 2010]. Note that the QR decomposition is unique up to the sign of the elements in the diagonal [e.g., Agrell et al. 2002]. Hence, one alternative to the problem of finding **Q** would be to compute the QR decomposition of both matrices (**H** and $\widehat{\mathbf{H}}$) and verify that the **R** triangular matrix in both cases was the same up to the negation of its columns.

2.2.1.8 Geometry of Unimodular Transformations

While the first two kinds of elementary unimodular operations, (i) and (ii), change the sign of the determinant of the lattice, they do not change its modulo. In other words, they can include reflections but the volume remains the same. The concept of operation (iii) is illustrated in Figure 2.2 for a 2D case and with $a = 2$.

It is easy to see that the determinant of the lattice remains unchanged after the later type of elementary operation too. It should be noted that if the restriction on $a \in \mathbb{Z}$ is dropped and a is allowed to be real, the volume of the associated region also remains unchanged. In fact, the volume is solely dependent on the length of the Gram–Schmidt orthogonalized vectors. However, it should be noted that when $a \notin \mathbb{Z}$, the new set of vectors *no longer* constitutes a basis for the lattice as the vector may no longer lie on the lattice.

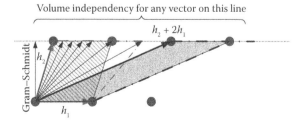

Figure 2.2 **The elementary operation that skews the fundamental region of a lattice preserves the determinant. The two shaded areas are the same.**

2.2.1.9 Shortest Vector and Successive Minima

Lattices have a shortest vector (and at least its symmetrical with the same norm). Many times one is interested in finding the shortest vectors that are also linearly independent (so that a vector and its symmetrical cannot be both considered). Hence, λ_i is the *i*th *successive minimum* of a lattice if λ_i is defined as the smallest real number which is the smallest radius of a sphere containing i pairwise independent vectors, all with norms smaller or equal to λ_i. The shortest vector clearly has norm λ_1.

2.2.2 Dual Lattice

Every lattice has a *dual lattice*[*] (the first being known as the *primal lattice*). The dual lattice is traditionally defined for real lattices, though the definition has also been extended to complex lattices [Ling et al. 2006]. In the real domain, given a primal lattice Λ with a basis $\mathbf{H}$, the dual lattice is defined as

$$\Lambda_D = \left\{ \mathbf{z} \in \mathbb{R} : \langle \mathbf{z}, \mathbf{x} \rangle \in \mathbb{Z}, \quad \forall \mathbf{x} \in \Lambda \right\}. \tag{2.13}$$

The dual lattice can be expressed in terms of the dual basis $\mathbf{H}^{(D)}$ as

$$\Lambda_D = \left\{ \mathbf{z} \in \mathbb{R} : \mathbf{z} = \underbrace{\left(\mathbf{H}^+ \right)^T}_{\mathbf{H}^{(D)}} \mathbf{x}, \quad \mathbf{x} \in \mathbb{Z}^n \right\}, \tag{2.14}$$

where $\mathbf{H}^{(D)}$ involves the Moore–Penrose pseudo-inverse (to be defined in Section 2.5.2).

$$\begin{aligned} \mathbf{H}^{(D)} &= (\mathbf{H}^+)^T = ((\mathbf{H}^T\mathbf{H})^{-1}\mathbf{H}^T)^T \\ &= \mathbf{H}((\mathbf{H}^T\mathbf{H})^{-1})^T = \mathbf{H}(\mathbf{H}^T\mathbf{H})^{-1}. \end{aligned} \tag{2.15}$$

Note that there is a unique dual lattice for each primal lattice. However, because a lattice holds an infinite number of bases, there is also an infinite number of bases for its dual, always observing $\mathbf{H}^{(D)} = (\mathbf{H}^+)^T$, in the case of real lattices, as given in Equation 2.14.

---

[*] The dual lattice appears in the literature also as the *polar lattice* or, more commonly, as the *reciprocal lattice*. All these names were already in use in 1971 [Cassels 1971, p. 24]. Since then, the name polar fell into disuse, though reciprocal is a name that is still common to be found in the literature.

Consider the case of full-rank real matrices. In fact, for $\mathbf{x}_1, \mathbf{x}_2 \in \mathbb{Z}^n$,

$$\langle \mathbf{z}, \mathbf{y} \rangle = \mathbf{z}^T \mathbf{x} = \underbrace{\left(\mathbf{H}^+\right)^T \mathbf{x}_1}_{\mathbf{z} \in \Lambda^{(D)}} \underbrace{\mathbf{H}\mathbf{x}_2}_{\mathbf{y} \in \Lambda} = \mathbf{x}_1^T \mathbf{H}^+ \mathbf{H}\mathbf{x}_2 = \mathbf{x}_1^T \mathbf{x}_2 \in \mathbb{Z}.$$

It is also possible to show that each point in the dual lattice can be written as an integer combination of the columns of $\mathbf{H}^{(D)}$. Denoting the rows of $\mathbf{H}^{-1}$ by $\mathbf{r}_1, \mathbf{r}_2, \ldots, \mathbf{r}_n$, for any point $\mathbf{z} \in \Lambda^{(D)}$ it is possible to write

$$\mathbf{z}^T = \mathbf{z}^T \mathbf{H}\mathbf{H}^{-1}$$
$$= \underbrace{(\mathbf{z}^T \mathbf{h}_1)}_{\in \mathbb{Z}}\mathbf{r}_1 + \underbrace{(\mathbf{z}^T \mathbf{h}_2)}_{\in \mathbb{Z}}\mathbf{r}_2 + \cdots + \underbrace{(\mathbf{z}^T \mathbf{h}_n)}_{\in \mathbb{Z}}\mathbf{r}_n, \tag{2.16}$$

which shows that the point in the dual lattice is defined by a linear combination of the rows of $\mathbf{H}^{-1}$, that is, a linear combination of the columns of $(\mathbf{H}^{-1})^T$. These arguments can be extended to the cases where the Moore–Penrose inverse is required and also to complex lattices.

One interesting relation between the two bases is that

$$(\mathbf{H}^{(D)})^T \mathbf{H} = \mathbf{I}, \tag{2.17}$$

which is equivalent to saying that $\langle \mathbf{h}_i, \mathbf{h}_j^{(D)} \rangle = \delta_{i,j}$, using the Kronecker delta.

The volumes of the primal and the dual lattice are related by

$$\text{vol}(\Lambda_D) = \frac{1}{\text{vol}(\Lambda)} \tag{2.18}$$

and their Gram matrices are related by

$$\mathbf{G}^{(D)} = (\mathbf{H}^{(D)})^T \mathbf{H}^{(D)} = ((\mathbf{H}^{-1})^T)^T (\mathbf{H}^{-1})^T$$
$$= \mathbf{H}^{-1}(\mathbf{H}^T)^{-1} = (\mathbf{H}^T \mathbf{H})^{-1} = \mathbf{G}^{-1}. \tag{2.19}$$

Obviously, the dual of the dual lattice is the primal lattice itself. The geometry of the dual lattice is closely related to the geometry of the primal lattice. The connection is that each point in the dual lattice defines a family of parallel $(n - 1)$-dimensional hyperplanes, where translates of an $(n - 1)$-dimensional sublattice lie. The union of those planes captures all the points of the primal lattice. This means that the shortest vector in the dual lattice will define the most distant $(n - 1)$-dimensional hyperplanes, whose union builds up the whole primal lattice. These hyperplanes can be interpreted as parallel layers and (as a consequence of being the ones farthest apart) are the densest ones in the lattice. In MIMO literature, the geometrical interpretation of the dual lattice as a tool for improving detection seems to have been first noted in [Agrell et al. 2002, p. 2207] for SD, and then in [Ling et al. 2006] for SIC [Ling and Mow 2009], though it is also implied in the detector in [Su and Kschischang 2007, p. 1944].

From the definition in Equation 2.13, for both Λ and $\Lambda^{(D)}$ in n dimensions, the inner product between some given point $\mathbf{z}$ in the dual lattice and any vector in the primal lattice is always an integer. Therefore,

$$\langle \mathbf{z}, \mathbf{x} \rangle \in \mathbb{Z}, \ \mathbf{z} \in \Lambda^{(D)}, \ \mathbf{x} \in \Lambda \ \Leftrightarrow \ \|\mathbf{z}\| \|\mathbf{x}\| \cos(\theta) = \|\mathbf{z}\| \mathrm{Proj}_{\mathbf{e}_z}(\mathbf{x}) \in \mathbb{Z}, \qquad (2.20)$$

where $\mathbf{e}_z = \mathbf{z}/\|\mathbf{z}\|$.

From Equation 2.20, it is then possible to define a family of parallel hyperplanes $\mathcal{P}(v)$, for $v \in \mathbb{Z}$, such that $\mathrm{Proj}_{\mathbf{e}_z}(\mathbf{x}) = \|\mathbf{z}\|^{-1} v$. These are planes in dimension $n-1$ with a distance $D = \|\mathbf{z}\|^{-1}$ between them, as illustrated in Figure 2.3. Note that vectors $\mathbf{a}_1$, $\mathbf{a}_2$, $\mathbf{a}_3$ all have the same projection onto the vector $\mathbf{z}$ that defines the set of parallel hyperplanes that is shown. v is then the index of the hyperplane with respect to the distance between hyperplanes, that is, $D = \|\mathbf{z}\|^{-1}$.

Figure 2.4 shows an example of two different partitions (i.e., a family of parallel hyperplanes) of a lattice associated with two different choices of vectors of the dual lattice. The example is set for

$$\mathbf{H} = \begin{bmatrix} 3/7 & -2/7 \\ -1/7 & 3/7 \end{bmatrix} \quad \text{and} \quad \mathbf{H}^{(D)} = \begin{bmatrix} 3 & 1 \\ 2 & 2 \end{bmatrix}.$$

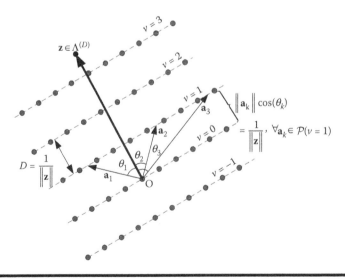

Figure 2.3 A primal lattice in n dimensions as the union of translates of a sublattice and these translates lie on $(n-1)$-dimensional hyperplanes.

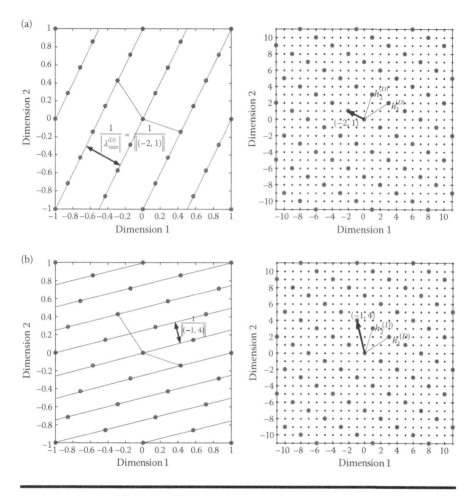

Figure 2.4 Identification of the hyperplanes in the primal lattice (on the left side) associated with a certain point in the dual lattice (on the right side). (a) Selection of (−2, 1) in the dual lattice. (b) Selection of (−1, 4) in the dual lattice.

2.3 Communication with Vectors

2.3.1 System Model

In MIMO SM with N_T transmit antennas and N_R receive antennas (with $N_R \geq N_T$), the relation between the transmitted (input) vector $\mathbf{x}_c = [x_{c,1}, x_{c,2}, \ldots, x_{c,N_T}]^T \in \mathbb{C}^{N_T \times 1}$ and the received (output) vector $\mathbf{y}_c = [y_{c,1}, y_{c,2}, \ldots, y_{c,N_R}]^T \in \mathbb{C}^{N_R \times 1}$ is modeled in the baseband as

$$\mathbf{y}_c = \mathbf{H}_c \mathbf{x}_c + \mathbf{n}_c, \qquad (2.21)$$

where $\mathbf{H}_c \in \mathbb{C}^{N_R \times N_T}$ is the channel matrix, with its entries h_{ij} representing the complex coefficient associated with the SISO link between the ith receive (Rx) antenna and the jth transmit (Tx) antenna, and with $h_{i,j} \sim \mathcal{N}_c(0,1)$, that is, taken from a zero-mean circularly symmetric complex Gaussian distribution with unitary variance (i.e., variance 1/2 in both the real and imaginary components). The phase of these elements is uniformly distributed in $[0 \ 2\pi]$, and their amplitude has a Rayleigh distribution. This corresponds to the i.i.d. (independent and identically distributed) Rayleigh fading channel model. The subscripts denote that the elements in the vectors and the entries in $\mathbf{H}_c$ are all complex variables. Furthermore, there is noise added to each entry of the received vector, modeled by the column vector $\mathbf{n}_c = [n_{c,1}, n_{c,2}, \ldots, n_{c,N_R}]^T \in \mathbb{C}^{N_R \times 1}$ with independent circularly symmetric complex Gaussian random variables taken from $\mathcal{N}_c(0, \sigma_n^2)$, that is, with zero average and variance σ_n^2 (corresponding to a variance $\sigma_n^2/2$ in both real and imaginary components). This noise model is often dubbed in MIMO literature as zero-mean spatially white (ZMSW) noise [e.g., Goldsmith 2005]. For independent input data, its covariance is $\mathbf{R}_x = E\{\mathbf{x}_c \mathbf{x}_c^H\} = \sigma_x^2 \mathbf{I}_n$. Similarly, the covariance of the independent noise vector is $\mathbf{R}_n = E\{\mathbf{n}_c \mathbf{n}_c^H\} = \sigma_n^2 \mathbf{I}_n$. Henceforth, the subscript in $\mathbf{I}_n$ will be abandoned.

It should now be clear that with integer input symbols x_i, any of these possible vectors is a point on the $\mathbb{Z}^n$ lattice. The effect of the channel is that of warping $\mathbb{Z}^n$ according to the linear transformation $\mathbf{H}_c$ (as illustrated in Figure 2.5).

Both phase shift keying (PSK) or quadrature amplitude modulation (QAM) constellations can be used in MIMO, however, only the latter lends itself for a lattice interpretation and most of the literature on MIMO SM concentrates on QAM, which is also the modulation that is considered in this chapter. Consequently, the input symbols in each transmit antenna are taken from a finite complex constellation $\mathcal{A}_c$, which is some M-ary QAM (quadrature amplitude modulation). The symbols have zero mean, so that $E\{\mathbf{x}_c\} = 0$. This complex constellation is constructed

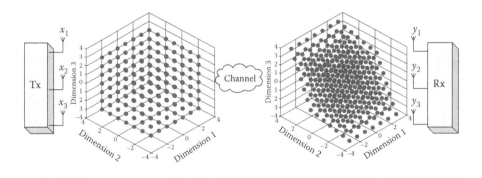

Figure 2.5 SM with real inputs. $\mathbb{Z}^n$ is transmitted and then skewed by the effect of the channel.

from the Cartesian product $\mathcal{A}_c = \mathcal{A} \times \mathcal{A}$ where $\mathcal{A}$ is the real alphabet

$$\mathcal{A} = \left\{ -(\sqrt{M} - 1)a, \ldots, -5a, -3a, -a, +a, +3a, +5a, + \cdots + (\sqrt{M} - 1)a \right\}. \quad (2.22)$$

Traditionally, $a = 1$, and the alphabet in each real dimension is

$$\mathcal{A} = \left\{ -(\sqrt{M} - 1), \ldots, -5, -3, -1, +1, +3, +5, + \cdots + (\sqrt{M} - 1) \right\}. \quad (2.23)$$

Without loss of generality, one can assume Rx filters with impulse response $h(t)$ normalized to $\int |h(t)|^2 \, dt = 1$, and therefore, the *average energy* of the complex symbols in $\mathcal{A}$ is given by

$$E_s = E\left[|x_{c,i}|^2 \right] = \frac{1}{M} \sum_{x_{c,i} \in \mathcal{A}} |x_{c,i}|^2 = \frac{1}{M} \sum_{x_{c,i} \in \mathcal{A}} (\Re x_{c,i})^2 + (\Im x_{c,i})^2, \quad (2.24)$$

which coincides with their *average power* ($E_s = \sigma_x^2$). Table 2.2 lists the values of E_s for the modulations used later in this work.

The "overall" SNR at the receiver is

$$\rho \triangleq \frac{E\left\{ \|\mathbf{y}\|^2 \right\}}{E\left\{ \|\mathbf{n}\|^2 \right\}} = \frac{E\left\{ \|\mathbf{H}_c \mathbf{x}\|^2 \right\}}{E\left\{ \|\mathbf{n}\|^2 \right\}} = \frac{E\left[\sum_{i=1}^{N_R} \sum_{j=1}^{N_T} |h_{c,i,j} x_j|^2 \right]}{E\left[\sum_{i=1}^{N_R} n_c^2 \right]}$$
$$= \frac{N_T N_R \sigma_x^2}{N_R \sigma_n^2} = N_T \frac{\sigma_x^2}{\sigma_n^2}, \quad (2.25)$$

which is actually the same as N_T times the SNR of a SISO channel. This comes from the fact that each antenna receives the incoming power from N_T antennas, while each receive antenna perceives the same amount of noise as in SISO. The result is valid on *average* and only when $h_{i,j} \sim \mathcal{N}_c(0,1)$, that is, each y_i receives the sum of N_T symbols weighted by unit power random variables ($E[|h_{ij}|^2] = 1$). Particular channel realizations will lead to different instantaneous SNRs. In this chapter, when assessing the performance of a receiver, the SER will be plotted against the SNR as defined in Equation 2.25. One

Table 2.2 Symbol Energy for the Used Modulations

	QPSK	16-QAM	64-QAM
E_s	2	10	42

other important metric that will appear in the capacity formula is the *SNR per transmit antenna*, which is

$$\rho_a = \frac{\rho}{N_T} = \frac{\sigma_x^2}{\sigma_n^2}. \tag{2.26}$$

This latter normalized SNR is in fact the same as E_s/N_0 (where N_0 is the unilateral power spectral density of the noise) because, assuming the Nyquist bandwidth and a *raised-cosine filter*,

$$\frac{\sigma_x^2}{\sigma_n^2} = \frac{E_s/T}{N_0 \cdot B} = \frac{E_s R}{N_0 B} = \frac{E_s}{N_0}. \tag{2.27}$$

Unlike what happens in SISO systems, in MIMO most performance results are given as a function of the SNR. However, some literature uses the average energy per bit,[*] $E_b = E_s/\log_2(M)$, and the unilateral spectral density of noise, N_0. Accordingly, the SNR given by Equation 2.25 is the same as

$$\rho = N_T \frac{E_s}{N_0} = N_T \log_2(M) \frac{E_b}{N_0}, \tag{2.28}$$

which allows comparisons across the two different approaches seen in the literature.

It is also worth mentioning that an equivalent model for SM assuming unit noise variance and unit E_s is also often found in the literature. Maintaining all that was said for the SNR, this model must be written as

$$\mathbf{y}_c = \sqrt{\frac{\rho}{N_T}} \mathbf{H}_c \mathbf{x}_c + \mathbf{n}_c \quad \text{or also as} \quad \mathbf{y}_c = \sqrt{\frac{E_s}{\sigma_n^2}} \mathbf{H}_c \mathbf{x}_c + \mathbf{n}_c, \tag{2.29}$$

where now the real alphabet in each dimension is

$$\mathcal{A} = \sqrt{\frac{3}{2(M-1)}} \left\{ -(\sqrt{M}-1), \ldots, -5, -3, -1, +1, +3, +5, + \cdots + (\sqrt{M}-1) \right\} \tag{2.30}$$

to assure that $E\{x_i^* x_i\} = 1$ and $E\{\mathbf{x}^H \mathbf{x}\} = N_T$.

The norms in Equation 2.25 are the Euclidian norm. Notice that for the case of matrices several norms may be defined [Stewart 1973, Chapter 4].

$$\|\mathbf{A}\|_F = \sqrt{\text{trace}(\mathbf{A}^H \mathbf{A})} = \sqrt{\text{trace}(\mathbf{A}\mathbf{A}^H)} = \sqrt{\sum_i \sum_j |a_{ij}^2|} \tag{2.31}$$

---

[*] The error rate is also sometimes given in the MIMO literature in terms of the bit error rate (BER), which is obtained from the SER taking into consideration the number of bits per complex symbol.

It is not difficult to see that in this norm

$$E\left\{\|\mathbf{H}_c\|_F^2\right\} = N_T N_R. \tag{2.32}$$

In this chapter, the uncorrelated channel is considered, as seen in the definition of $\mathbf{H}_c$. When correlation exists between the multipath components, a more general model is necessary and that implies a full characterization of the correlation matrix involving the cumbersome vectorization of $\mathbf{H}_c$ [see, e.g., Oestges and Clerckx 2007, Section 3.1.1]. The Kronecker model is a popular way of avoiding this, by decoupling the effect of correlation at the transmit side (characterized by $\mathbf{R}_{c,\text{Tx}}$) from effects at the receive side (characterized by $\mathbf{R}_{c,\text{Rx}}$). Then, the model consists only of matrix multiplications[*]

$$\mathbf{H}_c = R_{c,\text{Rx}}^{1/2} H_{c,\text{ind}} R_{c,\text{Tx}}^{1/2}. \tag{2.33}$$

The separation of Tx and Rx effects can be interpreted as if the multipath components at the receiver had "forgotten" about the effects of antenna coupling and scattering close to the Tx, which can be considered as separate from what happens in terms of antenna coupling and scattering close to the Rx.

For signals with *symbol time T* and *bandwidth B*, as a rule of thumb, the coherence time and the coherence bandwidth are given by

$$B_{coh} \approx \frac{1}{\text{rms delay spread}}, \quad \tau_{coh} \approx \frac{1}{\text{Doppler spread}}. \tag{2.34}$$

When assessing the performance of SM systems, in this chapter we will always consider the channel to be (i) *flat* (nonselective in the frequency domain), that is, $B < B_{coh}$, and (ii) *slow* (nonselective in time for a transmit vector), that is, $T < \tau_{coh}$. For typical values for the coherence time in different mobility scenarios in LTE, see [Ghosh et al. 2010, Section 2.4]. Typical descriptions of the wireless channels can be found in, for example, [Ghosh et al. 2010, Section 2.4], [Giannakis et al. 2007, Section 1.3] and a deeper discussion is offered by [Oestges and Clerckx 2007].

2.3.2 Real Equivalent Model

The model for SM was described in Equation 2.21 in terms of complex vector spaces, however it is not difficult to prove that, by stacking the real and complex parts of the vectors (respectively denoted by $\Re$ and $\Im$), and by appropriate con-

---

[*] The square is root of a matrix defined as $\mathbf{A}^{1/2} = \mathbf{U}\Sigma^{1/2}\mathbf{V}^H$, where $\mathbf{U}$ and $\mathbf{V}$ are the unitary matrices of the SVD of $\mathbf{A}$, with Σ the diagonal matrix made of the singular values of $\mathbf{A}$ [Biglieri and Taricco 2005, p. 18].

struction of a modified channel matrix, the problem can equivalently be described by means of real variables as

$$
\mathbf{y} = \mathbf{Hx} + \mathbf{n} \Leftrightarrow \begin{bmatrix} \Re(\mathbf{y}_c) \\ \Im(\mathbf{y}_c) \end{bmatrix} = \begin{bmatrix} \Re(\mathbf{H}_c) & -\Im(\mathbf{H}_c) \\ \Im(\mathbf{H}_c) & \Re(\mathbf{H}_c) \end{bmatrix} \begin{bmatrix} \Re(\mathbf{x}_c) \\ \Im(\mathbf{x}_c) \end{bmatrix} + \begin{bmatrix} \Re(\mathbf{n}_c) \\ \Im(\mathbf{n}_c) \end{bmatrix} \tag{2.35}
$$

with $\mathbf{y} = [y_1, y_2,..., y_{2N_R}]^T \in \mathbb{R}^{N_R \times 1}$, $\mathbf{x} = [x_1, x_2,..., x_{2N_T}]^T \in \mathbb{R}^{N_T \times 1}$, $\mathbf{n} = [n_1, n_2,..., n_{2N_R}]^T \in \mathbb{R}^{N_R \times 1}$, and $\mathbf{H} \in \mathbb{R}^{N_R \times N_T}$ (notice that the "c" subscripts in the variables will be dropped from now on). Expanding Equation 2.35, each component of $\mathbf{y}$ is

$$
\begin{cases}
y_i = \Re h_{i,1} \cdot \Re x_1 + \cdots + \Re h_{i,N_T} \cdot \Re x_{N_T} \\
\quad - \Im h_{i,1} \cdot \Im x_1 - \cdots - \Im h_{i,N_T} \cdot \Im x_{N_T} + \Re n_i, \quad i \le N_R \\
y_i = \Im h_{i-N_R,1} \cdot \Re x_1 + \cdots + \Im h_{i-N_R,N_T} \cdot \Re x_{N_T} \\
\quad + \Re h_{i-N_R,1} \cdot \Im x_1 + \cdots + \Re h_{i-N_R,N_T} \cdot \Im x_{N_T} + \Im n_{i-N_R}, \quad N_R < i \le 2N_R
\end{cases} \tag{2.36}
$$

and therefore these components have a χ^2 distribution with $2N_T$ degrees of freedom, before the noise is added.

The equivalent real model is the one that will be used throughout this work and, therefore, the transposition operator, T, will replace the Hermitian operator, H (conjugation followed by transposition). Moreover, orthogonal matrices play the role of singular matrices, for instance in the singular-value decomposition (SVD) or for performing orthogonal rotations on lattices. In this model, the discrete inputs in the elements of x that are considered in this chapter are taken from the real alphabet $\mathcal{A}$, as defined in Equation 2.23. In this model, both real and imaginary components of $\mathbf{n}_c = \Re(\mathbf{n}_c) + j\Im(\mathbf{n}_c)$ have variance

$$
\sigma_n^2 = \frac{1}{2} \frac{N_T E_s}{\rho}. \tag{2.37}
$$

Another consequence of this model is that, henceforth, full-rank real lattices will be considered to have n dimensions.

2.3.3 Capacity with Channel State Information at the Receiver

Assuming that there is perfect channel state information at the receiver (CSIR), that is, perfect knowledge of $\mathbf{H}$, it can be proven that the maximization of the mutual information between the input and output in MIMO amounts to maximizing

$\log_2(\det(\pi e \mathbf{R}_y))$. The output covariance is $\mathbf{R}_y = E\{\mathbf{yy}^H\} = \mathbf{HR}_x\mathbf{H}^H + \sigma_n^2\mathbf{I}_{N_R}$ and therefore the capacity is

$$C = \max_{\mathbf{R}_x} \log_2 \det\left(\mathbf{I} + \frac{\rho}{N_T}\mathbf{HR}_x\mathbf{H}^H\right), \tag{2.38}$$

where trace $(\mathbf{R}_x) = E\{\mathbf{x}^H\mathbf{x}\} < 1$ is the transmit power constraint in the optimization problem. When the channel *is* known at the transmitter (Tx), that is, there is CSIT, the input covariance can be built to match the channel. When CSIT does *not* exist, the input is made to have $\mathbf{R}_x = \sigma_x^2\mathbf{I}_{N_T}$, leading to a capacity

$$C = \log_2 \det\left(\mathbf{I} + \frac{\rho}{N_T}\mathbf{HH}^H\right), \tag{2.39}$$

where CSIR is assumed. This implies that the receiver is able to accurately estimate the channel matrix.

The matrix $\mathbf{HH}^H$ can be interpreted as the Gram matrix associated with the lattice generated by the *rows* of $\mathbf{H}$. Therefore, as shown in Section 2.2.1, $\mathbf{HH}^H$ is a semidefinite-positive matrix with $r = \text{rank}(\mathbf{H})$ eigenvalues, whose values are $\lambda_i = s_i^2$, that is, they are the square of corresponding r singular values s_i. Applying the SVD to $\mathbf{HH}^H$, and remembering that a Gram matrix is symmetric, its left and right unitary matrices in the SVD are the same. Hence, Equation 2.38 becomes

$$\log_2 \det\left(\mathbf{I} + \frac{\rho}{N_T}\mathbf{U\Sigma U}^H\right) = \log_2 \det\left(\mathbf{U}\left(\mathbf{I} + \frac{\rho}{N_T}\mathbf{\Sigma}\right)\mathbf{U}^H\right) \tag{2.40}$$

and

$$C = \sum_{k=1}^{r} \log_2\left(1 + \frac{\rho}{N_T}\lambda_i\right). \tag{2.41}$$

As known from Section 2.3.1, $\mathbf{H}$ is not deterministic but rather a random matrix. One should note that in the case of *slow fading* (when a code word does not span more than one coherence period of the channel), regardless the choice of rate and coding scheme, there will always be a nonzero probability that the rate is higher than the capacity of the channel. Hence, the capacity that the channel can commit to is *zero* [Tse and Viswanath 2005, p. 188]. Even in that case, one can speak of the *instantaneous capacity* $C(\mathbf{H})$, as a function of the current channel.

This chapter assumes a channel with a *block fading* model (i.e., when a code-word goes through many different and independent channel instances [Tse and Viswanath 2005, p. 199]). In this model, the channel remains constant over a certain duration (the duration of a *block*), only changing from block to block. In this case, by taking the average of over many instances of the channel coefficients, a channel capacity can be obtained by applying an expectation to Equation 2.41:

$$C_{\text{erg}} = E_{\lambda_i} \left\{ \sum_{k=1}^{r} \log_2 \left(1 + \frac{\rho}{N_T} \lambda_i \right) \right\}. \tag{2.42}$$

This expectation depends on the eigenvalues λ_i, which are independent random variables, each with a Wishart distribution [Tulino and Verdú 2004, Oestges and Clerckx 2007, Appendix B], as Teletar first noted in [Telatar 1999]. While the number of receive antennas is not explicit in the capacity formula, its effect is hidden in the rank r of $\mathbf{H}\mathbf{H}^H$, which is

$$r = \min(N_t, N_R). \tag{2.43}$$

It is not difficult to show [Tse and Viswanath 2005, Khan 2009, pp. 148–149, Section 8.2] that for low SNR, using the approximation $\log_2(1 + x) \approx x \log_2 e$ valid for small x, and noting that $E\{\lambda_i\} = \text{Trace}(\mathbf{H}\mathbf{H}^H)$, as seen in Equations 2.39 through 2.40, then the capacity in Equation 2.42 is

$$C_{\text{erg}} \approx N_R \rho \log_2 e. \tag{2.44}$$

This result shows that, for the low SNR, the number of transmit antennas does not play any role, because what is important is the fact that N_R receive antennas can coherently combine the incoming signals working as a "spatial matched filter." This result will be revisited in Section 2.5, when both linear receivers are shown to approach the matched filter when noise is high compared to the mutual interference between layers. In the high SNR regime, it is also not difficult to conclude that expression (2.42) becomes

$$C_{\text{erg}} \approx r \log_2 \left(\frac{\rho}{N_T} \right) + \sum_{i=1}^{r} E_{\lambda_i} \{ \log_2(\lambda_i) \}, \tag{2.45}$$

which, remembering Equation 2.43, shows the famous linear increase of the ergodic capacity with the minimum number of antennas on each side.

In STC a code word spans more than one transmit vector, as a matrix assumes the role of $\mathbf{x}_c$ in model (2.21) and several vectors may experience different channel realizations $\mathbf{H}_i$ over time. Moreover, using outer-codes the code words may experience enough channel realizations and the average of the capacities of the channels may be close to the ergodic capacity. In the case of *uncoded* SM, those situations do not exist and therefore the outage capacity and the outage probability are the concepts best suited to describe these systems.

In 2004, Guo, Verdú, and Shamai made a breakthrough in information theory, finding a very simple relation between mutual information and the minimum mean square error at a receiver. Debbah in [Sibille 2011, Chapter 2] shows how that approach eventually leads to an elegant (average) geometrical interpretation for the capacity of a MIMO system, much similar to that of the traditional geometrical interpretation of capacity of the binary symmetric channel, as first observed by Shannon. The determinant of the auto-correlation matrix of the transmitted symbols measures the volume of space associated with a MIMO code word. The determinant of the covariance matrix of the MMSE estimate measures the small volume around the received vector, where the signal is expected to lie with high probability. It can be proven that the MIMO capacity is

$$C = \log_2 \frac{\det(\mathbf{R}_{xx})}{\det(\mathbf{R}_{\mathrm{MMSE}})}, \tag{2.46}$$

which amounts to the *sphere packing problem* in lattices [Conway and Sloane 1999, Aste and Weaire 2000]. It is worth noting that Stoica et al. have independently arrived at the exact same lattice interpretation solely by signal processing considerations [Stoica et al. 2005].

2.4 Detection in MIMO Spatial Multiplexing

While in STC the central research problem is finding codes that maximize the code word pairwise distance and thus that minimizes pairwise error probability [e.g., Viterbo and Oggier 2004], the main research problem in SM in the last 10 years has been detecting $\mathbf{x}$ given the noisy observation $\mathbf{y}$. For that problem, the *maximum a posterior probability* (MAP) of $\mathbf{x}$ is

$$\mathbf{x} = \arg\max_{\mathbf{x} \in \mathcal{A}} P(\mathbf{x}|\mathbf{y}) = \arg\max_{\mathbf{x} \in \mathcal{A}} \frac{P(\mathbf{x}|\mathbf{y})P(\mathbf{x})}{P(\mathbf{y})}. \tag{2.47}$$

As all vectors $\mathbf{x}$ are equiprobable, $P(\mathbf{x}|\mathbf{y})$ is a sufficient statistics for the detection process. Therefore, MAP detection can be reduced to *maximum likelihood* (ML) without any performance loss. For the independent and identically distributed (i.i.d.) Rayleigh channel with i.i.d. transmitted symbols with $\mathbf{R}_x = \sigma_x^2 \mathbf{I}_n$ and $\mathbf{R}_n = \sigma_n^2 \mathbf{I}_n$, one has the N-dimensional probability distribution

$$
\begin{aligned}
P(\mathbf{x}|\mathbf{y}) &= \frac{1}{(2\pi\sigma_n^2)^{N/2}} \exp\left(-\frac{(\mathbf{y} - \mathbf{Hx})^T (\mathbf{y} - \mathbf{Hx})}{2\sigma_n^2}\right) \\
&= \frac{1}{(2\pi\sigma_n^2)^{N/2}} \exp\left(-\frac{\|\mathbf{y} - \mathbf{Hx}\|^2}{2\sigma_n^2}\right),
\end{aligned} \tag{2.48}
$$

and therefore the detection problem becomes that of minimizing the exponent of Equation 2.48:

$$\hat{\mathbf{x}}_{ML} = \arg\max_{\mathbf{x}\in\mathcal{A}}\left\{\left\|\mathbf{y} - \mathbf{H}\mathbf{x}\right\|^2\right\}. \tag{2.49}$$

This problem now has a clear *geometrical* interpretation: the optimal $\mathbf{x}$ (the one that best explains the observation $\mathbf{y}$) is the one that, among all possible input vectors, and after the linear transformation, generates the closest vector $\mathbf{H}\mathbf{x}$ (in the Euclidian sense) to the received vector $\mathbf{y}$. This problem is known in integer optimization as *integer least squares* and in lattice theory as CVP (as mentioned above): "given a target vector off the lattice, $\mathbf{y}$, which point in the lattice is the closest one?" The problem is exemplified in Figure 2.6 for the simple cases of $\mathbb{Z}$, $\mathbb{Z}^2$, and $\mathbb{Z}^3$ lattices (the CVP in these lattices is not NP-hard; for $\mathbb{Z}^n$ lattices the algorithmic detection complexity of the optimal detector is actually polynomial $\mathcal{O}(n^3)$, due to the orthogonal structure).

The solution of the CVP is equivalent to drawing a Voronoi cell around the target point and finding which single lattice point lies inside the region. Conversely, this is equivalent to having the lattice tiled by the Voronoi region and in selecting which region the target is. Obviously, computing the Voronoi region is also NP-hard.

Notice that the complex-valued lattice in a MIMO link with N_T transmit antennas and an M-QAM modulation will have M^{N_T} complex points within its border. In the equivalent real model the number is obviously the same, $(\sqrt{M})^{2N_T}$.

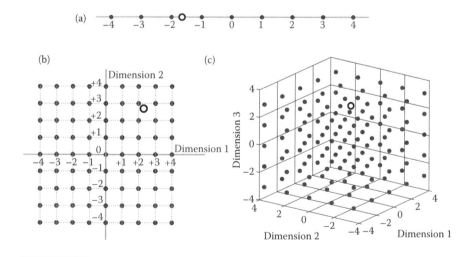

Figure 2.6 The CVP in (a) one, (b) two, and (c) three dimensions, given an off-lattice target point.

2.4.1 Optimal Detection and Complexity

The algorithmic complexity of the CVP is proven to be *NP-hard*, which, in the current state of understanding of the complexity of algorithms, places it in the worst tier in the hierarchy of complexity classes. One should not conclude from this that any hope of finding accurate solutions should be deemed unrealistic. In fact, very good approximations to the optimal solution can be found, especially when the number of dimensions is small [Hochba 1996]. As the number of dimensions grows, the complexity of the problem, measured as the number of operations, grows exponentially (this is what is known as "the curse of dimensionality" [Bishop 2006]); however, the complexity of some approximate detection techniques grows only polynomially. Usually, the number of operations (flops or algebraic operations) required by an algorithm is expressed in the "big O" notation[*] and in practice this suffices for comparing the complexity of algorithms, *when* it is feasible to test them. However, the complexity theory of algorithms is a vast and convoluted topic [e.g., Talbot and Welsh 2006], and the precise definition of each of the complexity classes falls beyond the scope of this work. Nonetheless, there are simpler and insightful working definitions describing these classes, for example, [Welsh 1988], [Cormen et al. 2001], and [Bertsimas and Weismantel 2005, Appendix B]. The most broad complexity classes are P, NP, NP-complete, and NP-hard and, according to what is known today, are related as shown in Figure 2.7 (if P ≠ NP, as is believed to be the most likely case).

The P class encompasses the problems that can be solved in polynomial time (making a correspondence between the number of operations and time). NP stands for *nondeterministic polynomial* complexity. It means that, if a certain *certificate* is provided

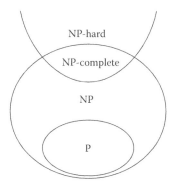

Figure 2.7   Complexity classes.

---

[*] For an input data of size n (e.g., n bits are necessary to represent the data, or, in the particular case of MIMO, n is the number of dimensions of the lattice), complexity $\mathcal{O}(f(n))$ means that the function $f(n)$ is an upper bound, up to a constant multiple factor, for the function of the number of operations as a function of n—a detailed description of this and other notations is given in [Epp 2011].

(i.e., a possible "solution" to the problem), it is then possible to verify in polynomial time if that certificate is a valid solution to the problem or not. This involves a mere yes/no answer because the problem is formulated as a *decision problem*. As every problem in P can also be posed as a decision problem (one just needs to use the solution as a certificate), then P ⊂ NP. The problems in the NP-complete class all share the property that if one of them is proven to be in P, then all the other problems in the class would also be in P, unless the polynomial hierarchy collapses. (It is worth mentioning that the vast majority of NP problems are in fact NP-complete.) Technically, these "entanglements" are proven through a "P time" *reduction* of one problem to the other. A simple definition of the NP-hard class unavoidably ends up vague: it consists of the problems that are "at least as hard as the NP-complete ones." NP-hard problems hold some property that, if solved in "P time," would make any problem in class NP-complete to be solvable in "P time," though they cannot be included in the NP-complete family. Obviously, given the definition, all problems in the NP-complete class are also NP-hard; however, there exist problems that are in the NP-hard class but not in NP-complete. A useful theorem states as follows: if an optimization problem has a *decision version* that is NP-complete, then the *optimization version* is NP-hard. In the CVP, one can think of the following decision version: "is there a lattice point at a distance shorter than some distance d from the target point?" Given a certificate, it is trivial to compute the distance and verify that such a point exists. The optimization version is the CVP itself: "what is the point at the shortest distance from the target point?" There are thousands of NP-hard problems [Bertsimas and Weismantel 2005, Appendix B], such as the *subset sum problem* (an example of a decision problem), the *knapsack problem* (a problem of combinatorial optimization, [Kreher 1999, Bertsimas and Weismantel 2005]), the *binary optimization problem*, the *traveling salesman problem*, or the CVP.

The CVP was first proven NP-hard by van Emde Boas in 1981, but the technicalities of that proof are considered cumbersome. In 2001, [Micciancio 2001] was able to show a reduction from the subset sum problem (known to be NP-hard) to a CVP in a lattice, which finally proved, in a very simple and elegant way, the NP-hardness of the CVP.

As is the case of many problems in the class NP-hard, being in this class does not mean that the problem cannot ever be solved in an optimal manner. When both the number of dimensions n and the modulation order M are low, a "brute force" approach is affordable. Furthermore, when ML-type detection is no longer possible, the challenge of finding suboptimal affordable solutions can be quite successful. The next section will describe how the geometry of lattice is closely related with the detection strategies used in MIMO.

2.5 Traditional Receivers

In the following, the most important type of MIMO receivers will be introduced with a geometric perspective in mind, which explains their performance loss in respect to the optimum detector. The linear receivers, which are the simplest ones, but also the ones having the worst symbol error rate (SER), are the first ones to be geometrically interpreted. Then, the OSIC technique is described, followed by the lattice-reduction-aided (LRA) approach, and finally the sphere-decoding concept is introduced.

2.5.1 Linear Receivers

Linear receivers consist of (i) a linear transformation $\mathbf{W}$ of the received vector which then is followed by (ii) a quantization to the symbol alphabet (also known as *slicing*). The linear transformation is a filter that can be designed with two different criteria, leading to the *zero-forcing* (ZF) detector or to the *minimum mean square error* (MMSE) detector. These receivers constitute the simplest set of (nonoptimal) receivers to be widely used for MIMO receivers. The detected solution $\hat{\mathbf{x}}$ given by these techniques is obtained by applying

$$\mathbf{x}_W = \mathbf{W}\mathbf{y}, \tag{2.50}$$

$$\hat{\mathbf{x}}_W = Q_{\mathbb{Z}}[\mathbf{x}_W], \tag{2.51}$$

where $Q_{\mathbb{Z}}[\cdot]$ denotes rounding to the nearest integer and the subscript W in $\hat{\mathbf{x}}_W$ indicates the filter design criterion: ZF or MMSE.

D-BLAST was the first technique to be proposed [Foschini 1996], followed by the more practical OSIC [Golden et al. 1999], but, as will be seen below, OSIC includes a linear inversion (either ZF or MMSE). As mentioned in Section 2.1.3, both ZF and MMSE techniques are well known in other contexts of detection and equalization in SISO (c.f. Table 2.1), but the formalism that is used in MIMO is the one that was first developed for multiuser detection [Verdú 1998].

In both types of linear receivers, the linear transformation $\mathbf{W}$ can be seen as a focusing process of the points in the received lattice back onto $\mathbb{Z}^n$ (or $\mathbb{C}^n$). This "backwards transformation" is of interest because it maps the received lattice back onto $\mathbb{Z}^n$, which lends itself to simple orthogonal *slicing*. This is the primary motivation for this particular design. In [Monteiro and Kschischang 2011], the concept of having a linear transformation as the first stage of a detection technique has been generalized to the concept of *focusing* a received lattice onto some other given lattice, whose geometric structure is also of interest.

Noticeably, despite the early use of both ZF and MMSE detectors, a thorough theoretical understanding of their performance seems not to have been pursued until [Jiang et al. 2011].

2.5.2 Zero-Forcing Detection

It is natural to think first of a solution to Equation 2.49 involving the linear transformation that *undoes* the linear transformation, which is obviously the inverse matrix.

The inversion of **H** is a trivial operation (e.g., by applying Gauss elimination to an extended matrix), but can only be defined for matrices with nonzero determinant (i.e., *invertible* or *nonsingular* matrices). Hence, the need for $N_T < N_R$. A geometrical interpretation for the cases with singular matrices is simple. A *singular* matrix that defines a lattice in $\mathbb{R}^n$ (the extension to $\mathbb{C}^n$ is always implied) performs a linear transformation whose outcome is a "flat" lattice that lies on a "flat subspace" of that space, that is, a lattice that does not fill all the dimensions in $\mathbb{R}^n$ and consequently can be fully described in a smaller dimensional space. For example, think of lattices that lie on a 2D plane or on a straight line embedded in an n-dimensional real space. In the space in which they are defined, these lattices have zero volume (i.e., zero determinant). Even so, an inverse correspondence to the original lattice seems still impossible. This would only not be possible if $N_R < N_T$, since the higher dimensional transmit lattice in N_T cannot, in general, be captured in a lower dimensional space.

The algebraic interpretation of the channel inversion problem of an $n \times n$ singular channel matrix **H** is related to its SVD $\mathbf{H} = \mathbf{U\Sigma V}^T$, where **U** and **V** are unitary matrices and $\mathbf{\Sigma} = \mathrm{diag}(\lambda_1, \lambda_2, \ldots, \lambda_r)$, the diagonal matrix with the singular values of **V**. The inverse is $\mathbf{H} = (\mathbf{V}^T)^{-1}\mathbf{\Sigma}^{-1}\mathbf{U}^{-1}$, or $\mathbf{H} = (\mathbf{V}^T)^{-1}\mathbf{\Sigma}^{-1}\mathbf{U}^{-1}$, or $\mathbf{H} = \mathbf{V\Sigma}^{-1}\mathbf{U}^T$, as both **U** and **V**. Since the inverse of the diagonal matrix is $\mathbf{\Sigma} = \mathrm{diag}(1/\lambda_1, 1/\lambda_2, \ldots, 1/\lambda_r)$, when there are only $r < n$ nonzero singular values (i.e., the rank is r), then $n - r$ singular values cannot be inverted in a finite domain.

The *pseudo-inverse* matrix, also known as the *Moore–Penrose*[*] *(inverse) matrix*, is the solution to the *normal equation* $\mathbf{H}^H\mathbf{y} = \mathbf{H}^H\mathbf{Hx}$, obtained from Equation 2.35. The straightforward solution to this equation is to make $(\mathbf{H}^H\mathbf{H})^{-1}\mathbf{H}^H\mathbf{y} = \mathbf{x}$, where $\mathbf{H}^H\mathbf{H}$ is invertible because it is positive definite (indeed, it corresponds to the Gram matrix of the lattice). The Moore–Penrose inverse of $\mathbf{H} \in \mathbb{C}^{N_R \times N_T}$, when $N_R > N_T$, *always* exists and is defined as

$$\mathbf{H}^+ = (\mathbf{H}^H\mathbf{H})^{-1}\mathbf{H}^H. \tag{2.52}$$

From Equations 2.50 and 2.51, the ZF receiver is formed by the pseudo-inverse matrix (the linear filter) followed by a quantization to the symbol alphabet by threshold decision, that is,

$$\mathbf{W}_{ZF} = (\mathbf{H}^H\mathbf{H})^{-1}\mathbf{H}^H, \quad \mathbf{x}_{ZF} = \mathbf{W}_{ZF}\mathbf{y}, \quad \hat{\mathbf{x}}_{ZF} = Q_A[\mathbf{x}_{ZF}], \tag{2.53}$$

and therefore

$$\hat{\mathbf{x}}_{ZF} = Q_A[\mathbf{W}_{ZF}(\mathbf{Hx} + \mathbf{n})] = Q_A[\mathbf{x} + \mathbf{W}_{ZF}\mathbf{n}]. \tag{2.54}$$

The filtered noise is transformed by $W_{ZF} = \mathbf{H}^+$, which constitutes a noise enhancement factor. The receiver structure is shown in Figure 2.8.

---

[*] Discovered first by E. H. Moore in 1920 and later rediscovered by Roger Penrose in [Penrose 1955].

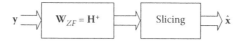

Figure 2.8 Zero-forcing receiver.

The detected vector $\hat{\mathbf{x}}_{ZF}$, as obtained from Equation 2.54, is in fact the solution to

$$\hat{\mathbf{x}}_{ZF} = \arg\min_{\mathbf{x}\in\mathbb{C}^{N_T}}\left\{\left\|\mathbf{y} - \mathbf{H}\mathbf{x}\right\|\right\}. \tag{2.55}$$

Comparing Equation 2.55 with Equation 2.49, one should note how the search is now made in the *continuous* domain $\mathbb{R}^n$ (or in $\mathbb{C}^n$, for complex lattices) instead of the discrete complex alphabet $\mathcal{A}$ (or $\mathcal{A}_r$). This is the origin of the suboptimality of the ZF receiver. As mentioned previously after the inverse transformation, all the points in the lattice are matched to the initial $\mathbb{Z}^n$. The orthogonal geometry of $\mathbb{Z}^n$ eliminates all the interference between the dimensions of the lattice, that is, between the MIMO layers; in a system with $N_T = N_R$, the ith antenna at the receiver "sees" the transmission from the corresponding ith antenna at the transmitter cleared from any interference from the other $N_T - 1$ antennas. If $N_T < N_R$, the same happens in the appropriate subdimensions of **y**. The name zero forcing is due to the fact that the interference in each antenna is *forced to zero*. In a Euclidian signal space, the geometrical interpretation is that ZF projects the received vector onto the space that is orthogonal to space where all the interferers lie (as illustrated in Figure 2.9).

This generalized pseudo-inverse transformation for singular matrices holds properties that are similar to the properties of the "true" inverse matrix of a nonsingular matrix (for an extensive description of the properties, see [Piziak and Odell 2007, Section 4.4]).

Any solution to Equation 2.49 involving the inversion of **H** will imply a number of operations that depends on the inversion methods used. The number of those operations is given in [Piziak and Odell 2007, p. 170] and is listed in Table 2.3. The common assumption in MIMO literature is that the number of operations involved

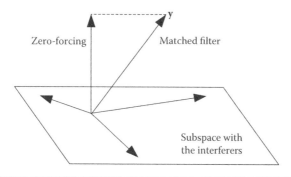

Figure 2.9 Geometric interpretation of ZF filtering in a signal space.

Table 2.3   Number of Operations Involved in Solving y = Hx by Several Methods Based on Channel Inversion (with a Square H)

	Additions and Subtractions	Multiplications and Divisions
Gauss elimination with back substitution	$\frac{1}{3}n^3 + \frac{1}{2}n^2 - \frac{5}{6}n$	$\frac{1}{3}n^3 + n^2 - \frac{1}{3}n$
Gauss–Jordan elimination (reduced Echelon form)	$\frac{1}{3}n^3 + \frac{1}{2}n^2 - \frac{5}{6}n$	$\frac{1}{3}n^3 + n^2 - \frac{1}{3}n$
Cramer's rule [Westlake 1968]	$(n + 1)!$	$(n + 1)!$
Inversion of **H**, when it is nonsingular	$n^3 - n^2$	$n^3 + n^2$

in ZF (or in any stage involving channel inversion) is $\mathcal{O}(n^3)$ and Table 2.3 corroborates that this is the case for all the inversion methods listed.

2.5.2.1 Geometry of Zero-Forcing Detection

As is mentioned in the previous section, ZF solves the CVP by relaxing it to a search in a continuous neighborhood instead of computing the distance between the received vector (also called the *target*) and every point in the lattice. The geometrical implication can be better understood thinking of the linear transformation of the hypercubic Voronoi regions of $\mathbb{Z}^n$ by **H**. The resulting regions are called the *ZF decision regions* and correspond to the space where a lattice point will be interpreted as being close to the lattice point associated with that region.

The decision regions associated with ZF criterion are simple to obtain as they are the fundamental region $\mathcal{R}(\mathbf{H})$, as defined in Equation 2.6. Because the lattices in MIMO are the Gaussian lattices defined in Section 2.3.1, the basis generated by a channel may have some highly correlated vectors. Geometrically, this corresponds to lattices with very narrow fundamental regions, which are generated by ill-conditioned matrices, that is, when one or more singular values are close to zero, and consequently the volume of the lattice vanishes. Figure 2.10 shows the ZF decision regions associated with the following equivalent bases:

$$\mathbf{H}_1 = \begin{bmatrix} 6 & 2 \\ 1 & 5 \end{bmatrix} \quad \text{and} \quad \mathbf{H}_2 = \begin{bmatrix} 6 & 8 \\ 1 & 6 \end{bmatrix}. \tag{2.56}$$

Let us concentrate in the case where the transmit point was the origin. The shaded areas indicate regions which will lead to wrong decisions when using the ZF

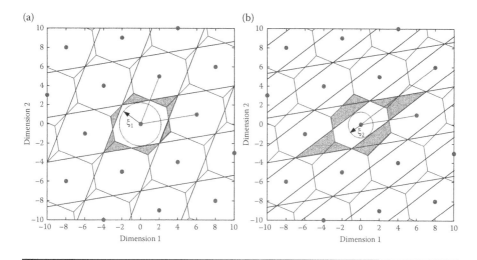

Figure 2.10 Decision regions associated with the two different bases of the same lattice. (a) H_1 and (b) H_2.

technique: either because the point is inside the Voronoi region and outside the ZF decision region or because the closest lattice point would be decided as being the origin while the Voronoi region shows that to be false. It is possible to observe in Figure 2.10 that different bases will output different decisions given a target point. For the examples at a given SNR, the SER with H_1 will be always lower, because the *coverage* of the MLD (i.e., Voronoi) regions is larger than in the case of basis H_2. The notion of coverage is essential to understand MIMO detection [Su and Kschischang 2007]. To simplify the operational meaning of coverage, [Ling 2011] was introduced the notion of *proximity factors* dependent on the notion of the largest sphere that can be fitted inside the region of coverage. These spheres are also shown in Figure 2.10 for the two basis, having *decoding radii* ξ_1 and ξ_2, respectively.

Note that a receiver with a better performance is one whose decision regions better approximate the shape of the regions associated with MLD, that is, with a better coverage of the Voronoi regions. This is the central concept in the design of novel detection techniques such as [Su and Kschischang 2007] and [Monteiro and Kschischang 2011].

2.5.2.2 Algebraic Analysis of ZF

It is also possible to explain the behavior of ZF analytically. One starts by noticing that the covariance matrix of the noise affecting the decisions after any linear transformation $\mathbf{W}$ is (for complex lattices)

$$\begin{aligned} \mathbf{R}_{n,\mathbf{W}} &= E\{(\mathbf{Wn})(\mathbf{Wn})^H\} = E\{\mathbf{Wnn}^H\mathbf{W}^H\} \\ &= \mathbf{WR}_n\mathbf{W}^H. \end{aligned} \tag{2.57}$$

The output SNR of ZF-detected vector (before slicing) at the ith layer is

$$\rho_{ZF,i} = \frac{\mathbf{R}_x}{\mathbf{R}_{n,ZF}} = \frac{\sigma_x^2 \mathbf{I}}{\left[\mathbf{W}_{ZF}\mathbf{R}_n\mathbf{W}_{ZF}^H\right]_{i,i}}, \quad 1 \le i \le n. \tag{2.58}$$

For $\mathbf{W}_{ZF}$ given by Equation 2.52 and for the model with $\mathbf{R}_x = \sigma_x^2 \mathbf{I}$ and with $\mathbf{R}_n = \sigma_n^2 \mathbf{I}$, this SNR becomes

$$
\begin{aligned}
\rho_{ZF,i} &= \frac{1}{\left[(\mathbf{H}^H\mathbf{H})^{-1}\mathbf{H}^H((\mathbf{H}^H\mathbf{H})^{-1}\mathbf{H}^H)^H\right]_{i,i}} \rho_a = \frac{1}{\left[(\mathbf{H}^H\mathbf{H})^{-1}\mathbf{H}^H\mathbf{H}((\mathbf{H}^H\mathbf{H})^{-1})^H\right]_{i,i}} \rho_a \\
&= \frac{1}{\left[(\mathbf{H}^H\mathbf{H})^{-1}\mathbf{H}^H\mathbf{H}(\mathbf{H}^{-1}(\mathbf{H}^H)^{-1})^H\right]_{i,i}} \rho_a = \frac{1}{\left[(\mathbf{H}^H\mathbf{H})^{-1}\mathbf{H}^H\underbrace{\mathbf{H}\mathbf{H}^{-1}}_{\mathbf{I}}(\mathbf{H}^H)^{-1}\right]_{i,i}} \rho_a \\
&\underbrace{\qquad\qquad\qquad\qquad}_{\mathbf{I}} \\
&= \frac{1}{[(\mathbf{H}^H\mathbf{H})^{-1}]_{i,i}} \rho_a, \quad 1 \le i \le n.
\end{aligned}
$$

$$\tag{2.59}$$

This expression relates the input SNR per transmit antenna with the output SNR at each receive antenna and quantifies the noise enhancement that explains the poor performance of ZF detection when deciding for $\hat{\mathbf{x}}_{ZF}$. One may note in Equation 2.59 that the denominator $(\mathbf{H}^H\mathbf{H})^{-1}$ is the inverse of the Gram matrix $\mathbf{G}$, that is, it is the Gram matrix of the dual lattice, according to Equation 2.19. When this Gram matrix is close to the identity matrix, that means the lattice is generated by some unitary matrix $\mathbf{U}$ (orthonormal in the real model), with $\mathbf{U}\mathbf{U}^H = \mathbf{U}^H\mathbf{U} = \mathbf{I}$. In this case, the interference between all layers is zero, and no noise enhancement will happen in ZF detection. However, when the lattice is not orthogonal, $\left[(\mathbf{H}^H\mathbf{H})^{-1}\right]_{i,i} = \mathbf{G}_{ii}^{(D)}$ corresponds to the quadratic norm of the generators of the dual vector. According to the interpretation given in Section 2.2.2, when these generators of the dual are large, the lattice will have a narrow separation between the parallel hyperplanes where the lattice points lie, and so have a small decoding radius. This brings together the algebraic analysis and the geometrical interpretation.

The diversity order collected by the ZF is $N_R - N_T + 1$, as known since the early papers on MIMO, however, the analytical proof of that was only later given in [Ma and Zhang 2008].

2.5.3 Minimum Mean-Squared Error Detection

The other (and more sophisticated) linear receiver aims at finding the filter that minimizes the mean-squared error between the estimated vector and the original vector, that is, the filter should be

$$\mathbf{W}_{MMSE} = \arg\min_{\mathbf{W}} E\left\{\|\mathbf{W}\mathbf{y} - \mathbf{x}\|^2\right\}. \tag{2.60}$$

This criterion does not aim at cancelling all the interference between layers as does ZF. Instead, the MMSE criterion takes into consideration both the interference *and* the noise in order to minimize the expected error. This minimization implies finding the point where the gradient of the objective function in Equation 2.60 is zero. There is however a fast track to finding this estimator by applying the *orthogonality principle*, well known in estimation theory and widely used in equalization problems in the ISI channel [Haykin 1996, Section 5.2, Fischer 2005, Sections 2.2.3 and 2.3.4, Madhow 2008, Section 5.6].* The optimum estimator for Equation 2.60 is the one that produces an error vector $\Delta = \mathbf{W}_{MMSE}\,\mathbf{y} - \mathbf{x}$ that is orthogonal to received signal, that is, the two vectors are uncorrelated (as illustrated in Figure 2.11).

The minimum norm $\|\Delta\|$ occurs when $\mathbf{W}_{MMSE}\mathbf{y} \perp \Delta$, that is

$$E\{(\mathbf{W}_{MMSE}\mathbf{y} - \mathbf{x})\mathbf{y}^H\} = 0. \tag{2.61}$$

Applying this principle, $\mathbf{W}_{MMSE}$ can be obtained from (using complex vectors)

$$
\begin{aligned}
&\mathbf{W}_{MMSE}E\{\mathbf{y}\mathbf{y}^H\} - E\{\mathbf{x}\mathbf{y}^H\} = 0 \\
&\Leftrightarrow \mathbf{W}_{MMSE}(\mathbf{H}\mathbf{R}_x\mathbf{H}^H + \mathbf{R}_n) - E\{\mathbf{x}(\mathbf{H}\mathbf{x})^H\} = 0 \\
&\Leftrightarrow \mathbf{W}_{MMSE}(\mathbf{R}_n + \mathbf{H}\mathbf{R}_x\mathbf{H}^H) - \mathbf{R}_x\mathbf{H}^H = 0 \\
&\Leftrightarrow \mathbf{W}_{MMSE} = \mathbf{R}_x\mathbf{H}^H(\mathbf{R}_n + \mathbf{H}\mathbf{R}_x\mathbf{H}^H)^{-1}.
\end{aligned}
\tag{2.62}
$$

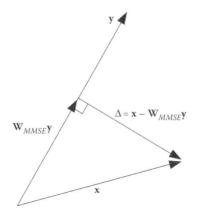

Figure 2.11 Orthogonality principle: the expected error is made orthogonal to the space where the solution lies.

---

* The principle is valid for in general estimation theory and can be derived in a Bayesian framework for linear estimation [Kay 1993, Chapter 12, Poor 1994, Sections V-C, VII-C1].

At this point most authors commonly invoke the *matrix inversion lemma*[*] [Kay 1993, p. 571, Haykin 1996, pp. 565–566], and immediately obtain from it one of the two possible formulas of the MMMSE filter. That path is cumbersome and eventually ends with an expression for $\mathbf{W}_{MMSE}$ that is not even the expression that is concluded from that derivation (although it will be proven later on that they are equivalent). In the following is presented a derivation of the filter involving much simpler algebra[†]:

$$
\begin{aligned}
\mathbf{W}_{MMSE} &= \mathbf{R}_x \mathbf{H}^H (\mathbf{R}_n + \mathbf{H} \mathbf{R}_x \mathbf{H}^H)^{-1} \\
&= \mathbf{R}_x \mathbf{H}^H (\mathbf{R}_n (\mathbf{I} + \mathbf{R}_n^{-1} \mathbf{H} \mathbf{R}_x \mathbf{H}^H))^{-1} \\
&= \mathbf{R}_x \mathbf{H}^H ((\mathbf{I} + \mathbf{R}_n^{-1} \mathbf{H} \mathbf{R}_x \mathbf{H}^H)^{-1} \mathbf{R}_n^{-1}).
\end{aligned}
\tag{2.63}
$$

For the correlation models that were considered, Equation 2.63 is reduced to

$$
\begin{aligned}
W_{MMSE} &= \sigma_x^2 \mathbf{I} \cdot \mathbf{H}^H \left(\sigma_n^2 \mathbf{I} \cdot \mathbf{H} \cdot \sigma_x^2 \cdot \mathbf{I} \cdot \mathbf{H}^H \right)^{-1} \\
&= \mathbf{H}^H \left(\mathbf{H} \mathbf{H}^H + \frac{\sigma_n^2}{\sigma_x^2} \mathbf{I} \right)^{-1} = \mathbf{H}^H \left(\mathbf{H} \mathbf{H}^H + \frac{1}{\rho_a} \mathbf{I} \right)^{-1}.
\end{aligned}
\tag{2.64}
$$

It should be highlighted that the final expression in Equation 2.64 is *not* the only one that appears in the literature. Just as often, one may encounter the following distinct version for the MMSE filter:

$$
\mathbf{W}_{MMSE} = \left(\mathbf{H}^H \mathbf{H} + \frac{1}{\rho_a} \mathbf{I} \right)^{-1} \mathbf{H}^H.
\tag{2.65}
$$

Expressions (2.64) and (2.65) are equivalent, although this is rarely mentioned in the literature. The equivalence is a consequence of the following matrix identity:

Theorem: $(\mathbf{AB} + \mathbf{I})^{-1} \mathbf{A} = \mathbf{A}(\mathbf{BA} + \mathbf{I})^{-1}$.

Proof. The identity[‡] $\mathbf{A}(\mathbf{BA} + \mathbf{I}) = (\mathbf{AB} + \mathbf{I})\mathbf{A}$ holds because

---

[*] The matrix inversion lemma, $(\mathbf{A} + \mathbf{BCD})^{-1} = \mathbf{A}^{-1} - \mathbf{A}^{-1}\mathbf{B}(\mathbf{DA}^{-1}\mathbf{C}^{-1})^{-1} \mathbf{DA}^{-1}$, is actually one of the several variants of the Woodbury identity [Petersen and Pedersen 2008, p. 17]. Moreover, the lemma is a particular case of the Hendersson–Searle formulas [Piziak and Odell 2007, Section 1.2.1].

[†] The steps taken here are mostly used when proving several of the Hendersson–Searle formulas.

[‡] This identity can be seen as a particular case of one of the Searle identities [Petersen and Pedersen 2008, p. 18].

$$\mathbf{A}(\mathbf{BA} + \mathbf{I}) = \mathbf{ABA} + \mathbf{A} = (\mathbf{AB} + \mathbf{I})\mathbf{A}$$

$$\Leftrightarrow \mathbf{A}(\mathbf{BA} + \mathbf{I})\underbrace{(\mathbf{BA} + \mathbf{I})}_{\text{new term}} = (\mathbf{AB} + \mathbf{I})\mathbf{A}\underbrace{(\mathbf{BA} + \mathbf{I})}_{\text{new term}}$$

$$\Leftrightarrow \mathbf{A} = (\mathbf{AB} + \mathbf{I})\mathbf{A}(\mathbf{BA} + \mathbf{I})(\mathbf{BA} + \mathbf{I})^{-1}(\mathbf{BA} + \mathbf{I})^{-1} \qquad (2.66)$$

$$\Leftrightarrow \underbrace{(\mathbf{AB} + \mathbf{I})^{-1}}_{\text{new}} \mathbf{A} = \underbrace{(\mathbf{AB} + \mathbf{I})^{-1}}_{\text{new}}(\mathbf{AB} + \mathbf{I})\mathbf{A}(\mathbf{BA} + \mathbf{I})^{-1}$$

$$\Leftrightarrow (\mathbf{AB} + \mathbf{I})^{-1}\mathbf{A} = \mathbf{A}(\mathbf{BA} + \mathbf{I})^{-1}.$$

Similarly to Equation 2.53, the filtering matrix $\mathbf{W}_{MMSE}$ is given by Equations 2.64 or by Equation 2.65,[*] and so the MMSE receiver can be described by

$$\mathbf{x}_{MMSE} = \mathbf{W}_{MMSE}\mathbf{y}, \quad \hat{\mathbf{x}}_{MMSE} = Q[\mathbf{x}_{MMSE}], \qquad (2.67)$$

with the block diagram as the one in Figure 2.12.

It should be mentioned that, as shown in [Hassibi 2000], expression (2.67) is in fact equivalent to Equation 2.53 if $\mathbf{H}$ is replaced by the extended matrix

$$\underline{\mathbf{H}} = \begin{bmatrix} \mathbf{H} \\ \sigma_n^2 \mathbf{I}_{2N_R} \end{bmatrix}. \qquad (2.68)$$

As it is often mentioned in the literature, a careful comparison of Equation 2.64 or Equation 2.65 with Equation 2.53 allows one to conclude that the MMSE filter tends to the ZF filter at high SNR. Therefore, one could expect a similar performance for both of them in the high SNR regime. However, it is well known (from very early on) that this is *not* true and this fact seems to be forgotten whenever such comment is made. It was only in 2011 that the existing gap between ZF and MMSE detection was characterized in [Jiang et al. 2011]. The authors finally proved (for the ideal Rayleigh channel) several other assertions taken for granted in the last decade for the ZF, MMSE, and SIC receivers based on ZF or MMSE filters. Furthermore, analytical expressions for the BER for the correlated channel have been devised in [Hong and Armada 2011]. In [Kim et al. 2008], the MMSE

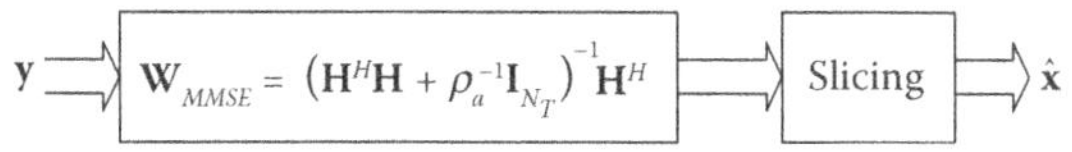

$$\mathbf{y} \rightarrow \boxed{\mathbf{W}_{MMSE} = \left(\mathbf{H}^H\mathbf{H} + \rho_a^{-1}\mathbf{I}_{N_T}\right)^{-1}\mathbf{H}^H} \rightarrow \boxed{\text{Slicing}} \rightarrow \hat{\mathbf{x}}$$

Figure 2.12   MMSE receiver.

---

[*] The simulations for the performance of the MMSE receivers to be presented at the end of this chapter make use of Equation 2.65 adapted for the real equivalent model.

receiver had already been analytically studied but only for low number of antennas (and also for the ideal Rayleigh fading channel).

For low SNR, the effect of interference is *less* important than the effect of the Gaussian noise, hence the MMSE filter tends to $\mathbf{H}^H$, which corresponds to the *matched filter* to that channel. In this case, the detection is treated as a maximum correlation problem.

The covariance matrix of the noise after the MMSE transformation is

$$\mathbf{R}_{n,MMSE} = \mathbf{W}_{MMSE}\mathbf{R}_n\mathbf{W}^H_{MMSE}, \tag{2.69}$$

and, similarly to Equations 2.57 through 2.59,

$$\begin{aligned}
\mathbf{R}_{n,MMSE} &= E\left\{(\mathbf{W}_{MMSE}\mathbf{n})(\mathbf{W}_{MMSE}\mathbf{n})^H\right\} \\
&= E\left\{\left((\mathbf{H}^H\mathbf{H} + \rho_a^{-1}\mathbf{I})^{-1}\mathbf{H}^H\mathbf{n}\right)\left((\mathbf{H}^H\mathbf{H} + \rho_a^{-1}\mathbf{I})^{-1}\mathbf{H}^H\mathbf{n}\right)^H\right\} \\
&= E\left\{(\mathbf{H}^H\mathbf{H} + \rho_a^{-1}\mathbf{I})^{-1}\mathbf{H}^H\mathbf{n}\mathbf{n}^H\mathbf{H}((\mathbf{H}^H\mathbf{H} + \rho_a^{-1}\mathbf{I})^{-1})^H\right\} \\
&= E\left\{(\mathbf{H}^H\mathbf{H} + \rho_a^{-1}\mathbf{I})^{-1}\mathbf{H}^H\mathbf{n}\mathbf{n}^H\mathbf{H}((\mathbf{H}^H\mathbf{H} + \rho_a^{-1}\mathbf{I})^{-1})^H\right\}. \tag{2.70}
\end{aligned}$$

It is possible to show that the output SNRs at the ith layer after the MMSE filter becomes

$$\rho_{MMSE,i} = \frac{1}{\left[\left(\mathbf{H}^H\mathbf{H} + \rho_a^{-1}\mathbf{I}\right)^{-1}\right]_{i,i}}\rho_a, \quad 1 \le i \le n. \tag{2.71}$$

It was seen before that the detection based in ZF is extremely problematic when its decision regions become too long and narrow. This happens for channels that are ill conditioned, with one or several of the eigenvalues very small in comparison to the others. This means that the (hyper-) ellipse associated with the linear transformation of an (hyper-) sphere is highly eccentric. In these cases, one may consider penalizing the solutions to the problem that would imply having a large norm for the detected $\hat{\mathbf{x}}$. One other way of interpreting the MMME solution is in the context of the optimization problem generated from a relaxation of Equation 2.49. Following the proposal in [Jaldén and Ottersten 2006], one may generalize the problem for binary symbols, to the problem with $\sqrt{M}$ symbols per dimension. In doing this, the problem becomes equivalent to

$$\mathbf{y}_{MMSE} = \arg\min_{\mathbf{x}\in\mathcal{A}}\left\{\|\mathbf{y} - \mathbf{H}\mathbf{x}\|^2 + \frac{1}{\rho_a}\|\mathbf{x}\|^2 - \frac{N_T E_s}{\rho_a}\right\}, \tag{2.72}$$

noting that for $\mathbf{x} \in \mathcal{A} \Rightarrow E\left\{\rho_a^{-1}\|\mathbf{x}\|^2 + \rho_a^{-1}(N_T E_s)\right\} = 0$. The MMSE criterion in Equation 2.72 is attained by relaxing the search in $\mathbf{x} \in \mathcal{A}$ to a search in the continuous space where $\mathbf{x} \in \mathbb{C}^n$. As the last term in Equation 2.72 does not involve $\mathbf{x}$, the minimization is also

$$\mathbf{y}_{MMSE} = \arg \min_{\mathbf{x} \in \mathbb{C}^n} \{ \underbrace{\|\mathbf{y} - \mathbf{Hx}\|^2}_{\substack{\text{(Squared)} \\ \text{Euclidean} \\ \text{distance}}} + \underbrace{\sigma_n^2 \|\mathbf{x}\|^2}_{\substack{\text{Penalization} \\ \text{on longer} \\ \text{solutions}}} \} \tag{2.73}$$

corresponding to the solution of the typical CVP but with a term that penalizes large $\|\mathbf{x}\|$ and is proportional to the energy of the noise. This explains why MMSE performs better than ZF for ill-conditioned channel realizations.

2.5.3.1 Projection Matrices

Denoting $\mathbf{H}_{\bar{j}}$ as the matrix obtained from $\mathbf{H}$ by deleting the (column) generator $\mathbf{h}_j$, it is indicated in [Jiang et al. 2011] that Equation 2.59 can also be written as

$$\begin{aligned} \rho_{ZF,j} &= \left[\mathbf{h}_j^H \mathbf{h}_j - \mathbf{h}_j^H \mathbf{H}_{\bar{j}} (\mathbf{H}_{\bar{j}}^H \mathbf{H}_{\bar{j}})^{-1} \mathbf{H}_{\bar{j}}^H \mathbf{h}_j \right] \rho_a \\ &= \mathbf{h}_j^H \left(\mathbf{I} - \mathbf{H}_{\bar{j}} \left(\mathbf{H}_{\bar{j}}^H \mathbf{H}_{\bar{j}} \right)^{-1} \mathbf{H}_{\bar{j}}^H \right) \mathbf{h}_j \rho_a \\ &= (\mathbf{h}_j^H \mathbf{P}_{\mathbf{H}_{\bar{j}}^\perp} \mathbf{h}_j) \rho_a. \end{aligned} \tag{2.74}$$

This expression can be better interpreted evoking the notion of *projection matrices*. A projection matrix is always (i) symmetric, (ii) idempotent (i.e., the successive application of a projection, twice or more times, does not change the result), and (iii) positive semidefinite [e.g., Meyer 2000, p. 386].

Consider a linear space spanned by the columns of $\mathbf{H}$, that is, span($\mathbf{H}$). The projection of a vector $\mathbf{a}$ onto that space is denoted as $\mathrm{Proj}_{\mathbf{H}}(\mathbf{a})$, the projection onto the space orthogonal to span($\mathbf{H}$) is denoted as $\mathrm{Proj}_{\mathbf{H}^\perp}(\mathbf{a})$, and they are given by [Piziak and Odell 2007, Section 8.3]:

$$\mathrm{Proj}_{\mathbf{H}}(\mathbf{a}) = \mathbf{P}_{\mathbf{H}} \mathbf{a} = \mathbf{H} \mathbf{H}^+ \mathbf{a}, \tag{2.75}$$

$$\mathrm{Proj}_{\mathbf{H}^\perp}(\mathbf{a}) = \mathbf{P}_{\mathbf{H}^\perp} \mathbf{a} = (\mathbf{I} - \mathbf{H} \mathbf{H}^+) \mathbf{a}. \tag{2.76}$$

From Equations 2.75 and 2.76, the projection onto the span($\mathbf{H}_{\bar{j}}$) is

$$\mathbf{P}_{\mathbf{H}_{\bar{j}}} = \mathbf{H}_{\bar{j}} \mathbf{H}_{\bar{j}}^+ \tag{2.77}$$

and the projection onto its orthogonal complement is

$$\mathbf{P}_{\mathbf{H}_{\bar{j}}^{\perp}} = \mathbf{I} - \mathbf{H}_{\bar{j}}\mathbf{H}_{\bar{j}}^{+}$$
$$= \mathbf{I} - \mathbf{H}_{\bar{j}}\left(\mathbf{H}_{\bar{j}}^{H}\mathbf{H}_{\bar{j}}\right)^{-1}\mathbf{H}_{\bar{j}}^{H}, \tag{2.78}$$

where the Moore–Penrose pseudo-inverse was used in the last line.

The factor $\mathbf{h}_{j}^{H}\mathbf{P}_{\mathbf{H}_{\bar{j}}^{\perp}}\mathbf{h}_{j}$ that appears in Equation 2.74 is a quadratic form, as defined in Equation 2.9, hence, $\mathbf{h}_{j}^{H}\mathbf{P}_{\mathbf{H}_{\bar{j}}^{\perp}}\mathbf{h}_{j} \geq 0$ as it corresponds to the norm of the projection of generator $\mathbf{h}_{j}$ onto space spanned by the remaining generators in the basis $\mathbf{H}_{\bar{j}}$.

Figure 2.13 shows the geometry of these projections in the same bidimensional example of Figure 2.10, with basis $\mathbf{H}_{2}$, given in Equation 2.56. One can observe that the factor $\mathbf{h}_{j}^{H}\mathbf{P}_{\mathbf{H}_{\bar{j}}^{\perp}}\mathbf{h}_{j}$ corresponds to the distance between parallel layers where the lattice points lie and it measures the separation between the decision thresholds for the jth layer, associated with $\mathbf{h}_{j}$. When that distance is ≥ 1, there is an SNR gain in that layer in respect to the SNR in the expected average layer.

It is worth mentioning that in the case with ZF, $\mathbf{h}_{j}^{H}\mathbf{P}_{\mathbf{H}_{\bar{j}}^{\perp}}\mathbf{h}_{j}$ has a χ^2 distribution with $2(N_R - N + 1)$ degrees of freedom [Jiang et al. 2011]. A similar expression to

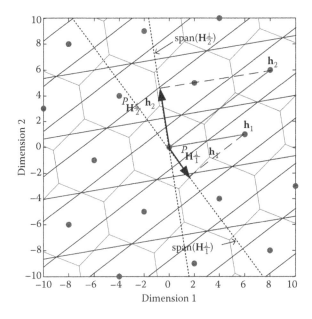

Figure 2.13   Geometry of the SNR relation factor in Equation 2.74. It includes the Voronoi regions and the ZF decision regions of the lattice $\Lambda(\mathbf{H}_2)$, as given in Equation 2.56.

Equation 2.74 can be obtained for MMSE if the factor $(\mathbf{H}_{\bar{j}}^{H}\mathbf{H}_{\bar{j}})^{-1}$ is replaced by $(\mathbf{H}_{\bar{j}}^{H}\mathbf{H}_{\bar{j}} + \rho_{a}^{-1}\mathbf{I})^{-1}$ in Equation 2.74.

2.5.4 Geometry of Optimally Ordered Successive Interference Cancellation

As mentioned in Chapter 1, D-BLAST was the first scheme to be proposed by Foschini. Given its detection complexity, V-BLAST ended up being the standard architecture for SM. The detection algorithm first proposed in [Wolniansky et al. 1998] and [Golden et al. 1999] uses the principles of SIC, already known in ISI control and MUD and is known as the V-BLAST detector (as mentioned, in this chapter the name V-BLAST is identified with that particular detection method). The general principle of SIC is that an initial "best" layer is detected and then, assuming that the symbol was correctly detected, the interference caused by that symbol is replicated and subtracted from all the other layers. The procedure is then applied to the "next best" layer: one symbol more is detected, its interference recreated and then subtracted from the remaining ones.

One important question that arises is the one of determining the order of detection of the N_R antennas. For a MIMO $n \times n$ system, one has to find the optimum permutation $\Pi(k)$ of the column indexes $\{1,2, \ldots, n\}$ that minimizes the SER amid all the $n!$ possible permutations. An exhaustive search over all the permutations would rapidly become unbearable as n increases. The optimal solution to this problem was found early on in [Wolniansky et al. 1998] and [Golden et al. 1999], in the first implementations of the V-BLAST detector.

The optimal criterion at each stage is to select the layer that less emphasizes the noise power after a ZF or an MMSE filter. Consider the following example (for one an alternative example see [Windpassinger 2004]) with input data $\mathbf{x} = [1 \ \ 1 \ \ 1]^{T}$ and a noise vector that does not induce an error in the MLD sense, that is, that does not take the point out of its Voronoi region:

$$\mathbf{y} = \begin{bmatrix} -0.3 \\ 2.4 \\ -1.28 \end{bmatrix} = \underbrace{\begin{bmatrix} -0.97 & 0.48 & 0.31 \\ 1.35 & 0.12 & 1.43 \\ -1.04 & 1.2 & -1.94 \end{bmatrix}}_{\mathbf{H}} \begin{bmatrix} x_1 \\ x_2 \\ x_2 \end{bmatrix} + \underbrace{\begin{bmatrix} 0.5 \\ -0.5 \\ 0.5 \end{bmatrix}}_{\mathbf{n}}. \tag{2.79}$$

Considering that ZF is used,

$$\mathbf{W}_{ZF}^{(1)} = \mathbf{H}^{+} = \begin{bmatrix} -1.0296 & 0.2954 & 0.3823 \\ 0.5980 & 0.8239 & 0.5118 \\ 0.9219 & 0.3512 & -0.4039 \end{bmatrix} \begin{matrix} \leftarrow \|\mathbf{w}_1(1,:)\|^2 = 1.2936 \\ \leftarrow \|\mathbf{w}_1(2,:)\|^2 = 1.2983 \\ \leftarrow \|\mathbf{w}_1(3,:)\|^2 = 1.1363, \end{matrix}$$

where the energy of each row in the filtering matrix is indicated on the right.

The lowest noise enhancement factor is the one associated with the third row (i.e., the third layer), so this third symbol is decided via a decision threshold to the alphabet as* $\hat{\mathbf{x}}(3) = Q_A[W_{ZF}^{(1)}\mathbf{y}] = 1$. Next, the effect of that symbol, after, is subtracted from the other layers:

$$\mathbf{y}^{(2)} = \mathbf{y} - \mathbf{H}(:,3)\,\hat{\mathbf{x}}(3) = \mathbf{y} - \begin{bmatrix} 0.31 \\ 1.43 \\ -1.94 \end{bmatrix} 1 = \begin{bmatrix} 0.01 \\ 0.97 \\ 0.66 \end{bmatrix}.$$

The third generator vector in the channel matrix is now nulled:

$$\mathbf{H}^{(2)} = \begin{bmatrix} -0.97 & 0.48 & 0 \\ 1.35 & 0.12 & 0 \\ -1.04 & 1.2 & 0 \end{bmatrix},$$

and the associated ZF matrix is now

$$\mathbf{W}_{ZF}^{(2)} = \begin{bmatrix} -0.2185 & 0.6045 & 0.0270 \\ 0.0837 & 0.6279 & 0.7371 \\ 0 & 0 & 0 \end{bmatrix} \begin{matrix} \leftarrow \|\mathbf{w}_2(1,:)\|^2 = 0.4139 \\ \\ \leftarrow \|\mathbf{w}_2(2,:)\|^2 = 0.9446. \end{matrix}$$

At this stage one observes that the first row is the one that less enhances the noise and thus $\mathbf{x}(1)$ (i.e., the element transmitted in the first layer) is now decided as $\hat{\mathbf{x}}(1) = Q[W_{ZF}^{(2)}\mathbf{y}^{(2)}] = 1$ and its interference subtracted from $\mathbf{y}^{(2)}$:

$$\mathbf{y}^{(3)} = \mathbf{y}^{(2)} - \mathbf{H}(:,1)\,\hat{\mathbf{x}}(1) = \begin{bmatrix} 0.01 \\ 0.97 \\ 0.66 \end{bmatrix} - \begin{bmatrix} -0.97 \\ 1.35 \\ -1.04 \end{bmatrix} 1 = \begin{bmatrix} 0.98 \\ -0.38 \\ 1.70 \end{bmatrix}.$$

Now, the first generator of **H** is zeroed

$$\mathbf{H}^{(3)} = \begin{bmatrix} 0 & 0.48 & 0 \\ 0 & 0.12 & 0 \\ 0 & 1.2 & 0 \end{bmatrix}$$

---

* The notation is slightly abused, as the quantization is applied only to the third element of $W_{ZF}^{(1)}\mathbf{y}$. Moreover, some Matlab will be used to denote matrices rows and columns.

and the corresponding pseudo-inverse is

$$\mathbf{W}_{ZF}^{(3)} = \begin{bmatrix} 0 & 0 & 0 \\ 0.2849 & 0.0712 & 0.7123 \\ 0 & 0 & 0 \end{bmatrix} \leftarrow \left\| \mathbf{w}_2(2,:) \right\|^2 = 0.5936.$$

Finally, $\hat{\mathbf{x}}(2) = Q[W_{ZF}^{(3)}\mathbf{y}^{(3)}] = 1$, that is once again a correctly detected symbol.

The ordering strategy described in this example by means of algebraic arguments is in fact is a direct application of a fact that, until the paper by [Wolniansky et al. 1998], was known as a rule of thumb in the MUD literature, but whose optimality had never been proven. The rule consists of selecting at each step the vector that minimizes noise enhancement and in [Wolniansky et al. 1998] the authors pointed out that the SNR at layer i is

$$\rho_{i,OSIC} = \frac{E[x_i^2]}{\sigma_n^2 \sum_{l=1}^{N_T} w_i^2} = \frac{E[x_i^2]}{\sigma_n^2 \left\| \mathbf{w}_i \right\|^2}, \tag{2.80}$$

and used Equation 2.80 to justify the optimality of the criterion.

In the following, it is shown that further insight concerning this optimization process can be enlightened, and indeed proven to be optimal, if a geometric perspective is applied using the projection tools introduced in Section 2.5.3.1 and using the geometric ideas of the Babai nearest plane algorithm [Babai 1986, Micciancio and Goldwasser 2002, Chapter 2] in algorithmic number theory, which corresponds to SIC in MIMO, as first noticed by [Agrell et al. 2002] and [Mow 2003].

To minimize the error probability when deciding layer j, the generator vector $\mathbf{h}_j$ has to be selected at any given decision step k, with $k \in \{1, 2, \ldots, n\}$, should be the vector that maximizes the projection onto the orthogonal space to the space spanned by the matrix that remains after that same vector is taken out from $\mathbf{H}$.

The initial step is to find the column vector $\mathbf{h}_1$ that, when removed from $\mathbf{H}$, transforms $\mathbf{H}$ into $\mathbf{H}_{\bar{1}}$ (as $\mathbf{H}_{\bar{j}}$ denotes the matrix that is obtained from $\mathbf{H}$ after removing column j). $\mathbf{H}_{\bar{1}}$ is the generator of an $(n-1)$-dimensional lattice Λ_{n-1}. Hence, the original lattice can be written in the form

$$\Lambda = \Lambda_{n-1} + i\mathbf{h}_1, \quad i \in \mathbb{Z}, \tag{2.81}$$

signifying that Λ can be created from the *union* of translates of the Λ_{n-1} sublattice.

Once a decision is produced for one layer, the subsequent step is to repeat the process, now in the sublattice with basis $\mathbf{H}_{\bar{j}}$, that is, by removing generator $\mathbf{h}_j$ from the set. The process repeats itself until a decision is made in a one-dimensional lattice, corresponding to the decision of the last layer to be detected.

Figure 2.14 depicts SIC applied to a lattice partitioned as in Equation 2.81. In a first stage, the nearest hyperplane is found and a decision for the layer associated to $\mathbf{h}_j$ is produced. In a second stage, depicted at the bottom of Figure 2.14, the same procedure is applied but now conducted in the sublattice Λ_{n-1}.

Figure 2.15 shows the SIC decision region for the origin of the lattice with basis

$$\mathbf{H} = \begin{bmatrix} 3 & 1 \\ 1 & 3 \end{bmatrix}. \tag{2.82}$$

The example shows a target point located in a region where SIC outputs an erroneous decision. The first SIC step in the example in Figure 2.15 is to select which plane is the nearest one to the target point. In the example, SIC would decide for plane 1 while the Voronoi region indicates that the correct point lies in plane 2.

The diversity attained by SIC is $N_R - N_T + 1$ and sorting the layers does not contribute to any improvement in this respect, as recently proven in [Jiang et al. 2011]. Sorting can only yield a power gain in SM detection.

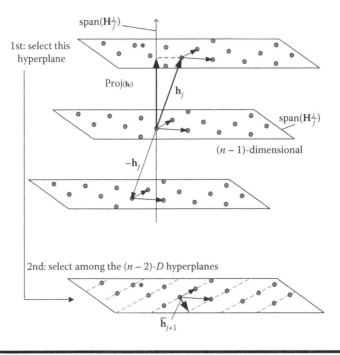

**Figure 2.14   The nearest plane algorithm with sorting. Choosing the *j*th genera-
tor vector that maximizes the distance between parallel hyperplanes. The lattice
is the union of such translates.**

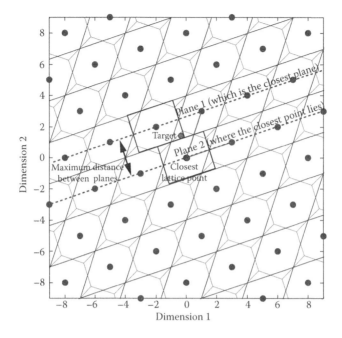

Figure 2.15 **Error events in SIC. Plane 1 is selected because it is the closest plane, however, the closest lattice point lies in plane 2. The SIC decision region for the origin is shown.**

The decision regions associated to SIC are hyper-rectangular and it is not difficult to perceive (e.g., from Figures 2.14 and 2.15) that these decision regions are unequivocally defined by the Gram–Schmidt vectors on the basis of the lattice [e.g., Ling 2011].

The fastest implementation of the original OSIC idea was provided in [Shang and Xia 2008] and was made cubic in n, that is, with complexity $\mathcal{O}(n^3)$, nevertheless other $\mathcal{O}(n^3)$ algorithms were known for OSIC much before (c.f. [Windpassinger 2004, p. 39] and references therein). Ling et al. also proposed an OSIC algorithm with $\mathcal{O}(n^3)$ complexity based on the geometric insights offered by the dual lattice [Ling et al. 2006, 2009]. In doing that, the same optimal ordering known for OSIC [Golden et al. 1999] is proven and the same performance is attained without needing a matrix inversion for each layer to be detected. This approach makes use of the shortest vector in the dual basis at each detection step.

One can now formalize SIC in a very concise manner; the hth index for the permutation is then selected from

$$\Pi(k) = \arg\max_{j \in \mathcal{A}_k} \left\{ \left\langle \mathbf{h}_j^T, \mathrm{Proj}_{\mathbf{H}_{\bar{j}}^{\perp}}(\mathbf{h}_j) \right\rangle \right\}, \tag{2.83}$$

where A_k is the set of columns that have not been chosen yet. From Section 2.5.3.1, Equation 2.83 becomes

$$\Pi(k) = \underset{j \in A_k}{\arg\max} \left\{ b_j^T \left(\mathbf{I} - \mathbf{H}_{k,j}^T (\mathbf{H}_{k,j}^T \mathbf{H}_{k,j})^{-1} \mathbf{H}_{k,j}^T \right) b_j \right\}, \qquad (2.84)$$

which is a very concise expression that summarizes the entire OSIC with optimal ordering [Trujillo et al. 2009]. Starting with $A_1 = \{1,2, \ldots, n\}$ (i.e., with all the columns of **H**), the set A_k is reduced by one element each time a column is selected, and continues until only one is left. Although concise, this formulation for finding the permutation $\Pi(k)$ does not lead to a practical implementation.

There is however a very elegant way of finding $\Pi(k)$ remembering that the distance between hyperplanes in the primal lattice is established by the lattice points in the dual (as proved in Chapter 2). Selecting the smallest basis vector in the dual basis ensures that the decision for that layer will be made from selecting between the most distant hyperplanes associated to that *specific* basis. Nonetheless, it is important to highlight that these are *not* necessarily the most distant hyperplanes in the lattice. This observation confirms why there is room for improving a receiver based on the OSIC principle.

It is thus natural to look for short vectors in the dual lattice rather than searching among the generators constituting the basis. Shorter vectors in the dual lattice would maximize the distance between the parallel hyperplanes and thus minimize erroneous decisions. Finding shorter vectors in the dual lattice is accomplished by means of lattice-reduction-aided (LRA) techniques, which will be presented later. Lattice reduction provides an equivalent basis with shorter (and more orthogonal) generator vectors.

It is noteworthy that the geometric interpretation presented in this section also sheds light on the finding by Taherzadeh et al. that reducing the dual matrix is preferable to reducing the primal basis [Taherzadeh et al. 2007, Taherzadeh 2008].

2.5.4.1 Gram–Schmidt Orthogonalization and QR Decomposition

The Gram–Schmidt (GS) orthogonalization (Algorithm 2.1) is a well-known method that takes one set of generating vectors of space and obtains another set of vectors that span the same space but which are all mutually orthogonal. Notice that although the new basis spans the same *continuous* real (or complex) space, it does *not* span the lattice. In general, the GS vectors are not members of the lattice and therefore cannot be members of any of its bases. Finding a basis that is close to orthogonal while still spanning the same *discrete* space is the much more difficult problem of *lattice reduction*, which will be described in Section 2.5.5.

The Gram–Schmidt vectors process can create a matrix $\mathbf{Q}$, with all columns mutually orthogonal, that is, $\mathbf{Q}^H\mathbf{Q} = \text{diag}(\mathbf{q}_1^H\mathbf{q}_1,...,\mathbf{q}_n^H\mathbf{q}_n)$. However, it is possible to make its column vectors orthonormal, that is, $\mathbf{Q}^H\mathbf{Q} = \mathbf{I}$ and $\|\mathbf{q}_i\| = 1$. The two sets of vectors of the two versions are obviously related by the respective norms. The matrices that perform the transformation of the original matrices to the orthogonal or orthonormal forms are triangular in each of the two cases. The relation between the two triangular matrices is less obvious, though important in the MIMO context.

The orthogonal version is (i) relevant in lattice reduction techniques such as the LLL algorithm, and (ii) an essential tool in the interpretation of SIC. However, the orthonormal form of GS orthogonalization corresponds to the QR decomposition and is (i) much used in sorted or unsorted OSIC detection, and (ii) central to SD.

Algorithm 2.1 computes the set of orthogonal vectors as

$$\bar{\mathbf{h}}_j = \mathbf{h}_j - \sum_{k=1}^{j-1} \frac{\langle h_j, \bar{h}_k \rangle}{\langle \bar{h}_k, \bar{h}_k \rangle} \bar{h}_k = \mathbf{h}_j - \sum_{k=1}^{j-1} \mu_{j,k} \bar{\mathbf{h}}_k. \tag{2.85}$$

In matrix form, the original column vectors can be related with the GS vectors by

$$\begin{bmatrix} \mathbf{h}_1 \ \mathbf{h}_2 \ \mathbf{h}_3 \cdots \mathbf{h}_n \end{bmatrix} = \underbrace{\begin{bmatrix} \bar{\mathbf{h}}_1 \ \bar{\mathbf{h}}_2 \ \bar{\mathbf{h}}_3 \cdots \bar{\mathbf{h}}_n \end{bmatrix}}_{\substack{\text{Orthogonal but} \\ \text{not orthonormal} \\ \text{columns}}} \begin{bmatrix} 1 & \mu_{2,1} & \mu_{3,1} & \cdots & \mu_{n,1} \\ 0 & 1 & \mu_{3,2} & \cdots & \mu_{n,2} \\ 0 & 0 & 1 & \cdots & \mu_{n,3} \\ 0 & 0 & 0 & \ddots & \vdots \\ 0 & 0 & 0 & 0 & 1 \end{bmatrix}. \tag{2.86}$$

Algorithm 2.1: Gram–Schmidt Orthogonalization

Input: linearly independent vectors $\mathbf{h}_1, \mathbf{h}_2,...,\mathbf{h}_n \in \mathbb{R}^n$
Output: Orthogonal basis $\bar{\mathbf{H}} = [\bar{\mathbf{h}}_1, \bar{\mathbf{h}}_2,...,\bar{\mathbf{h}}_n] \in \mathbb{R}^n$ and coefficients $\mu_{j,k} \in \mathbb{R}$
$\bar{\mathbf{h}}_1 = \mathbf{h}_1$
for $j = 2{:}n$
$\quad \bar{\mathbf{h}}_j = \mathbf{h}_j$
$\quad$ for $k = 1{:}j-1$
$\quad\quad \mu_{j,k} = \dfrac{h_j \cdot \bar{h}_k}{\bar{h}_k \cdot \bar{h}_k}$
$\quad\quad \bar{\mathbf{h}}_j = \mathbf{h}_j - \sum_{k=1}^{n} \mu_{j,k} \cdot \bar{\mathbf{h}}_k$
$\quad$ end
end

Note that $\det(\mathbf{H}) = \det(\bar{\mathbf{H}})$, as the upper triangular (u.t.) matrix has unit determinant. An orthonormal basis can also be constructed from the GS vectors if they are normalized:

$$\mathbf{q}_j = \frac{\bar{\mathbf{h}}_j}{\|\bar{\mathbf{h}}_j\|}, \quad j = 1, 2, \ldots, n. \tag{2.87}$$

GS orthogonalization can also be used to compute the QR decomposition of a channel matrix as

$$\mathbf{H} = \mathbf{QR}, \tag{2.88}$$

with $\mathbf{Q}$ orthogonal and $\mathbf{R}$ upper triangular (u.t).

To compute Equation 2.88, one starts by computing the GS vectors using

$$\bar{\mathbf{h}}_j = \mathbf{h}_j - \sum_{k=1}^{N} \langle \mathbf{h}_j, \mathbf{q}_k \rangle, \quad j = 1, 2, \ldots, n. \tag{2.89}$$

Moreover, each of the n vectors of the original basis $\mathbf{H}$ can be expressed in terms of the orthonormal vectors $\mathbf{q}_j$ as

$$\mathbf{h}_j = \langle \mathbf{h}_j, \mathbf{q}_1 \rangle \mathbf{q}_1 + \cdots + \langle \mathbf{h}_j, \mathbf{q}_{j-1} \rangle \mathbf{q}_{j-1} + \|\bar{\mathbf{h}}_j\| \mathbf{q}_j. \tag{2.90}$$

This relation can be conveniently written in matrix form as

$$[\mathbf{h}_1\ \mathbf{h}_2\ \mathbf{h}_3 \cdots \mathbf{h}_n] = \underbrace{\left[\mathbf{q}_1\ \mathbf{q}_2\ \mathbf{q}_3 \cdots \mathbf{q}_n\right]}_{\substack{Q \text{ with} \\ \text{orthonormal columns}}} \begin{bmatrix} \|\bar{\mathbf{h}}_1\| & \langle \mathbf{h}_2, \mathbf{q}_1 \rangle & \langle \mathbf{h}_3, \mathbf{q}_1 \rangle & & \langle \mathbf{h}_n, \mathbf{q}_1 \rangle \\ 0 & \|\bar{\mathbf{h}}_2\| & \langle \mathbf{h}_3, \mathbf{q}_2 \rangle & & \langle \mathbf{h}_n, \mathbf{q}_2 \rangle \\ 0 & 0 & \|\bar{\mathbf{h}}_3\| & \cdots & \langle \mathbf{h}_n, \mathbf{q}_3 \rangle \\ 0 & 0 & 0 & \ddots & \vdots \\ 0 & 0 & 0 & 0 & \|\bar{\mathbf{h}}_n\| \end{bmatrix}. \tag{2.91}$$

Since $\det(\mathbf{Q}) = 1$, the volume of the lattice is the product of the orthogonal vectors

$$\mathrm{vol}(\Lambda) = \prod_{i=1}^{n} \|\bar{\mathbf{h}}\|, \tag{2.92}$$

which corresponds to the volume of the hyper-rectangular decision regions in SIC.

A comparison of Equation 2.85 with Equation 2.89 reveals that

$$\mu_{j,k}\overline{b}_k = \left\langle \mathbf{h}_j, \mathbf{q}_k \right\rangle = \mathrm{Proj}_{\mathbf{q}_k}(\mathbf{h}_j), \tag{2.93}$$

each of which is the projection of the original generator vector onto the one-dimensional space spanned either by $\mathbf{q}_k$ or $\mathbf{h}_k$.

2.5.5 Lattice-Reduction-Aided Detection

As was exemplified in Figure 2.10, the two bases given by Equation 2.56 generate the same lattice but their fundamental regions have different coverage. To maximize the coverage of the MLD region, one is interested in bases with vectors that are both short and close to orthogonal, which is called a *reduced basis*. Figure 2.16 shows a lattice with a rather "skewed" basis and a reduced basis.

It was seen in Chapter 2 that two different bases of a lattice are related by a unimodular transformation **M**. In particular, the two basis in Equation 2.56 are related by

$$\mathbf{H}_1 = \mathbf{H}_2\mathbf{M} \Leftrightarrow \begin{bmatrix} 6 & 2 \\ 1 & 5 \end{bmatrix} = \begin{bmatrix} 6 & 8 \\ 1 & 6 \end{bmatrix}\begin{bmatrix} 1 & 1 \\ 0 & 1 \end{bmatrix}, \tag{2.94}$$

and in this case it is easy to see that det $(\mathbf{M}) = 1$.

As observed in Section 2.5.2.1, it is preferable to invert a well-conditioned channel matrix, and therefore having a more orthogonal basis contributes to a smaller noise enhancement factor whenever a ZF or an MMSE filter is applied (standalone or included in the OSIC stages). In LRA receivers, a preprocessing stage is introduced before the detection algorithm, as shown in Figure 2.17.

The application of lattice-reduction-aided (LRA) techniques to MIMO detection was first brought to light in [Yao and Wornell 2002], and since then the research in LR applications to MIMO has boomed not only for the detection but also for precoding in the BC (as mentioned in Chapter 1). These authors

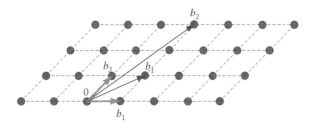

Figure 2.16 A reduced basis (with short vectors) and a skewed basis (with longer vectors) for the same lattice.

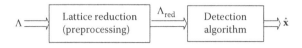

Figure 2.17 MIMO detection with lattice-reduction preprocessing.

applied the Lenstra Lenstra Lovász reduction (LLL, also sometimes denoted as L³) [Yao 2003] to reduce the channel matrix. In 2007, Seysen's reduction was simultaneously rediscovered for MIMO in [Seethaler et al. 2007] and in [Niu and Lu 2007]. This technique is based on the simultaneous reduction of both the primal and dual basis. One other important lattice reduction approach (LR) that delivers a more reduced basis than the others is the Hermite–Korkin–Zolotarev (HKZ) reduction [e.g., Banihashemi 1997, Chapter 3]; however, HKZ has not been used in MIMO until recently in [Zhang et al. 2012], probably due to its high complexity. The quality of the output of an LR algorithm can be measured by the *orthogonality defect*, defined as [e.g., Micciancio and Goldwasser 2002, p. 131]

$$OD(\mathbf{H}) = \frac{\prod_{i=1}^{n}\|\mathbf{h}_i\|}{\det(\mathbf{H})}. \tag{2.95}$$

Shorter generator vectors correspond to a lower orthogonality defect. Clearly, $OD(\mathbf{H}) \geq 1$, with equality attained only by the $\mathbb{Z}^n$ lattice.

An overview of the applications of lattice reduction techniques in MIMO (including SM and BC) exists in [Wübben et al. 2011]. LRA detection achieves the maximum diversity available in SM, as proved in [Taherzadeh et al. 2007]* for the case of LLL reduction. (Seysen's algorithm and HKZ also achieve that maximum diversity since, on average, they output bases even closer to orthogonal bases.)

The idea is that the system model can be rewritten as

$$\mathbf{y} = \mathbf{H}\mathbf{x} + \mathbf{n} \Leftrightarrow \underbrace{\mathbf{H}\mathbf{M}}_{\mathbf{H}_{red}}\underbrace{\mathbf{M}^{-1}\mathbf{x}}_{\mathbf{z}} + \mathbf{n} \Leftrightarrow \mathbf{H}_{red}\mathbf{z} + \mathbf{n}. \tag{2.96}$$

In this model, $\mathbf{z}$ is a modified data vector that can be detected with a lower SER than would $\mathbf{x}$ without LR. This is true regardless of the type of receiver that follows

---

* The same result was also proved in [Ma and Zhang 2008] for the complex-LLL algorithm (CLLL), where the real equivalent model is not used and LLL is applied directly to the complex lattice.

the LR preprocessing (usually ZF, MMSE, or OSIC). The original data vector can then be recovered from $\mathbf{z}$ noting that

$$\mathbf{z} = \mathbf{M}^{-1}\mathbf{x} \Rightarrow \mathbf{x} = \mathbf{Mz}. \tag{2.97}$$

Because M-QAM constellations and their PAM equivalent alphabets are defined without the origin and have nonunitary distance between the symbols (c.f. Section 2.3.1), in order to apply the lattice tools as in Equations 2.96 and 2.97, it is necessary to make a translation of the constellation, creating the modified received vector

$$\mathbf{y}_{\text{red}} = \frac{1}{2}(\mathbf{y} + \mathbf{H} \cdot \mathbf{1}) = \frac{1}{2}(\mathbf{Hx} + \mathbf{n} + \mathbf{H} \cdot \mathbf{1}), \tag{2.98}$$

where $\mathbf{1}$ is the column vector of n elements all equal to 1. This latter modification (2.98) applies to real lattices only and its extension to complex lattices is described in the next section.

Now, in the case of a ZF criterion,

$$\mathbf{z} = \mathbf{H}_{\text{red}}^{+}\mathbf{y}_{\text{red}}, \tag{2.99}$$

and in performing

$$\hat{\mathbf{x}} = 2\mathbf{M}\underbrace{Q_{\mathbb{Z}}(\mathbf{z})}_{\hat{\mathbf{z}}} - \mathbf{1}, \tag{2.100}$$

the symbol $\hat{\mathbf{z}}$ is detected and put back in the alphabet $\mathcal{A}$.

The LLL algorithm was first derived for integer lattices and then applied to real lattices. It has also been shown that a complex-LLL (CLLL) algorithm can be defined and that by applying it directly to a complex $\mathbf{H}_c$ (see the next section), the complexity becomes half of the one involved in the application of Algorithm 2.2 to the real equivalent lattice [Mow 2004, Gan et al. 2009]. The LLL algorithm can be seen as an extension to higher dimensions of Gauss's algorithm [Micciancio and Goldwasser 2002, p. 28], which operates in two dimensions only (c.f. example in Figure 2.18). It is also noteworthy that the LLL algorithm can be derived by making appropriate changes to the GS orthogonalization (Algorithm 2.1), as shown in [Fischer 2010] and is also closely related to sorting in OSIC [Ling and Mow 2009].

Ling et al. proved that the complexity of the LLL reduction is $\mathcal{O}(n^4 \log n)$ not only for integer lattices [e.g., Park et al. 2011] but also proved that for Gaussian lattices [Ling et al. 2013]. They further proposed a change to the algorithm that maintains a similar performance while having complexity $\mathcal{O}(n^3 \log(n))$. It was shown in [Jaldén

Algorithm 2.2: Lenstra Lenstra Lovász (LLL)

Input: a basis $\mathbf{H}$ with generator vectors $\mathbf{h}_1, \mathbf{h}_2, \ldots, \mathbf{h}_n \in \mathbb{R}^n$ in its columns
Output: a ζ-LLL reduced basis $\mathbf{H}_{\text{red}}$ $\mathbf{H}_{\text{red}}$, a unimodular matrix $\mathbf{M}$

1: (Preliminaries)

   Compute the GS orthogonal vectors $\bar{\mathbf{h}}_1, \bar{\mathbf{h}}_2, \ldots, \bar{\mathbf{h}}_n$ using Algorithm 2.1

   Set $\mathbf{M} = \mathbf{I}_n$

2: (Reduction step)

   for $i = 2$ to n

      for $j = i - 1$ down to 1

   $$\mathbf{h}_i = \mathbf{h}_i - \left\lceil \frac{\langle \mathbf{h}_i, \bar{\mathbf{h}}_j \rangle}{\langle \bar{\mathbf{h}}_j, \bar{\mathbf{h}}_j \rangle} \right\rfloor \mathbf{h}_j$$

   $$\mathbf{m}_i = \mathbf{m}_i - \left\lceil \frac{\langle \mathbf{h}_i, \bar{\mathbf{h}}_j \rangle}{\langle \bar{\mathbf{h}}_j, \bar{\mathbf{h}}_j \rangle} \right\rfloor \mathbf{m}_j$$

   end

   recompute the GS orthogonal vectors $\bar{\mathbf{h}}_1, \bar{\mathbf{h}}_2, \ldots, \bar{\mathbf{h}}_n$ using Algorithm 2.1

3: (Swap step)

   if there is i such that $\zeta \left\| \bar{\mathbf{h}}_i \right\|^2 > \left\| \bar{\mathbf{h}}_{i+1} + \mu_{i+1,i} \bar{\mathbf{h}}_i \right\|^2$ then

      swap columns $\mathbf{h}_i$ and $\mathbf{h}_{i+1}$

      swap columns $\mathbf{m}_i$ and $\mathbf{m}_{i+1}$

      go to 1

   end

4: return $\mathbf{h}_1, \mathbf{h}_2, \ldots, \mathbf{h}_n$ and $\mathbf{M}$

et al. 2008] that for some instances of lattices, the complexity of the LLL algorithm for noninteger matrices is not polynomial but that probability tends to zero.

2.5.5.1 Complex Lattice Reduction

Lattice reduction can also process directly complex numbers (i.e., the constellation symbols in the antennas) rather than using the real-valued equivalent model introduced in Section 2.3.2. Let us now reconsider a complex-valued transmission model

$$\mathbf{y}_c = \mathbf{H}_c \mathbf{s}_c + \mathbf{n}_c, \tag{2.101}$$

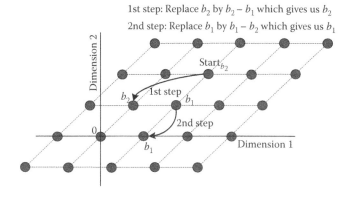

Figure 2.18 The Gauss algorithm (i.e., LLL in 2D).

such as in Equation 2.21, where $\mathbf{H}_c$ is the $N_R \times N_T$ *complex* channel matrix, $\mathbf{s}$ is the $N_T \times 1$ *complex* column vector whose elements are the complex-valued symbols transmitted on the N_T antennas and $\mathbf{n}$ is the $N_R \times 1$ complex column vector. Assuming that the transmitted symbols are drawn from a constellation isomorphic to a subset of the Gaussian integers (denoted as $\mathbb{Z}[j] = \{a + bj : a, b \in \mathbb{Z}\}$), through shifting and scaling operations (e.g., M-QAM constellations) one can write

$$\mathbf{s}_c = \alpha \left[\mathbf{x}_c - \left(\frac{1}{2} + \frac{1}{2}j \right) \mathbf{I}_{N_T} \right], \tag{2.102}$$

where $\mathbf{x}_c \in \mathbb{Z}^{N_T}[j]$, α represents the factor used for energy normalization, and $\mathbf{I}_{N_T}$ is the $N_T \times N_T$ identity matrix. One can rewrite Equation 2.101 as

$$\begin{aligned}
\mathbf{y}_c &= \mathbf{H}_c \alpha \left[\mathbf{x}_c - \left(\frac{1}{2} + \frac{1}{2}j \right) \mathbf{I}_{N_T} \right] + \mathbf{n}_c \\
&= \alpha \mathbf{H}_c \mathbf{x}_c - \left(\frac{1}{2} + \frac{1}{2}j \right) \alpha \mathbf{H}_c \mathbf{I}_{N_T} + \mathbf{n}_c \\
&= \bar{\mathbf{H}}_c \mathbf{x}_c - \left(\frac{1}{2} + \frac{1}{2}j \right) \bar{\mathbf{H}}_c + \mathbf{n}_c,
\end{aligned} \tag{2.103}$$

where

$$\bar{\mathbf{H}}_c = \alpha \mathbf{H}_c. \tag{2.104}$$

A complex-valued lattice in the N_R-dimensional complex space $\mathbb{C}^{N_R}$ is therefore defined as

$$\Lambda_c = \left\{ \sum_{i=1}^{N_T} x_i \bar{\mathbf{h}}_i \; : \; x_i \in \mathbb{Z}[j], \bar{\mathbf{h}}_i \in \mathbb{C}^{N_R} \right\}, \tag{2.105}$$

or in matrix form,

$$\Lambda_c = \left\{ \bar{\mathbf{H}}\mathbf{x} \; : \; \mathbf{x}_c \in \mathbb{Z}^{N_T}[j], \bar{\mathbf{H}}_c \in \mathbb{C}^{N_R \times N_T} \right\}, \tag{2.106}$$

From Equation 2.103, it is easy to realize that, apart from the shifting operation, the received vectors correspond to noisy versions of the lattice points defined by the basis $\bar{\mathbf{H}}_c$. One is interested in the estimate $\hat{\mathbf{w}}_c = \bar{\mathbf{H}}_c \hat{\mathbf{x}}_c$ which is closest to the shifted received vector $\|\mathbf{y}_c - \hat{\mathbf{w}}_c\| \le \|\mathbf{y}_c - \mathbf{w}_c\|, \forall \mathbf{w}_c \in \Lambda_c$.

Let $\tilde{\mathbf{H}}_c$ be a new basis for the lattice obtained by right multiplying the original basis by a unimodular matrix $\mathbf{U}$ containing only Gaussian integers, that is,

$$\tilde{\mathbf{H}}_c = \bar{\mathbf{H}}_c \mathbf{U}, \tag{2.107}$$

the received vector can be rewritten as

$$\mathbf{y}_c = \tilde{\mathbf{H}}_c \mathbf{U}^{-1} \mathbf{x}_c - \left(\frac{1}{2} + \frac{1}{2}j \right) \tilde{\mathbf{H}}_c \mathbf{U}^{-1} + \mathbf{n}_c = \tilde{\mathbf{H}}_c \mathbf{z}_c - \left(\frac{1}{2} + \frac{1}{2}j \right) \tilde{\mathbf{H}}_c \mathbf{U}^{-1} + \mathbf{n}_c, \tag{2.108}$$

with $\mathbf{z}_c = \mathbf{U}^{-1} \mathbf{x}_c$. Because $\mathbf{U}$ is unimodular, the elements of $\mathbf{z}_c$ are also Gaussian integers. If the new basis has a smaller orthogonality defect, the performance of the detector gets improved. The complex LLL algorithm (CLLL) was proposed in [Gan et al. 2009] and outperforms the real LLL both, attaining smaller orthogonality defects while further reducing the complexity. It can also be combined with the linear receivers and with OSIC to accomplish the decoding. Note that the MMSE criterion can be applied as if it was a ZF detector by using an extended system model approach [Wübben et al. 2003, 2004]. The extended channel matrix and the extended received vector are (considering $\alpha = 1$ for simplicity)

$$\underline{\mathbf{H}}_c = \begin{bmatrix} \mathbf{H}_c \\ \sqrt{2}\sigma \mathbf{I}_{N_T} \end{bmatrix} \quad \text{and} \quad \underline{\mathbf{y}}_c = \begin{bmatrix} \mathbf{y}_c \\ \mathbf{0}_{N_T \times 1} \end{bmatrix}, \tag{2.109}$$

where $\mathbf{0}_{N_T \times 1}$ is an all zeros size N_T column vector (note that in Equation 2.109 one has a complex model while in Equation 2.68 the real complex model was considered, hence the $\sqrt{2}$). As suggested in [Wübben et al. 2004], one can apply LLL reduction to the extended model and in this case the reduced basis $\tilde{\underline{\mathbf{H}}}_c$ can be "QR decomposed" as

$$\tilde{\underline{\mathbf{H}}}_c = \mathbf{QR}. \tag{2.110}$$

The extended received vector is multiplied by $\mathbf{Q}^T$ and added to $\left(\frac{1}{2} + \frac{1}{2}j\right)$ $\mathbf{RU}^{-1}\mathbf{I}_{N_T}$ to allow rounding over $\mathbb{Z}[j]$, resulting in

$$\hat{\mathbf{y}}_c = \mathbf{Q}^T \underline{\mathbf{y}}_c + \left(\frac{1}{2} + \frac{1}{2}j\right)\mathbf{RU}^{-1}\mathbf{I}_{N_T} = \mathbf{Rz}_c + \mathbf{Q}^T\left[\begin{array}{c}\mathbf{n}_c \\ -\sqrt{2}\sigma\mathbf{I}_{N_T}\mathbf{x}_c\end{array}\right], \tag{2.111}$$

where the second term takes into account both noise and residual interference.

Since $\mathbf{R}$ is triangular, one can start by computing the estimate of $\hat{z}_{N_T}$ as $\hat{y}_{N_T}/r_{N_T,N_T}$ (where $\hat{y}_i$ and $r_{i,j}$ represent elements of $\hat{\mathbf{y}}_c$ and $\mathbf{R}$, respectively, dropping the c subscript) and then proceed to obtain the remaining elements through the successive application of

$$\hat{z}_i = \text{round}\left(\frac{\hat{y}_i - \sum_{j=i+1}^{N_T} r_{i,j}\hat{z}_j}{r_{i,i}}\right). \tag{2.112}$$

The function "round()" corresponds to standard rounding to the nearest Gaussian integers as proposed in the conventional SIC algorithm [Babai 1986]. The final transmitted vector estimate can then be achieved through shifting and rescaling as

$$\hat{\mathbf{s}}_c = \alpha\left(\mathbf{U}\hat{\mathbf{z}}_c - \left(\frac{1}{2} + \frac{1}{2}j\right)\mathbf{I}_{N_T}\right). \tag{2.113}$$

As the decoding is performed assuming an infinite lattice, it is necessary to restrict the estimate to a valid constellation vector (this can be accomplished applying the restriction to $\mathbf{U}\hat{\mathbf{z}}_c$).

2.5.6 Randomized Lattice Decoding

Lattice-reduction-aided decoding can achieve full diversity in MIMO transmissions, although its performance gap to ML decoding depends on the detection algorithm employed and on the dimension N_T [Ling 2011, Liu et al. 2011]. In fact, while LLL lattice-reduction-aided MMSE detection can achieve near ML performance for small MIMO systems [Wübben et al. 2004], the performance gap widens as the dimension increases [Liu et al. 2011]. To narrow down this gap, a randomized version of SIC was proposed in [Liu et al. 2011], and based on [Klein 2000], leading to impressive performance results. In the following, a description of this algorithm will be presented.

In the SIC algorithm described in the previous section, standard rounding to the nearest Gaussian integers was applied in Equation 2.112. Instead, we can replace the standard rounding by Klein's randomized rounding [Klein 2000] (see Algorithm 2.3) and implement a randomized lattice decoder. In this case, we compute the elements $\hat{z}_i$ using

$$\hat{z}_i = \text{randround}_{c_i}\left(\frac{\hat{y}_i - \sum_{j=i+1}^{N_T} r_{i,j}\hat{z}_j}{r_{i,i}}\right). \tag{2.114}$$

where function $\text{randround}_c(r)$ rounds the real and imaginary parts of r (denoted as r_{re} and r_{im}) to integers q_{re} and q_{im}, respectively, according to the discrete Gaussian distribution defined as

$$P(Q = q_{re}) = \frac{e^{-c(r_{re}-q_{re})}}{\sum_{q=-\infty}^{+\infty} e^{-c(r_{re}-q_{re})}} \tag{2.115}$$

for the real part and

$$P(Q = q_{im}) = \frac{e^{-c(r_{im}-q_{im})}}{\sum_{q=-\infty}^{+\infty} e^{-c(r_{im}-q_{im})}} \tag{2.116}$$

for the imaginary part, and the parameter c_i is computed as

$$c_i = \frac{\log(\rho)}{\min_j\left(r_{j,j}^2\right)} r_{i,i}^2 \tag{2.117}$$

with ρ being another parameter whose optimum value is obtained from

$$K = (e\rho)^{\frac{4N_T}{\rho}}, \tag{2.118}$$

where K is the number of candidate lattice points that are considered in the algorithm. The candidate list is built by repeating K times the procedure for computing a lattice point with Equation 2.114. If the transmission is uncoded, the final estimate will then correspond to the closest of the K lattice point candidates. Due to the randomized nature of the algorithm, to avoid the possibility of the final estimate being further away than the one produced by the MMSE-SIC decoder, one of the K candidates should be obtained through standard rounding. The final transmitted vector estimate is computed using Equation 2.113.

Algorithm 2.3: Randomized Decoding

Inputs: $\mathbf{y}_c$, $\underline{\mathbf{H}}_c$, K, δ, ρ (the variable $1/2 < \delta \leq 1$ selects the complexity performance tradeoff of the CLLL algorithm)

Output: $\hat{\mathbf{S}}$

1      $\tilde{\mathbf{H}}_c$, $\mathbf{U} \leftarrow$ Complex_LLL($\underline{\mathbf{H}}_c$, δ)

2      $\mathbf{Q}$, $\mathbf{R} \leftarrow$ QR_Decomposition($\tilde{\mathbf{H}}_c$)

3      $\hat{\mathbf{y}}_c \leftarrow \mathbf{Q}^T \mathbf{y}_c + (1/2 + 1/2\,j)\mathbf{RU}^{-1}\mathbf{I}_{N_T}$

4   **for** $k = 1$ **to** K **do**

5   **for** $i = 1$ **to** N_T **do**

6   **if** $k = = 1$

7
$$\hat{z}_i^1 \leftarrow Round\left(\hat{y}_i - \sum_{j=i+1}^{N_T} r_{i,j}\hat{z}_j^1 \Big/ r_{i,i} \right)$$

8   **else**

9
$$c_i \leftarrow \frac{\log(\rho)}{\min_j(r_{j,j}^2)} r_{i,i}^2$$

$$\check{z}_i^k \leftarrow randRound_{c_i}\left(\hat{y}_i - \sum_{j=i+1}^{N_T} r_{i,j}\check{z}_j^k \Big/ r_{i,i} \right)$$

10   **end if**

11   **end for**

12   **end for**

13      $\hat{\mathbf{z}}_c \leftarrow$ Closest_Vector$\left(\mathbf{y}_c + \left(1/2 + 1/2\,j\right)\mathbf{H}_c, \hat{\mathbf{z}}^1,\ldots, \hat{\mathbf{z}}^K\right)$

14      $\hat{\mathbf{s}}_c \leftarrow \alpha\left(\mathbf{U}\hat{\mathbf{z}}_c - \left(\frac{1}{2} + \frac{1}{2}\,j\right)\mathbf{I}_{N_T} \right)$

15   Return $\hat{\mathbf{s}}_c$

2.5.7 Sphere Decoding

SD is an exact detection method (i.e., it achieves the same performance as MLD) with a complexity that, on average, is much lower than MLD. The idea is that a rigid rotation $\mathbf{Q}$ can be applied to the ensemble $\{\Lambda, \mathbf{y}\}$, for which the CVP needs to be solved, so that the lattice can be described by an equivalent lattice in u.t. form. The u.t. property allows describing the norm of any lattice point to be detected as a sum that can be computed incrementally, taking into consideration the cumulative effect of each vector components. Consider now that an upper bound (UB) on the norms is established. The u.t. property of the basis allows all the possible values in the last component of the data vector, $x(n)$, to be detected. As the norm can be computed as a sum of "ordered" contributions, if some of the tested values in $x(n)$ generates a total vector norm that is larger than the UB, then those values of $x(n)$ need not be considered further as possible values in the solution. This procedure can be extended to the next layer $x(n-1)$, where only the possible values of $x(n)$ are considered. In conclusion, finding vectors with norm smaller than the UB is a problem that can be solved by expanding and pruning a tree that represents the lattice points. All these ideas can be converted into the CVP, if the lattice is shifted to the target **y**.

A sphere decoder has the structure shown in Figure 2.19. After traversing the tree with a particular symbol enumeration, the MLD solution is always found if the initial radius that is chosen is large enough to contain a lattice point inside the hypersphere. Figure 2.20 gives an example tree associated with 3×3 antenna and 4-PAM, showing the branches that have been expanded at each tree level.

As mentioned previously, one can define an UB for the radius (or, equivalently, for the squared radius) of the sphere around the received point, that is,

$$\left\| \mathbf{y} - \mathbf{Hx} \right\|^2 \leq \xi^2, \tag{2.119}$$

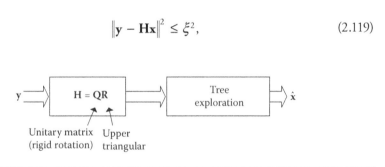

Figure 2.19   Receiver based on SD.

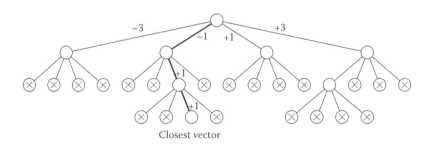

Figure 2.20   Tree exploration of a tree with three layers, considering a 4-PAM alphabet.

$$\left\| \mathbf{y} - [\mathbf{Q}_1 \ \mathbf{Q}_2] \begin{bmatrix} \mathbf{R} \\ \mathbf{0} \end{bmatrix} \mathbf{x} \right\|^2 \leq \xi^2. \tag{2.120}$$

Applying the inverse rotation $\mathbf{Q}^H$ (remembering that $\mathbf{Q}$ is unitary, or orthogonal in the real case),

$$\left\| \begin{bmatrix} \mathbf{Q}_1^H \\ \mathbf{Q}_2^H \end{bmatrix} \mathbf{y} - \begin{bmatrix} \mathbf{R} \\ \mathbf{0} \end{bmatrix} \mathbf{x} \right\|^2 \leq \xi^2 \Leftrightarrow \left\| Q_1^H \mathbf{y} - \mathbf{R}\mathbf{x} \right\|^2 + \left\| \mathbf{Q}_2^H \mathbf{y} \right\|^2 \leq \xi^2. \tag{2.121}$$

If one defines $\mathbf{y}' = Q_1^H \mathbf{y}$ and $\xi'^2 = \xi'^2 - \left\| \mathbf{Q}_2^H \mathbf{y} \right\|^2$, then the CVP can be written as

$$\left\| \mathbf{y}' - \mathbf{R}\mathbf{x} \right\|^2 \leq \xi^2. \tag{2.122}$$

Finally, remembering that $\mathbf{R}$ is u.t., the problem can be written as the sum

$$\sum_{i=1}^{m} \left(y_i' - \sum_{j=i}^{m} r_{ij} x(j) \right)^2 \leq \xi^2. \tag{2.123}$$

The complexity of SD is usually measured by the number of nodes that need to be visited in the tree until the MLD solution is found. The fact that the complexity is a random variable is a limitation of SD. To circumvent this problem, it is possible to expand only say, K branches at each level of the tree, as is the case in the *K-best receivers* and its variations [Detert 2007, Okawado et al. 2008, Roger et al. 2010], which will be a central tool in Chapter 10. Nonetheless, there are different approaches as to how the tree should be traversed. Historically, the Fincke–Pohst method [Su 2005] was the first to be used, followed by the more efficient Schnorr–Euchner node enumeration [Agrell et al. 2002], which attains the same performance, while expanding a smaller number of tree branches. Su showed in [Su and Wassell 2005] and [Su 2005] that a dramatically more efficient exploration of the tree could be made if not only the channel is taken into consideration when enumerating the symbols in the alphabet, but also the particular target point was also taken into account during that sorting process. Furthermore, Su's ordered traversal of the tree also eliminated the need for an initial radius and provides automatic *boundary control* for spherical lattice codes [Su et al. 2009]. Recently, simplifications to the Fincke–Pohst and Schnorr–Euchner's enumerations have been proposed, eliminating about 75% of the operations previously required [Ghasemmehdi and Agrell 2011].

The average complexity of SD is exponential [Hassibi and Vikalo 2005], given by $\mathcal{O}(M^{\alpha N_T})$ with $0 \leq \alpha \leq 1$ [Jaldén and Ottersten 2005], however, for low-dimensional lattices, that number is affordable. A celebrated improvement to SD was the development of fixed-complexity sphere decoding (FCSD) in [Barbero and Thompson 2008]. FCSD splits the tree exploration into two: one where all valid branches are further expanded, and a second phase, conducted for the remaining layers, where only one branch is expanded from any node. Recently, an automatic adjustment of the switching point was proposed in [Lai et al. 2011]. The performance of FCSD was described analytically in [Jaldén et al. 2009].

SD is typically used to obtain the MLD performance curves for the more challenging configurations such as 4×4 with 64-QAM. The SD described here is a simple implementation of the Fincke–Pohst enumeration, such as the one given in [Hassibi and Vikalo 2005] or [Bölcskei et al. 2006, Chapter 15], where the output of the algorithm is not just one point but the set of all points inside the defined sphere. The implementation of this SD is given in Algorithm 2.4.

Algorithm 2.4: Sphere Decoding (Fincke–Pohst)

Input: $\mathbf{Q}$, orthogonal, $\mathbf{R}$, upper triangular, a target vector $\mathbf{y} = \mathbf{Q}_1 \mathbf{x}$, radius ξ

Output: an MLD solution to the CVP, $\hat{\mathbf{x}}_{SD}$

1: Set $k = m$, $\xi_m'^2 = \xi^2 - \left\| \mathbf{Q}_2 \mathbf{x} \right\|^2$, $\mathbf{y}_{m|m+1} = \mathbf{y}_m$

2: (Bounds for s_k) Set $\mathrm{UB}(x_k) = \left\lfloor \dfrac{\xi_k' + y_{k|k+1}'}{r_{kk}} \right\rfloor$, $x_k = \left\lceil \dfrac{-\xi_k' + y_{k|k+1}'}{r_{kk}} \right\rceil - 1$

3: (Increase x_k) $x_k = x_k + 1$

4: (Increase k) $k = k + 1$

   if $k = m + 1$

      Terminate (no lattice point found)

   else go to 3

5: (Decrease k)

   if $k = 1$

      go to 6

   else

      $k = k - 1$

      $y_{k|k+1}' = y_k' - \displaystyle\sum_{j=k+1}^{m} r_{kj} x_j$

$$\xi_k'^2 = \xi_{k+1}'^2 - (y_{k+1|k+2} - r_{k+1,k+1}x_{k+1})^2$$
go to 2

6: (Solution found)

return $\hat{\mathbf{x}}_{SD} = \mathbf{x}$

go to 3

The algorithm starts by detection of the last element in **x**. Note that the subscript in $y_{k|k+1}'$ denotes the symbol y_k', in the kth layer, incorporating the effect of the layers already detected, in agreement with Equation 2.123. In this algorithm the option was made to denote the elements in a vector by x_n instead of $x(n)$, in order to accommodate notation such as $y_{k|k+1}'$ that also reflects the updates of a vector over time or the updates of the radius over time.

2.6 Performance Comparison

In this chapter, one could find descriptions and insights about the most popular techniques for detecting MIMO signals, including complexity comparisons. This chapter ends with a graphical comparison of the performance of some of the algorithms described, namely, the ZF, MMSE, OSIC, SD (i.e., the same as MLD performance), LRA combined with ZF, MMSE, and with OSIC. The results of one receiver recently proposed in [Monteiro and Kschischang 2011] is also plotted in Figures 2.21 and 2.22.

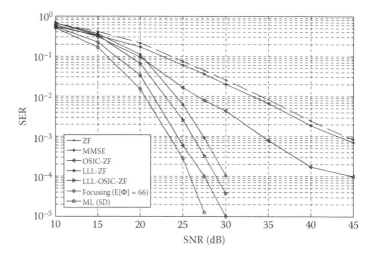

Figure 2.21 Detection in $n = 8$ real dimensions (4×4 antennas) with 16-QAM. ($E[\Phi]$ is the average of a complexity metric associated with the novel receiver.)

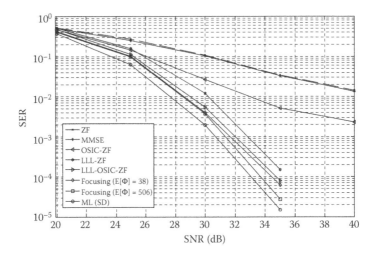

Figure 2.22 Detection in $n = 8$ **real dimensions (4 × 4 antennas) with 64-QAM. (**$E[\Phi]$ **is the average of a complexity metric associated with the novel receiver.)**

Acknowledgments

This work was partially supported by FCT (Fundação para a Ciência e Tecnologia) through project PEst-OE/EEI/LA0008/2013. F. Monteiro also received scholarships from the Gulbenkian Foundation and FCT (in Portugal), and grants from the Royal Academy of Engineering, the Cambridge Philosophical Society, The Computer Laboratory of the University of Cambridge, and Fitzwilliam College (in the UK).

References

Agrell, E. and Eriksson, T. Optimization of lattices for quantization, *IEEE Transactions on Information Theory*, 44(5), 1814–1828, September 1998.

Agrell, E., Eriksson, T., Vardy, A., and Zeger, K. Closest point in lattices, *IEEE Transactions on Information Theory*, 48(8), 2201–2214, August 2002.

Ahmadi, S. An overview of next-generation mobile WiMAX technology, *IEEE Communications Magazine*, 47(6), 84–98, June 2009.

Alamouti, S.M. A simple transmit diversity for technique wireless communications, *IEEE Journal on Selected Areas on Communications*, 16(8), 1451–1458, October 1998.

Anderson, J.A. Lattices, in *Discrete Mathematics with Combinatorics*. Upper Saddle River, New Jersey, USA: Pearson/Prentice Hall, 2nd ed., 2003, Chapters 9.2–9.3, pp. 357–371.

Aste, T. and Weaire, D. Packings and kisses in high dimensions, in *The Persuit of the Perfect Packing*. Bristol, UK: Institute of Physics Publishing, 2000, Chapter 12, pp. 113–118.

Ayanoglu, E., Larsson, E.G., and Karipidis, E. Computational complexity of decoding orthogonal space–time block codes, *IEEE Transactions on Communications*, 59(4), 936–941, April 2011.

Babai, L. On Lovász' lattice reduction and the nearest lattice, *Combinatoria*, 6(1), 1–13, January 1986.

Bai, D., Nguyen, H., Kim, T., and Kang, I. LTE-Advanced modem design: Challenges and perspectives, *IEEE Communications Magazine*, 50(2), 178–186, February 2012.

Banerjee, U. Unimodular matrices, in *Loop Transforms for Restructuring Compilers*. Dordrecht, The Netherlands: Kluwer Academic Press, 1993, Chapter 2.

Banihashemi, A.H. Decoding complexity and trellis structure of lattices, PhD thesis, University of Waterloo, Waterloo, Ontario, Canada, 1997.

Barbero, L.G. and Thompson, J.S., Fixing the complexity of the sphere decoder for MIMO detection, *Transactions on Wireless Communications*, 7(6), 2131–2142, June 2008.

Bertsimas, D. and Weismantel, R. *Optimization over Integers*. Belmont, MA, USA: Dynamic Ideas, 2005.

Biggs, N.L. *Discrete Mathematics*. Oxford, UK: Oxford University Press, 2002.

Biglieri, E. and Taricco, G. *Transmission and Reception with Multiple Antennas: Theoretical Foundations*. Hanover, Massachusetts, USA: now Publishers, 2004.

Bishop, C.M. The curse of dimensionality, in *Pattern Recognition and Machine Learning*. New York, NY: Springer, 2006, Chapter 1.4, pp. 33–38.

van Emde Boas, P. Another NP-complete partition problem and the complexity of computing short vectors in a lattice, Mathematisch Instituut, Amsterdam, The Netherlands, Report 81-04, April 1981.

Bölcskei, H., Gesbert, D., Papadias, C.B., and van der Veen, A.-J. *Space–Time Wireless Systems—From Array Processing to MIMO Communications*. Cambridge, UK: Cambridge University Press, 2006.

Boutros, J., Viterbo, E., Rastello, C., and Belfiore, J.-C. Good lattice constellations for both Rayleigh fading and Gaussian channels, *IEEE Transactions on Information Theory*, 42(2), 502–518, March 1996.

Cassels, J. *An Introduction to the Geometry of Numbers*, 2nd ed. Berlin, Germany: Springer, 1971.

Conway, J.H., and Sloane, N.J. *Sphere Packings, Lattices and Groups*, 3rd ed. New York, New York, USA: Springer, 1999.

Cormen, T.H., Leiserson, C.E., Rivest, R.L., and Stein, C. NP-completeness, in *Introduction to Algorithms*, 2nd ed. Cambridge, Massachusetts, USA: MIT Press, 2001, Chapter 34.

Detert, T. An efficient fixed complexity QRD-M algorithm for MIMO-OFDM using per-survivor slicing, in *Proc. of ISWCS'07, 4th International Symposium on Wireless Communication Systems*, Trondheim, October 2007, pp. 572–576.

Dubois, E. The sampling and reconstruction of time-varying imagery with applications in video systems, *Proceedings of the IEEE*, 73(4), 502–522, April 1985.

Edelman, A., Arias, T.A., and Smith, S.T. The geometry of algorithms with orthogonality constraints, *SIAM Journal on Matrix Analysis and Applications (SIMAX)*, 20(2), 303–353, July 1998.

El-Hajjar, M. and Hanzo, L. Multifunctional MIMO systems: A combined diversity and multiplexing design perspective, *IEEE Wireless Communications*, 17(2), 73–79, April 2010.

Epp, S.S. Analysis of algorithm efficiency, in *Discrete Mathematics with Applications*, 4th ed. Boston, Massachusetts, USA: Brookes/Cole-Cengage, 2011, Chapter 11.

Erez, U. and Zamir, R. Achieving 1/2 log (1+SNR) on the AWGN channel with lattice encoding and decoding, *IEEE Transactions on Information Theory*, 50(10), 2293–2314, October 2004.

Feng, C., Silva, D., and Kschischang, F.R. An algebraic approach to physical-layer network coding, in *Proc. of ISIT'10—The Inter. Symp. on Information Theory*, Austin, TX, USA, June 2010, pp. 1017–1021.

Fincke, U. and Pohst, M. Improved methods for calculating vectors of short length in a lattice, including a complexity analysis, *Mathematics of Computation*, 44(170), 463–471, April 1985.

Fischer, R. *Precoding and Signal Shaping for Digital Transmission*. Chichester, UK: John Wiley & Sons, 2005.

Fischer, R. From Gram–Schmidt orthogonalization via sorting and quantization to lattice reduction, in *Proc. of the 6th Joint Workshop on Coding and Communications (JWCC)*, Santo Stefano Belbo, Italy, 2010, pp. 13–17.

Forney, G.D. Coset codes—Part I: introduction and geometrical classification, *IEEE Transactions on Information Theory*, 34(5), 1123–1151, September 1988.

Foschini, G.J. Layered space–time architecture for wireless communication in a fading environment when using multi-element antennas, *Bell Labs Technical Journal*, pp. 41–59, Autumn 1996.

Galbraith, S. *Mathematics of Public Key Cryptography*. Cambridge, UK: Cambridge University Press, 2013, Chapters 18,19,20, and 22.

El Gamal, H., Caire, G., and Damen, M.O. Lattice coding and decoding achieve the optimal diversity–multiplexing tradeoff of MIMO channels, *IEEE Transactions on Information Theory*, 50(6), 968–985, June 2004.

Gan, Y.H., Ling, C., and Mow, W.H. Complex lattice reduction algorithm for low-complexity full-diversity MIMO detection, *IEEE Transactions in Signal Processing*, 57(7), 2701–2710, July 2009.

Ghasemmehdi, A. and Agrell, E. Faster recursions in sphere decoding, *IEEE Transactions on Information Theory*, 57(6), 3530–3536, June 2011.

Ghosh, A., Zhang, J., Andrews, J.G., and Muhamed, R. *Fundamentals of LTE*. Boston, Massachusetts, USA: Prentice-Hall, 2010.

Giannakis, G., Liu, Z., Ma, X., and Zhou, S. *Space-Time Coding for Broadband Wireless Communications*. Hoboken, New Jersey, USA: Wiley, 2007.

Golden, G.D., Foschini, C.J., Valenzuela, R.A., and Wolniansky, P.W. Detection algorithm and initial laboratory results using V-BLAST space–time communication architecture, *IET Electronics Letters*, 35(1), January 1999.

Goldsmith, A. *Wireless Communications*. Cambridge, UK: Cambridge University Press, 2005.

Gower, J.C. Procrustes methods, *Wiley Interdisciplinary Reviews: Computational Statistics*, 2, 503–508, July/August 2010.

Gower, J.C. and Dijksterhuis, G.B. *Procrustes Problems*. Oxford, UK: Oxford University Press, 2004.

Hassibi, B. An efficient square-root algorithm for BLAST, in *Proc. of ICASSP '00—IEEE Inter. Conf. on Acoustics, Speech, and Signal Processing*, Vol. 2, Istanbul, Turkey, 2000, pp. II737–II740.

Hassibi, B. and Vikalo, H. On the sphere-decoding algorithm I. Expected complexity, *IEEE Transactions on signal Processing*, 53(8), 2806–2818, August 2005.

Haykin, S. *Adaptive Filter Theory*, 3rd ed. Upper Saddle River, New Jersey, EUA: Prentice-Hall, 1996.

Heath, R.W. and Paulraj, A.J. Switching between diversity and multiplexing in MIMO systems, *IEEE Transactions on Communications*, 53(6), 962–968, June 2005.

Higham, N.H. Matrix nearness problems and applications, in *Applications of Matrix Theory*, M. J. C. Gover and S. Barnett, Eds. Oxford, UK: Oxford University Press, 1989, pp. 1–27.

Hochba, D.S. Ed., *Approximation Algorithms for NP-Hard Problems*. Boston, Massachusetts, USA: Course Technology/PWS Publishing Company, 1996.

Hochwald, B.M. and ten Brink, S. Achieving near-capacity on a multiple-antenna channel, *IEEE Transactions on Information Theory*, 51(3), 389–399, March 2003.

Hong, L. and Armada, A.G. Bit error rate performance of MIMO MMSE receivers in correlated Rayleigh flat-fading channels, *IEEE Transactions on Vehicular Technology*, 60(1), 313–317, January 2011.

Jafarkhani, H. *Space-Time Coding—Theory and Practice*. Cambridge, UK: Cambridge University Press, 2005, Chapter 9, pp. 221–234.

Jaldén, J. and Ottersten, B. On the complexity of sphere decoding in digital communications, *IEEE Transactions on Signal Processing*, 53(4), 1474–1484, April 2005.

Jaldén, J. and Ottersten, B. High diversity detection using semidefinite relaxation, in *40th Asilomar Conference on Signals, Systems and Computers, Pacific Grove*, California, USA, 2006, pp. 2082–2086.

Jaldén, J., Seethaler, D., and Matz, G. Worst- and average-case complexity of LLL lattice reduction in MIMO wireless systems, in *Proc. of ICASSP'08—The IEEE Inter. Conf. on Acoustics, Speech and Signal Processing*, Las Vegas, NV, USA, April 2008, pp. 2685–2688.

Jaldén, J., Barbero, L.G., Ottersten, B., and Thompson, J.S. The error probability of the fixed-complexity sphere decoder, *IEEE Transactions on Signal Processing*, 57(7), 2711–2720, July 2009.

Jiang, Y., Varanasi, M.K., and Li, J. Performance analysis of ZF and MMSE equalizers for MIMO systems: An in-depth study of the high SNR regime, *IEEE Transactions on Information Theory*, 57(4), 2008–2026, April 2011.

Kannan, R. Algorithmic geometry of numbers, *Annual Review of Computer Science*, 2, 231–267, June 1987.

Kay, S.M. *Fundamentals of Statistical Signal Processing: Estimation Theory*, Vol. I. Upper Saddle River, New Jersey, USA: Prentice-Hall, 1993.

Khan, F. *LTE for 4G Mobile Broadband: Air Interface Technologies and Performance*. Cambridge, UK: Cambridge University Press, 2009.

Kim, N., Lee, Y., and Park, H. Performance analysis of MIMO system with linear MMSE receiver, *IEEE Transactions on Wireless Communications*, 7(11), 4474–4478, November 2008.

Klein, P. Finding the closest lattice vector when it's unusually close, in *Proc. ACM-SIAM Symposium on Discrete Algorithms (SODA)*, San Francisco, CA, 2000, pp. 937–941.

Kreher, D.L. and Stinson, D.R. *Combinatorial Algorithms: Generation, Enumeration, and Search*. Boca Raton, FL: CRC Press, 1999.

Lai, K.-C., Huang, C.-C., and Jia, J.-J. Variation of the fixed-complexity sphere decoder, *IEEE Communications Letters*, 15(9), 1001–1003, September 2011.

Larsson, E.G., MIMO detection methods: How they work, *IEEE Signal Processing Magazine*, 26(3), 91–95, May 2009.

Larsson, E.G. and Stoica, P. *Space-Time Block Coding for Wireless Communications*. Cambridge, UK: Cambridge University Press, 2003.

Lenstra, H.W. Lattices, in *Algorithmic Number Theory*, J. P. Buhler and P. Stevenhagen, Eds. Cambridge, UK: Cambridge University Press, 2008, pp. 127–181.

Lidl, R. and Pilz, G. Lattices, in *Applied Abstract Algebra*, 2nd ed. New York, New York, USA: Springer, 2nd ed., 2010, Chapter 1, pp. 1–53.

Ling, C. On the proximity factors of lattice reduction-aided decoding, *IEEE Transactions on Signal Processing*, 59(6), 2795–2808, June 2011.

Ling, C. and Mow, W.H. A unified view of sorting in lattice reduction: From V-BLAST to LLL and beyond, in *Proceedings of the IEEE Information Theory Workshop*, Taormina, Italy, 2009, pp. 529–533.

Ling, C., Gan, L., and Mow, W.H. A dual-lattice view of V-BLAST detection, in *Proc. of ITW' 06, The IEEE Information Theory Workshop*, Chengdu, China, October 2006, pp. 478–482.

Ling, C., Mow, W.H., and Gan, L. Dual-lattice ordering and partial lattice reduction for SIC-based MIMO detection, *IEEE Journal in Selected Topics in Signal Processing*, 3(6), 975–985, December 2009.

Ling, C., Mow, W.H., and Howgrave-Graham, N. Reduced and fixed-complexity variants of the LLL algorithm for communications, *IEEE Transactions on Communications*, 61, 1040–1050, March 2013.

Liu, S., Ling, C., and Stehlé, D. Decoding by sampling: A randomized lattice algorithm for bounded distance decoding, *IEEE Transactions on Information Theory*, 57(9), 5933–5945, Sep. 2011.

Lovász, L. *An Algorithmic Theory of Numbers, Graphs and Convexity*. Philadelphia, Pennsylvania, USA: Society for Industrial and Applied Mathematics (SIAM), 1986, Chapter 1, pp. 15–38.

Lozano, A. and Jindal, N. Transmit diversity vs. spatial multiplexing in modern MIMO systems, *IEEE Transactions on Wireless Communications*, 9(1), 186–197, January 2010.

Ma, X. and Zhang, W. Performance analysis for MIMO systems with lattice-reduction aided linear equalization, *IEEE Transactions on Communications*, 56(2), 309–318, February 2008.

Madhow, U. *Fundamentals of Digital Communication*. Cambridge, UK: Cambridge University Press, 2008.

Magarini, M. Spatial loading in V-BLAST systems with limited feedback and ZF-OSIC detection, in *Proceedings of the 4th International Symposium on Wireless Communication Systems (ISWCS)*, Trondheim, Norway, 2007, pp. 350–354.

Meyer, C.D. *Matrix Analysis and Applied Linear Algebra*. Philadelphia, PA: Society for Industrial and Applied Mathematics (SIAM), 2000.

Micciancio, D. The hardness of the shortest vector problem with preprocessing, *IEEE Transactions on Information Theory*, 47(3), 1212–1215, March 2001.

Micciancio, D. and Goldwasser, S. *Complexity of Lattice Problems—A Cryptographic Perspective*. Norwell, Massachusetts, USA: Kluwer Academic Publishers, 2002.

Micciancio, D. and Regev, O. Lattice-based cryptography, in *Post-Quantum Cryptography*, D.J. Bernstein, J. Buchmann, and E. Dahmen, Eds. Berlin, Germany: Springer, 2009, pp. 146–191.

Monteiro, F.A. and Kschischang, F.R. Trellis detection for random lattices, in *Proceedings of ISWCS'11—The 8th International Symposium on Wireless Communication Systems 2011*, Aachen, Germany, November 2011.

Mow, W.H. Maximum likelihood sequence estimation from the lattice viewpoint, MPhil thesis, Chinese University of Hong Kong, Hong Kong, Hong Kong, 1991.

Mow, W.H. Universal lattice decoding: Principles and recent advances, *Wireless Communications and Mobile Computing*, 3, 553–569, March 2003.

Mow, H.W. Universal lattice decoding: A review and some recent results, in *Proc. of the IEEE Int. Conf. on Comm. (ICC)*, 5, 2842–2846, Paris, France, June 2004.

Nazer, B. and Gastpar, M. Reliable physical layer network coding, *Proceedings of the IEEE*, 99(3), 438–460, March 2011.

Nguyen, P.Q and Micciancio, D. Entries on lattice, shortest vector problem, closest vector problem, and lattice based cryptography, in *Encyclopedia of Cryptography and Security*, Henk C. A. van Tilborg, Ed. New York, New York, USA: Springer, 2005, pp. 345–349.

Nguyen, P.Q and Vallée, B. Eds., *The LLL Algorithm*. Berlin, Germany: Springer, 2010.

Niu, J. and Lu, I-T. A new lattice-reduction-based receiver for MIMO systems, in *In Proc. of CISS—41st Annual Conf. on Information Sciences and Systems*, Baltimore, Maryland, March 2007, pp. 499–504.

Oestges, C. and Clerckx, B. *MIMO Wireless Communications: From Real-World Propagation to Space–Time Code Design*. Oxford, UK: Academic Press/Elsevier, 2007.

Oggier, F. Algebraic methods for channel coding, PhD thesis, École Polytechnique Fédérale de Lausanne, Lausanne, Switzerland, 2005.

Oggier, F., Rekaya, G., Belfiore, J.-C., and Viterbo, E. Perfect space-time block codes, *IEEE Transactions on Information Theory*, 52(9), 3885–3902, September 2006.

Okawado, A., Matsumoto, R., and Uyematsu, T. Near ML detection using Dijkstra's algorithm with bounded list size over MIMO channels, in *Proc. of ISIT'08—IEEE Inter. Symp. on Information Theory*, Toronto, Canada, July 2008, pp. 2022–2025.

Park, J., Chun, J., and Luk, F.T. Lattice reduction aided MMSE decision feedback equalizers, *IEEE Transactions on Signal Processing*, 59(1), 436–441, January 2011.

Parkvall, S., Furuskär, A., and Dahlman, E. Evolution of LTE toward IMT-advanced, *IEEE Communications Magazine*, 49(2), 84–91, February 2011.

Paulraj, A., Nabar, R., and Gore, D. *Introduction to Space–Time Wireless Communications*. Cambridge, Cambridge: Cambridge University Press, 2003.

Paz, A. and Schnorr, C.P. Approximating integer lattices by lattices with cycle factor groups, in *Proc. of the 14th Inter. Conf. on Automata, Languages and Programming*, Karlsruhe, Germany, LNCS 267, July 1987, pp. 386–393.

Penrose, R. A generalized inverse for matrixes, *Proc. of the Cambridge Philosophical Society*, Vol. 51, pp. 406–413, 1955.

Perahia, E. and Stacey, R. *Next Generation Wireless LANs—Throughput, Robustness, and Reliability in 802.11n*. Cambridge, UK: Cambridge University Press, 2008.

Petersen, K.B and Pedersen, M.S. 2008, February *The Matrix Cookbook*. http://matrixcookbook.com

Piziak, R. and Odell, P.L. *Matrix Theory—From Generalized Inverses to Jordan Form*. Boca Raton, FL: Chapman & Hall - CRC, 2007.

Poor, H.V. *An Introduction to Signal Detection and Estimation*, 2nd ed. USA: Springer, 1994.

Roger, S., Gonzalez, A., Almenar, V., and Vidal, A.M. Limassol, Cyprus, March 2010.

Schnorr, C. P. and Euchner, M. Lattice basis reduction: Improved practical algorithms and solving subset sum problems, *Mathematical Programming*, 66, 191–199, 1994.

Schrijver, A. *Theory of Linear and Integer Programming*. Chischester, UK: John Wiley & Sons, 1986, Chapters 4, 5, 6.

Seethaler, D., Matz, G., and Hlawatsch, F. Low-complexity MIMO data detection using Seysen's lattice reduction algorithm, in *Proc. of ICASSP'07—IEEE Inter. Conf. on Acoustics, Speech and Signal Processing*, Honolulu, Hawaii, USA, April 2007, pp. 15–20.

Shang, Y. and Xia, X.-G. An improved fast recursive algorithm for V-BLAST with optimal ordered detections, in *Proc of ICC'08—IEEE Inter. Conference on Communications*, Beijing, China, May 2008, pp. 756–760.

Sibille, A., Oestges, C., and Zanella, A. Eds., *MIMO: From Theory to Implementation*. Amsterdam, Netherlands: Academic Press, 2011.

Siegel, C.L. *Lectures on the Geometry of Numbers*. Berlin, Germany: Springer, 1989, Chapter 5.

Sloan, I.H and Joe, S. *Lattice Methods for Multiple Integration*. Oxford, UK: Oxford University Press, 1994.

Stewart, G.W. *Introduction to Matrix Computations*. London, UK: Academic Press, 1973.

Stoica, P., Jiang, Y., and Li, J. On MIMO channel capacity: An intuitive discussion, *IEEE Signal Processing Magazine*, 22(3), 83–84, May 2005.

Su, K. Detection and decoding of signals transmitted over linear MIMO channels, PhD thesis, University of Cambridge, Cambridge, UK, 2005.

Su, K. and Kschischang, F.R. Coset-based lattice detection for MIMO systems, in *Proc. of ISIT'07 - IEEE Inter. Symp. on Information Theory*, Nice, France, June 2007, pp. 1941–1945.

Su, K. and Wassell, I.J. A new ordering for efficient sphere decoding, in *IEEE Inter. Conf. on Communications*, 3, Seoul, Korea, 2005, pp. 1906–1910.

Su, K., Berenguer, I., Wassell, I.J., and Wang, X. Efficient maximum-likelihood decoding of spherical lattice codes, *IEEE Transactions on Communications*, 57(8), 2290–2300, August 2009.

Taherzadeh, M. Lattice-based precoding and decoding in MIMO fading systems, PhD thesis, University of Waterloo, Waterloo, Ontario, Canada, 2008.

Taherzadeh, M., Mobasher, A., and Khandani, A.K. Communication over MIMO broadcast channels using lattice-basis reduction, *IEEE Transactions on Information Theory*, 53(12), 4567–4582, December 2007.

Talbot, J. and D. Welsh, *Complexity and Cryptography: An Introduction*. Cambridge, UK: Cambridge University Press, 2006.

Tarokh, V., Seshadri, N., and Calderbank, A.R. Space-time codes for high data rate wireless communication: Performance criterion and code construction, *IEEE Transactions on Information Theory*, 44(2), 744–765, March 1998.

Tarokh, V., Jafarkhani, H., and Calderbank, A.R. Space-time block codes from orthogonal designs, *IEEE Transactions on Information Theory*, 45(5), 1456–1467, July 1999.

Telatar, Ì.E. Capacity of multiple-antenna Gaussian channels, *European Transactions on Telecommunications*, 10(6), 585–595, November-December 1999.

Trolin, M. Lattices with many cycles are dense, in *Proc. of the 21th Inter. Conf. on Theoretical Aspects of Computer Science (STACS)*, Montpellier, France, LNCS 2996, March 2004, pp. 370–381.

Trujillo, R.A., Garcia, V.M., Vidal, A.M., Roger, S., and Gonzalez, A. A gradient-based ordering for MIMO decoding, in *Proc. of the 9th IEEE Inter. Symp. on Signal Processing and Information Technology (ISSPIT)*, Ajman, United Arab Emirates, 2009, pp. 5–8.

Tse, D. and Viswanath, P. *Fundamentals of Wireless Communication*. Cambridge, UK: Cambridge University Press, 2005.

Tulino, A.M. and Verdú, S. *Random Matrix Theory and Wireless Communications*. Delft, Netherlands: Now, 2004, Chapter 2, pp. 24–38.

Verdú, S. *Multiuser Detection*. Cambridge, UK: Cambridge University Press, 1998.

Viterbo, E. and Boutros, J. A universal lattice code decoder for fading channels, *IEEE Transactions on Information Theory*, 45(5), 1639–1642, July 1999.

Viterbo, E. and Oggier, F. *Algebraic Number Theory and Code Design for Rayleigh Fadign Channels*. Hanover, MA, USA: now Publishers, 2004.

Welsh, D. Computational complexity, in *Codes and Cryptography*. Oxford, UK: Oxford University Press, 1988, Chapter 9, pp. 143–148.

Westlake, J. *Handbook of Numerical Matrix Inversion and Solution of Linear Equations*. Chichester, UK: John Wiley & Sons, 1968, Chapter 2.6.

Windpassinger, C. Detection and precoding for multiple input multiple output channels, PhD thesis, University of Erlangen-Nürnberg, Erlangen, Germany, 2004.

Wolniansky, P.W., Foschini, G.J., Golden, G.D., and Valenzuela, R.A. V-BLAST: An architecture for realizing very high data rates over the rich-scattering wireless channel, in *Proc. of URSI Int. Symposium on*, Pisa, Italy, September 1998, pp. 295–300.

Wübben, D., Böhnke, R., Kühn, V., and Kammeyer, K.D. MMSE extension of V-BLAST based on sorted QR decomposition, in *IEEE Proceedings of Vehicular Technology Conference (VTC)*, Orlando, Florida, USA, October 2003.

Wübben, D., Böhnke, R., Kühn, V., and Kammeyer, K. D. Near-maximum-likelihood detection of MIMO systems using MMSE-based lattice reduction, in *Proc. IEEE Int. Conf. Communications (ICC)*, Vol. 2, pp. 798–802, Paris, France, June 2004.

Wübben, D., Seethaler, D., Jaldén, J., and Matz, G. Lattice reduction, *IEEE Signal Processing Magazine*, 28(3), 70–91, May 2011.

Yao, H. *Efficient Signal, Code and Receiver Designs for MIMO Communications Systems*, Massachusetts Institute of Technology, Cambridge, MA, USA, PhD thesis, 2003.

Yao, H. and Wornell, G.W. Lattice-reduction-aided detectors for MIMO communication systems, in *Proc. of GLOBECOM' 02—IEEE Global Telecommunications Conference*, Taipei, Taiwan, 2002, pp. 424–428.

Zamir, R. On lattice quantization noise, *Transactions on Information Theory*, 42(4), 1152–1159, July 1996.

Zhang, W., Ma, X., Gestner, B., and Anderson, D. Designing low-complexity equalizers for wireless systems, *IEEE Communications Magazine*, 47(1), 56–62, January 2009.

Zhang, W,. Qiao, S., and Wei, Y. HKZ and Minkowski reduction algorithms for lattice-reduction-aided MIMO detection, *IEEE Transactions on Signal Processing*, 60, 5963–5976, Nov. 2012.

Zheng, L. and Tse, D.N.C. Diversity and multiplexing: A fundamental tradeoff in multiple antenna channels, *IEEE Transactions on Information Theory*, 49(5), 1073–1096, May 2003.

Chapter 3

Precoding for MIMO

Dzevdan Kapetanović and Fredrik Rusek

Contents

3.1 Introduction to Precoding

3.1.1 SU-MIMO

In this chapter, we focus on the single-user MIMO (SU-MIMO) scenario, with short remarks on the closely related multi-user MIMO (MU-MIMO) broadcast channel. We state the model here again for convenience:

$$\mathbf{y} = \mathbf{Hx} + \mathbf{n}. \tag{3.1}$$

In Equation 3.1, $\mathbf{y}$ is the $N_R \times 1$ received vector, $\mathbf{H}$ an $N_R \times N_T$ matrix that represents the MIMO channel and $\mathbf{x}$ is the $N_T \times 1$ vector of transmitted data symbols. Throughout the chapter, white Gaussian noise is assumed, hence $\mathbf{n}$ is an $N_R \times 1$ vector of complex Gaussian noise variables $\mathbf{n} \sim \mathcal{CN}(\mathbf{0}, N_0 \mathbf{I}_{N_R \times N_R})$.

3.1.1.1 Channel State Information

The MIMO channel depends on the environment, which determines how the channel changes over time and frequency. As soon as the channel variations are small, characterized by the coherence time T_C and the coherence bandwidth B_C, it is possible to track the channel and obtain reliable CSI. To obtain CSI at the receiver, a popular method is to send known pilot symbols from the transmitter to the receiver [Hassibi and Hochwald 2003]. Decoding them at the receiver makes it possible to obtain an estimate $\hat{\mathbf{H}}$ of the channel $\mathbf{H}$, which can be made accurate by devoting more resources to the training phase. The amount of pilot data that needs to be transmitted is a well-researched topic and has been analyzed in, for example [Hassibi and Hochwald 2003]. Note that transmitting pilot symbols, estimating the channel based on the pilot observations and using the estimate as if it is correct in the subsequent data detection phase, is not the optimal approach as it is inferior to performing a noncoherent detection. The ultimate limits of noncoherent detection have been studied in [Zheng and Tse 2002]. However, pilot transmission yields significantly less computational complexity at the receiver side. In order for the transmitter to obtain $\mathbf{H}$, two common techniques are

1. *Feedback*: In this approach, the estimated channel $\hat{\mathbf{H}}$ is sent from the receiver to the transmitter on a feedback link. This feedback inherently gives rise to some delay δ. In order for $\hat{\mathbf{H}}$ to be reliable at the transmitter, we must have $\delta \ll T_C$. If the channel varies rapidly, this approach requires more frequent estimates $\hat{\mathbf{H}}$ and feedback.

2. *Channel Reciprocity*: This technique uses the assumption that the estimated channel from the transmitter to the receiver is the same as the channel from the receiver to the transmitter. Problems with this technique include calibration issues as well as the fact that the forward and backward channels are not necessarily close in time and frequency [Paulraj et al. 2003].
3. Despite the practical difficulties in obtaining perfect CSI, in situations when the channel varies slowly and the feedback link has sufficient capacity, the perfect CSI assumption is assumed to hold throughout this chapter.

3.1.1.2 Information Rate

An important performance measure for a MIMO system is its mutual information, which determines the achievable information rates for the MIMO channel. Assume that the MIMO channel does not change from one slot to another (quasistatic channel). Let $p_{\mathbf{X}}(\mathbf{x})$ denote the joint pdf/pmf[*] of the vector $\mathbf{X} = [X_1, \ldots, X_{N_T}]^T$ of N_T random variables.[†] It is assumed that $\mathbf{X}$ has zero mean $\mathbb{E}\{\mathbf{X}\} = \mathbf{0}_{N_T}$ and covariance matrix $\mathbf{R}_{\mathbf{X}} = \mathbb{E}\{\mathbf{X}\mathbf{X}^H\}$. The mutual information $\mathcal{I}(\mathbf{Y}; \mathbf{X})$ between the input $\mathbf{x}$ and output $\mathbf{y}$ for a MIMO channel is defined as

$$\mathcal{I}(\mathbf{Y}; \mathbf{X}) \overset{\Delta}{=} \mathcal{H}(\mathbf{Y}) - \mathcal{H}(\mathbf{Y} \mid \mathbf{X}), \qquad (3.2)$$

where $\mathcal{H}(\cdot)$ is the differential entropy operator [Cover and Thomas 2006]

$$\mathcal{H}(\mathbf{Y}) = -\int_{\mathbf{y}} p_{\mathbf{Y}}(\mathbf{y}) \log_2(p_{\mathbf{Y}}(\mathbf{y})) d\mathbf{y}.$$

$\mathcal{I}(\mathbf{Y}; \mathbf{X})$ is the number of bits that can be carried by $\mathbf{X}$ through $\mathbf{H}$, given the specified pdf $p_{\mathbf{X}}(\mathbf{x})$.

Definition 3.1

The information rate $I(\mathbf{H}, p_x) = \mathcal{I}(\mathbf{Y}; \mathbf{X})$ is the maximum number of bits per channel use that can be carried error free through the MIMO channel $\mathbf{H}$, *given the pdf $p_{\mathbf{X}}(\mathbf{x})$.*

If one maximizes $I(\mathbf{H}, p_{\mathbf{X}})$ over the pdf $p_{\mathbf{X}}(\mathbf{x})$, but keeps the correlation matrix $\mathbf{R}_{\mathbf{X}}$ fixed, one obtains the *constrained capacity* for the MIMO channel.

---

[*] Probability density function or probability mass function.
[†] Capital and bold symbols denote vectors of random variables, except for matrices which always are in capital and bold.

Definition 3.2

The constrained capacity for a MIMO channel is

$$C(\mathbf{H},\mathbf{R_X}) = \sup_{p_{\mathbf{X}}(\mathbf{x}):\mathbb{E}\{\mathbf{XX}^H\}=\mathbf{R_X}} I(\mathbf{H},p_{\mathbf{X}}).$$

Soon we will present a closed-form expression for the constrained capacity of the MIMO channel. Finally, maximizing $C(\mathbf{H}, \mathbf{R_X})$ over $\mathbf{R_X}$ yields the *capacity* for a MIMO channel. This maximization is valid only if there is a constraint on $\mathbf{R_X}$, and the average transmit power constraint is commonly used.

Definition 3.3

The capacity for a MIMO channel is

$$\hat{C}(P_0,\mathbf{H}) = \max_{\mathbf{R_X}:\mathrm{tr}(\mathbf{R_X})\le P_0} C(\mathbf{H},\mathbf{R_X}).$$

Note that the information rate per channel use is given by the mutual information between two sequences of random variables, $\mathbf{Y}$ and $\mathbf{X}$.

Telatar derived exact analytical expressions for the constrained capacity $C(\mathbf{H}, \mathbf{R_X})$ and the capacity $\hat{C}(P_0,\mathbf{H})$ of a MIMO system [Telatar 1999]. The constrained capacity of a MIMO system is given by

$$C(\mathbf{H},\mathbf{R_X}) \overset{\Delta}{=} \max_{p_{\mathbf{X}}(\mathbf{x}):\mathbb{E}\{\mathbf{XX}^H\}=\mathbf{R_X}} I(\mathbf{H},p_{\mathbf{X}}) = \log_2 \det\left(\mathbf{I}_{N_T} + \frac{1}{N_0}\mathbf{HR_X}\mathbf{H}^H\right). \quad (3.3)$$

The constrained capacity in Equation 3.3 is attained by a multivariate Gaussian distribution on $\mathbf{X}$, with the correlation matrix $\mathbf{R_X}$. The capacity is obtained by subsequent maximization of Equation 3.3 as in Definition 3.3. The solution to this optimization is the well-known waterfilling technique. Let $\mathbf{H} = \mathbf{USV}^H$ be the SVD decomposition of the channel $\mathbf{H}$ and $\mathbf{R_X} = \mathbf{Q\Sigma Q}^T$ be the eigenvalue decomposition of $\mathbf{R_X}$. Further, let $\sigma_{j,j}$ be the diagonal elements of $\mathbf{\Sigma}$ and $s_{j,j}$ the diagonal elements of $\mathbf{S}$, respectively. The optimization in Definition 3.3 can be shown to be equivalent to

$$\hat{C}(P_0,\mathbf{H}) = \max_{\sum_{j=1}^{r}\sigma_{j,j}=P_0} \sum_{k=1}^{r} \log_2\left(1 + \frac{\sigma_{j,j}s_{j,j}^2}{N_0}\right), \quad (3.4)$$

where r is the rank of the channel $\mathbf{H}$. The unitary matrix $\mathbf{Q}$ is equal to $\mathbf{V}$. The solution to Equation 3.4 is

$$\sigma_{j,j}^{\text{opt}} = \left(\mu - \frac{N_0}{s_{j,j}} \right)_+ , \tag{3.5}$$

where one defines

$$x_+ = \begin{cases} x & x \geq 0 \\ 0 & x < 0 \end{cases}$$

for a number x, and μ is such that the power constraint is satisfied. Hence, to achieve the rate in Equation 3.4, the transmitted vector $\mathbf{x}$ is constructed as $\mathbf{x} = \mathbf{V}\sqrt{\Sigma}\mathbf{a}$, where $\mathbf{a}$ is a zero mean circularly symmetric complex Gaussian (ZMCSCG) with $\mathbf{R_A} = \mathbf{I}_{N_T \times N_T}$. This transforms the linear channel in Equation 3.1 into a set of parallel channels,

$$y_k = s_{k,k}\sigma_{k,k}^{\text{opt}}a_k + n_k, \quad k = 1,\ldots,r. \tag{3.6}$$

Thus, optimal transmission over the linear channel occurs over its eigenmodes $\{s_{j,j}\}$. Note, however, that this is only true if the data can be a multivariate Gaussian. As soon as the symbols x_j in $\mathbf{x}$ are drawn from a discrete constellation, which will be the main focus later in this chapter, signaling as in Equation 3.6 is not optimal!

Transmitting at bit rates in Equations 3.3 and 3.4, the probability of detecting an erroneous message at the receiver can be made arbitrarily close to 0 with long data blocks, in theory. Assume that the transmitter wants to convey one out of 2^k different messages to the receiver. A bit pattern $\boldsymbol{b}$ of k bits is used to represent each message. This bit pattern is represented by a sequence of vectors $\{\mathbf{x}_1, \ldots, \mathbf{x}_n\}$, which are sent through $\mathbf{H}$ in n different channel uses. The rate R of the system is defined as $R \overset{\Delta}{=} k/n$ bits/channel use. Assuming that channel is used indefinitely, that is, $k, n \to \infty$ but with a fixed ratio $R = k/n$, it is possible to recover the transmitted message with error probability tending to zero as long as $R < \hat{C}(P_0, \mathbf{H})$, if the encoding of the 2^k messages to the sequence of n vectors is done in an optimal fashion.

However, signaling exactly at this rate in practice is impossible for several reasons. First of all, it requires that the transmitted symbols x_j are taken from a Gaussian alphabet, which is not very practical. Moreover, the number of messages, k, has to be infinite (in theory), that is, the transmission has to occur for an indefinite amount of time. Infinitely many vectors $\mathbf{x}_i$, $i = 1, \ldots$, need to be transmitted, and the receiver has to receive the whole signal $\mathbf{y}_i$, $i = 1, \ldots$, in order to make optimal detection. Still, it is possible to come close to these rates by modern coding

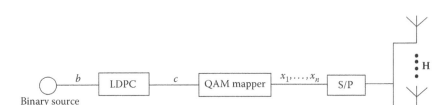

Figure 3.1 **A practical transmission system that can come close to the rates in Definitions 3.1 through 3.3. The bit sequence b is encoded into a much longer bit sequence c by an LDPC encoder. The bits in c are mapped onto a QAM constellation, which results in a sequence of vectors $\{x_1, \ldots, x_n\}$, each with N_T symbols. After the serial to parallel converter S/P, each vector is transmitted from the antenna array across the channel H.**

systems. A popular method [Brink et al. 2004] is to code the bit stream **b** with, for example, a low-density parity-check (LDPC) code, into a new bit sequence **c** of length $m > k$. Thus, the rate of the encoder is $R_c = k/m$. Next, the bits in **c** are mapped onto a discrete alphabet $\mathcal{X}$ (e.g., quadrature amplitude modulation—QAM) of cardinality $|\mathcal{X}| = M$. This creates a sequence of $n = m/N_T \log_2(M)$ symbol vectors $\mathbf{x}_j^T = [x_{j,1}, \ldots, x_{j,N_T}], j = 1, \ldots, n$. The sequence of vectors is passed through a serial to parallel converter and then transmitted from the antenna array. Hence, there are $n = m/N_T \log_2(M)$ channel uses, and the total rate of the system is $R = R_c \log_2(M) N_T$ bits per channel use. This transmitter is shown in Figure 3.1. At the receiver, an iterative decoding algorithm is applied, that iterates between decoding the MIMO channel and the LDPC encoder, in accordance with the Turbo principle [Hagenauer 1997].

As soon as the alphabet for the symbols x_j is constrained to be discrete, the rates in Equations 3.3 and 3.4 can never be reached exactly. Instead, the limit is $I(\mathbf{H}, p_x)$ in Definition 3.1. To achieve high rates, a large QAM alphabet is necessary. Beside a large QAM alphabet, long code words need to be produced by the encoder in order to reach $I(\mathbf{H}, p_x)$ and thereby approach $C(\mathbf{H}, \mathbf{R}_x)$. For an LDPC encoder, the needed block lengths can be as large as 10^5 [Brink et al. 2004; Lu et al. 2004; Bennatan and Burshtein 2006]. For some time-critical applications, it is of interest to send short code words and have less latency at the receiver side. This will inevitably lead to an error probability that is bounded away from 0. Furthermore, the alphabet $\mathcal{X}$ is in practice discrete. Thus, minimizing error probabilities or quantities related to it, without taking advanced codes into account, is a practically important topic.

3.1.2 Precoding for Linear Channels

From the previous discussion, it follows that the interest is to find a discrete set of vectors that can achieve the highest mutual information rate. However, this is also a nontractable problem if no constraints are put on the discrete set, except for the obvious energy constraint. A widely used technique to generate the discrete set of vectors, which starts to lend analytical tractability, is to construct the symbol vector **x** as

$$\mathbf{x} = \mathcal{L}(\mathbf{a}), \tag{3.7}$$

where **a** is a data vector that comes from a well-defined B-dimensional discrete alphabet $\mathcal{A}^B$, and $\mathcal{L}$ is a certain function/mapping. Note here that the mapping $\mathcal{L}$ is $\mathcal{L} : \mathbb{C}^B \to \mathbb{C}^{N_T}$, where $B \neq N_T$ can hold. In practice, the alphabet $\mathcal{A}$ is an M-QAM alphabet,

$$\mathcal{A} \overset{\Delta}{=} \{\kappa(z_r + iz_i) : z_r, z_i \in \{(-\sqrt{M} + 1)/2, \dots, (\sqrt{M} - 1)/2\}\},$$

where κ is a normalizing constant such that the average energy in $\mathcal{A}$ is one.

From now on, we will always let $\mathcal{A}$ be the QAM alphabet.

In general, to find the optimal mapping $\mathcal{L}$ for a certain receiver structure and performance measure (e.g., information rate) is a very tough problem. The mappings can be divided into two classes: linear mappings and nonlinear mappings. The latter often give rise to higher complexity (either encoding or decoding complexity), but can in general perform better than linear mappings. However, linear mappings always have a linear encoding complexity, while the decoding complexity depends on the receiver. We will now describe these two classes of mappings.

3.1.2.1 Linear Precoding

When $\mathcal{L}$ is a linear map, it can be represented by a matrix equation, that is, $\mathbf{x} = \mathbf{Fa}$ for some matrix **F**. A linear map is called a *linear precoder*. Here, **a** is a $B \times 1$ vector, and **F** is $N_T \times B$. Since **a** is discrete and structured, so will **x** be. We have already seen an application of linear precoding. To achieve the capacity $\hat{C}(P_0, \mathbf{H})$ in Equation 3.4, the vector **x** is constructed as $\mathbf{x} = \mathbf{V\Sigma a}$, where $\mathbf{a} \sim \mathcal{CN}(\mathbf{0}_{N_T}, \mathbf{I}_{N_T \times N_T})$ (i.e., $B = N_T$). Hence, in this case, $\mathbf{F} = \mathbf{V\Sigma}$. In general, since both the transmitter and receiver have perfect channel knowledge, it is possible to optimize over a linear transformation **F** of the data symbols **a**, in order to improve a performance measure imposed on Equation 3.1. Hence, a more general linear model than Equation 3.1 arises from this consideration:

$$\mathbf{y} = \mathbf{HFa} + \mathbf{n}. \tag{3.8}$$

As in Equation 3.1, $\mathbf{n} \sim \mathcal{CN}(\mathbf{0}_{N_R}, \mathbf{I}_{N_R \times N_R})$. Since $\mathbf{x} = \mathbf{Fa}$, $\mathbf{R_X} = \mathbb{E}\{\mathbf{Faa}^H\mathbf{F}^H\} = \mathbf{FR_A F}^H$, where the last equality follows from the linearity of the expectation operator $\mathbb{E}\{\cdot\}$ and the fact that **F** is not stochastic. Assuming uncorrelated QAM symbols (a practical assumption), $\mathbf{RA} = \mathbf{I}_{B \times B}$ and $\mathbf{RX} = \mathbf{FF}^H$. In this case, $\mathbf{F}$ determines the correlation matrix of $\mathbf{x}$. The average transmit energy constraint is now

$$\text{tr}(\mathbf{FF}^H) \leq P_0. \tag{3.9}$$

3.1.2.2 Nonlinear Precoding and the Multiuser Broadcast Channel

If $\mathcal{L}$ is a nonlinear function, then we obtain a *nonlinear precoder*. A common nonlinear precoding technique is *vector perturbation* [He and Salehi 2008, Ryan et al. 2009, Razi et al. 2010], which perturbs the data vector $\mathbf{a}$ with another vector $\mathbf{p}$ that comes from a lattice, in order to reduce the transmit energy. The data symbols a_j are assumed to belong to a bounded region in the complex-valued plane. Usually, this region is the cube $\mathcal{K} = \{a: |\text{Re}\{a\}| < 0.5, |\text{Im}\{a\}| < 0.5\}$, that is, $a_j \in \mathcal{K}$ and $\mathbf{a} \in \mathcal{K}^B$, the B dimensional cube. Next, a vector $\mathbf{s}$ is constructed as

$$\mathbf{s} = \mathbf{H}^+(\mathbf{a} + \mathbf{p}), \tag{3.10}$$

where $\mathbf{H}^+$ is the Moore–Penrose pseudo inverse of $\mathbf{H}$ and $\mathbf{p}$ is the solution to

$$\mathbf{p} = \arg \min_{\mathbf{q} \in \mathbb{Z}[i]^B} \| \mathbf{H}^+(\mathbf{a} + \mathbf{q}) \|^2 . \tag{3.11}$$

In Equation 3.11, $\mathbb{Z}[i]^B$ denotes the set of B dimensional Gaussian integer vectors. The transmitted vector $\mathbf{x}$ is then

$$\mathbf{x} = \sqrt{\frac{P_0}{\kappa(\mathbf{H})}} \mathbf{s}, \tag{3.12}$$

where $\kappa(\mathbf{H})$ is the average energy

$$\kappa(\mathbf{H}) = \mathbb{E}_A\{\| \mathbf{s} \|^2\} \tag{3.13}$$

with respect to the data vector $\mathbf{a}$. Hence, the received signal is

$$\mathbf{y} = \sqrt{\frac{P_0}{\kappa(\mathbf{H})}} (\mathbf{a} + \mathbf{p}) + \mathbf{n}. \tag{3.14}$$

Decoding $\mathbf{a}$ is now a simple matter. Since $\mathbf{p} \in \mathbb{Z}[i]^B$ and $\mathbf{a} \in \mathcal{K}^B$, the lattice vectors $\mathbf{p}$ translate the cube $\mathcal{K}^B$ so that it tiles the complex-valued B dimensional space $\mathbb{C}^B$; that is, the translated cubes cover $\mathbb{C}^B$ and they do not intersect. Let $\hat{a}_j$ denote the jth decoded data symbol. Then

$$\hat{a}_j = y_j \operatorname{mod}\mathcal{K}, \qquad (3.15)$$

where mod $\mathcal{K}$ means that y_j is translated to $\mathcal{K}$, that is, $y_j - z_j \in \mathcal{K}$ for a (unique) z_j. Thus, a simple modulo operation for each received stream y_j recovers a_j. Note that the decoded $\hat{a}_j$ is corrupted by a modulo Gaussian noise, $n_k \operatorname{mod} \mathcal{K}$.

Vector perturbation finds applications in MIMO broadcast channels as well [Hochwald et al. 2005]. A MIMO broadcast channel, with single-antenna users, is modeled as in Equation 3.1, where each entry in the vector **y** is the received symbol at a certain user. Each entry i, x_i, in the vector **x** is the symbol intended for user i. Hence, in contrast to SU-MIMO, all other symbols except x_i act as interference to user i. By applying vector perturbation at the transmitter, it is readily seen from Equation 3.14 that no interference occurs at the user terminals.

Vector perturbation gives rise to a simple decoding method, by inverting the channel and translating the data vector **a** to reduce the transmit energy. Moreover, it is shown in [Hochwald et al. 2005] that vector perturbation comes close to the sum capacity of a MIMO broadcast system. In [Taherzadeh et al. 2007], it is shown that vector perturbation achieves the maximum diversity in MIMO broadcast channels. The main bottleneck of vector perturbation is the computational complexity needed to find the optimal **p** in Equation 3.11, which is a well-known NP-hard problem [Razi et al. 2010]. Hence, an NP-hard problem needs to be solved online for every realization of **H** and **a**. Suboptimal low-complexity implementations of vector perturbation exist [Windpassinger et al. 2004] that achieve performance close to the original vector perturbation.

3.1.3 Construction of Linear Precoders

The low encoding complexity of linear precoders is very desirable for practical applications. For this reason, linear precoders have been an active area of research throughout the history of MIMO communications and are, for example, incorporated in the long-term evolution (LTE) standard [Dahlman et al. 2011]. However, depending on the receiver and the different performance measures of interest, different optimal precoders are obtained. We will now review some linear precoding techniques for different receiver structures.

3.1.3.1 Optimal Linear Precoders for Linear Receivers

The Wiener filter is the optimal linear receiver for many performance measures of interest. Employing this filter at the receiver, the next task is to find linear precoders that maximize different performance measures. A thorough investigation of this problem is performed in [Palomar and Jiang 2007], [Palomar et al. 2003], and [Scaglione et al. 2002]. The optimization problems that arise are efficiently solved with majorization techniques, and it turns out that the optimum precoder can be derived in a relatively easy fashion. Let $N \leq \min(N_R, N_T)$. If an arbitrary objective function $f(e_{1,1}, \ldots, e_{N,N})$, where $e_{1,1}, \ldots, e_{N,N}$ are the MMSE values across each received antenna after the Wiener filter, is increasing in its arguments and minimized when its arguments are sorted in decreasing order, the solution to the optimization problem

$$\min_{F} f(\{e_{j,j}\})$$

subject to

$$e_{j,j} \leq \rho_j, \quad j = 1, \ldots, N \qquad (3.16)$$

$$\mathrm{tr}(\mathbf{FF}^H) \leq P_0$$

is of the form $\mathbf{F} = \mathbf{V}\mathrm{diag}(\sqrt{\mathbf{p}})\mathbf{Q}$. Here, $\mathbf{V}$ is the right unitary matrix of $\mathbf{H}$, $\mathbf{Q}$ is a unitary matrix such that $e_{j,j} = \rho_j, j = 1, \ldots, N$, ($\mathbf{Q}$ can be obtained by a rather simple algorithm, Palomar and Jiang 2007, Algorithm 2.2), and $\mathrm{diag}(\sqrt{\mathbf{p}})$ is a diagonal matrix with the vector $\sqrt{\mathbf{p}}$ on its main diagonal.

It turns out that the BER function is convex as soon as it is below a certain threshold $\approx 10^{-3}$. Thus, the problem in Equation 3.16 becomes a convex optimization problem, that is, minimizing the BER is a convex problem that can be solved efficiently with convex optimization techniques.

Instead, if the interest is to maximize the mutual information, the problem reduces to minimizing the determinant of the MMSE matrix [Palomar and Jiang 2007], and the optimal $\mathbf{F}$ has $\mathbf{Q} = \mathbf{I}_{N_T, N_T}$ and $p_i = (\mu - \lambda_{\mathbf{H},i}^{-1})_+$, where μ is such that $\sum_{j=1}^{N} p_j = P_0$ holds. Hence, it is thus possible to derive closed-form solutions to the optimal precoder in the case of simple functions f, and also optimal numerical solutions when f is convex.

3.1.3.2 Optimal Linear Precoders for the ML Receiver

The Wiener filter is after all a suboptimal receiver, which thus gives a suboptimal performance of a MIMO system. If instead the optimal ML decoding rule is employed at the receiver, the analysis of the optimal linear precoders is significantly tougher for discrete alphabets. For coded systems, the main goal is to maximize the information rates in Section 3.1.2, which describe the performance of the ML decoder. Henceforth, in all sections, we assume that the ML decoder is employed at the receiver. Note that since $\mathbf{x} = \mathbf{Fa}$, $p_{\mathbf{X}}(\mathbf{x}) = p_{\mathbf{A}}(\mathbf{a}) = 1/|\mathcal{A}^B|$ and $\mathbf{R}_{\mathbf{X}} = \mathbf{FF}^H$. The information rate $I(\mathbf{H}, p_{\mathbf{X}})$ in Definition 3.1 can be denoted as $I(\mathbf{H}, \mathbf{F})$, where the optimization variables are explicit. The information rate optimal precoder is found by solving

$$\max_{\mathbf{F}} I(\mathbf{H},\mathbf{F})$$

$$\text{subject to} \qquad\qquad (3.17)$$

$$\text{tr}(\mathbf{F}\mathbf{F}^{H}) \leq P_{0}.$$

Note that it is trivial to solve Equation 3.17 with $\hat{C}(P_{0},\mathbf{H})$ as objective function, that is, when the alphabet $\mathcal{A}$ is Gaussian: the optimum linear precoder then performs waterfilling. However, since the alphabet $\mathcal{A}$ is discrete, there is no closed-form expression available for the objective function $I(\mathbf{H}, \mathbf{F})$. Finding the precoder $\mathbf{F}$ that solves Equation 3.17 is a challenging problem. In [Perez-Cruz et al. 2010], the Karush–Kuhn–Tucker (KKT) conditions were derived for Equation 3.17, which produced a fixed point equation for the optimal $\mathbf{F}$. Based on this equation, an iterative optimization technique was developed that produced precoders providing high information rate. However, the problem with this approach is that the iterative optimization technique is not guaranteed to converge to the optimum, since Equation 3.17 is a nonconvex problem in $\mathbf{F}$.

In [Lozano et al. 2006], an optimal $\mathbf{F}$ adhering to a diagonal structure is derived. Combining this with the result in [Payaro and Palomar 2009], that the optimal $\mathbf{F}$ is such that its SVD factorization $\mathbf{F} = \mathbf{U}_{\mathbf{F}}\,\mathbf{P}_{\mathbf{F}}\,\mathbf{V}_{\mathbf{F}}$ satisfies $\mathbf{U}_{\mathbf{F}} = \mathbf{V}$, where $\mathbf{H} = \mathbf{U}\mathbf{S}\mathbf{V}^{H}$ is the SVD of $\mathbf{H}$, it is concluded that the optimal $\mathbf{F}$ is known up the right unitary matrix in its SVD factorization. Furthermore, in [Perez-Cruz et al. 2010], it was shown that in the low SNR-regime, the right unitary matrix of the optimal precoder is the identity matrix. Thus, for low SNRs, the optimal precoder is of the form $\mathbf{F} = \mathbf{V}_{\mathbf{H}}\mathbf{P}_{\mathbf{F}}$, with $\mathbf{P}_{\mathbf{F}}$ constructed as in [Lozano et al. 2006]. The work in [Lozano et al. 2006] showed that the diagonal elements in $\mathbf{P}_{\mathbf{F}}$ are obtained by a procedure named as *mercury waterfilling*, which is the analogue of waterfilling for Gaussian alphabets, but now for discrete alphabets instead. In contrast to classical waterfilling for Gaussian alphabets, mercury waterfilling amounts to first pouring mercury in tubes corresponding to each data stream; the mercury is poured up to a certain precalculated level. Thereafter, water with a certain volume is poured on top of the mercury until the water level in all tubes is the same. The height of the water in each tube is the power to be allocated to each data stream. A graphical illustration of this process, as well as analysis, is presented in [Lozano et al. 2006]. A recent advance in [Xiao et al. 2011] shows that $I(\mathbf{H},\mathbf{F})$ is concave over the Gram matrix $\mathbf{G} = \mathbf{F}^{H}\mathbf{H}^{H}\mathbf{H}\mathbf{F}$. This enables construction of an algorithm that converges to the optimal $\mathbf{F}$. However, this algorithm is of very high complexity and is not very feasible for large QAM constellations and MIMO dimensions.

The work in [Perez-Cruz et al. 2010] showed an interesting connection between information rate and the minimum distance of the received signal vectors. Since the bit error rate (BER) at the receiver is decreasing with the minimum distance of the received signal vectors for high SNRs, it is desirable to maximize the minimum distance between the received vectors. Theorem 3.4 in [Perez-Cruz et al. 2010] shows that for high SNRs, the precoder solving Equation 3.17 is the one maximizing the minimum distance. Thus, for high SNRs, the solution to Equation 3.17 is obtained by optimizing the minimum distance. Hence, an interesting connection exists among the three well-known optimization criteria: the minimum distance, BER, and information rate. The precoder that minimizes the BER at high SNRs maximizes the data rate and the minimum distance at the same time!

3.1.3.3 Linear Precoders for Minimizing the BER

Owing to difficulty in finding the precoder that maximizes the mutual information rate for moderate SNR, an alternative is to minimize the BER. This is also directly applicable to systems not using advanced error correcting codes. Since the BER expression for an ML decoder cannot be put in closed form, different approximations of it are minimized, such as the Chernoff upper bound. In [Lokesh et al. 2008], a general expression for the precoder that minimizes the BER was presented. Similar results are derived in [Payaro and Palomar 2009]. However, these expressions are given in terms of unknown matrices, and to determine these matrices is a nontractable task in dimensions higher than two. Works such as [Lee et al. 2007], [Vu and Paulraj 2006], [Scaglione et al. 2002], [Mohammed et al. 2011], and [Jian et al. 2005] present suboptimal constructions to minimize the BER. In general, finding the precoder that minimizes the BER is a nontractable problem, and approximations to the BER are made, which relax the problem into a tractable one.

3.1.3.4 Linear Precoders without CSI at the Transmitter

When the transmitter has no knowledge about the channel coefficients **H**, the construction of precoders is of a different nature than before. In many cases, the receiver has the capability to obtain a good enough estimate of the channel. One alternative then is to feed back the CSI to the transmitter, as mentioned in Section 3.1.1.1, so that the transmitter has a channel to work with. However, due to the inherent delay, and in the case of low rate feedback links, this method is not viable. Instead, in this scenario, it is desirable that the receiver only feeds back a small amount of information to the transmitter, which is sufficient for determining the precoder at the transmitter. In [Jindal 2006], it was shown that for MIMO broadcast* channel with a zero forcing precoder (ZF) at the base station, the number of feedback bits required increases linearly with the SNR.

Usually, the transmitter is equipped with an already static, finite collection of precoders, a *precoder codebook*, and the receiver only feeds back a string of bits across the MIMO channel, that represent the position of the precoder in the codebook that the transmitter should use. This is known as limited feedback precoding, and is a part of the LTE standard [Dahlman et al. 2011]. Hence, the art of limited feedback precoding is in designing the finite precoder codebook. Many different techniques exist for

---

* MIMO broadcast is also refered to as MU-MIMO.

this purpose. In [Love and Heath 2005], precoders with k orthogonal columns are designed, where $k < N_T$. It is shown that the optimal such precoder, for many performance measures, has its k columns isotropically (i.e., "evenly") distributed across the unitary space $\mathcal{U}(N_T, k)$ of $N_T \times k$ matrices with k orthogonal columns. Hence, the optimal codebook should consist of precoders that are evenly spread across $\mathcal{U}(N_T, k)$. By constructing different distance measures between the subspaces that each such precoder spans, it is possible to construct codebooks of different sizes containing evenly spread precoders. Beside orthogonal precoding, other methods exist that feedback a few bits representing different elements of a precoder that optimizes, for example, the minimum distance [Ghaderipoor and Tellambura 2006].

3.1.4 Linear Precoders for Maximizing the Minimum Distance

Owing to the difficulty in finding a precoder that minimizes the BER, a common method is to minimize a quantity directly related to the BER. As discussed earlier, the minimum distance is the dominant factor in the BER for an ML decoder at high SNRs, and the precoder that maximizes it not only minimizes the BER, but also maximizes the information rate. There have been many attempts to construct precoders that increase the minimum distance, see, for example, [Payaro and Palomar 2009], [Vu and Paulraj 2006], [Scaglione et al. 2002], [Bergman 2009], among many others. All of these attempt to produce suboptimal constructions for moderate dimensions of the MIMO system. However, in [Collin et al. 2007], the precoder that maximizes the minimum distance for MIMO channels with two data streams ($B = 2$ and $N_R \geq 2$) and 4-QAM alphabet was found. It is shown that there are essentially only two different precoder "structures" that are optimal. With "structure," it is meant that the mathematical expression for the precoder takes on two different forms, but the precoder itself changes continuously with the channel, c.f. [Collin et al. 2007]. For any MIMO channel $\mathbf{H}$ with a ratio of its singular values that is above a certain threshold, one of these structures is always optimal, while for a channel with a ratio below the threshold, the other structure is optimal. Thus, in a way, the optimal precoder behaves in a discrete fashion. Later, this analysis was extended to 16-QAM alphabets in [Ngo et al. 2009], where it was found that there are only eight different precoder structures. Hence, these results hint upon a discrete structure of the optimal precoder. This chapter will reveal this optimal structure and provide new insights into the design of linear precoders $\mathbf{F}$ that maximize the minimum distance of the received signal vectors. The analysis for MIMO systems with two data streams is covered in Section 3.2.

3.1.5 Optimization Problem

Let $D_{\min}^2(\mathbf{HF}, \mathcal{A})$ denote the minimum distance of the received signal points at the receiver. The problem of finding the $N_T \times B$ precoder $\mathbf{F}$ that maximizes the minimum distance can be formulated as

$$\mathbf{F}_{\text{opt}} = \arg\max_{\mathbf{F}} D_{\min}^2(\mathbf{HF}, \mathcal{A})$$

$$\text{subject to} \qquad (3.18)$$

$$\text{tr}(\mathbf{FF}^H) \leq P_0.$$

Let $\mathbf{e} = \mathbf{a} - \mathbf{a} \in \mathcal{E}^B$, where $\mathcal{E}$ is the difference set of the alphabet $\mathcal{A}$. We now have $D_{\min}^2(\mathbf{HF}, \mathcal{A}) = \min_{\mathbf{e} \neq 0_B} \mathbf{e}^H \mathbf{F}^H \mathbf{H}^H \mathbf{HFe}$. Define

$$\mathbf{G} \triangleq \mathbf{F}^H \mathbf{H}^H \mathbf{HF} \qquad (3.19)$$

to be the *Gram matrix* of $\mathbf{HF}$. Then Equation 3.18 becomes

$$\mathbf{F}_{\text{opt}} = \arg\max_{\mathbf{F}} \min_{\mathbf{e} \neq 0_B, \mathbf{e} \in \mathcal{E}^B} \mathbf{e}^H \mathbf{Ge}$$

$$\text{subject to} \qquad (3.20)$$

$$\text{tr}(\mathbf{FF}^H) \leq P_0.$$

It is readily seen that the optimal $\mathbf{F}_{\text{opt}}$ is such that it minimizes $\text{tr}(\mathbf{FF}^H)$ subject to a fixed constraint on the minimum distance, for example, $\min_{\mathbf{e} \neq 0_B, \mathbf{e} \in \mathcal{E}^B} \mathbf{e}^H \mathbf{Ge} \geq 1$. Thus, we can rewrite Equation 3.20 as

$$\mathbf{F}_{\text{opt}} = \arg\min_{\mathbf{F}} \text{tr}(\mathbf{FF}^H)$$

$$\text{subject to}$$

$$\min_{\mathbf{e} \neq 0_B, \mathbf{e} \in \mathcal{E}^B} \mathbf{e}^H \mathbf{Ge} \geq 1 \qquad (3.21)$$

$$\mathbf{G} = \mathbf{F}^H \mathbf{H}^H \mathbf{HF}.$$

Let $\mathbf{H} = \mathbf{USV}^H$ be the singular-value decomposition of $\mathbf{H}$. From the definition of $\mathbf{G}$ in Equation 3.19, it follows that $\mathbf{U}$ has no impact on $D_{\min}^2(\mathbf{HF}, \mathcal{A})$, and can therefore be removed at the receiver. Furthermore, the matrix $\mathbf{V}$ can be absorbed into $\mathbf{F}$ without changing the value of the objective function in Equation 3.21. Only the $N \times N$ diagonal submatrix in $\mathbf{S}$, that contains the singular values, is of interest, since the other elements are zero. Hence, an equivalent model to Equation 3.1 arises:

$$\mathbf{y} = \mathbf{SFa} + \mathbf{n}, \qquad (3.22)$$

where $\mathbf{S}$ is an $N \times N$ diagonal matrix with nonzero diagonal entries, $\mathbf{F}$ is an $N \times B$ matrix subject to $\text{tr}(\mathbf{F}^H \mathbf{F}) \leq P_0$, and $\mathbf{a}$ a $B \times 1$ vector. Further, $\mathbf{y}$ is now $N \times 1$ and so is $\mathbf{n}$. In total, the system in Equation 3.1, where $\mathbf{x} = \mathbf{Fa}$, can without loss of generality be reduced to Equation 3.22. $\mathbf{G}$ is now $\mathbf{G} = \mathbf{F}^H \mathbf{S}^2 \mathbf{F}$, a $B \times B$ matrix. It is further assumed that $B \leq N$, that is, the number of spatially multiplexed data streams is determined by the rank. Thus, we can now rewrite Equation 3.21 as

$$\mathbf{F}_{\text{opt}} = \arg\min_{\mathbf{F}} \text{tr}(\mathbf{F}\mathbf{F}^H)$$

$$\text{subject to}$$

$$\min_{\mathbf{e} \neq \mathbf{0}_B, \mathbf{e} \in \mathcal{E}^B} \mathbf{e}^H \mathbf{G} \mathbf{e} \geq 1 \tag{3.23}$$

$$\mathbf{G} = \mathbf{F}^H \mathbf{S}^2 \mathbf{F}.$$

In [Payaro and Palomar 2009], it was shown that solving Equation 3.23 is an NP-hard problem. Thus, at first sight, this problem seems mathematically intractable, and finding a solution to it online amounts to solving an NP-hard problem. Therefore, the first natural approach is to construct suboptimal solutions to Equation 3.23, as was done in [Scaglione et al. 2002] and [Vrigneau et al. 2008]. This is the rationale behind Section 3.1.7.

Assume that the Gram matrix $\mathbf{G}$ in Equation 3.23 is given. We will now find the $\mathbf{F}$ that solves Equation 3.23 for the given $\mathbf{G}$. Since $\mathbf{G}$ is Hermitian, and thus a normal matrix, its eigendecomposition is

$$\mathbf{G} = \mathbf{Q}\mathbf{D}\mathbf{Q}^H. \tag{3.24}$$

The factorization in Equation 3.24 is unique if the diagonal elements of $\mathbf{D}$ are ordered in a decreasing order. From the definition of $\mathbf{G}$ and Equation 3.24, we see that $\mathbf{F} = \mathbf{S}^{-1}(\sqrt{\mathbf{D}}\,\mathbf{0}_{B,N-B})^T \mathbf{Q}^H$ is a precoder such that $\mathbf{F}^H \mathbf{S}^2 \mathbf{F} = \mathbf{G}$. The $\mathbf{0}_{B,N-B}$ matrix is a $B \times N - B$ zero matrix, accounting for the case when $B < N$. Next, we prove that this $\mathbf{F}$ has the lowest energy of all possible $\mathbf{G}$ satisfying $\mathbf{G} = \mathbf{F}^H \mathbf{S}^2 \mathbf{F}$.

Theorem 3.1

Let $\mathbf{G} = \mathbf{Q}\mathbf{D}\mathbf{Q}^H$ *where the diagonal elements of* $\mathbf{D}$ *are ordered in decreasing order. Then, of all* $\mathbf{F}$ *satisfying* $\mathbf{G} = \mathbf{F}^H \mathbf{S}^2 \mathbf{F}$,

$$\mathbf{F} = \mathbf{S}^{-1}(\sqrt{\mathbf{D}}\,\mathbf{0}_{B,N-B})^T \mathbf{Q}^H \tag{3.25}$$

is the one with least energy $\text{tr}(\mathbf{F}^H\mathbf{F})$.

Proof. Combining $\mathbf{G} = \mathbf{F}^H \mathbf{S}^2 \mathbf{F}$ and Equation 3.24 we get $\mathbf{F}^H \mathbf{S}^2 \mathbf{F} = \mathbf{Q}\mathbf{D}\mathbf{Q}^H$. Rewriting, we have

$$\mathbf{Q}^H \mathbf{F}^H \mathbf{S}_H^2 \mathbf{F} \mathbf{Q} = \mathbf{D}. \tag{3.26}$$

Assume that $\mathbf{F}^H\mathbf{R}\mathbf{F}$, where $\mathbf{R}$ is a positive semidefinite matrix, is equal to a diagonal matrix Λ, where the diagonal elements in Λ are in decreasing order (in our case Λ is $B \times B$). Then, in [Palomar and Jiang 2007, Lemma 3.16], it is proved that we can always choose $\mathbf{F} = \mathbf{V}_\mathbf{R}\mathbf{D}_\mathbf{R}^{-1/2}\sqrt{\Lambda}$, where $\mathbf{V}_\mathbf{R}$ contains the B eigenvectors of $\mathbf{R}$ corresponding to the B largest eigenvalues of $\mathbf{R}$ and $\mathbf{D}_\mathbf{R}$ contains the B largest eigenvalues of $\mathbf{R}$, respectively, in order to minimize tr($\mathbf{F}^H\mathbf{F}$). Hence in our case, we choose $\mathbf{F}\mathbf{Q} = \mathbf{S}^{-1}(\mathbf{I}_{\mathbf{B}\times\mathbf{B}}\,\mathbf{0}_{B,N-B})^T\sqrt{\mathbf{D}}$ which gives $\mathbf{F} = \mathbf{S}^{-1}(\sqrt{\mathbf{D}}\,\mathbf{0}_{B,N-B})^T\mathbf{Q}^H$ and completes the proof.

3.1.6 Suboptimal Constructions

Numerous suboptimal constructions of $\mathbf{F}$ exist for maximizing the received minimum distance. One of the first suboptimal approaches is presented in [Scaglione et al. 2002]. Therein, it is noted that

$$D_{\min}^2(\mathbf{H}\mathbf{F},\mathcal{A}) \geq \lambda_N(\mathbf{G}) \min_{\mathbf{a},\hat{\mathbf{a}}\in\mathcal{A},\mathbf{a}\neq\hat{\mathbf{a}}} \|\mathbf{a} - \hat{\mathbf{a}}\|^2,$$

where $\lambda_B(.)$ is the B-th largest eigenvalue of its argument (since B is the dimension of the Gram matrix $\mathbf{G}$, it is the smallest eigenvalue in this case). Hence, by constructing $\mathbf{F}$ such that the smallest singular value of $\mathbf{H}\mathbf{F}$ is maximized, the lower bound on the minimum distance is maximized. It is shown that the optimal F takes on the form

$$\mathbf{F} = \mathbf{V}\mathbf{D},$$

where as before $\mathbf{V}$ denotes the right unitary matrix in the SVD decomposition of $\mathbf{H}$. Although providing performance superior to no precoding, this method has been outperformed by later suboptimal constructions.

In [Vrigneau et al. 2008], a suboptimal construction based on the optimal two-dimensional precoder in [Collin et al. 2007] is presented. The precoder in [Collin et al. 2007] is only optimal for 4-QAM alphabets. The idea is to pair off the largest and smallest singular values of the channel, and apply the optimal two-dimensional precoder for each pair. Namely, let $s_{(i)}$ denote the i:th largest singular values of $\mathbf{H}$. The singular values are paired as $(s_{(1)}, s_{(N)})$, $(s_{(2)}, s_{(N-1)})$, ..., $(s_{(N/2)}, s_{(N/2+1)})$. Let $\mathbf{S}_{(i)}$ denote the i:th pair. The optimal precoder $\mathbf{F}_i$ is constructed for each pair $\mathbf{S}_{(i)}$, $1 \leq i \leq N/2$, and scaled such that the minimum distance for each channel $\mathbf{S}_{(i)}\mathbf{F}_i$ is the same. The performance of this precoding will be presented in later sections.

Another approach for constructing precoders is presented in [Bergman 2009]. Therein, a lattice theoretic approach is undertaken to construct precoders with large minimum distances. By using densest lattice packings in different dimensions, and optimizing the lattice basis, precoders with large minimum distances were found. It was shown with bounds and numerical observations that, in general, the performance of lattice precoders is close to the optimum. However, the precoder construction is heuristic, and it is not known whether densest lattice packings are optimal or not. Section 3.2 will present a proof that, for MIMO systems with two input streams, well-known lattices are indeed optimal for precoding with large discrete alphabets. Recent works in [Kapetanović et al. 2012] shows that densest lattice packings are not always optimal for a higher number of input streams. Further suboptimal constructions can be found in [Rusek and Kapetanović 2009], [Mohammed et al. 2011], [Ngo et al. 2010, 2013], and [Kapetanović and Rusek 2011].

3.1.7 Summary of Precoding Techniques

In Table 3.1, we present current knowledge at hand when it comes to optimal linear precoding. The complexity of a decoder is measured by the conventional ordo notation, that is, the complexity is proportional to the argument of $\mathcal{O}(\cdot)$. Complexity of preprocessing before decoding, that is, matrix/vector operations, is not considered. For the MMSE receiver, the decoding complexity does not depend on the alphabet size $|\mathcal{A}|$, and for Gaussian alphabets, decoding complexity is not really a definable concept (Gaussian alphabets are approximated by discrete ones in practice).

As seen in the table, for the optimal ML receiver, the optimal precoder construction is found in [Xiao et al. 2011], which produces an optimal precoder maximizing the mutual information. However, due to the high computational cost in generating these precoders at the transmitter, it is of interest to study alternative methods, especially in the high/low SNR regime. For low SNRs, the solution is given by mercury waterfilling. For high SNRs, the optimal precoder is the one that maximizes the minimum distance of the received signal points. Moreover, this precoder is also useful for delay-sensitive applications not using advanced coding techniques. In Table 3.2, we list the current knowledge at hand for precoders maximizing the minimum distance.

Table 3.1 Summary of Current Results on Optimal Linear Precoders

Receiver	SNR	Constellation	Optimal Precoder	Decoding Complexity
ML	Any	Gaussian	Waterfilling	NA
ML	Low	Discrete	Mercury waterfilling [Lozano et al. 2006]	$\mathcal{O}(B|\mathcal{A}|)$
ML	High	Discrete	Maximum min. dist. precoders	$\mathcal{O}(|\mathcal{A}|^B)$
ML	Any	Discrete	Precoding in [Xiao et al. 2011]	$\mathcal{O}(|\mathcal{A}|^B)$
Linear (MMSE)	Any	Any	Constructions in [Palomar and Jiang 2007]	$\mathcal{O}(B)$

Note: The dimension of the MIMO system is arbitrary.

Table 3.2   Current Status on Precoders Maximizing the Minimum Distance

No. of Input Streams B	Constellation Size	Optimal Precoder
2	4	Precoders in [Collin et al. 2007]
2	16	Precoders in [Ngo et al. 2009]
2	>16	Unknown
2	Infinite	Presented in Section 3.2
>2	Small to moderate	Unknown
>2	Infinite	Precoders in [Kapetanović et al. 2012]

Note that for infinite discrete alphabets (e.g., an infinite QAM alphabet), the optimal precoders are known by work presented in Section 3.2 and in [Kapetanović et al. 2012]. The infinity assumption comes from using lattice theoretic tools for characterization of the optimal solution, while in practice, as will be demonstrated in Section 3.2, this can be relaxed to large alphabets ($\geq$16-QAM for MIMO systems with two input streams). The main result is that there are a discrete number of optimal Gram matrices **G**, from which the optimal precoder is obtained by means of the construction in Theorem 3.1. This implies that for small variations of the MIMO channel, the Gram matrix of the optimal precoder does not change. Hence, the Gram matrices can be enumerated offline for different sets of MIMO channels, and used for online precoder construction. This can be efficiently accomplished in the case of two input data streams ($B = 2$), while for a larger number of streams, efficient enumeration of the Gram matrices as well as identifying which ones are optimal for different channel sets remains a topic for future research.

3.2 Precoding from a Lattice Point of View

This section will explore the connection between optimal minimum distance precoders and lattice theory. Hence, the minimum distance problem will be viewed from a lattice theoretic perspective, and this will enable us to explain the structures observed in [Collin et al. 2007] and [Ngo et al. 2009] for $B = 2$ data streams. For convenience, this section only focuses on the case $N = B = 2$, with the remark that the results also hold for all $N \times N$ MIMO channels with two input data streams, that is, $N \times B$ precoders with $B = 2$. This follows from the expression of the optimal precoder in Theorem 3.1, which shows that if $B < N$, the optimal precoder is an $N \times 2$ matrix with an $(N - 2) \times 2$ zero matrix in the last rows. Hence, effectively, this corresponds to employing a 2×2 precoder across the strongest eigenmodes of the MIMO channel.

3.2.1 Lattices

We now present a brief account on lattice theory (c.f. Section 2.2.1 in Chapter 2) needed for our purposes. All matrices and vectors in this section are assumed to be real-valued. This covers complex-valued matrices and vectors too, since any complex-valued matrix A is isomorphic to a real-valued matrix $\mathbf{A}_r$ through the transformation

$$\mathbf{A}_r = \begin{bmatrix} \mathrm{Re}\{\mathbf{A}\} & \mathrm{Im}\{\mathbf{A}\} \\ -\mathrm{Im}\{\mathbf{A}\} & \mathrm{Re}\{\mathbf{A}\} \end{bmatrix} \tag{3.27}$$

and similarly

$$\mathbf{s}_r = \begin{bmatrix} \mathrm{Re}\{\mathbf{s}\} \\ \mathrm{Im}\{\mathbf{s}\} \end{bmatrix} \tag{3.28}$$

for complex-valued vectors $\mathbf{s}$, where, $\mathrm{Re}\{\cdot\}$ and $\mathrm{Im}\{\cdot\}$ denote the real and imaginary parts of a matrix/vector, respectively.

Let $\mathbf{L} \in \mathbb{R}^{N \times N}$ and let the columns of $\mathbf{L}$ be denoted by $\mathbf{l}_1, \ldots, \mathbf{l}_N$. A lattice $\Lambda_\mathbf{L}$ is the set of points

$$\Lambda_\mathbf{L} = \{\mathbf{L}\mathbf{u} : \mathbf{u} \in \mathbb{Z}^N\}. \tag{3.29}$$

In Equation 3.29, $\mathbf{u}$ is an integer vector and $\mathbf{L}$ is called a *generator matrix* for the lattice $\Lambda_\mathbf{L}$. The *squared minimum distance* of $\Lambda_\mathbf{L}$ is defined as

$$D^2_{\min}(\mathbf{L}) = \min_{\mathbf{u} \neq \mathbf{v}} \| \mathbf{L}(\mathbf{u} - \mathbf{v}) \|^2 = \min_{\mathbf{e} \neq \mathbf{0}_N} \| \mathbf{L}\mathbf{e} \|^2 = \min_{\mathbf{e} \neq \mathbf{0}_N} \mathbf{e}^T \mathbf{G}_\mathbf{L} \mathbf{e},$$

where $\mathbf{u}$, $\mathbf{v}$ and $\mathbf{e} = \mathbf{u} - \mathbf{v}$ are integer vectors and $\mathbf{G}_\mathbf{L}$ is the Gram matrix for the lattice $\Lambda_\mathbf{L}$. The *fundamental volume* is $\mathrm{Vol}(\Lambda_\mathbf{L}) = |\det(\mathbf{L})|$, that is, it is the volume spanned by $\mathbf{l}_1, \ldots, \mathbf{l}_N$. Let $\mathbf{p}_j$ denote a lattice point in $\Lambda_\mathbf{L}$. A Voronoi region around a lattice point $\mathbf{p}_j$ is the set $\mathcal{V}_{\mathbf{p}_j}(\Lambda_\mathbf{L}) = \{\mathbf{w} : \| \mathbf{w} - \mathbf{p}_j \| \leq \| \mathbf{p}_k - \mathbf{w} \|, \mathbf{p}_k \in \Lambda_\mathbf{L}\}$. Due to the symmetry of a lattice, it holds that $\mathcal{V}_{\mathbf{p}_j}(\Lambda_\mathbf{L}) = \mathbf{p}_j + \mathcal{V}(\Lambda_\mathbf{L})$, where the Voronoi region around $\mathbf{0}_N$ is denoted by $\mathcal{V}(\Lambda_\mathbf{L})$.

As can be seen from the definition of $\Lambda_\mathbf{L}$, the column vectors $\mathbf{l}_1, \ldots, \mathbf{l}_N$ form a *basis* for the lattice. There are infinitely many bases for a lattice. Assume that $\mathbf{L}'$ is another basis for $\Lambda_\mathbf{L}$. It holds that $\mathbf{L}' = \mathbf{L}\mathbf{Z}$, where $\mathbf{Z}$ is a unimodular matrix, that is, $\mathbf{Z}$ has integer entries and $\det(\mathbf{Z}) = \pm 1$ [Conway and Sloane 1999]. Hence, the generator matrix $\mathbf{L}'$ generates the same lattice as $\mathbf{L}$, that is, $\Lambda_\mathbf{L} \equiv \Lambda_\mathbf{L}'$ where $\equiv$ denotes equality between sets. Two Gram matrices $\mathbf{G}_{\mathbf{L}_1} = \mathbf{L}_1^T \mathbf{L}_1$ and $\mathbf{G}_{\mathbf{L}_2} = \mathbf{L}_2^T \mathbf{L}_2$ are *isometric* if there exists a unimodular $\mathbf{Z}$ and a constant c such that $\mathbf{G}_{\mathbf{L}_1} = c\mathbf{Z}^T \mathbf{G}_{\mathbf{L}_2} \mathbf{Z}$. Geometrically, this means that $\mathbf{L}_1$ and $\mathbf{L}_2$ are the same lattice up to rotation and scaling of the basis

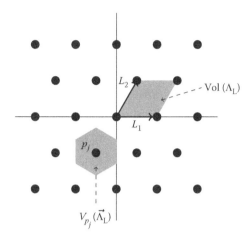

Figure 3.2 The hexagonal lattice depicted with a geometrical description of the introduced lattice quantities.

vectors. Figure 3.2 shows a geometrical visualization of a hexagonal lattice and the definitions above.

From the definition of the different lattice measures, it follows that

$$D_{\min}^2(\mathbf{WLZ}) = D_{\min}^2(\mathbf{L}), \tag{3.30}$$

where $\mathbf{W}$ is any orthogonal matrix. Similarly, $\mathrm{Vol}(\Lambda_{\mathbf{WLZ}}) = \mathrm{Vol}(\Lambda_{\mathbf{L}})$.

A number of lattices are especially interesting and have been given formal names in the literature. In particular, the densest lattices in the sense that they maximize the quotient $D_{\min}^2(\Lambda)/\mathrm{Vol}(\Lambda)$ are of interest. In 2, 4, 6, and 8 dimensions, the densest lattices are the hexagonal A_2, Schläfli D_4, E_6, and the Gosset E_8 lattices, respectively [Conway and Sloane 1999]. Apart from these 4, we will also make use of the two-dimensional square lattice Z_2.

3.2.2 Introduction to Lattice Precoding

To study the minimum distance problem from a lattice point of view, the QAM alphabet $\mathcal{A}$ has to be infinite, that is, $\mathbf{a} \in \mathcal{A}^B = \mathbb{Z}^B[i]$, the set of B-dimensional Gaussian integer vectors. We will use this assumption for the purpose of the analysis, while practically, as will be demonstrated by simulations, this corresponds to a large enough QAM alphabet. Hence, the error vectors $\mathbf{e}$ are B dimensional Gaussian integer vectors. From now on $\mathcal{A}$ will not be explicitly written out, since it is implicit that it is equal to $\mathbb{Z}[i]$. Thus, $D_{\min}^2(\mathbf{SF}, \mathcal{A})$ will be denoted as $D_{\min}^2(\mathbf{SF})$. Since there are infinitely many error vectors $\mathbf{e}$, the $B \times B$ Gram matrix $\mathbf{G}$ must have rank B in order for the inequalities $\mathbf{e}^H\mathbf{G}\mathbf{e} \geq 1$ to hold. If not, then the minimum distance $D_{\min}^2(\mathbf{SF}) = \mathbf{e}^H\mathbf{Ge}$ is arbitrarily close to 0, since $\mathbf{e}$ can be arbitrarily close to the eigenvectors that corresponds to zero eigenvalues. Thus, $\mathbf{G}$ is a positive-definite matrix, and $N \geq B$ must hold (the reader is reminded that the precoder $\mathbf{F}$ in Equation 3.22 has dimensions $N \times B$, and $\mathbf{G} = \mathbf{F}^H \mathbf{S}^2 \mathbf{F}$).

Using notions from Section 3.2.1, we start by reformulating (3.23) as a lattice problem. Let $\mathbf{M} = \mathbf{SF}$ be the lattice generator matrix at the receiver, which, as described above, must have full rank. $\mathbf{M}$ can be factorized as $\mathbf{M} = \mathbf{WBZ}$, where $\mathbf{W}$ is a unitary/orthogonal matrix, $\mathbf{B}$ is an $N \times N$ matrix and $\mathbf{Z}$ is a unimodular matrix. The lattice structure of $\mathbf{M}$ is determined by the matrix $\mathbf{B}$, while $\mathbf{Z}$ is the basis through which the lattice is represented. The matrix $\mathbf{W}$ is merely a rotation of the lattice, but plays an important role in the optimization to follow. With this factorization of $\mathbf{M}$, it follows that $\mathbf{F}$ can be written as

$$\mathbf{F} = \mathbf{S}^{-1}\mathbf{M} = \mathbf{S}^{-1}\mathbf{WBZ}. \tag{3.31}$$

Hence, Equation 3.23 can be formulated as

$$\min_{\mathbf{W},\mathbf{B},\mathbf{Z}} \mathrm{tr}(\mathbf{Z}^H\mathbf{B}^H\mathbf{W}^H\mathbf{S}^{-2}\mathbf{WBZ})$$
$$\text{subject to } D_{\min}^2(\mathbf{WBZ}) = 1. \tag{3.32}$$

For completeness, we shall separate between two cases: (i) Real-valued precoding, where all quantities in Equation 3.32 are real-valued, and (ii) complex-valued precoding, where all quantities, except $\mathbf{S}$, are complex valued.

From Theorem 3.1, it follows that the optimization over $\mathbf{W}$ is straightforward once $\mathbf{BZ}$ is fixed: the optimal $\mathbf{W}$ equals the left unitary matrix of $\mathbf{BZ}$. This leaves us with the optimization of $\mathbf{B}$ and $\mathbf{Z}$, and we shall start with $\mathbf{B}$ in Section 3.2.3.1, while optimization over $\mathbf{Z}$ is treated in Section 3.2.3.2.

3.2.3 Optimal Two-Dimensional Lattice Precoders

In [Bergman 2009] and [Forney Jr. and Wei 1989], it is proposed to design $\mathbf{F}$ based on dense lattice packings. A lattice-based construction implicitly assumes that the signal constellation is a finite but sufficiently "large" set of lattice points, and the idea is that if the received constellation points $\mathbf{SFa}$'s are arranged as a dense lattice packing, the minimum distance is expected to be "good." However, no exact results on optimality have been presented in either of these articles.

To gain some insight into the problem, let us examine some simple special cases. First, we rewrite Equation 3.32 in its equivalent form

$$\max_{\mathbf{F}} D_{\min}^2(\mathbf{SF})$$
$$\text{subject to } \mathrm{tr}(\mathbf{FF}^H) \leq P_0. \tag{3.33}$$

The formulation in Equation 3.33 turns out to be easier to analyze numerically. In real-valued precoding, some specific instances of the problem in Equation 3.33 can be viewed geometrically. Assume that $\text{tr}(\mathbf{F}\mathbf{F}^H) = 4$ and the elements of the input $\mathbf{a}$ are identically and independently distributed (i.i.d.) random variables. Normalize $\mathbf{S}$ to have $s_{2,2} = 1$, which only scales the optimal solution to Equation 3.33 with a constant, so that changing $\mathbf{S}$ corresponds to varying the value of $s_{1,1}$. Since there are only four real-valued elements in $\mathbf{F}$, and they are bounded by the energy constraint, it is possible to determine the optimal $\mathbf{F}$ to Equation 3.33 for some carefully chosen value of $s_{1,1}$, say, by empirical means. When $\mathbf{S} = \mathbf{I}$ (i.e., $s_{1,1} = 1$), one optimal solution to Equation 3.33 is $\mathbf{F} = \mathbf{I}$, while another one is

$$
\mathbf{F} = \begin{pmatrix} 1 & 0.5 \\ 0 & \sqrt{3/4} \end{pmatrix},
$$

which spans a hexagonal lattice. However, as soon as $\mathbf{S}$ deviates from $\mathbf{I}$ (even with a very small change, say, $s_{1,1} = 1.01$), the optimal $\mathbf{F}$ is unique (up to sign changes in the columns) and it gives rise to an $\mathbf{SF}$ that is a generator matrix for the hexagonal lattice. Varying $s_{1,1}$ further, the optimal $\mathbf{F}$ changes in a continuous way, while the received lattice $\mathbf{SF}$ remains the same (up to scaling). This behavior continues until $s_{1,1}$ reaches a certain value, for which the optimal $\mathbf{F}$ suddenly changes in a discontinuous way, resulting in a discontinuous change in $\mathbf{SF}$. However, surprisingly, $\mathbf{SF}$ still spans a hexagonal lattice, in spite of its subtle changes!

Figure 3.3 depicts such a behavior by plotting as vectors the columns of the optimal $\mathbf{F}$ and the corresponding $\mathbf{SF}$ for three different $\mathbf{S}$ with $s_{1,1} = 1.5$, 2.7 and 2.8, respectively. The received constellation points $\mathbf{SFa}$ are shown as discrete points. The optimal $\mathbf{F}$ changes continuously as $s_{1,1}$ increases from 1.5 to 2.7, and the columns of $\mathbf{SF}$ are simply being scaled and always span the same hexagonal lattice (up to scaling). When $s_{1,1}$ further increases from 2.7 to 2.8, there is a discontinuous change in the elements of the optimal $\mathbf{F}$. The columns of $\mathbf{SF}$ also change discontinuously, but they still span the hexagonal lattice (up to scaling and rotation). This intriguing behavior of the optimal precoder poses a challenging puzzle, and the aim of this section is to resolve this puzzle.

Although the following results are derived for infinite constellations, by using lattice theory, the results are applicable to "large" QAM constellations. In the numerical result section, we shall investigate how "large" a QAM constellation is sufficient for the presented results to be fruitfully applied. With the solution at hand, we are able to answer questions, such as the following:

▪ Is there a general underlying structure of the precoding optimization problem (3.32)?
▪ Under what conditions, does the solution to Equation 3.32 vary with the channel matrix $\mathbf{S}$ in a continuous (respectively, discrete) manner?

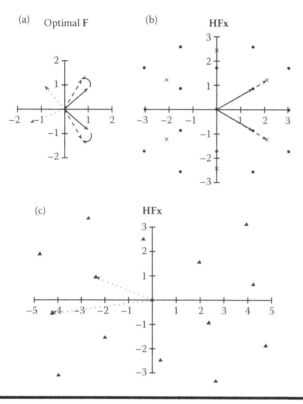

Figure 3.3 **Visualization of the solution to the precoding optimization problem in Equation 3.33 when *E* is *Z*, and *H* is a diagonal channel matrix S = diag([$s_{1,1}$ 1]). (a) Columns of the optimal real-valued precoding matrix *F* are depicted. Three different S are considered: $s_{1,1} = 1.5$ (solid line arrow); $s_{1,1} = 2.7$ (dashed line arrow); $s_{1,1} = 2.8$ (dotted line arrow). Columns of the same matrix are plotted as arrows with the same line style. (b) Columns of the matrices SF and their corresponding received constellation points SFa's for $h_{1,1} = 1.5$ (solid line arrows, filled circles) and for $s_{1,1} = 2.7$ (dashed line arrows, crosses) are plotted. (c) Columns of the matrices SF and their corresponding received constellation points SFa for $s_{1,1} = 2.8$ (dotted line arrows, filled triangles) are plotted.**

∎ Is it possible to offline construct a codebook of optimal precoders so that there is no need to perform any online optimization?

The answers to these questions are that there is indeed a profound structure in the solution of Equation 3.32. Remarkably, there is a single precoder structure which is optimal, and it organizes the received constellation points as a hexagonal lattice for real-valued **F**'s, and as a Schläfli lattice for complex-valued **F**'s. However, the basis through which the lattice **SF** is observed changes (up to scaling) in a discrete fashion when **S** changes. This implies that Equation 3.32 is actually a discrete optimization problem and not a continuous one.

3.2.3.1 Optimal Precoding Lattices

In this section, the optimal lattice **B** for the real-valued and the complex-valued cases is derived.

For the real-valued case, the main result is:

Theorem 3.2

*For any non-singular channel matrix **S**, the optimal lattice **B** in Equation 3.32 is the hexagonal lattice, that is,*

$$
\mathbf{B} = \begin{bmatrix} 1 & \dfrac{1}{2} \\ 0 & \dfrac{\sqrt{3}}{2} \end{bmatrix}.
$$

Proof. First, the constraint in Equation 3.32 will be made more manageable. It follows from Equation 3.30 that $D_{\min}^2(\mathbf{WBZ}) = D_{\min}^2(\mathbf{B})$. Let $\mathbf{b}_1, \mathbf{b}_2 \in \mathbb{R}^2$ be the columns of **B** and assume that $\|\mathbf{b}_1\| \le \|\mathbf{b}_2\|$. In 1801, C.F. Gauss noted [Gauss 1981] that if $\mathbf{b}_1$ and $\mathbf{b}_2$ fulfill, $|\mathbf{b}_2 \cdot \mathbf{b}_1| \le \|\mathbf{b}_1\|^2/2$, where "·" is the scalar product between vectors, then $D_{\min}^2(\mathbf{B}) = \|\mathbf{b}_1\|^2$. Given $\mathbf{b}_1$, the set of all $\mathbf{b}_2$ satisfying the inequality is the *minimum distance region* of $\mathbf{b}_1$. Figure 3.4 depicts this region geometrically.

$\mathbf{b}_1$ and $\mathbf{b}_2$ are actually the *shortest basis* for the lattice, since $\|\mathbf{b}_1\|$ is the length of the shortest vector in the lattice, and it can be shown that $\|\mathbf{b}_2\|$ is the length of the next shortest vector in the lattice. Hence, by putting $\|\mathbf{b}_1\| = 1$ and letting $\mathbf{b}_2$ be

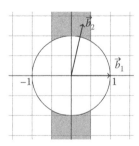

Figure 3.4 The minimum distance region of b_1 is shaded. All b_2 inside the shaded region generate a lattice, spanned by b_1 and b_2, with a minimum distance equal to the length of b_1.

any vector in the minimum distance region of $\mathbf{b}_1$, the matrix $\mathbf{B}$ will be a generator matrix for any lattice in the plane with unit minimum distance.

Let $r = \|\mathbf{b}_2\|$. The constraint $D_{\min}^2(\mathbf{WBZ}) = 1$ can be written as $r \geq 1$ and $|\cos(\phi)| \leq 1/2r$ where ϕ is the angle between $\mathbf{b}_1$ and $\mathbf{b}_2$. Hence, $\mathbf{WB}$ can be written as

$$\mathbf{WB} = \begin{pmatrix} \sin(\alpha) & r\sin(\alpha \pm \phi) \\ \cos(\alpha) & r\cos(\alpha \pm \phi) \end{pmatrix}. \tag{3.34}$$

Optimization (3.32) can now be formulated over α, ϕ and r:

$$\min_{\alpha,\phi,r} \mathrm{tr}(\mathbf{Z}^H \mathbf{B}^H \mathbf{W}^H \mathbf{S}^{-2} \mathbf{WBZ}) \quad \text{subject to} \quad r \geq 1, |\cos(\phi)| \leq 1/2r. \tag{3.35}$$

It follows that the intervals for α and ϕ are $0 \leq \alpha \leq 2\pi, |\phi| \leq \cos^{-1}(1/2r)$.

Let $s_{1,1}, s_{2,2}$ be the diagonal elements of $\mathbf{S}$ and assume $s_{1,1} \geq s_{2,2}$. For notational convenience, we let z_{jj} be the elements in $\mathbf{Z}$. Define $s \stackrel{\Delta}{=} s_{2,2}/s_{1,1}$ and

$$a \stackrel{\Delta}{=} \frac{z_{11}^2 + z_{12}^2}{\|\mathbf{Z}\|^2} \quad b \stackrel{\Delta}{=} \frac{z_{11}z_{21} + z_{12}z_{22}}{\|\mathbf{Z}\|^2} \quad c \stackrel{\Delta}{=} \frac{1 + s^2}{2}. \tag{3.36}$$

To obtain easier expressions, we scale the objective function (3.35) with $1/s_{2,2} \|\mathbf{Z}\|^2$ which has no impact on the solution, and by doing so we get the following objective function:

$$\begin{aligned} f(\alpha,\phi,r) &\stackrel{\Delta}{=} \mathrm{tr}(\mathbf{Z}^H \mathbf{B}^H \mathbf{W}^H \mathbf{S}^{-2} \mathbf{WBZ})/s_{2,2} \|\mathbf{Z}\|^2 \\ &= c(a + r^2(1 - a) + 2br\cos(\phi)) + (1 - c)(a\cos(2\alpha) \\ &\quad + (1 - a)r^2\cos(2\alpha + 2\phi) + 2br\cos(2\alpha + \phi)). \end{aligned} \tag{3.37}$$

Since $0 \leq s \leq 1$, it follows that $1/2 \leq c \leq 1$.

First, we minimize $f(\alpha,\phi,r)$ over α by making use of the following Lemma.

Lemma 3.1

Let $g(x) = \sum_{j=1}^n a_j \cos(x + \theta_j)$ for some real-valued constants $\{a_j\}$ and $\{\theta_j\}$. It holds that

$$\min_x g(x) = -\sqrt{\sum_{j=1,k=1}^n a_j a_k \cos(\theta_j - \theta_k)}. \tag{3.38}$$

Proof. Rewrite $g(x)$ as $g(x) = \text{Re}\{\sum_{j=1}^{n} a_j e^{i(x+\theta_j)}\} = \text{Re}\{e^{ix}(\sum_{j=1}^{n} a_j e^{i\theta_j})\} = \mathcal{R}\{e^{ix}z\}$, where $z \triangleq \sum_j a_j e^{i\theta_j}$. The minimum occurs when z is rotated to the negative part of the real axis, that is, $x = \pi - \beta$, and the minimum value is then equal to $-|z|$. This gives expression (3.38).

Applying Lemma 3.1 to Equation 3.37 in order to minimize over α, we get

$$h(\phi, r) \triangleq \min_{\alpha} f(\alpha, \phi, r) = c(a + r^2(1 - a) + 2rb\cos(\phi))$$
$$+ (c - 1)[a^2 + r^4(1 - a)^2 + 4r^2b^2 + 2r^2a(1 - a)\cos(2\phi)$$
$$+ 4rb(a + r^2(1 - a))\cos(\phi)]^{1/2}.$$

Using the identity $\cos(2\phi) = 2\cos^2(\phi) - 1$ and defining $t \triangleq \cos(\phi)$, we get

$$q(t, r) \triangleq h(\cos^{-1}(t), r) = c(a + r^2(1 - a) + 2rbt) + (c - 1)[a^2 + r^4(1 - a)^2$$
$$+ 4r^2b^2 - 2r^2a(1 - a) + 4r^2a(1 - a)t^2 + 4rb(a + r^2(1 - a))t]^{1/2}.$$

$$(3.39)$$

From the definition of t, it follows that $-1/2r \leq t \leq 1/2r$. It can be verified that $q(t, r)$ is a concave function in t. This implies that the minimum of $h(t, r)$ over t is attained at one of the two endpoints $t = \pm 1/2r$. For these values, and with the variable substitution $\rho = r^2$, we get

$$l_\pm(\rho) \triangleq q(\pm 1/2r, r)$$
$$= c(a + \rho(1 - a) \pm b) + (c - 1)[a^2 + \rho^2(1 - a)^2 + 4b^2\rho$$
$$- 2\rho a(1 - a) + a(1 - a) \pm 2b(a + \rho(1 - a))]^{1/2}, \qquad (3.40)$$

where $\rho \geq 1$. $l_+(\rho)$ has "$+$" instead of $\pm$ and $l_-(\rho)$ has "$-$." The functions $l_\pm(\rho)$ are both concave in ρ. Now, since $l_\pm(\rho)$ is the objective function of Equation 3.35, it follows that it must always be positive. Therefore, the minimizer must be $\rho = 1$, which gives that $r = 1$ in Equation 3.39. This implies that the minimum over t in Equation 3.39 occurs at $t = \pm 1/2$, which corresponds to $\phi \in \{\pm \pi/3, \pm 2\pi/3\}$ in Equation 3.37. This shows that the minimum of $f(\alpha, \phi, r)$ in Equation 3.37 occurs at $r = 1$ and $\phi \in \{\pm \pi/3, \pm 2\pi/3\}$. Inserting these values in the generator matrix **B**, one obtains the generator matrix for the hexagonal lattice as stated in the Theorem. This completes the proof.

While the real-valued case is interesting for theoretical purposes, the complex-valued case is more important for practical MIMO or OFDM applications. Nevertheless, the real-valued result has immediate applications to precoding for mitigation of I/Q imbalance in scalar complex-valued channels.

For the complex-valued case, the main result is:

Theorem 3.3

For any nonsingular channel matrix **S**, *the optimal lattice* **B** *in Equation 3.32 is the complex representation of the Schläfli lattice, that is,*

$$
\mathbf{B} = \begin{bmatrix} 1 & \dfrac{1}{\pm 1 \pm i} \\[2mm] & \dfrac{1}{2} \\[2mm] 0 & \pm\dfrac{1}{\sqrt{2}} \end{bmatrix}.
$$

Proof. It turns out that there is a similar minimum distance preserving condition for complex-valued **B** as for real-valued ones. In [Yao and Wornell 2002], the authors prove that if $\|\mathbf{b}_1\| \le \|\mathbf{b}_2\|$ and

$$
|\mathrm{Re}\{\mathbf{b}_1^H \mathbf{b}_2\}| \le \frac{1}{2} \quad \text{and} \quad |\mathcal{I}\{\mathbf{b}_1^H \mathbf{b}_2\}| \le \frac{1}{2}, \tag{3.41}
$$

then $D_{\min}^2(\mathbf{B}) = \|\mathbf{b}_1\|^2$. The matrix **W** is now

$$
\mathbf{W} = \begin{pmatrix} e^{i(\phi_1 - \gamma_1)} & 0 \\ 0 & e^{i(\phi_3 - \gamma_1)} \end{pmatrix} \begin{pmatrix} \sin(\alpha)e^{-i\phi_1} & \cos(\alpha)e^{-i\phi_2} \\ \cos(\alpha)e^{-i\phi_3} & -\sin(\alpha)e^{-i\phi_4} \end{pmatrix} \tag{3.42}
$$

and **B** is

$$
\mathbf{B} = \begin{pmatrix} re^{i\gamma_1} & \sin(\omega)e^{i\gamma_2} \\ 0 & \cos(\omega)e^{i\gamma_3} \end{pmatrix}. \tag{3.43}
$$

Hence, **WB** becomes

$$
\mathbf{WB} = \begin{pmatrix} r\sin(\alpha) & \sin(\alpha)\sin(\omega)e^{i\theta_1} + \cos(\alpha)\cos(\omega)e^{i\theta_2} \\ r\cos(\alpha) & \cos(\alpha)\sin(\omega)e^{i\theta_1} - \sin(\alpha)\cos(\omega)e^{i\theta_2} \end{pmatrix}, \tag{3.44}
$$

where $\phi_1 - \phi_2 \equiv \phi_3 - \phi_4 \pmod{2\pi}$, $\theta_1 = \gamma_2 - \gamma_1$ and $\theta_2 = \gamma_3 - \gamma_1 + \phi_1 - \phi_2$. Conditions (3.41) become

$$
|\mathrm{Re}\{\sin(\omega)e^{-i\theta_1}\}| \le \frac{1}{2r} \quad \text{and} \quad |\mathcal{I}\{\sin(\omega)e^{-i\theta_1}\}| \le \frac{1}{2r}, \tag{3.45}
$$

where $r \geq 1$. Define $f(\alpha,\omega,\theta_1,\theta_2,r) \triangleq \text{tr}(\mathbf{Z}^H \mathbf{B}^H \mathbf{W}^H \mathbf{S}^{-2} \mathbf{W} \mathbf{B} \mathbf{Z})/s_{2,2}$. We have

$$
\begin{aligned}
f(\alpha,\omega,\theta_1,\theta_2,r) = & \ c[r^2(|z_{11}|^2 + |z_{12}|^2) + |z_{21}|^2 + |z_{22}|^2 \\
& + 2\mathcal{R}\{(rz_{11}z_{21}^H + rz_{12}z_{22}^H)\sin(\omega)e^{-i\theta_1}\}] + (1-c)[r^2(|z_{11}|^2 + |z_{12}|^2) \\
& - (|z_{21}|^2 + |z_{22}|^2)\cos(2\omega) + 2\text{Re}\{(rz_{11}z_{21}^H \\
& + rz_{12}z_{22}^H)\sin(\omega)e^{-i\theta_1}\}]\cos(2\alpha) \\
& - (1-c)[(|z_{21}|^2 + |z_{22}|^2)\sin(2\omega)\cos(\theta_1 - \theta_2) \\
& + 2\mathcal{R}((rz_{11}z_{21}^H + rz_{12}z_{22}^H)\cos(\omega)e^{-i\theta_2})]\sin(2\alpha),
\end{aligned}
\tag{3.46}
$$

where $c = (1 + (s_{2,2}/s_{1,1})^2)/2$. First, we minimize over α. It is seen that f depends on α as

$$
\begin{aligned}
f(\alpha,\omega,\theta_1,\theta_2,r) &= a_1 + a_2\cos(2\alpha) + a_3\sin(2\alpha) \\
&= a_1 + \sqrt{a_2^2 + a_3^2}\left(\frac{a_2}{\sqrt{a_2^2 + a_3^2}}\cos(2\alpha) + \frac{a_3}{\sqrt{a_2^2 + a_3^2}}\sin(2\alpha)\right) \\
&= a_1 + \sqrt{a_2^2 + a_3^2}(\sin(\psi)\cos(2\alpha) + \cos(\psi)\sin(2\alpha)) \\
&= a_1 + \sqrt{a_2^2 + a_3^2}\sin(2\alpha + \psi),
\end{aligned}
\tag{3.47}
$$

where the constants a_1, a_2, and a_3 are easily read from Equation 3.46 and ψ is such that $\sin(\psi) = a_2/\sqrt{a_2^2 + a_3^2}$. The minimum of Equation 3.47 over α occurs at $\alpha = -\pi/4 - \psi/2$, which gives $f(-\pi/4 - \psi/2,\omega,\theta_1,\theta_2,r) = a_1 - \sqrt{a_2^2 + a_3^2}$. Since only a_3 depends on θ_2, minimizing f over θ_2 implies maximizing a_3^2 over θ_2. We have

$$
\begin{aligned}
a_3 &= -(1-c)[(|z_{21}|^2 + |z_{22}|^2)\sin(2\omega)\cos(\theta_1 - \theta_2) \\
& \quad + 2\text{Re}((rz_{11}z_{21}^H + rz_{12}z_{22}^H)\cos(\omega)e^{-i\theta_2})] \\
&= -(1-c)\mathcal{R}\{e^{-i\theta_2}((|z_{21}|^2 + |z_{22}|^2)\sin(2\omega)e^{i\theta_1} + 2\cos(\omega)(rz_{11}z_{21}^H + rz_{12}z_{22}^H))\}.
\end{aligned}
$$

It follows that the maximizing θ_2 is such that $e^{i\theta_2}$ rotates the expression it multiplies to the real axis. We get,

$$
\begin{aligned}
\min_{\theta_2} f(-\pi/4 - \psi/2,\theta_1,\theta_2,\omega,r) &= l(\theta_1,\omega,r) \\
&= c[r^2(|z_{11}|^2 + |z_{12}|^2) + |z_{21}|^2 + |z_{22}|^2 + 2\text{Re}\{\sin(\omega)e^{-i\theta_1}(rz_{11}z_{21}^H + rz_{12}z_{22}^H)\}] \\
&\quad + (c-1)[(r^2(|z_{11}|^2 + |z_{12}|^2) + |z_{21}|^2 + |z_{22}|^2 \\
&\quad + 2\text{Re}\{\sin(\omega)e^{-i\theta_1}(rz_{11}z_{21}^H + rz_{12}z_{22}^H)\})^2 - 4\cos^2(\omega)|\det(\mathbf{Z})|^2]^{1/2}.
\end{aligned}
\tag{3.48}
$$

As in the real-valued case, it can easily be shown that the expression in Equation 3.48 is concave in $\sin(\omega)$. Thus, the minimum is attained at the endpoints of $\sin(\omega)$. The constraints in Equation 3.49 can be written as $|\sin(\omega)\cos(\theta_1)| \leq 1/2r$ and $|\sin(\omega)\sin(\theta_1)| \leq 1/2r$. Assume $|\sin(\theta_1)| \leq |\cos(\theta_1)|$. It follows that the interval for $\sin(\omega)$ is $-1/(2r\cos(\theta_1)) \leq \sin(\omega) \leq 1/(2r\cos(\theta_1))$, while the interval for θ_1 is $-\pi/4 \leq \theta_1 \leq \pi/4$. Inserting either one of these endpoints for $\sin(\omega)$ in Equation 3.48 and using the trigonometric identity $1/\cos^2(x) = 1 + \tan^2(x)$, we get that l takes on the following form:

$$l(\theta_1, r) = c(b_1 + b_2 \tan(\theta_1))$$
$$+ (c-1)\left[(b_1 + b_2 \tan(\theta_1))^2 + \frac{|\det(\mathbf{Z})|^2}{r^2}\tan^2(\theta_1) \right.$$
$$\left. + |\det(\mathbf{Z})|^2 \ (4 - 1/r^2) \right], \qquad (3.49)$$

where b_1 and b_2 are constants with respect to θ_1. Again, it is clear that Equation 3.49 is concave in $\tan(\theta_1)$, and thus the minimum is attained at one of the endpoints of θ_1, which are $-\pi/4$ and $\pi/4$. If we instead assumed that $|\sin(\theta_1)| \geq |\cos(\theta_1)|$, the only difference is that $\tan(\theta_1)$ becomes $\cos(\theta_1)$ and $\pi/4 \leq \theta_1 \leq 3\pi/4$. This gives rise to the same behavior of $l(\theta_1, r)$ and thus the same results are obtained.

To recap, we showed that the minimum for $l(\theta_1, \omega, r)$ in Equation 3.48 over θ_1, ω occurs when $\theta_1 = \pm \pi/4$ and at the endpoints for $\sin(\omega)$, which are then $\sin(\omega) = \pm 1/(2r\cos(\theta_1)) = \pm 1/\sqrt{2}r$. We now continue by inserting this expression for $\sin(\omega)e^{-i\theta_1}$ in Equation 3.48 and obtain a one-dimensional function in $\rho = r^2$ of the form

$$l_1(\rho) = k_1 + k_2\rho + (c-1)\sqrt{k_3\rho^2 + k_4\rho + k_5 + |\det(\mathbf{Z})|^2 \ (2/\rho - 4)}, \quad (3.50)$$

where the k_j are constants with regard to ρ and with k_3 positive. If we instead study the function $l_2(\rho) = k_1 + k_2\rho + (c-1)\sqrt{k_3\rho^2 + k_4\rho + k_5 - 2|\det(\mathbf{Z})|^2}$, it follows from the same concavity arguments as before that $l_2(\rho)$ is a concave function and thus the minimum is attained at the endpoints, which are $\rho = 1$ and $\rho = \infty$. From the concavity of $l_2(\rho)$ it follows that if the minimum is attained at ∞, then the minimum value is $-\infty$, which is impossible since the trace function is always positive; thus the minimum of $l_2(\rho)$ must be attained at $\rho = 1$. Now comparing $l_2(\rho)$ with $l(\rho)$, the only difference is the term $|\det(\mathbf{Z})|^2 \ (2/\rho - 4)$ in the square root, with maximum value of $2|\det(\mathbf{Z})|^2$ attained at $\rho = 1$; hence $l_2(1) = l_1(1)$. Since $c - 1$ is always nonpositive, it follows that $l_2(\rho) \leq l_1(\rho)$ for $\rho \geq 1$, which gives that the minimum of $l_1(\rho)$ occurs when $\rho = r = 1$ (because the minimum of $l_2(\rho)$ occurs for $\rho = 1$).

We have now showed that the minimum of $l(\theta_1, \omega, r)$ in Equation 3.48 occurs for $\theta_1 = \pm\pi/4$, $\sin(\omega) = \pm1/\sqrt{2}r$, $r = 1$. Inserting these values into the lattice generator $\mathbf{B}$ in Equation 3.43, we arrive at the following optimal lattice generator:

$$
\mathbf{B} = \begin{pmatrix} 1 & \dfrac{\pm1 \pm i}{2} \\ 0 & \pm\dfrac{1}{\sqrt{2}} \end{pmatrix}.
\tag{3.51}
$$

Extending $\mathbf{B}$ to its real-valued representation by means of Equation 3.27, it holds that for each realization of $\pm$ as $+$ or $-$, that $\mathbf{B}_r$ is a generator matrix for the Schläfli lattice D4.

Hence, by "real-valued representation," it is meant that if the transformation Equation 3.27 is performed on $\mathbf{B}$ in Theorem 3.3, the Schläfli lattice in four real-valued dimensions results. Its real-valued generator matrix is

$$
D_4 = \begin{pmatrix} 1 & 1 & 0 & 0 \\ 1 & -1 & 0 & 0 \\ 0 & 1 & -1 & 0 \\ 0 & 0 & 1 & -1 \end{pmatrix}.
\tag{3.52}
$$

To summarize, the two-dimensional minimum distance optimal precoder for "large" input constellations is always an instance of the hexagonal or the Schläfli lattice for real-valued and complex-valued precoding, respectively.

3.2.3.2 Optimal **Z** Matrix

Since $\mathbf{B}$ is now known, it remains to find the optimal basis matrix $\mathbf{Z}$ in order to solve Equation 3.32. This section describes the core idea of the algorithms that find the optimal real-valued and complex-valued $\mathbf{Z}$, respectively. A complete MATLAB code for the algorithms can be found at *www.eit.lth.se/goto/Zalgorithm*.

By inserting the optimal real-valued $\mathbf{B}$ and $\mathbf{W}$ into Equation 3.32, the optimization Equation 3.32 is equivalent to[*]

$$
\mathbf{Z} = \arg\min_{\mathbf{Z}}\mu_\pm^r(\mathbf{Z}),
$$

---

[*] The optimization over $\mathbf{W}$ is treated in the proofs of Theorems 2 and 3.

where

$$\mu_\pm^r(\mathbf{Z}) \triangleq s_{2,2}(z_{11}^2 + z_{12}^2 + z_{21}^2 + z_{22}^2)l_\pm(1)$$
$$= c[z_{11}^2 + z_{12}^2 + z_{21}^2 + z_{22}^2 \pm (z_{11}z_{21} + z_{12}z_{22})]$$
$$+ (c-1)[(z_{11}^2 + z_{12}^2)^2 + (z_{21}^2 + z_{22}^2)^2 + 4(z_{11}z_{21} + z_{12}z_{22})^2$$
$$- (z_{11}^2 + z_{12}^2)(z_{21}^2 + z_{22}^2) \pm 2(z_{11}z_{21} + z_{12}z_{22})(z_{11}^2 + z_{12}^2 + z_{21}^2 + z_{22}^2)]^{1/2}.$$

$$(3.53)$$

In the complex-valued case, we have the following optimization:

$$\mathbf{Z} = \arg\min_{\mathbf{Z}}\mu_\pm^c(\mathbf{Z}),$$

where

$$\mu_\pm^c(\mathbf{Z}) \triangleq c(\| \mathbf{Z} \|^2 + \mathcal{R}\{(\pm 1 \pm i)(z_{11}z_{21}^H + z_{12}z_{22}^H)\})$$
$$+ (c-1)\sqrt{(\| \mathbf{Z} \|^2 + \mathcal{R}\{(\pm 1 \pm i)(z_{11}z_{21}^H + z_{12}z_{22}^H)\})^2 - 2}. \quad (3.54)$$

The $\pm$ signs in both Equations 3.53 and 3.54 can be absorbed into the elements of $\mathbf{Z}$, without changing the unimodularity of $\mathbf{Z}$. Define $\beta_r \triangleq z_{11}^2 + z_{12}^2 + z_{21}^2 + z_{22}^2 - (z_{11}z_{21} + z_{12}z_{22})$ and $\beta_c \triangleq |z_{11}|^2 + |z_{12}|^2 + |z_{21}|^2 + |z_{22}|^2 + \mathrm{Re}\{(1 + i)(z_{11}z_{21}^H + z_{12}z_{22}^H)\}$, where we do not explicitly denote the dependency of β_r and β_c on $\mathbf{Z}$. Since $|\det(\mathbf{Z})| = 1$, Equations 3.53 and 3.54 become

$$\mu^r(\beta_r) = c\beta_r + (c-1)\sqrt{\beta_r^2 - 3} \quad (3.55)$$

and

$$\mu^c(\beta_c) = c\beta_c + (c-1)\sqrt{\beta_c^2 - 2}, \quad (3.56)$$

respectively. The difference between Equations 3.55 and 3.53 (similarly between Equations 3.56 and 3.54) is that the former only depends on one variable, that implicitly depends on the elements $\{z_{ij}\}$, while the latter is directly expressed in the elements $\{z_{ij}\}$. Deriving the optimal β_r and β_c does not produce the optimal elements $\{z_{ij}\}$, however, it can provide easier optimality conditions for $\{z_{ij}\}$. If we for the moment drop the constraint that β_r has to be integer-valued, the function μ^r ($\mathbf{Z}$) in Equation 3.55 will be minimized over β_r. It can be verified that $\mu^r(\mathbf{Z})$ is a convex function. Differentiating $\mu(\beta_r)$ with respect to β and setting the derivative to 0 gives that $\beta_{r,\mathrm{opt}} = \sqrt{(3c^2/2c - 1)}$ is the optimal point. Since $\mu^r(\mathbf{Z})$ is convex, the minimum of $\mu(\beta_r)$ over unimodular matrices can only occur at two specific matrices. Either it is the $\mathbf{Z}$ that produces the largest β_r smaller than $\beta_{r,\mathrm{opt}}$, or it is the $\mathbf{Z}$ that produces the smallest β_r larger than $\beta_{r,\mathrm{opt}}$. A similar analysis can be applied to the complex-valued Equation 3.56, and it follows that the largest β_c smaller, or smallest β_c larger, than $\beta_{c,opt} = \sqrt{2c}/\sqrt{2c-1}$ is optimal. Hence, in the real-valued case, an algorithm can be developed that traverses unimodular $\mathbf{Z}$'s and stops when two matrices $\mathbf{Z}_1$ and $\mathbf{Z}_2$ are found, such that $\mathbf{Z}_1$ gives the β_r that equals the largest integer smaller than $\beta_{r,opt}$, and $\mathbf{Z}_2$ gives the β_r that equals the smallest integer larger than $\beta_{r,opt}$. An algorithm for the complex-valued case works in the same way. For the purpose of clarity of this chapter and the fact that the algorithms are ad-hoc, we omit the implementation details and refer to *www.eit.lth.se/goto/Zalgorithm,* where the MATLAB code for both algorithms can be found.

Since we now know that solving Equation 3.32 is a discrete optimization problem, it is of interest to see how often the solution changes with varying $\mathbf{S}$. Figure 3.5 shows the ratio $s_{1,1}/s_{2,2}$ on the x-axis, and the markers show the ratios where $\mathbf{Z}$ changes. As seen, the same solution can be used for a wide interval.

3.2.3.3 Applications

In this section, we consider a number of practical applications of the optimal minimum distance lattice-based precoder and make comparisons to other schemes. As discussed in Section 3.1.3.2, minimum distance-based precoders are asymptotically optimal in the high SNR regime, but minimum distance plays little role at low SNR, so significant performance gains cannot be expected there.

Consider first the 2×2 channel studied in [Perez-Cruz et al. 2010],

$$\mathbf{S} = \begin{bmatrix} \sqrt{3} & 0 \\ 0 & 1 \end{bmatrix}. \tag{3.58}$$

In [Perez-Cruz et al. 2010], this channel was studied at asymptotically high SNR for binary baseband alphabets with real-valued precoding. The objective was

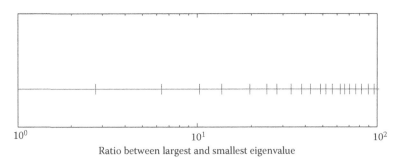

Ratio between largest and smallest eigenvalue

Figure 3.5 Change in Z with respect to the ratio $s_{1,1}/s_{2,2}$. The solution to Equation 3.32 is constant for all S with a ratio between any two consecutive markers. The scale on the x-axis is logarithmic.

to find the real-valued precoder **F** that maximizes the mutual information. For high SNR, it is known that the optimal mutual information precoder converges to the optimal minimum distance precoder, and the numerical optimization framework in [Perez-Cruz et al. 2010] thus produced the optimal minimum distance precoder. The precoder is of the following simple form:

$$\mathbf{F} = \begin{bmatrix} \sqrt{2} & \sqrt{2} \\ -\sqrt{2} & \sqrt{2} \end{bmatrix}. \tag{3.59}$$

It can be verified by standard techniques that the combined channel–precoder matrix **SF** is an instance of the hexagonal lattice—which is precisely the result if an infinite lattice constellation was used. For such a lattice constellation, the strength of the results in Theorems 3.2 and 3.3 is that no numerical optimization of the precoder is necessary since it is known *a priori* that the hexagonal lattice *must* be the solution, and it only remains to find the optimal basis matrix **Z** according to the algorithm mentioned in Section 3.2.3.2. By doing so, we find that the optimal **Z** for asymptotically large constellations coincides with the basis matrix that is built into Equation 3.59. Altogether, for the particular channel (3.58) studied in [Perez-Cruz et al. 2010], a "large" constellation means binary and it is known beforehand what structure the solution must have.

In Figure 3.6, we continue to study the channel (3.58), but now by evaluating its mutual information that is achieved by 4-QAM inputs when the complex-valued minimum distance optimal precoder for large constellations is used. As comparisons, plots of the achieved mutual information for (1) no precoding at all, that is, **F** = **I**, (2) mercury waterfilling from [Lozano et al. 2006], and (3) capacity achieved by Gaussian inputs and waterfilling are presented. The performance of the optimal mutual information precoder coincides with that of mercury waterfilling in the low SNR regime, while it coincides with that of the minimum distance precoder in the high SNR regime. As can be seen, there is a 2 dB gain offered by the minimum distance precoder over uncoded systems and mercury waterfilling at high SNR. At low SNR, the mercury waterfilling policy is optimal and outperforms the minimum distance precoder.

For the channel in Equation 3.58, we observed that the large constellation assumption made in this section was not very critical as it produced the same result as a binary input constellation does. This is, however, not true in general, and it is necessary to investigate the impact of the cardinality of the input constellation. Consider diagonal channel matrices **H** where each diagonal element is a zero-mean, unit-variance, circulary symmetric complex Gaussian random variable ($\mathcal{CN}(0,1)$). The average mutual information, against SNR, is computed for 4-QAM and 16-QAM input constellations for (1) the minimum distance optimal precoder for large constellations, (2) minimum distance optimal precoders for the particular constellations used, and (3) no precoder. The average is evaluated over 10^6 channel realizations by straightforward Monte Carlo simulation. For 4-QAM

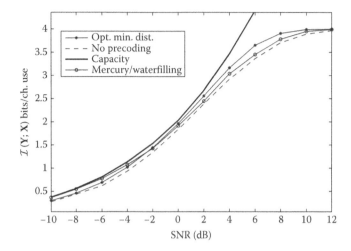

Figure 3.6 Mutual information for the channel (3.58) studied in Perez-Cruz et al. (2010) with 4-QAM inputs under different settings. The solid heavy line shows the capacity with waterfilling, the curve marked with asterixes shows the ensuing mutual information from the lattice precoder in this section and the curve marked with circles shows the mercury waterfilling mutual information. The bottom line is the no precoding case.

and 16-QAM, the minimum distance optimal precoders have been reported in [Collin et al. 2007] and [Ngo et al. 2009], while the optimal precoder for 64-QAM has so far not been reported in the literature which is the reason why we do not go beyond 16-QAM. The results are shown in Figure 3.7.

The uppermost heavy solid line corresponds to the average capacity of the channel achieved by Gaussian inputs with waterfilling. The lower set of curves corresponds to 4-QAM, while the upper corresponds to 16-QAM. Within each set of curves, the lower curve (without markers) shows the no precoder case, the middle curve (marked with asterixes) is the performance of the precoder constructed from a large constellation assumptions, and the upper curve (marked with circles) is the performance of the precoder explicitly constructed for the input constellation used. For 4-QAM inputs, a small loss of the large constellation construction can be seen, while for 16-QAM the ensuing mutual information from a large constellation assumption is virtually indistinguishable from that of a construction explicitly made for 16-QAM. Hence, it can be concluded from this example that a 16-QAM input constellation can be replaced by an infinite lattice constellation without appreciably affecting the results. This greatly simplifies the precoder optimization problem since lattice theoretic tools can be applied.

In Figure 3.8, we turn our attention toward the error probability of 2 × 2 MIMO systems with (1) the minimum distance optimal precoder for large constellations, (2) minimum distance optimal precoders for the particular constellations used, and (3)

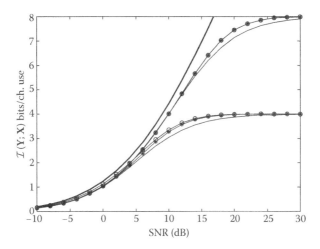

Figure 3.7 Average mutual information for random diagonal channels with 4-QAM (bottom set) and 16-QAM (upper set). The heavy solid line is the capacity with waterfilling. Within each set, the line marked with circles shows the performance of a precoder constructed explicitly for the input constellation used, and the curve marked with asterixes shows the performance of the precoder constructed from an infinite lattice constellation assumption. These two curves are virtually identical for 16-QAM. The bottom line within each set corresponds to the no precoding case.

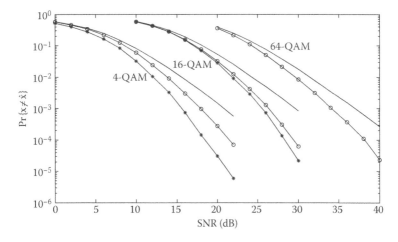

Figure 3.8 ML receiver tests of various precoders with 4-QAM, 16-QAM, and 64-QAM. Within each set, the rightmost curve is the no precoding case, the middle curve is the precoder constructed from an infinite lattice constellation assumption, and the leftmost curve is the performance of a precoder constructed explicitly for the input constellation used (not present for 64-QAM).

no precoding. 4-QAM, 16-QAM, and 64-QAM input constellations, together with a maximum likelihood detector, are considered. The lines marked with circles correspond to the minimum distance optimal precoder for large constellations, the lines marked with squares correpond to the optimal precoder designed for the particular input constellations used, and the unmarked lines correspond to the no-precoder case. As can be seen, there is a large gain from explicitly taking the input constellation into account for 4-QAM. However, for 16-QAM inputs, this gain reduces significantly, so that the precoder designed for large constellations performs close to optimal. For 64-QAM, the gap to the optimal precoder designed explicitly for 64-QAM cannot be determined. However, given the large reduction of the gap between the 4-QAM and 16-QAM cases, it is expected that the gap for 64-QAM is minor, so that the precoder designed for large constellations is virtually optimal.

As a final example, an OFDM system with N subcarriers $\{H_k\}_{k=1}^N$ is investigated. For simplicity, all subcarriers are assumed to be independent zero-mean, unit-variance, circulary symmetric complex Gaussian random variables ($\mathcal{CN}(0,1)$). In practice, adjacent carriers are strongly correlated but for the transceiver system to be considered, N is large and such correlations are immaterial. The approach taken in [Vrigneau et al. 2008] is pursued, but now with the 2×2 minimum distance optimal precoder constructed from the large constellation assumption. This precoder is then used as a building block to construct precoders of larger dimension. The N subcarriers are first grouped into $N/2$ pairs. The particular pairing used in [Vrigneau et al. 2008] is to combine the strongest subcarrier with the weakest subcarrier, the second strongest with the second weakest, and so on. Let $\{\tilde{H}_k\}_{k=1}^N$ denote the subcarriers $\{H_k\}_{k=1}^N$, but sorted according to their strengths so that $|\tilde{H}_1| \geq |\tilde{H}_1| \geq \cdots \geq |\tilde{H}_N|$. We have $N/2$ independent transmissions

$$\mathbf{y}_k = \begin{bmatrix} \tilde{H}_k & 0 \\ 0 & \tilde{H}_{N-k+1} \end{bmatrix} \mathbf{F}_k a_k + \mathbf{n}_k = \mathbf{S}_k \mathbf{F}_k a_k + \mathbf{n}_k, \quad 1 \leq k \leq N/2$$

and we need to construct $N/2$ precoders $\{\mathbf{F}_k\}_{k=1}^{N/2}$. A total energy of $NP/2$ is assumed, and we allocate a fraction γ_k to $\mathbf{F}_k$ under the constraint that $\Sigma \gamma_k = NP/2$. The power allocation policy undertaken here is that all channel–precoder pairs $\mathbf{S}_k \mathbf{F}_k$ should have equal minimum distances. We can find the precoders according to this policy as follows:

■ Design $\{\mathbf{F}_k\}_{k=1}^{N/2}$ according to the constraint $\text{Tr}(\mathbf{F}_k^H \mathbf{F}_k) = 1$.
■ From lattice theory, it is guaranteed that the minimum distance for each channel–precoder pair equals the length of the shortest vector of the lattice spanned by $\mathbf{S}_k \mathbf{F}_k$. Let D_k^2 denote the minimum distance.
■ The power allocation that equalizes all minimum distance is proportional to $\gamma_k \propto \dfrac{1}{D_k^2}$ and the overall power constraint $\Sigma \gamma_k = NP/2$ finally yields the set of precoders.

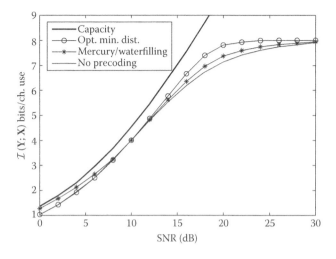

Figure 3.9 **Average mutual information per sub-carrier pair with 16-QAM inputs under different settings. The solid heavy line shows the capacity with waterfilling, the curve marked with circles shows the ensuing mutual information from the lattice precoder in this section and the curve marked with asterixes shows the mercury/waterfilling mutual information. The bottom line is the no precoding case.**

The ensuing average mutual information of this strategy is compared with the no-precoder case, mercury waterfilling, and the capacity of the channel. The input constellation is 16-QAM in all cases (except for the capacity case where it is complex Gaussian). The results are shown in Figure 3.9. Note that the average mutual information per channel–precoder pair is plotted. The top heavy solid curve is the average capacity of the channel, the curve marked by circles is the system based on the minimum distance optimal precoder described above, the curve marked with asterixes is the mercury waterfilling system, and the bottom curve shows the performance of the no-precoder case. As in the previous examples, there are no gains at low-moderate SNR by the minimum distance optimal precoder, while the gains are significant at high SNR. Note that the mercury waterfilling is close to optimal at low SNR, while it suffers from large penalties at high SNR.

3.2.4 Current Status and Future Research

The results in this chapter show that in two dimensions, whether it is real-valued or complex-valued precoding, the optimal precoder produces well-known lattice structures at the receiver. Theoretically, this result only holds for an infinite alphabet assumption. However, as confirmed by the simulation results in this chapter, for practical channels these results hold for 16-QAM or larger alphabets. Current research in this topic has revealed optimal lattice precoders in higher dimensions, as mentioned in Section 3.1.7. The analytical techniques used for two-dimensional precoding in this section are not easily extendable to higher dimensions. New ways to look at the problem were developed in [Kapetanović et al. 2012] to characterize optimal lattices for precoding. Namely, it is shown in [Kapetanović et al. 2012] that optimal precoders in any dimension give rise to perfect lattices at the receiver, that is, the optimal precoder is always generated by a perfect lattice. An elaborated analysis of these results can be found in [Kapetanović et al. 2012].

For small QAM alphabets, it can be beneficial to use a rank-deficient precoder which avoids data transmission across weak eigenmodes of the channel **H**. For two-dimensional MIMO systems, the result in Section 3.2 can be combined with the optimal rank one precoder in [Collin et al. 2007] for 4-QAM alphabets and [Ngo et al. 2009] for 16-QAM, and further performance enhancements can be obtained. However, for larger dimensions of the MIMO channel, rank-deficient precoders are not always instances of lattices as presented in [Kapetanović et al. 2012], and their structure is not known. Thus, future research into this area should develop efficient methods that find optimal or close to optimal rank-deficient precoders.

References

Bennatan, A. and D. Burshtein, Design and analysis of nonbinary LDPC codes for arbitrary discrete-memoryless channels, *IEEE Trans. Inf. Theory*, 52(2), 549Â–583, Feb. 2006.

Bergman, S., *Bit loading and precoding for MIMO communication systems*, PhD thesis, Signal Processing Laboratory, Royal Institute of Technology, Stockholm, May 2009.

Brink, S., G. Kramer, and A. Ashikhmin, Design of low-density parity-check codes for modulation and detection, *IEEE Trans. Commun.*, 52(4), 670–678, Apr. 2004.

Collin, L., O. Berder, P. Rostaing, and G. Burel, Optimal minimum-distance based precoder for MIMO spatial multiplexing systems, *IEEE Trans. Signal Process.*, 52(3), 617–627, Mar. 2007.

Conway, J. H. and N.J.A. Sloane, *Sphere Packings, Lattices and Groups*, Springer-Verlag, New York, 1999.

Cover, T. M. and J. A. Thomas, *Elements of Information Theory*, John Wiley & Sons, Inc., Hoboken, NJ, 2006.

Dahlman, E., S. Parkvall, and Johan Sköld, *4G LTE/LTE-Advanced for Mobile Broadband*, Academic Press, London, 2011.

Forney, G. D. Jr. and L.-F.Wei, Multidimensional constellations. I. Introduction, figures of merit, and generalized cross constellations, *IEEE J. Select. Areas Commun.*, 7(6), 877–892, Aug. 1989.

Gauss, C.F., *Disquisitiones Arithmeticae. Leipzig 1801*. German translation: Untersuchungen über die hohere Arithmetik. Springer, Berlin 1889. (Reprint: Chelsea, New York, 1981.)

Ghaderipoor, A. and C. Tellambura, Minimum distance-based limited-feedback precoder for MIMO spatial multiplexing systems, *in Proc. of the IEEE Vehicular Tech. Conf. (VTC)*, Montreal, Canada, Sep. 2006.

Hagenauer, J., The turbo principle: Tutorial introduction and state of the art, in *Proc. of theInt. Symp. Turbo Codes*, pp. 1–11, ENST de Bretagne, France, Sep. 1997.

Hassibi, B. and B. M. Hochwald, How much training is needed in multiple-antenna wireless links?, *IEEE Trans. Inf. Theory*, 49(4), 951–963, Apr. 2003.

He, J. and M. Salehi, A lattice precoding scheme for flat-fading MIMO channels, in *Proc. IEEE MILCOM*, San Diego, CA, Nov. 2008.

Hochwald, B. M., C. B. Peel, and A. L. Swindlehurst, A vector-perturbation technique for near-capacity multiantenna communicationpart II: Perturbation, *IEEE Trans. Commun.*, 53(5), 537—544, May 2005.

Jian, Y., J. Li, and W. W. Hager, Uniform channel decomposition for MIMO communications, *IEEE Trans. Signal Process.*,53(11), 4283–4294, Nov. 2005.

Jindal, N., MIMO broadcast channels with finite-rate feedback, *IEEE Trans. Inf. Theory*, 52(11), 5045–5060, Nov. 2006.

Kapetanovic', D. and F. Rusek, Linear precoders for parallell Gaussian channels with low decoding complexity, *in Proc. of the IEEE Vehicular Tech. Conf.*, San Francisco, CA, Sep. 2011.

Kapetanović, D., H. V. Cheng, W. H. Mow, and F. Rusek, A lattice-theoretic characterization of optimal minimum-distance linear precoders, arXiv, available at http://arxiv.org/abs/1204.1933, 2012.

Lee, H., S. Park, and I. Lee, A new MIMO beamforming technique based on rotation transformations, in *Proc. of theIEEE Int. Conf. Comm. (ICC)*, Glasgow, Scotland, Jun. 2007.

Lokesh, S. S., A. Kumar, and M. Agrawal, Structure of an optimum linear precoder and its application to ML equalizer, *IEEE Trans. Signal Process.*, 56(8), Aug. 2008.

Love, D. J. and R. W. Heath, Limited feedback unitary precoding for spatial multiplexing systems, *IEEE Trans. Inf. Theory*, 51(8), 2967–2976, Aug. 2005.

Lozano, A., A. M. Tulino, and S. Verdu, Mercury/waterfilling: Optimum power allocation with arbitrary input constellations, *IEEE Trans. Inf. Theory*, 52(7), 3033–3051, Jul. 2006.

Lu, B., G. Yue, and X. Wang, Performance analysis and design optimization of LDPC-coded MIMO OFDM systems, *IEEE Trans. Signal Process.*, 52(2), pp. 348–361, Feb. 2004.

Mohammed, S. K., E. Viterbo, Y. Hong, and A. Chockalingam, MIMO precoding with X- and Y-codes, *IEEE Trans. Inf. Theory*, 57(6), 3542–3566, Jun. 2011.

Ngo, Q. T., O. Berder, B. Vrigneau, and O. Sentieys, Minimum distance based precoder for MIMO-OFDM systems using a 16-QAM modulation, in *Proc. of the IEEE Int. Conf. Comm. (ICC)*, Dresden, Germany, Jun. 2009.

Ngo, Q. T., O. Berder, and P. Scalart, 3-D minimum Euclidean distance based sub-optimal precoder for MIMO spatial multiplexing systems, in *Proc. of the IEEE Int. Conf. Comm. (ICC)*, Cape Town, South Africa, May 2010.

Ngo, Q.-T., O. Berder, and P. Scalart, General minimum Euclidean distance based precoder for MIMO wireless systems, *EURASIP J. Adv. Sig. Process.*, March 2013.

Palomar, D. P. and Y. Jiang, MIMO transceiver design via majorization theory, *Foundations and Trends in Communication and Information Theory*, 3(4–5), 2007.

Palomar, D. P., J. M. Cioffi, and M. A. Lagunas, Joint Tx-Rx beam-forming design for multicarrier MIMO channels: A unified framework for convex optimization, *IEEE Trans. Signal Process.*, 51(9), 2381–2401, Sep. 2003.

Paulraj, A., R. Nabar, and D. Gore, *Introduction to Space–Time Wireless Communications*, Cambridge University Press, Cambridge, UK, 2003.

Payaro, M. and D. P. Palomar, On optimal precoding in linear vector Gaussian channels with arbitrary input distribution,in *Proc. of the IEEE Int. Symp. on Inform. Theory (ISIT)*, Seoul, Jun. 2009.

Perez-Cruz, F., M. R. D. Rodrigues, and S. Verdu, MIMO Gaussian channels with arbitrary inputs: Optimal precoding and power allocation, *IEEE Trans. Inf. Theory*, 56(3), 1070–1084, Mar. 2010.

Razi, A., D. J. Ryan, I. B. Collings, and J. Yuan, Sum rates, rate allocation, and user scheduling for multi-user MIMO vector perturbation precoding, *IEEE Trans. Wireless Commun.*, 9(1), 356–365, Jan. 2010.

Rusek, F. and D. Kapetanović, Design of close to optimal euclidean distance MIMO-Precoders, *IEEE Int. Symp. Inf. Theory*, Seoul, Korea, June. 2009.

Ryan, D. J., I. B. Collings, I. V. L. Clarkson, and R. W. Heath, Performance of vector perturbation multiuser (MIMO) systems with limited feedback, *IEEE Trans. Commun.*, 57(9), 2633–2644, September 2009.

Scaglione, A., P. Stoica, S. Barbarossa, G. B. Giannakis, and H. Sampath, Optimal designs for space-time linear precoders and decoders, *IEEE Trans. Signal Process.*, 50(5), 1051–1064, May 2002.

Taherzadeh, M. A., Mobasher, and A. K. Khandani, Communication over MIMO broadcast channels using lattice-basis reduction, *IEEE Trans. Inf. Theory*, 53(12), December 2007.

Telatar, I., Capacity of multi-antenna Gaussian channels, *European Trans. Tel.*, 10(6), 585–595, Nov.-Dec. 1999.

Vrigneau, B. et al. Extension of the MIMO precoder based on the minimum Euclidean distance: A cross-form matrix, *IEEE J. Select. Top. Signal Process.*, 2(2), 135–146, April 2008.

Vu, M. and A. Paulraj, Optimal linear precoders for MIMO wireless correlated channels with nonzero mean in space-time coded systems, *IEEE Trans. Signal Process.*, 54(6), 2318–2332, June 2006.

Windpassinger, C., R. F. H. Fischer, and J. B. Huber, Lattice-reductionaided broadcast precoding, *IEEE Trans. Commun.*, 52(12), 2057–2060, December 2004.

Xiao, C., Y. R. Zheng, and Z. Ding, Globally optimal linear precoders for finite alphabet signals over complex vector Gaussian channels, *IEEE Trans. Signal Process.*, 59(7), 3301–3314, July 2011.

Yao H. and G. W. Wornell, Lattice-reduction-aided detectors for MIMO communication systems, in *Proc. of the IEEE Global Telecomm. Conf. (GLOBECOM)*, Taipei, Nov. 2002.

Zheng L. and D. N. C. Tse, Communication on the Grassmann manifold: A geometric approach to the noncoherent multiple-antenna channel, *IEEE Trans. Inf. Theory*, 48(2), 359–383, Feb. 2002.

Chapter 4

MIMO Optimized for OFDM

Nuno Souto and Francisco A. Monteiro

Contents

The ever-increasing demand of high transmission rates in broadband wireless systems poses major design challenges especially due to the severe time dispersion occurring in mobile radio channels. Emergent radio systems often resort to block transmission techniques with frequency domain equalization (FDE) methods in order to sustain the envisioned data rates. Owing to the capability of converting a frequency selective channel into several parallel flat fading channels allied to their implementation simplicity both at the transmitter and receiver, orthogonal frequency-division multiplexing (OFDM) schemes have become the most popular and were incorporated into numerous communications standards, for example, digital video broadcast (DVB) and 4G systems, namely, LTE-Advanced [3GPP 2012] and WirelessMan-Advanced [IEEE 2011a]. Because of its popularity, OFDM has been frequently combined with the use of multiple-input multiple-output (MIMO) techniques in order to improve its performance and/or capacity. As a consequence, 4G specifications such as 3GPP LTE-Advanced already include a wide variety of MIMO transmission modes [3GPP 2012; 3GPP 2013a] supporting both single-user MIMO (SU-MIMO) and multiuser MIMO (MU-MIMO) in uplink and downlink as well as coordinated multipoint (CoMP) transmission and reception.

This chapter describes the use of OFDM with MIMO schemes in severely time-dispersive channels. Section 4.1 presents a general explanation of MIMO-OFDM signal representation and transmitter implementations as well as the combination with coding. Section 4.2 presents the application of several conventional MIMO detection methods and channel decoders in OFDM systems also taking into account the problem of channel estimation. Section 4.3 describes the use of iterative receiver configurations while the use of lattice-based receivers is explained in Section 4.4. In Section 4.5, it is shown how signal space diversity based on complex rotation matrices (CRM) can provide good performance gains in MIMO-OFDM systems without sacrificing spectral efficiency.

4.1 System Characterization

4.1.1 OFDM Signaling

4.1.1.1 SISO-OFDM

We will start this section by considering a SISO multicarrier transmission where a stream of N symbols is divided into N parallel streams, each mapped to a different subcarrier. Each of the N symbols is transmitted with a duration N times larger than what it would be if they were transmitted sequentially or, equivalently, the transmission rate of each subchannel is N times lower than in a single carrier transmission. In this case, the corresponding complex envelope can be written as

$$s(t) = \sum_{k=0}^{N-1} S_k r(t) \exp(j2\pi kFt), \tag{4.1}$$

where F is the subcarrier spacing, S_k is the modulated symbol mapped onto the kth subcarrier and $r(t)$ is the transmitted pulse. Applying the Fourier transform to this expression allows us to obtain the equivalent frequency domain representation as

$$S(f) = \sum_{k=0}^{N-1} S_k R(f - kF), \tag{4.2}$$

where $R(f)$ is the Fourier transform of $r(t)$. In order to avoid intercarrier interference (ICI), it is necessary to ensure that $R(f)$ obeys the orthogonality condition

$$\int_{-\infty}^{+\infty} R(f - kF)R^*(f - k'F)df = 0, \quad \forall k,k' \in \{0,\ldots,N-1\} : k \neq k'. \tag{4.3}$$

Although a conventional frequency domain multiplexing (FDM) transmission where $R(f)$ has an associated bilateral bandwidth smaller than F verifies the orthogonality condition, it is also possible to fulfill it even when the bandwidths of the translated $R(f)$ overlap [Marques da Silva et al. 2010]. This happens, for example, when we employ

$$R(f) = \text{sinc}\left(\frac{f}{F}\right), \tag{4.4}$$

which corresponds to a rectangular pulse with duration $1/F$ in the time domain, that is

$$r(t) = F \cdot \text{rect}(t \cdot F). \tag{4.5}$$

This type of FDM transmission where the total bandwidth is divided into several parallel overlapping subbands allows a more efficient use of the bandwidth and is referred to as OFDM. According to Equation 4.1, an OFDM transmission could be simply implemented as N parallel single carrier transmissions, each working with a different subcarrier frequency $f_k = f_c + kF$, where $k = 0,\ldots, N-1$ and f_c is the frequency of the first subcarrier. However, this approach is not practical for large N, due to the increasing number of oscillators and multipliers required. A simpler implementation can be obtained using a sampled version of Equation 4.1 followed by a digital-to-analog conversion (DAC) and a reconstruction filter applied to the in-phase and quadrature components of the samples. To understand how working with a sampled version of Equation 4.1 allows a simpler construction of the signal, let us assume a sampling frequency of $F_a = NF$. In this case, we can write the sequence of samples of Equation 4.1 as

$$s\left(t = \frac{n}{NF}\right) = \sum_{k=0}^{N-1} S_k r_n \exp\left(j2\pi k \frac{n}{N}\right)$$
$$= N r_n s_n,$$

(4.6)

where $r_n \triangleq r(t = n/NF)$,

$$s_n \triangleq \frac{1}{N}\sum_{k=0}^{N-1} S_k \exp\left(j2\pi k \frac{n}{N}\right) = \mathrm{IDFT}\{S_k\},$$

(4.7)

and IDFT is the inverse discrete Fourier transform (in this case of sequence S_k). Note that, according to Equations 4.6 and 4.7, apart from the scalar factor N, the sampled version of the complex envelope $s(t)$ can be simply obtained by multiplying the samples of the transmission pulse $r(t)$ with the IDFT of the block of N-modulated data symbols S_k, which can be efficiently implemented using an inverse fast Fourier transform (IFFT) algorithm. The corresponding implementation is illustrated in Figure 4.1. The "Add CP" block represents the insertion of the cyclic prefix (CP), which will be explained further ahead while the "D/A" blocks correspond to a DAC.

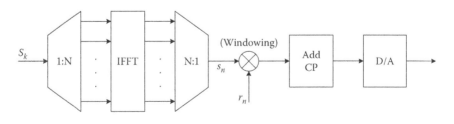

Figure 4.1   OFDM transmitter implementation using IFFT.

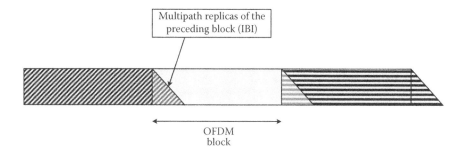

Figure 4.2   Interblock interference caused by the multipath propagation.

The use of the minimum sampling rate, $F_a = NF$, complicates the implementation of the reconstruction filter required for generating the wave shape $s(t)$, and thus oversampling is often applied. A higher sampling rate can be simply obtained by using N_{null} null subcarriers in each block, that is, using $S_k = 0$ for $k = (N - N_{null})/2$, ...,$(N + N_{null})/2 - 1$, which corresponds to an oversampling factor of

$$M_{os} = \frac{N}{N - N_{null}}. \tag{4.8}$$

Apart from the details of signal construction, the main idea of OFDM is to convert a large bandwidth channel into several parallel lower bandwidth subchannels. Since in a typical wireless environment, a large bandwidth channel will often undergo frequency selective fading, which results in intersymbol interference (ISI), the use of lower bandwidth subchannels allows each of them to experience flat fading only, thus avoiding the ISI problem inside an OFDM block. Still, as shown in Figure 4.2, it can exist ISI caused by the multipath propagation of the preceding block, which is often referred to as interblock interference (IBI). To cope with this problem, a guard period, T_G, with duration longer than the channel delay spread, T_m, is usually added to the beginning of each OFDM block in the time domain as shown in Figure 4.3. Although the guard period can simply be an empty slot (denoted as zero

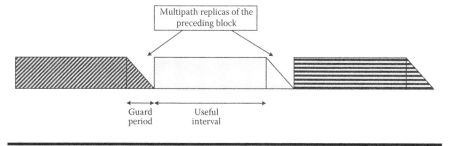

Figure 4.3   Use of an all-zero guard period to avoid IBI.

padding—ZP), in order to simplify the equalization in the frequency domain at the receiver, a repetition of the last N_{CP} samples of the OFDM block is often used instead. This allows the linear convolution of the channel with the OFDM block to be seen as a circular convolution. To avoid IBI, we must use $N_{CP} \geq L - 1$, with $L = F_a T_m$ being the length of the channel impulse response. This type of guard period is denoted as cyclic prefix (CP) and is illustrated in Figure 4.4. Besides ZP- and CP-assisted schemes, it is also possible to use pseudonoise (PN) symbol sequences.

Taking into consideration the structure of the OFDM signal, after transmission through the channel, the time domain samples, $y_n (n = -N_{CP}, \ldots, N - 1)$, of the received OFDM block can be written as

$$y_n = \sum_{l=0}^{L-1} h_l s_{n-l} + v_n, \qquad (4.9)$$

where h_l is the time domain channel coefficients (which we assume time invariant inside the block) and v_n represents independent and identically distributed Gaussian noise samples. Since only the CP samples will contain IBI and these are discarded at the receiver, we can neglect the preceding OFDM block and rewrite Equation 4.9 in a convenient matrix format as

$$\mathbf{y} = \Omega \mathbf{C}_{in} \mathbf{s} + \mathbf{v}, \qquad (4.10)$$

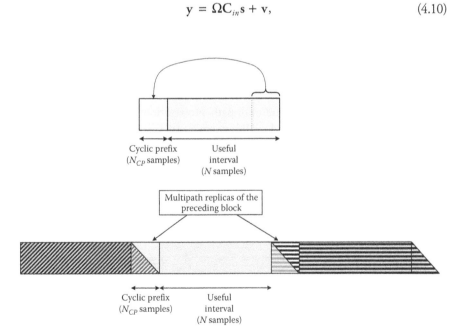

Figure 4.4   Use of CP to avoid IBI.

where **y** and **v** are length $N + N_{CP}$ column vectors containing respectively y_n and v_n as elements and Ω is a $(N + N_{CP}) \times (N + N_{CP})$ matrix defined as

$$
\Omega = \begin{bmatrix}
h_0 & 0 & \cdots & & & & 0 \\
h_1 & h_0 & 0 & \cdots & & & 0 \\
& \ddots & \ddots & & & & \vdots \\
0 & & \cdots & 0 & h_{L-1} & \cdots & h_0 & 0 \\
0 & & & \cdots & 0 & h_{L-1} & \cdots & h_0
\end{bmatrix}.
\tag{4.11}
$$

The length N column vector **s** contains the time domain-transmitted samples $s_j (j = 0,\ldots, N-1)$ as elements, which, following Equation 4.7, can also be written as

$$
\mathbf{s} = \frac{1}{N} \mathbf{F}^H \mathbf{S},
\tag{4.12}
$$

where **S** is a length N column vector whose elements are the modulated symbols S_k ($k = 0,\ldots, N-1$), and **F** represents the $N \times N$ discrete Fourier transform (DFT) matrix defined as

$$
\mathbf{F} = \begin{bmatrix}
1 & 1 & 1 & \cdots & 1 \\
1 & \omega^1 & \omega^2 & \cdots & \omega^{(N-1)} \\
1 & \omega^2 & \omega^4 & \cdots & \omega^{2(N-1)} \\
\vdots & \vdots & \vdots & \ddots & \vdots \\
1 & \omega^{(N-1)} & \omega^{(N-1)2} & \cdots & \omega^{(N-1)(N-1)}
\end{bmatrix},
\tag{4.13}
$$

with ω being an Nth primitive root of unity, $\omega = \exp(-j2\pi/N)$. Note that in Equation 4.12 $(\cdot)^H$ denotes the conjugate transpose operation, which, when applied on the scaled DFT matrix $(1/N)\mathbf{F}$, converts it into an IDFT operation.

CP insertion is accomplished in Equation 4.10 through the $(N + N_{CP}) \times N$ matrix $\mathbf{C}_{in}$ defined as

$$
\mathbf{C}_{in} = \begin{bmatrix}
\mathbf{0}_{N_{CP} \times (N - N_{CP})} & \mathbf{I}_{N_{CP}} \\
\mathbf{I}_N
\end{bmatrix},
\tag{4.14}
$$

with $\mathbf{0}_{N_{CP} \times (N - N_{CP})}$ representing a size $N_{CP} \times (N - N_{CP})$ matrix full of zeros and $\mathbf{I}_N$ denoting an $N \times N$ identity matrix. The CP removal at the receiver can be performed using an $N \times (N + N_{CP})$ matrix $\mathbf{C}_{rm}$ defined as

$$
\mathbf{C}_{rm} = [\mathbf{0}_{N \times N_{CP}} \quad \mathbf{I}_N],
\tag{4.15}
$$

followed by the conversion of the samples to the frequency domain using a DFT (which can be efficiently implemented using an FFT) according to

$$
\begin{aligned}
\mathbf{R} &= \mathbf{F}\mathbf{C}_{rm}\mathbf{y} \\
&= \frac{1}{N}\mathbf{F}\mathbf{C}_{rm}\mathbf{\Omega}\mathbf{C}_{in}\mathbf{F}^{H}\mathbf{S} + \mathbf{F}\mathbf{C}_{rm}\mathbf{v} \\
&= \frac{1}{N}\mathbf{F}\widehat{\mathbf{\Omega}}\mathbf{F}^{H}\mathbf{S} + \mathbf{N},
\end{aligned}
\tag{4.16}
$$

where

$$
\mathbf{N} = \mathbf{F}\mathbf{C}_{rm}\mathbf{v}.
\tag{4.17}
$$

In Equation 4.16, size $N \times N$ matrix $\widehat{\mathbf{\Omega}} = \mathbf{C}_{rm}\mathbf{\Omega}\mathbf{C}_{in}$ represents the combination of the channel with the CP insertion and removal operations. It is simple to verify that $\widehat{\mathbf{\Omega}}$ is a circulant matrix that can be written as

$$
\widehat{\mathbf{\Omega}} = \mathbf{C}_{rm}\mathbf{\Omega}\mathbf{C}_{in} =
\begin{bmatrix}
h_0 & 0 & \cdots & & h_{L-1} & \cdots & h_2 & h_1 \\
h_1 & h_0 & 0 & \cdots & & h_{L-1} & \cdots & h_2 \\
& \ddots & \ddots & & & & & \vdots \\
0 & & \cdots & 0 & h_{L-1} & \cdots & h_0 & 0 \\
0 & & & \cdots & 0 & h_{L-1} & \cdots & h_0
\end{bmatrix}
\tag{4.18}
$$

$$
= circ\{h_0, 0, \ldots, 0, h_{L-1}, \ldots, h_1\}.
$$

Therefore, we can apply the well-known decomposition of circulant matrices into the product of a conjugate transpose DFT matrix, a diagonal matrix, and a DFT matrix and write

$$
\widehat{\mathbf{\Omega}} = \frac{1}{N}\mathbf{F}^{H}\mathbf{H}\mathbf{F},
\tag{4.19}
$$

where

$$
\begin{aligned}
\mathbf{H} &= diag\{\mathbf{F}\widehat{\mathbf{\Omega}}_{,1}\} \\
&=
\begin{bmatrix}
H_0 & 0 & \cdots & 0 \\
0 & H_1 & \ddots & \vdots \\
\vdots & \ddots & \ddots & 0 \\
0 & \cdots & 0 & H_{N-1}
\end{bmatrix},
\end{aligned}
\tag{4.20}
$$

with

$$H_k = \text{DFT}\{h_n\} = \sum_{n=0}^{L-1} h_n \exp\left(-j2\pi\frac{kn}{N}\right).$$
(4.21)

In Equation 4.20, $\hat{\mathbf{\Omega}}_{:1}$ represents the first column of $\hat{\mathbf{\Omega}}$. Note that in Equation 4.21 we took into account that since the length of the channel impulse response is L, we have $h_n = 0$ for $n \geq L$. Bearing in mind that the DFT matrix defined in Equation 4.13 (which is not in a unitary form) follows the following property:

$$\mathbf{F}\mathbf{F}^H = N\mathbf{I}_N,$$
(4.22)

then, inserting Equation 4.19 into Equation 4.16 results in

$$\mathbf{R} = \mathbf{H}\mathbf{S} + \mathbf{N}.$$
(4.23)

Since $\mathbf{H}$ is diagonal, the frequency domain samples in $\mathbf{R}$ can thus be simply expressed as

$$R_k = H_k S_k + N_k,$$
(4.24)

which means that in the frequency domain, the channel in each subcarrier is equivalent to flat fading, represented by the single multiplicative coefficient H_k, and no ICI exists. This allows very simple equalization at the receiver end since it can be accomplished with a simple multiplication by $1/H_k$. The processing just described, apart from the equalization, is illustrated in Figure 4.5. It is important to note that although OFDM is robust in frequency selective channels and can be implemented with low complexity both in the emitter and receiver, it still has some weaknesses. For example, being similar to a transmission in a flat fading channel, its uncoded performance is rather poor. Furthermore, its high peak-to-average

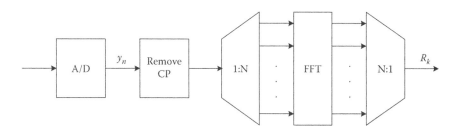

Figure 4.5   **OFDM receiver implementation using FFT.**

power ratio (PAR) leads to difficulties in transmitter amplification and receiver analog-to-digital conversion.

4.1.1.2 MIMO-OFDM

Let us now characterize the OFDM signal when combined with MIMO schemes with M_{tx} transmitting antennas and N_{rx} receiving antennas. In this case, the transmitter can be implemented using M_{tx} parallel chains similar to the one presented in Figure 4.1. After the transmission through the channel, the time domain samples, y_t^n $(t = -N_{CP}, \ldots, N - 1)$, of the received OFDM block in each receiver antenna n $(n = 1, \ldots, M_{tx})$ can be written as

$$y_t^n = \sum_{m=1}^{M_{tx}} \sum_{l=0}^{L-1} h_l^{m,n} s_{t-l}^m + v_t^n, \tag{4.25}$$

where $h_l^{m,n}$ represents the lth time domain channel coefficient between transmit antenna m and receive antenna n, s_l^m is the lth time domain sample transmitted using antenna m and v_t^n represents independent and identically distributed Gaussian noise samples in receive antenna n. Considering the use of a CP between adjacent OFDM blocks, we can rewrite Equation 4.25 in a matrix format as

$$\mathbf{y}^n = \sum_{m=1}^{M_{tx}} \mathbf{\Omega}^{m,n} \mathbf{C}_{in} \mathbf{s}^m + \mathbf{v}^n, \tag{4.26}$$

where $\mathbf{y}^n$ and $\mathbf{v}^n$ are length $N + N_{CP}$ column vectors containing respectively y_t^n and v_t^n as elements, $\mathbf{s}^m$ contains the time domain transmitted samples s_l^m and $\mathbf{\Omega}$ is a $(N + N_{CP}) \times (N + N_{CP})$ matrix defined as Equation 4.11 containing the time domain channel coefficients $h_l^{m,n}$ between transmit antenna m and receive antenna n.

The received frequency domain samples for antenna n can be computed through multiplication by the pilot removal matrix $\mathbf{C}_{rm}$ defined in Equation 4.15 and by the DFT matrix, that is, using

$$\mathbf{R}^n = \mathbf{F}\mathbf{C}_{rm}\mathbf{y}^n. \tag{4.27}$$

Using the same derivation presented for the SISO scheme, it is simple to verify that the frequency domain samples for each subcarrier k and receive antenna n can be expressed as

$$R_k^n = \sum_{m=1}^{M_{tx}} H_k^{m,n} S_k^m + N_k^n. \tag{4.28}$$

According to this expression, the transmission in each subcarrier is equivalent to a MIMO transmission in a flat fading channel without ICI, allowing the direct application of well-known MIMO detection methods as will be detailed in some of the following sections.

4.1.2 Transmitter Structure

Figure 4.6 shows a typical configuration for a MIMO-OFDM baseband transmitter chain. According to this scheme, an information stream is first encoded, interleaved, and mapped onto the constellation symbols from a specific modulation (e.g., QPSK, M − QAM). If spatial multiplexing is being used, the sequence of modulated symbols is split into several smaller streams that are transmitted simultaneously by M_{tx} antennas. Conversely, if transmit diversity is being used, the sequence of modulated symbols can be coded using a space–time block code (STBC) like the Alamouti scheme for two antennas [Alamouti 1998]. When STBCs are employed in OFDM systems, the time domain is often replaced by the frequency domain and the resulting codes are referred to as space–frequency block codes (SFBC). In both cases, spatial multiplexing and transmit diversity modes, pilot symbols can be inserted into the parallel-modulated symbol sequences before conversion to the time domain using IDFTs. These pilot symbols can be used for channel estimation purposes, as will be explained in Section 4.2.3.

The sequences of symbols are then converted into the time domain through $\left\{s_{i,l}^m, i = 0,1,\ldots,N-1\right\} = \text{IDFT}\left\{S_{k,l}^m, k = 0,1,\ldots,N-1\right\}$, where $S_{k,l}^m$ is the symbol transmitted by the kth subcarrier of the lth OFDM block using antenna m. Although not detailed in the chain of Figure 4.6, as explained in Section 4.1.1, before being transmitted, windowing is applied to the time domain samples $s_{i,l}^m$, followed by CP insertion (with $s_{-i,l}^m = s_{N-i,l}^m$, $i = 1,\ldots,N_{CP}$), DAC, and reconstruction filtering.

The transmit chain just described corresponds to a unicast transmission but can be easily extended to broadcast schemes, as presented in [Souto et al. 2008] for the case of MIMO-OFDM systems combined with hierarchical modulations,

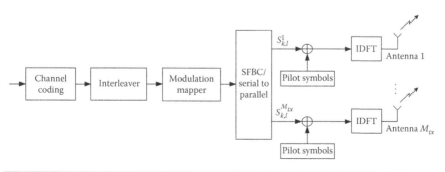

Figure 4.6   Generic baseband MIMO-OFDM transmitter chain.

where multiple data streams with different error protection levels are transmitted in parallel.

4.1.3 Channel Coding

As explained in Section 4.1.1, OFDM allows us to convert a frequency selective channel into several parallel flat fading channels. Although this solves the ISI problem in severely time-dispersive channels, OFDM does not take advantage of frequency diversity and, therefore, some subcarriers may experience deep fades thus compromising the reliable detection of the respective symbols. As a result, the performance of OFDM is rather poor when compared with other schemes like single carrier block transmission [Marques da Silva et al. 2010]. To circumvent this problem, OFDM is usually combined with the use of channel codes. As examples of codes used in OFDM systems, we have tail-biting convolutional codes that are specified for LTE-Advanced [3GPP 2013b] and WirelessMan-Advanced [IEEE 2011a] (for control channels), convolutional turbo codes (CTC), which are specified for 4G systems, namely, LTE-Advanced [3GPP 2013b] and WirelessMan-Advanced [IEEE 2011a], block turbo codes, which are used as optional coding in mobile WiMax [IEEE 2006], and low-density parity check codes (LDPC), which have been incorporated into WiMax [IEEE 2006], DVB-T2 [ETSI 2009b], DVB-S2 [ETSI 2009a], and IEEE Std 802.11n-2009 [IEEE 2011b]. Owing to their importance nowadays, we will briefly describe CTCs and LDPC codes in the following sections.

4.1.3.1 Convolutional Turbo Encoder

A CTC is formed by the parallel concatenation of recursive systematic convolutional (RSC) codes. As an example, Figure 4.7 portrays the turbo encoder employed in LTE-Advanced [3GPP 2013b], which is composed by two RSC encoders and an interleaver represented by the permutation function $\Pi_1(\cdot)$. The same information block is applied at the input of both constituent encoders but one of them sees an interleaved version of the block. This way, although both encoders are processing the same information bits, the output code words are different. According to the encoder of Figure 4.7, if the input block has size N_b, the interleaver will have the same size and the output code word will be composed of N_b systematic bits and $2N_b$ parity bits generated by the two constituent encoders. However, after encoding the input sequence, both constituent encoders are terminated into the all-zeros state as detailed in [3GPP 2013b], thus generating 12 additional tail bits. The overall coding rate is thus slightly lower than 1/3. Nevertheless, similar to the convolutional codes, the output of the turbo encoder can be punctured, usually the parity bits, to increase the coding rate.

If the memory of each component encoder e (in Figure 4.7 $e = \{1,2\}$) is v, the feed-forward and feedback connections can be defined as $\mathbf{g}_e^f = \left[g_{e,0}^f \cdots g_{e,v}^f \right]$ and $\mathbf{g}_e^b = \left[g_{e,0}^b \cdots g_{e,v}^b \right]$, respectively, with $g_{e,i}^{f/b} = \{0,1\}$. For the specific case of Figure 4.7, and since both constituent encoders are identical, we have

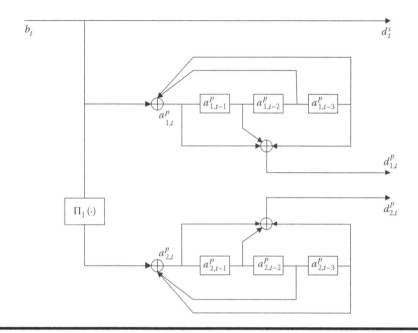

Figure 4.7 LTE-Advanced turbo encoder scheme.

$\mathbf{g}_e^f = [1 \quad 1 \quad 0 \quad 1]$ and $\mathbf{g}_e^b = [1 \quad 0 \quad 1 \quad 1]$. Considering that the input sequence of bits is $\mathbf{b} = \left[b_1 \ldots b_{N_b} \right]$, the systematic and parity sequences will be

$$
\begin{cases}
d_t^s = b_t \\
d_{e,t}^p = \displaystyle\sum_{i=0}^{v} g_{e,i}^f a_{e,t-i}
\end{cases}.
\tag{4.29}
$$

In Equation 4.29, the coefficients $a_{e,t}$ are expressed as

$$
a_{e,t} = b_{\Pi_{e-1}^{-1}(t)} \oplus \sum_{i=1}^{v} g_{e,i}^b a_{e,t-i},
\tag{4.30}
$$

where $b_{\Pi_{e-1}^{-1}(t)}$ is the tth bit at the input of encoder e (after the interleaver). The interleaver is denoted by the permutation function $\Pi_e(\cdot)$. For the first encoder, since there is no interleaver, it simply encodes the input sequence directly, which is equivalent to having a permutation function $\Pi_0(t) = t$. In all these expressions, the sums are modulo-2 additions.

Although the interleaver can be a block pseudorandom interleaver, other types are possible as described in Heegard and [Wicker 1999]. The information sequence fed to the second component encoder is reordered in such a way that it becomes decorrelated from the original input sequence as much as possible. In fact, the use of such

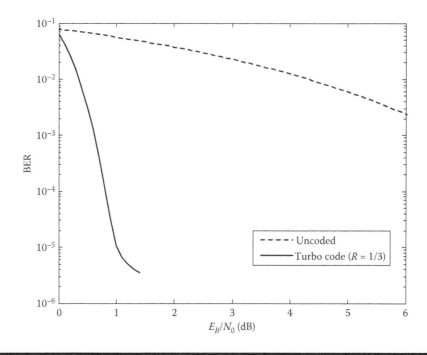

Figure 4.8   BER performance of 3GPP rate-1/3 turbo code [3GPP 2013b] in AWGN.

type of concatenation of two convolutional encoders separated by an interleaver has the objective of producing a set of code words where very few of them have low weight. This does not necessarily mean that the resulting code will have a particularly large free distance but at least the number of nearest neighbors of each code word will be substantially reduced. The substantially improved weight distribution of the resulting code words is one of the main reasons for the good performances of turbo codes. This weight distribution is also responsible for the typical turbo code performance curve in additive white Gaussian noise (AWGN), which can be seen in Figure 4.8. These curves are usually divided in two main zones. The first is located in the low SNR region and is characterized by an abrupt decrease in the bit error rate (BER), which is mostly caused by the existence of a low number of nearest neighbors surrounding each code word. The second zone is located in higher SNR regions and is characterized by a slower descent of the performance curve. This is caused by the not particularly high free distance of the turbo code due to the few low-weight code words.

4.1.3.2 LDPC Codes

LDPC codes were initially proposed by Robert Gallager in 1962, in [Gallager 1962], who showed that with a careful design of the parity check matrix, they could achieve near Shannon limit performance using iterative probabilistic-based

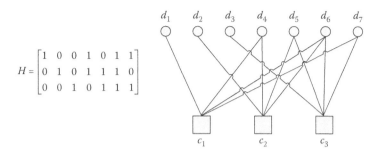

Figure 4.9 Hamming code (7,4) parity check matrix and respective Tanner graph.

---

decoders. However, the high complexity associated with the encoding and decoding processes made them computationally impractical at the time, and for 30 years, they did not attract the attention of the research community. After the advent of the turbo codes, the interest in these codes reappeared with the rediscovery of Gallager codes in [MacKay and Neal 1995, 1996] who showed that they could achieve similar performances. For this reason, LDPCs have been incorporated into several standards: WiMax [IEEE 2006], DVB-T2 [ETSI 2009b], DVB-S2 [ETSI 2009a, IEEE Std 802.11n-2009] [IEEE 2011b], and so on.

LDPC codes are (n,k) linear block codes (although it is also possible to build LDPC convolutional codes) defined through very sparse $(n - k) \times n$ parity check matrices $\mathbf{H}^*$ that satisfy

$$\mathbf{H} \cdot \mathbf{d}^T = 0 \tag{4.31}$$

for any code word $\mathbf{d} = [d_0 \quad d_1 \quad \dots \quad d_{k-1}]$. Each line of matrix $\mathbf{H}$ represents an equation that must be verified by a sequence $\mathbf{d}$ in order to be a valid code word. For example, regarding a (7,4) Hamming code whose parity check matrix is displayed in Figure 4.9, its parity check equations can be written as

$$\begin{cases} c_1 = d_1 \oplus d_4 \oplus d_6 \oplus d_7 \\ c_2 = d_2 \oplus d_4 \oplus d_5 \oplus d_6, \\ c_3 = d_3 \oplus d_5 \oplus d_6 \oplus d_7 \end{cases} \tag{4.32}$$

where $\oplus$ represents modulo-2 addition. In this case, $\mathbf{d}$ is a code word as long as $c_1 = c_2 = c_3 = 0$.

---

* Although in the previous sections and in most of the chapters, we use $\mathbf{H}$ for representing the channel matrix, in this subsection, we use symbol $\mathbf{H}$ for representing parity check matrices due to its widespread use in the literature. However, this representation is only required in this subsection and in Section 4.2.2 and thus there is no risk of confusion in the remainder of the chapter.

Since LDPC codes are defined through a parity check matrix $\mathbf{H}$, it is necessary to obtain the $k \times n$ generator matrix $\mathbf{G}$ from this one in order to perform the encoding process using

$$\mathbf{d} = \mathbf{b} \cdot \mathbf{G}. \tag{4.33}$$

The traditional method used in block codes consists of transforming matrix $\mathbf{H}$ into a systematic form using Gaussian elimination method and column permutations resulting in

$$\mathbf{H}' = [\mathbf{P} \;\vdots\; \mathbf{I}_{n-k}], \tag{4.34}$$

where $\mathbf{P}$ is a $(n - k) \times k$ parity submatrix and $\mathbf{I}_{n-k}$ is the $(n - k) \times (n - k)$ identity matrix. From Equation 4.34, it is simple to obtain the generator matrix as

$$\mathbf{G} = [\mathbf{I}_k \;\vdots\; \mathbf{P}^T]. \tag{4.35}$$

Although this method seems straightforward, it does not take into account the sparseness of the original matrix $\mathbf{H}$ and, as such, the resulting generator matrix $\mathbf{G}$ is likely to be dense. As a consequence, the encoding complexity can become significant, especially for large matrices (which are usually associated with better performances). Other lower-complexity approaches have been proposed, which take advantage of the sparseness of $\mathbf{H}$. For example, as proposed in [Richardson and Urbanke 2001b], it is possible to transform $\mathbf{H}$ into an almost lower triangular form using only row and column permutations, thus preserving the sparseness of the matrix. Then, Gaussian elimination is applied to the rows that do not match the triangular form. These rows are called the *gap* since they will be associated with the high-density part (higher complexity) of the encoding process. The presence of the other low-density rows allows the use of back substitution for the corresponding parity bits.

The LDPC code specified for WiMax [IEEE 2006] is defined using a parity check matrix $\mathbf{H}$ built from circulant submatrices and allows the encoding process to be efficiently implemented using back substitution with a gap of $n/24$.

In [Tanner 1981], the use of bipartite graphs, named Tanner graphs, was proposed as a simple approach for characterizing a (n,k) linear code. As an example, Figure 4.9 shows the Tanner graph associated with a $(7,4)$ Hamming code. The graph is composed of two types of nodes: parity check nodes (c_j) and variable nodes (d_i). Each parity check node is associated with a parity check equation while each variable node is associated with a code bit. The construction of the graphs is simple. Each parity check node (c_j) is connected to all the variable nodes (d_i) that are involved in the respective parity check equation (according to Equation 4.32).

From the Tanner graph, it is possible to know the degree of each node by simply counting the number of connections to that node. Since LDPC codes are defined through very sparse matrices, the degrees of the nodes are much smaller than n.

If the degrees of all variable nodes are equal and the same happens to the degrees of all parity check nodes, then the LDPC code is called regular, otherwise it is an irregular LDPC code [Richardson and Urbanke 2001a]. Tanner graphs are useful to find the length of closed cycles. For example, in Figure 4.9, the existence of a length four closed cycle associated with path $c_1 \rightarrow d_4 \rightarrow c_2 \rightarrow d_6 \rightarrow c_1$ is visible. When constructing LDPC codes, it is important to avoid short-length cycles, which are associated with worse performances of the decoder algorithm.

4.2 Conventional Receivers

4.2.1 Zero-Forcing, Minimum Mean Squared Error, and Maximum Likelihood Detector

In this section, we will focus on conventional OFDM receivers assuming that spatial multiplexing is being used. As previously explained, OFDM allows simple equalization in the frequency domain where the channel can be seen as flat fading. Therefore, the use of conventional MIMO decoders in an OFDM receiver with N_{rx} receiving antennas is straightforward as we show in Figure 4.10. According to the figure, the signal, which is considered to be sampled and with the cyclic prefix removed, is converted into the frequency domain after an appropriate size-N DFT operation. If the cyclic prefix is longer than the overall channel impulse response, the resulting sequence received in antenna n can be expressed as

$$R_{k,l}^n = \sum_{m=1}^{M_{tx}} S_{k,l}^m H_{k,l}^{m,n} + N_{k,l}^n, \tag{4.36}$$

with $H_{k,l}^{m,n}$ denoting the overall channel frequency response between transmit antenna m and receiving antenna n for the kth frequency of the lth time block and $N_{k,l}^m$ denoting the corresponding channel noise sample. The sequences of samples (Equation 4.36) enter the MIMO equalizer (spatial demultiplexer block), which separates the streams transmitted simultaneously by the multiple antennas, while the other blocks basically reverse the operations applied at the transmitter.

In the following, we will briefly describe conventional methods that can be used in the spatial demultiplexer block for separating the simultaneous transmitted streams. Details about the channel estimator block will be discussed in Section 4.2.3.

The most direct decoding technique is the zero-forcing (ZF), which simply uses a straight channel matrix inversion

$$\hat{S}_{k,l} = \hat{H}_{k,l}^H \cdot (\hat{H}_{k,l} \, \hat{H}_{k,l}^H)^{-1} R_{k,l}, \tag{4.37}$$

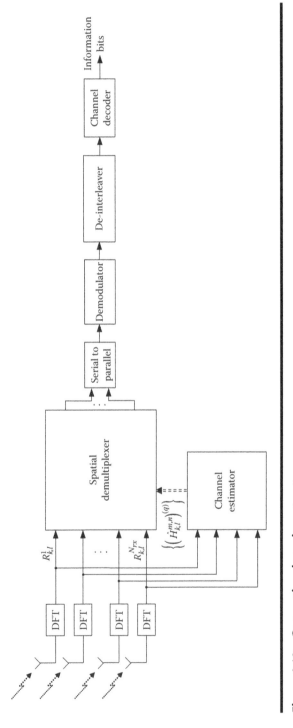

Figure 4.10 Conventional receiver structure.

where k is the subcarrier index, l is the OFDM, $\hat{S}_{k,l}$ is the $M_{tx} \times 1$ estimated transmitted signal vector with one different transmit antenna in each position, $\hat{H}_{k,l}$ is the $N_{rx} \times M_{tx}$ channel matrix estimate with each column representing a different transmit antenna and each line representing a different receive antenna, and $R_{k,l}$ is the $N_{rx} \times 1$ received signal vector with one different receive antenna in each position. The ZF approach can result in poor performance for some channel conditions and, therefore, a minimum mean squared error (MMSE) linear receiver is often used instead. The MMSE estimate is computed so that the minimum squared error, $E\left[|s - \hat{s}|^2\right]$, is minimized. In [Kay 1993], it is shown that the MMSE estimate is given by

$$\hat{S}_{k,l} = \hat{H}_{k,l}^H \cdot \left(\hat{H}_{k,l}\hat{H}_{k,l}^H + \sigma^2 I\right)^{-1} R_{k,l}, \tag{4.38}$$

where σ^2 is the noise variance. Note that the ZF estimate (Equation 4.37) can be obtained from Equation 4.38 by setting $\sigma^2 = 0$.

A well-known alternative detector for MIMO systems is the vertical Bell Labs layered space–time (V-BLAST) architecture [Foschini 1996]. In this scheme, the data bits to be transmitted are divided equally between the transmitting antennas and are then coded, modulated, and transmitted independently, as seen in Chapter 2. The decoder employs a relatively low-complexity approach based on nulling and cancelation, which is an equivalent technique to the successive interference cancelation often used for multiuser detection in CDMA systems. The difference is that the first performs the separation in the spatial channel domain while the latter works on the code domain. A brief description of this method is given next. First, the receiver performs a QR factorization [Friedberg et al. 2002] of the estimate of the channel matrix $\hat{H}_{k,l}$, that is

$$\hat{H}_{k,l} = QR, \tag{4.39}$$

where R is an $M_{tx} \times M_{tx}$ upper triangular matrix ($r_{m,n} = 0$, $m > n$) with nonnegative real-valued elements in the diagonal and Q is an $M_{rx} \times M_{tx}$ matrix composed of orthonormal column vectors (and thus $Q^H Q = I_{M_{tx}}, I_{M_{tx}}$ is the $M_{tx} \times M_{tx}$ identity matrix). Multiplying the received vector by the Hermitian of Q results in

$$\begin{aligned} y = Q^H R_{k,l} &= Q^H (QRS_{k,l} + N_{k,l}) \\ &= RS_{k,l} + v, \end{aligned} \tag{4.40}$$

where $v = Q^H N_{k,l}$ is a column of independent complex Gaussian variables. In [Foschini 1996], it was shown that instead of multiplying y by R^{-1} (completing a zero-forcing operation), the transmitted symbols can be better estimated through quantization and cancelation. This operation is performed taking into account the upper triangular structure of $\mathbf{R}$, which allows a direct estimation of the M_{tx}th transmitted symbol from the last element of $\mathbf{y}$ according to

$$y_{M_{tx}} = r_{M_{tx},M_{tx}} S_{k,l}^{M_{tx}} + v_{M_{tx}} \Rightarrow \hat{S}_{k,l}^{M_{tx}} = Q\left\{\frac{y_{M_{tx}}}{r_{M_{tx},M_{tx}}}\right\}, \qquad (4.41)$$

where $Q\{\}$ represents a quantization operation, which depends on the modulation in use. In the case of element $M_{tx} - 1$ of $\mathbf{y}$, it will suffer from interference of symbol M_{tx}, which has to be canceled for obtaining an estimate of the respective transmitted symbol using

$$y_{M_{tx}-1} = r_{M_{tx}-1,M_{tx}-1} S_{k,l}^{M_{tx}-1} + r_{M_{tx}-1,M_{tx}} S_{k,l}^{M_{tx}} + v_{M_{tx}-1}$$

$$\Rightarrow \hat{S}_{k,l}^{M_{tx}-1} = \frac{1}{r_{M_{tx}-1,M_{tx}-1}}\left(y_{M_{tx}-1} - r_{M_{tx}-1,M_{tx}} \hat{S}_{k,l}^{M_{tx}}\right). \qquad (4.42)$$

This approach can then proceed successively for the other elements of $\mathbf{y}$ according to

$$\hat{s}_m = \frac{1}{r_{m,m}}\left(y_m - \sum_{i=m+1}^{M_{tx}} r_{m,i} \hat{S}_{k,li}^i\right), \quad m = 1 \ldots M_{tx}. \qquad (4.43)$$

Note that the order of the nulling and cancelation operations affects the performance of this decoder, and, therefore, a permutation specifying the order in which the symbols are extracted has to be applied before these steps are executed, as explained in [Foschini 1996].

Another alternative decoding method is the optimal maximum likelihood detector (MLD), which compares the received signal to all the possible transmitted combination of symbols according to

$$\hat{\mathbf{S}}_{k,l} = \arg\min_{\mathbf{S}_{k,l}} \left\|\mathbf{R}_{k,l} - \hat{\mathbf{H}}_{k,l}\mathbf{S}_{k,l}\right\|. \qquad (4.44)$$

This method can become extremely complex as the number of antennas increases or when high-order modulations are employed. The MLD generates hard decisions for the modulated symbols but, when coding is used, it is important to work with soft decisions so as to avoid a performance loss of the channel decoder. Therefore, one can use other alternatives to the MLD such as a maximum likelihood-based soft output (MLSO) decoder. In the MLSO decoder, we use the following estimate for each symbol:

$$\hat{S}_{k,l}^m = E\left[S_{k,l}^m\big|\mathbf{R}_{k,l}\right]$$

$$= \sum_{s_i \in \Lambda} s_i \cdot P\left(S_{k,l}^m = s_i\big|\mathbf{R}_{k,l}\right)$$

$$= \sum_{s_i \in \Lambda} s_i \cdot \frac{P\left(S_{k,l}^m = s_i\right)}{p(\mathbf{R}_{k,l})} p\left(\mathbf{R}_{k,l}\big|S_{k,l}^m = s_i\right), \qquad (4.45)$$

where s_i corresponds to a constellation symbol from the modulation alphabet Λ, $E[\cdot]$ is the expected value, $P(\cdot)$ represents a probability, and $p(\cdot)$ is a probability density function (PDF). Considering equiprobable symbols, we have $P\left(S_{k,l}^m = s_i\right) = 1/M$, where M is the constellation size. The PDF values required in Equation 4.45 can be computed as

$$p\left(\mathbf{R}_{k,l}\big|S_{k,l}^m = s_i\right) = \frac{1}{M^{M_{tx}-1}} \sum_{S_{k,l}^{\text{interf}} \in \Lambda^{M_{tx}-1}} p\left(\mathbf{R}_{k,l}\big|S_{k,l}^m = s_i, S_{k,l}^{\text{interf}}\right), \qquad (4.46)$$

with

$$p\left(\mathbf{R}_{k,l}\big|S_{k,l}^m = s_i, S_{k,l}^{\text{interf}}\right) = \frac{1}{(2\pi\sigma^2)^{N_{rx}}} \exp\left[\sum_{n=1}^{N_{rx}} -\frac{\left|R_{k,l}^n - \hat{\mathbf{H}}_{k,l}(n,:)\cdot\mathbf{s}\right|^2}{2\sigma^2}\right], \qquad (4.47)$$

where $S_{k,l}^{\text{interf}}$ is a $(M_{tx} - 1) \times 1$ vector representing a possible combination of symbols transmitted simultaneously by all antennas except antenna m, s is an $M_{tx} \times 1$ vector composed by $S_{k,l}^{\text{interf}}$ and s_i and $\hat{\mathbf{H}}_{k,l}(n,:)$ is the nth line of channel matrix $\hat{\mathbf{H}}_{k,l}$.

4.2.2 Channel Decoder

4.2.2.1 Turbo Decoder

Although the code word weight distribution of the turbo code is important to achieve good performances, it is also necessary to have a decoder algorithm that is not excessively complex to be implemented. In fact, this is one of the key ideas behind the success of these codes. Let us consider a turbo code composed of two-component RSC (like the one of Figure 4.7). In theory, the turbo code can be modeled using a single Markov process but due to the existence of the interleaver at the input of one of the constituent codes, this representation is extremely complex and does not allow the direct implementation of computationally tractable algorithms. Consequently, a maximum likelihood decoder for turbo codes cannot be employed and some suboptimal decoder must be used in its place.

Instead of modeling the whole code using a single Markov process, each of the constituent codes can be associated with an individual Markov process and both can be linked by an interleaver. A trellis-based decoding algorithm can be used for estimating each of the Markov processes and, to improve the estimates, these two processes can exchange information iteratively. A basic turbo decoder accomplishing this is shown in Figure 4.11, where $\hat{b}$ denotes the information bits sequence estimate, $\hat{b}_\Pi$ the respective interleaved sequence, $\mathbf{y}^s$ the systematic observations, and $\mathbf{y}^{p,1}$ and $\mathbf{y}^{p,2}$ the two parity observations sequences. The two soft-input soft-output blocks implement a trellis decoding algorithm for the respective component RSC encoders. Each of these blocks has three inputs and one output. Two of the inputs correspond to the systematic and parity observations, both weighted by the channel reliability factor L_c ($L_c = 4E_c/N_0$). The third input is the *a priori* information obtained from the other decoder, which is represented in the form

$$L_{ap}(b_t) = \log \frac{\text{Prob}(b_t = +1)}{\text{Prob}(b_t = -1)}. \tag{4.48}$$

The output of the soft-input soft-output decoders is the sequence of log-likelihood ratios (LLRs) of the information bits, which can be written as the sum of three terms:

$$L(b_t) = \log \frac{\text{Prob}(b_t = +1|\mathbf{y})}{\text{Prob}(b_t = -1|\mathbf{y})} = L_c y_t^s + L_{ap}(b_t) + L_{1|2}^e(b_t). \tag{4.49}$$

In this expression, $L_{1|2}^e(b_t)$ is the extrinsic information produced by either decoder 1 or 2. While the first two terms are inputs to the component decoders, this third term represents new information derived in the decoder, which is used as *a priori* information, $L_{ap}(b_t)$, by the other decoder. To obtain $L_{1|2}^e(b_t)$, it is only necessary to subtract the weighted systematic observation and the *a priori* information

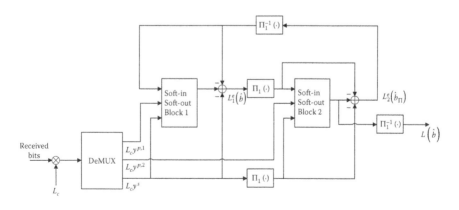

Figure 4.11   Turbo decoder block diagram.

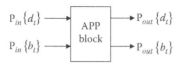

Figure 4.12   Soft-input soft-output APP module.

from the output (Equation 4.49) of the decoder. After passing this information to the interleaver (or de-interleaver, depending on the target decoder), the other decoder uses this extrinsic information as new *a priori* information and computes new estimates for the log-likelihoods of the information bits. The exchange of extrinsic information between the two decoders can proceed until the information sequence is correct (checked using CRC) or until a certain number of iterations is completed. The output of the second decoder is then interleaved and some decision function can be applied for obtaining the final estimates of the information bits. This decoding process can be easily extended to multiple turbo codes as presented in [Divsalar and Pollara 1995].

Each soft-input soft-output decoder can be implemented using a trellis decoding algorithm like the optimal maximum *a posterori* (MAP) algorithm [Bahl et al. 1974], its simplification max-log-MAP [Robertson et al. 1995], or the soft output Viterbi algorithm (SOVA) [Hagenauer and Hoher 1989]. In [Robertson et al. 1995], these three algorithms were compared and it was concluded that the MAP algorithm clearly has the best performance but at the cost of a substantial complexity. The SOVA exhibits a slight performance degradation compared with the max-log-MAP but has a lower decoding complexity. Nevertheless, it was shown in [Fossorier et al. 1998] that the SOVA can become equivalent to the max-log-MAP algorithm after a simple modification. The MAP algorithm can also be used to compute *a posteriori* probabilities of the encoder output symbols, as was shown in [Benedetto and Montorsi 1997] allowing the implementation of a soft-input soft-output module with four ports, as shown in Figure 4.12. The two input ports correspond to the *a priori* probabilities of the information symbols, $P_{in}\{b_t\}$, and the probabilities of the coded symbols, $P_{in}\{d_t\}$. The other two ports output the probabilities associated with the input symbols, $P_{out}\{b_t\}$ and with the coded symbols, $P_{out}\{d_t\}$. This soft-input soft-output module capable of outputting encoded symbols probabilities is useful in several applications where feedback from the turbo decoder to other blocks of the receiver is required, as for example, when implementing iterative joint estimation schemes.

4.2.2.2 LDPC Decoder—Sum-Product Algorithm

Tanner graphs are useful for implementing iterative probabilistic decoding algorithms where the different types of nodes exchange messages (probabilities) according to the graph connections. In the following, we will briefly describe the iterative sum-product algorithm (also known as belief propagation algorithm) assuming binary codes and working with log likelihood ratios (LLRs). The sum-product algorithm is a message passing-based algorithm with two types of messages:

■ Message from variable node d_j to parity check node c_i, which takes into account all the messages coming from the other check nodes connected to d_j, with the exception of the message coming from the target check node c_i:

$$L(q_{ij}) = L(p_j) + \sum_{\substack{i'=1 \\ i' \neq i}}^{n-k} h_{i'j} L(r_{i'j}).$$ (4.50)

■ Message from parity check node c_i to variable node d_j, which takes into account all the messages coming from the variable nodes connected to c_i, with the exception of the message coming from the target variable node d_j:

$$L(r_{ij}) = 2 \operatorname{atanh}\left[-\prod_{\substack{j'=1 \\ j' \neq j \wedge h_{ij'} \neq 0}}^{n} \tanh(-L(q_{ij'})/2) \right].$$ (4.51)

In these expressions, h_{ij} represents an element of parity check matrix $\mathbf{H}^*$. Furthermore, the following logarithms of probability ratios are used:

$$L(p_j) = \log\left(\frac{p_j^1}{1 - p_j^1} \right),$$ (4.52)

$$L(q_{ij}) = \log\left(\frac{q_{ij}^1}{1 - q_{ij}^1} \right),$$ (4.53)

$$L(r_{ij}) = \log\left(\frac{r_{ij}^1}{1 - r_{ij}^1} \right),$$ (4.54)

where p_j^1 is the likelihood probability associated with bit d_j being 1, that is

$$p_j^1 = p(y_j | d_j = 1).$$ (4.55)

In AWGN, Equation 4.52 can be written as

$$L(p_j) = \frac{2}{\sigma^2} y_j = \frac{2}{N_0/2 \cdot E_c} y_j = \frac{4 \cdot E_c}{N_0} y_j = L_c y_j,$$ (4.56)

---

[*] In this subsection, we are repeating the use of symbol, **H**, for representing the parity check matrix as in Section 4.1.3.

with L_c corresponding to the reliability factor defined as

$$L_c = \frac{2}{\sigma^2} = \frac{2}{(N_0/2 \cdot E_c)} = \frac{4 \cdot E_c}{N_0} \tag{4.57}$$

and

$$E_c = R \cdot E_b \tag{4.58}$$

(R is the code rate and E_b is the bit energy).

In Equation 4.51, the following hyperbolic functions are used:

$$\tanh(x) = \frac{e^{2x} - 1}{e^{2x} + 1} \tag{4.59}$$

and

$$\operatorname{atanh}(x) = \frac{1}{2}\log\frac{1 + x}{1 - x}, \quad |x| < 1. \tag{4.60}$$

The algorithm starts with $L(r_{ij}) = 0$ for all i and j, followed by the computation of $L(q_{ij})$ using Equation 4.50 and then $L(r_{ij})$ using Equation 4.51. The *a posteriori* log probability ratios for each variable node d_j can then be computed as

$$L(d_j) = \log\frac{\operatorname{Prob}\left(d_i = 1 \middle| \text{observations}\right)}{\operatorname{Prob}\left(d_i = 0 \middle| \text{observations}\right)} = L(p_j) + \sum_{i=1}^{n-k} h_{ij}L(r_{ij}). \tag{4.61}$$

If the Tanner graph did not have closed cycles, these *a posteriori* log probabilities would correspond to the exact ones. With the existence of cycles, it is necessary to repeat the computation of $L(q_{ij})$ and $L(r_{ij})$ for several iterations and the estimates computed using Equation 4.61 will be only approximations. These approximations are accurate as long as the cycles have long lengths.

In the computation of the message from parity check node c_i to variable node d_j, it is possible to avoid the use of hyperbolic functions as well as of the products of terms present in Equation 4.51 by using the following approximation:

$$L(r_{ij}) = (-1)^{\gamma_i} \min_{\substack{j'=1,\dots,n \\ j' \neq j \wedge h_{ij'} \neq 0}} \left(\left|L(q_{ij'})\right|\right) \prod_{\substack{j'=1 \\ j' \neq j \wedge h_{ij'} \neq 0}}^{n} \operatorname{sign}(L(q_{ij'})), \tag{4.62}$$

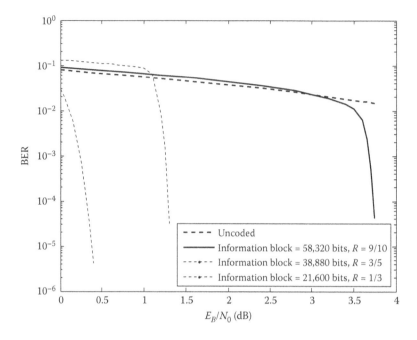

Figure 4.13   BER performance of several DVB-S2 [ETSI 2009a] LDPC codes.

where γ_i is the degree of check node i and sign (.) represents the sign function defined as

$$\text{sign}(x) = \begin{cases} 1, & x > 0 \\ 0, & x = 0 \\ -1, & x < 0 \end{cases} \qquad (4.63)$$

When using this approximation, the algorithm is named min-sum algorithm.

Figure 4.13 presents some simulation results obtained using DVB-S2 [ETSI 2009a] LDPC codes and employing the sum-product algorithm with a maximum of 50 iterations. Different coding rates are shown. It is visible that very high coding gains can be obtained using these codes.

4.2.3 Channel Estimation

In wireless communications, the propagation conditions lead to channels that distort the amplitude and phase of the transmitted signal. This distortion has to be estimated and tracked when performing coherent detection at the receiver. As an alternative, it is possible to design systems that employ noncoherent detection [Hanzo et al. 2010]. However, although the use of noncoherent systems eliminates the need for channel estimation, the cost is a significant performance degradation. Therefore, most mobile communications systems employ coherent detection and require that the amplitude and phase distortions caused by the channel be correctly estimated to avoid severe performance degradation.

Typically, the channel estimates are obtained with the help of training symbols that are multiplexed with the data symbols, either in the time domain or in the frequency domain [Cavers 1991, Hoher et al. 1997, Sanzi and Speidel 2000, Deneire et al. 2001], though the later approach is more common in OFDM systems. Regarding the transmission of pilot symbols in the frequency domain, it is possible to insert them into all the subcarriers of an OFDM symbol (block-type pilot arrangement) or periodically inserting the pilot tones inside some OFDM symbols (comb-type pilot arrangement) [Coleri et al. 2002]. In 3GPP LTE-Advanced, the later approach is used [3GPP 2013a]. The following description concerns the use of pilot symbols multiplexed with data in the frequency domain (inserted according to the transmitter scheme of Figure 4.6) and is valid for both block-type and comb-type arrangements. The assumed frame structure is shown in Figure 4.14. According to this structure, in a MIMO-OFDM system with N carriers, pilot symbols are multiplexed with data symbols using a spacing of ΔN_T OFDM blocks in the time domain and ΔN_F subcarriers in the frequency domain. To avoid interference between pilots of different transmitting antennas, frequency-division multiplexing (FDM) is often employed for the pilots, which means that pilot symbols cannot be transmitted over the same subcarrier in different antennas. Data symbols

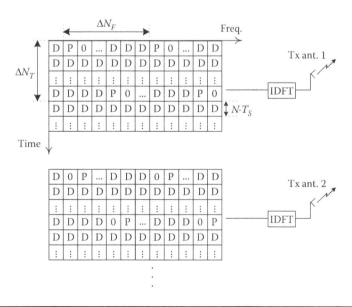

Figure 4.14 Frame structure for a MIMO-OFDM transmission with data multiplexed pilots (P—pilot symbol, D—data symbol, Ts—symbol duration).

are not transmitted on subcarriers reserved for pilots in any antenna and, therefore, the minimum spacing allowed in the frequency domain is $(\Delta N_F)_{min} = M_{tx}$.

With the pilot tones present in the transmitted signal, it is possible to use several different channel estimation techniques. As an example, we describe a typical approach where, to obtain the frequency channel response estimates for each transmitting/receiving antenna pair, the receiver applies the following steps:

1. The channel estimate between transmit antenna m and receive antenna n for each pilot symbol position is simply computed as

$$\tilde{H}_{k,l}^{m,n} = \frac{\left(S_{k,l}^{m,Pilot}\right)^*}{\left|S_{k,l}^{m,Pilot}\right|^2} R_{k,l}^n, \qquad (4.64)$$

where $S_{k,l}^{m,Pilot}$ corresponds to a pilot symbol transmitted in the kth subcarrier of the lth OFDM block using antenna m. The specific indexes k and l that correspond to pilot symbols depend on the periodicity of pilot insertion, that is, it depends on the values defined for ΔN_T and ΔN_F.

2. Channel estimates for the same subcarrier k, transmit antenna m, and receive antenna n but in time domain positions (index l) that do not carry a pilot symbol can be obtained through interpolation using a finite impulse response (FIR) filter with length W as follows:

$$\hat{H}_{k,l+t}^{m,n} = \sum_{j=-\lfloor (W-1)/2 \rfloor}^{\lfloor W/2 \rfloor} h_t^j \tilde{H}_{k,l+j\cdot\Delta N_T}^{m,n}, \qquad (4.65)$$

where t is the OFDM block index relative to the last one carrying a pilot (which is block with index l) and h_t^j are the interpolation coefficients of the estimation filter, which depend on the channel estimation algorithm employed. There are several proposed algorithms in the literature like the optimal Wiener filter interpolator [Cavers 1991] or the lowpass sinc interpolator [Kim et al. 1997]. Note that interpolation can also be applied between different OFDM symbols with pilots.

4.3 Iterative Receivers

4.3.1 Receiver Structure

A typical receiver is composed of several signal processing blocks dedicated to specific tasks. Examples of these tasks are channel estimation, MIMO decoding, equalization, multiuser detection, channel decoding, and so on. This division in several separate processing blocks is necessary since a joint processing of all these tasks, although optimal, is too complex to implement. Nevertheless, if these blocks work with soft decisions, the performance of the receiver can be improved if the first processing blocks make use of feedback information derived in the last blocks. This cyclic exchange of information between the various processing blocks can occur for several iterations although after a certain number of iterations, the gains can become negligible. By analogy with turbo codes, this type of techniques is usually referred to as turbo-processing techniques. One interesting application of these techniques is turbo equalization [Douillard et al. 1995] where equalization is combined with channel decoding. The task of the equalizer is to compensate inter-symbolic interference (ISI) present in frequency selective channels. The optimal equalizer is a trellis-based detector employing maximum likelihood sequence esti-mation (MLSE) algorithm similar to the Viterbi decoder although other algorithms are available. Iterating the equalization and channel decoding task can result in substantial improvements in the error rate.

In [Moher 1998], [Reed 1998], and [Wang and Poor 1999], the turbo prin-ciple was used for joint multiuser detection and channel decoding for obtaining near single user performance. The multiuser detector is implemented as a soft-input soft-output that uses *a priori* information from the previous channel decoder itera-tion to improve its performance and produces extrinsic information for the turbo decoders. Another possible application is turbo estimation [Valenti 2001] where the channel is reestimated after each iteration of the turbo code using the estimated transmitted symbols as additional pilots. In this section, we will explain how the turbo-processing concept can be applied to MIMO-OFDM receivers.

Figure 4.15 presents the basic structure of an iterative receiver capable of jointly performing channel estimation and data detection through iterative processing. N_{rx} receiving antennas are assumed to be used. The details about the channel estima-tion task will be discussed in the following section.

The sequence of steps applied to the received sequence is the same as the con-ventional receiver from Figure 4.10 up to the channel decoder block. In the iterative receiver, the channel decoder has two outputs: one is the estimated information sequence and the other is the sequence of LLR estimates of the code symbols (see Figure 4.12 and the description provided in Section 4.2.2). These LLRs go through the decision device, which outputs either soft-decision or hard-decision estimates of the code symbols, and enter the transmitted signal rebuilder, which performs the same operations of the transmitter (interleaving, modulation, and conversion of serial to parallel streams). The reconstructed symbol sequences can then be used for a refinement of the channel estimates and for improvement of the spatial demultiplex-ing task for the succeeding iteration. This can be accomplished using an interference canceler (IC) inside the spatial demultiplexer block, which can be represented as

$$\left(R_{k,l}^{n,m}\right)^{(q)} = R_{k,l}^{n} - \sum_{\substack{m'=1 \\ m' \neq m}}^{M_{tx}} \left(\hat{S}_{k,l}^{m'}\right)^{(q-1)} \hat{H}_{k,l}^{m',n}, \tag{4.66}$$

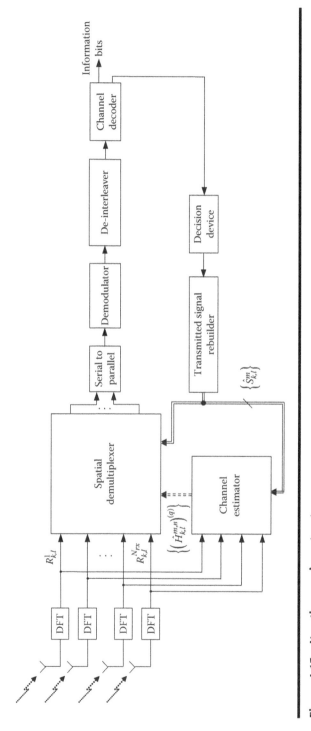

Figure 4.15 Iterative receiver structure.

where q is the iteration number and $\left(\hat{S}_{k,l}^m\right)^{(q-1)}$ represents the data symbols estimates of the previous iteration for transmit antenna m, subcarrier k, and OFDM block l.

4.3.2 Iterative Channel Estimation

In this section, we consider again the same receiver structure of the previous section but devote our attention to the task of estimating the amplitude and phase of the channel. Two pilot transmission schemes will be considered: data multiplexed pilots, which were already discussed in Section 4.2.3 and implicit pilots.

4.3.2.1 Data Multiplexed Pilots

When using an iterative receiver in a MIMO-OFDM transmission based on data multiplexed pilots, in the first iteration the channel estimates are computed as described in Section 4.2.3. However, after the first iteration, the data estimates can also be used as additional pilots for channel estimation refinement [Valenti 2001]. In this case, the respective channel estimates are computed as

$$\left(\hat{H}_{k,l}^{m,n}\right)^{(q)} = \frac{R_{k,l}^n \left(\hat{S}_{k,l}^m\right)^{(q-1)*}}{\left|\left(\hat{S}_{k,l}^m\right)^{(q-1)}\right|^2}. \tag{4.67}$$

4.3.2.2 Implicit Pilots

The transmission of training symbols multiplexed with data, either in the time domain or in the frequency domain, can result in an inefficient use of the available bandwidth, especially when the channel impulse response is very long. Therefore, it is often desirable to reduce the overheads required for channel estimation purposes. A promising method for overcoming this problem is to employ implicit pilots, also known as pilot embedding or superimposed pilots, which are added to the data block instead of being multiplexed with it [Farhang 1995, Ho et al. 2001, Zhu et al. 2003, Lugo et al. 2004]. This means that we can significantly increase the pilots' density, while keeping the system capacity. In fact, we can even have a pilot for each data symbol. However, the interference levels between the data symbols and pilots might be high. This means that the channel estimates are corrupted by the data signal, leading to irreducible noise floors (i.e., the channel estimates cannot be improved beyond a given level, even without channel noise). Moreover, there is also interference on the data symbols due to the pilots, leading to performance degradation. Therefore, this approach usually requires extra signal processing at the receiver to reduce the cross interference between pilots and data. In the following, we will describe how this type of pilot transmission can be used in MIMO-OFDM systems.

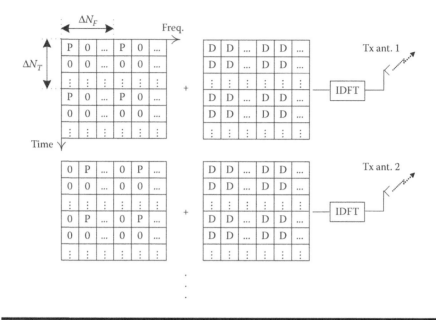

Figure 4.16 Frame structure for MIMO-OFDM transmission with implicit pilots (P—pilot symbol, D—data symbol).

The frame structure and pilot insertion process are shown in Figure 4.16. According to this structure, the implicit pilots are generated using a grid with a spacing of ΔN_T symbols in the time domain and ΔN_F symbols in the frequency domain. Similar to the data multiplexed pilots method, to avoid interference between implicit pilots of different transmitting antennas, FDM can be employed for the pilots, so that pilot symbols in different antennas are not transmitted over the same subcarrier. In this case, the minimum allowed spacing in the frequency domain is $(\Delta N_F)_{\min} = M_{tx}$ and the transmitted sequences are given by

$$X_{k,l}^m = S_{k,l}^m + S_{k,l}^{m,Pilot},\qquad(4.68)$$

where $S_{k,l}^{m,Pilot}$ is the implicit pilot transmitted over the kth subcarrier, in the lth OFDM block using antenna m. The resulting sequences are converted into the time domain through the usual process, $\left\{x_{i,l}^m, i = 0,1,\ldots,N-1\right\} = \mathrm{IDFT}\left\{X_{k,l}^m, k = 0,1,\ldots,N-1\right\}$, before being transmitted.

To reduce the mutual interference between pilots and data, which will be particularly high in MIMO systems with implicit pilots, and achieve reliable channel estimation and data detection, we can modify the previous iterative receiver structure according to Figure 4.17. The main difference from the previous receiver lies in the addition of the "remove pilots" and "remove data" processing blocks.

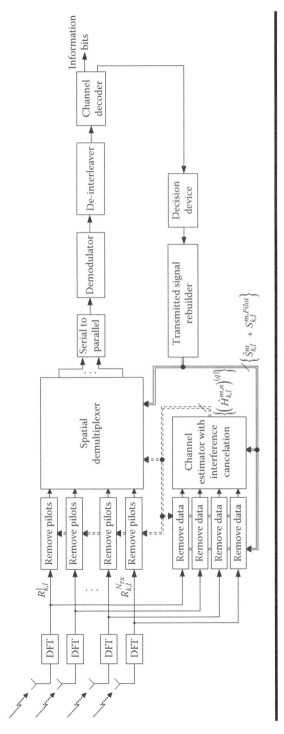

Figure 4.17 Iterative receiver structure for implicit pilot transmission.

In this case, after conversion to the frequency domain, the received sampled sequence can be expressed as

$$R_{k,l}^n = \sum_{m=1}^{M_{tx}} \left(S_{k,l}^m + S_{k,l}^{m,Pilot} \right) H_{k,l}^{m,n} + N_{k,l}^n. \tag{4.69}$$

The pilot symbols are removed from the sequence according to

$$\left(Y_{k,l}^n \right)^{(q)} = R_{k,l}^n - \sum_{m=1}^{M_{tx}} S_{k,l}^{m,Pilot} \left(\hat{H}_{k,l}^{m,n} \right)^{(q)}, \tag{4.70}$$

where $\left(\hat{H}_{k,l}^{m,n} \right)^{(q)}$ is the channel frequency response estimate and q is the current iteration. Considering the pilot distribution of Figure 4.16 where only one of the antennas can transmit a pilot over each subcarrier, the summation in Equation 4.70 has only one term. The sequences of samples (Equation 4.70) enter the spatial demultiplexer block, which can apply any of the methods discussed in Section 4.2.1. With the existence of implicit pilots, in each iteration, the receiver can apply the following steps in order to obtain the estimates of the frequency channel response for each transmitting/receiving antenna pair:

1. Data symbols estimates are removed from the pilots using

$$\left(\tilde{R}_{k,l}^n \right)^{(q)} = R_{k,l}^n - \sum_{m'=1}^{M_{tx}} \left(\hat{S}_{k,l}^{m'} \right)^{(q-1)} \left(\hat{H}_{k,l}^{m,n} \right)^{(q-1)}, \tag{4.71}$$

where $\left(\hat{S}_{k,l}^{m'} \right)^{(q-1)}$ and $\left(\hat{H}_{k,l}^{m,n} \right)^{(q-1)}$ are the data and channel response estimates of the previous iteration. This step can only be applied after the first iteration. In the first iteration, we set $\left(\tilde{R}_{k,l}^n \right)^{(1)} = R_{k,l}^n$.

2. The channel frequency response estimates is computed using a moving average with size W as follows:

$$\left(\hat{H}_{k,l}^{m,n} \right)^{(q)} = \frac{1}{W} \sum_{l'=l-\lfloor W/2 \rfloor}^{l+\lceil W/2 \rceil -1} \frac{\left(\tilde{R}_{k,l'}^n \right)^{(q-1)}}{S_{k,l'}^{m,Pilot}}. \tag{4.72}$$

3. After the first iteration, if a fully dense pilot distribution is not employed (i.e., $\Delta N_F \neq 1 \vee \Delta N_T \neq 1$), the data estimates can also be used as additional pilots for channel estimation refinement.

4. These channel estimates can be enhanced by ensuring that the corresponding impulse response has a duration N_G. This is accomplished

by computing the time domain impulse response of Equation 4.72 through $\left\{\left(\tilde{h}_{i,l}^{m,n}\right)^{(q)};i = 0,1,...,N-1\right\} = \text{DFT}\left\{\left(\tilde{H}_{k,l}^{m,n}\right)^{(q)};k = 0,1,...,N-1\right\}$, followed by the truncation of this sequence according to $\left\{\left(\hat{h}_{i,l}^{m,n}\right)^{(q)} = w_i\left(\tilde{h}_{i,l}^{m,n}\right)^{(q)};i = 0,1,...,N-1\right\}$ with $w_i = 1$ if the ith time domain sample is inside the cyclic prefix duration and $w_i = 0$ otherwise. The final frequency response estimates are then simply computed using $\left\{\left(\hat{H}_{k,l}^{m,n}\right)^{(q)};k = 0,1,...,N-1\right\} = \text{DFT}\left\{\left(\hat{h}_{i,l}^{m,n}\right)^{(q)};i = 0,1,...,N-1\right\}$.

One of the advantages of using implicit pilots is that it allows us to significantly increase the density of pilots without sacrificing system capacity. In fact, we can have a pilot for each data symbol, which can be important for fast-fading channels. As an example, Figures 4.18 and 4.19 show the performances of a 16-HQAM (nonuniform hierarchical QAM modulation with parameter $k_1 = 0.4$, as defined in [Marques da Silva et al. 2010] transmission for different speeds employing data multiplexed pilots ($\Delta N_F = 7$, $\Delta N_T = 7$, $E\left[\left|S_{k,l}^{m,Pilot}\right|^2\right]/E\left[\left|S_{k,l}^m\right|^2\right] = 0\,\text{dB}$) and implicit pilots ($\Delta N_F = 2$, $\Delta N_T = 1$, $E\left[\left|S_{k,l}^{m,Pilot}\right|^2\right]/E\left[\left|S_{k,l}^m\right|^2\right] = -4.5\,\text{dB}$). These results were obtained using an UTRA LTE-Advanced-based Monte Carlo simulator for a 10 MHz bandwidth

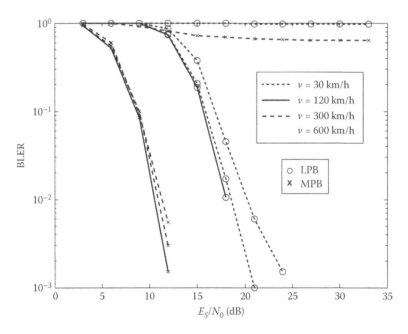

Figure 4.18 Performance of a 2×2 MIMO 16-HQAM ($k_1 = 0.4$) transmission with data multiplexed pilots ($\Delta N_F = 7$, $\Delta N_T = 7$) for different speeds.

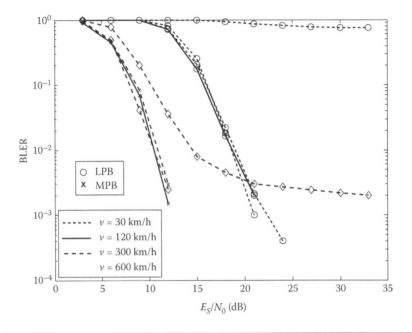

Figure 4.19 Performance of a 2×2 MIMO 16-HQAM ($k_1 = 0.4$) transmission with implicit pilots ($\Delta N_F = 2$, $\Delta N_T = 1$, $E\left[\left|S_{k,l}^{m,Pilot}\right|^2\right]\Big/E\left[\left|S_{k,l}^m\right|^2\right] = -4.5\,\text{dB}$) for different speeds.

configuration whose corresponding parameters are shown in Table 4.1. The channel impulse response is based on the Vehicular A environment [from ETSI, 1998] with Rayleigh fading adopted for the different paths. The channel encoders were rate-1/2 turbo codes based on two identical recursive convolutional codes as defined in [3GPP 2013b] with individual data block sizes of 4024 bits.

Table 4.1 Simulation Parameters for 10 MHz Bandwidth

Transmission BW	10 MHz
CP length (samples)	72
FFT size	1024
Number of occupied subcarriers	600
Subframe duration (ms)	0.5
Subcarrier spacing (kHz)	15
OFDM symbols per subframe	7

Note that the modulation employed, 16-HQAM, is a hierarchical one, allowing the transmission of two data streams with different error protection levels, as explained in [Marques da Silva et al. 2010]. Therefore, in the graph legends, MPB curves refer to the most protected bits and LPB corresponds to the least protected bits. According to the results, the performance for the data multiplexed transmission is almost insensitive to speeds up to 300 km/h, being visible only a small degradation in the performance of the LPB at these speeds. For higher velocities, the performance quickly degrades for both bit streams, becoming completely unreliable. With the use of implicit pilots, it is clear that the performance becomes more robust. In this case, the results are also practically insensitive to velocity for both streams until $v = 300$ km/h. For $v = 600$ km/h, the performance degrades substantially but it is still possible to obtain BLERs below 1% for the MPB, which means that even at such speeds it would be possible for a user to extract the basic information stream (e.g., a low-quality video).

4.4 Lattice-Reduction-Aided Receivers

4.4.1 Soft Metrics from LRA Receivers

As it was described in Chapter 2 for both real and complex lattices, there are several relatively low-complexity detection methods that can be applied to MIMO and those detection techniques can be easily extended to MIMO-OFDM. However, most of those methods often perform quite far from ML detection and since the use of ML detectors is only viable for a reduced number of antennas and small signal constellations, other solutions must be applied. As described in Chapter 2, one of the most promising approaches is the use of lattice reduction using for example the LLL algorithm [Lenstra et al. 1982] combined with either a linear equalizer, a successive interference cancelation (SIC) equalizer [Babai 1986], or a randomized version of SIC [Liu et al. 2011]. As explained in Chapter 2, lattice reduction is one of the most useful tools of lattice theory that allows us to find a new basis with improved mathematical properties that result in a gain in performance when a low-complexity detection method is applied.

In order to apply a lattice-reduction-aided receiver in a MIMO-OFDM system, we can start by writing the $N_{rx} \times 1$ received vector **Y** for a subcarrier frequency as

$$\mathbf{Y} = \bar{\mathbf{H}}\mathbf{S} + \mathbf{N}, \tag{4.73}$$

where $\bar{\mathbf{H}}$ is the $N_{rx} \times M_{tx}$ channel matrix for that frequency with each column representing a different transmit antenna and each line representing a different receive antenna, **S** is the $M_{tx} \times 1$ column vector whose elements are the complex-valued symbols transmitted on the M_{tx} antennas and **N** is the $N_{rx} \times 1$ column vector containing the noise samples at each receiving antenna. Similar to the procedure described in Chapter 2, if the constellation of transmitted symbols can be mapped to a subset of the Gaussian integers space $\mathbb{Z}^{M_{tx}}[j]$ through shifting and scaling (e.g., M-QAM constellations), we can write

$$S = \alpha\left[\mathbf{x} - \left(\frac{1}{2} + \frac{1}{2}j\right)\mathbf{I}_{M_{tx}}\right], \tag{4.74}$$

where $\mathbf{x} \in \mathbb{Z}^{M_{tx}}[j]$, α represents the factor used for energy normalization, and $\mathbf{I}_{M_{tx}}$ denotes an $M_{tx} \times M_{tx}$ identity matrix. In this case, a lattice basis can be represented by matrix $\mathbf{H}$ defined as

$$\mathbf{H} = \alpha\bar{\mathbf{H}}. \tag{4.75}$$

The transmitted vector estimate can then be achieved through shifting and rescaling as

$$\hat{\mathbf{S}} = \alpha\left(\mathbf{U}\hat{\mathbf{z}} - \left(\frac{1}{2} + \frac{1}{2}j\right)\mathbf{I}_{M_{tx}}\right), \tag{4.76}$$

where $\mathbf{U}$ is a unimodular matrix (with $|\det \mathbf{U}| = 1$) containing only Gaussian integers that allows us to obtain a new improved basis $\tilde{\mathbf{H}}$ (namely, shorter basis vectors and better orthogonality) using

$$\tilde{\mathbf{H}} = \mathbf{HU}, \tag{4.77}$$

and $\hat{\mathbf{z}}$ is an $M_{tx} \times 1$ column vector whose elements can be computed using the rounding and subtraction procedure of the SIC algorithm [Babai 1986], as described in Chapter 2. As also detailed in Chapter 2, instead of the ZF criterion, we can use the MMSE criterion by applying the SIC algorithm to the extended channel matrix and the extended received vector defined as

$$\underline{\mathbf{H}} = \begin{bmatrix} \mathbf{H} \\ \sqrt{2}\sigma\mathbf{I}_{M_{tx}} \end{bmatrix}$$

and

$$\underline{\mathbf{Y}} = \begin{bmatrix} \mathbf{Y} \\ \mathbf{0}_{M_{tx} \times 1} \end{bmatrix}, \tag{4.78}$$

where $\mathbf{0}_{M_{tx} \times 1}$ is an all zeros size-M_{tx} column vector. Furthermore, to narrow the performance gap to the ML decoder, instead of the standard rounding to the nearest Gaussian integers employed in the SIC algorithm, we can replace it by Klein's randomized rounding [Klein 2000] (c.f. Chapter 2) and implement a randomized lattice decoder.

As explained in Section 4.1.3, due to its poor uncoded performance, OFDM systems are usually combined with coding. However, to avoid a performance loss in the channel decoding task, the MIMO detector should be able to output LLRs for the demodulated bits. The use of randomized rounding in the randomized lattice decoder eases the computation of the required LLRs. Indeed, these can be obtained as

$$
\begin{aligned}
\lambda_i &= \log\left(\frac{p\left(c_i = 1 | \mathbf{y}\right)}{p\left(c_i = 0 | \mathbf{y}\right)}\right) \\
&= \log\left(\frac{\sum_{\mathbf{s} \in \Upsilon_{c_i=1}} p\left(\mathbf{Y} | \mathbf{S}\right)}{\sum_{\mathbf{s} \in \Upsilon_{c_i=0}} p\left(\mathbf{Y} | \mathbf{S}\right)}\right),
\end{aligned}
\tag{4.79}
$$

where $p(c_i | \mathbf{Y})$ is the probability of coded bit c_i conditioned on $\mathbf{Y}$, $\Upsilon_{c_i=1}$ ($\Upsilon_{c_i=0}$) represents the set of all possible transmitted vectors from the candidates' list containing $c_i = 1$ ($c_i = 0$) and $p(\mathbf{Y} | \mathbf{S})$ is the likelihood probability, which can be computed using

$$
p\left(\mathbf{Y} | \mathbf{S}\right) = \frac{1}{\left(2\pi\sigma\right)^{N_{rx}}} \exp\left(-\frac{1}{2\sigma^2}\left\|\mathbf{Y} - \bar{\mathbf{H}}\mathbf{S}\right\|^2\right).
\tag{4.80}
$$

4.4.2 Performance Results

To compare randomized lattice decoding against some of the MIMO-OFDM receivers described previously in this chapter, we present in Figure 4.20 the performance results for a 4×4 MIMO configuration, which were obtained using an UTRA LTE-Advanced-based Monte Carlo simulator for a 10 MHz bandwidth configuration whose corresponding parameters are the same as shown in Table 4.1 of Section 4.3.2. The channel impulse response is based on the Vehicular A environment [from ETSI, 1998] with Rayleigh fading adopted for the different paths. The channel encoders were rate-1/2 turbo codes based on two identical recursive convolutional codes as defined in [3GPP 2013b] with individual data block sizes of 4196 bits. At the receiver, 12 turbo decoding iterations are applied, but in the case of the iterative receiver, we used four receiver iterations with three turbo decoding iterations per receiver iteration (giving still a total of 12 turbo decoding iterations).

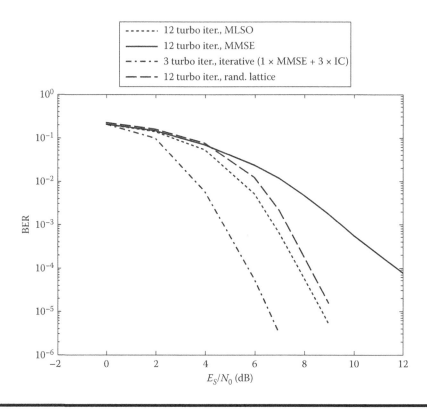

Figure 4.20 **Performance of a 4 × 4 MIMO QPSK transmission with *R* = 1/2 turbo coding and different receivers.**

In the case of the randomized lattice decoder, the MMSE criterion was employed combined with a candidate list size of $K = 25$. Perfect channel estimation and synchronization was assumed.

We can see that the iterative receiver achieves the best performance as it is the only one where the channel decoder also contributes to the MIMO decoding task. Regarding the non-iterative receivers, as expected, the best performance is achieved by the MLSO followed closely by the randomized lattice decoder. However, the MMSE receiver shows a considerable performance gap compared with the MLSO.

Although the randomized lattice decoder can achieve performances close to ML decoding, due to the random nature of the sampling, unnecessary complexity is often added (repetition of sampled points) and some performance loss can occur as a result of some points with low probabilities on the first elements being missed. To avoid these issues, a deterministic version of the algorithm, referred to as derandomized sampling algorithm, has been recently proposed in [Wang and Ling 2012].

4.5 MIMO-OFDM with Signal Space Diversity

4.5.1 Complex Rotation Matrices

When powerful channel coding schemes are employed, OFDM schemes can have excellent performance. However, the required code rate must be low, reducing the system's spectral efficiency. However, for uncoded systems or when high rate codes are employed, the performance of OFDM systems can be very poor. In this case, we can associate a given symbol to different subcarriers so as to take advantage of the diversity effects inherent to a severely frequency selective channel, which is typical in mobile communication environments. This alternative diversity technique is often named signal space diversity as proposed in [Boutros and Viterbo 1998] and is accomplished without additional power or bandwidth required. A simple way of doing this is to employ real rotation matrices (RRM) [Rainish 1996], which allows significant gains. Unfortunately, RRM were only designed to spread a symbol over two subcarriers, which is accomplished using

$$\mathbf{X} = \mathbf{A}_{RRM} \cdot \mathbf{S},\tag{4.81}$$

with

$$\mathbf{A}_{RRM} = \begin{bmatrix} \cos(\varphi) & \sin(\varphi) \\ -\sin(\varphi) & \cos(\varphi) \end{bmatrix},\tag{4.82}$$

and $\mathbf{S}$ being a 2×1 vector containing two modulated symbols.

Spreading a symbol over a larger number of subcarriers can be accomplished using the Hadamard matrix (HM) adopted in fully loaded multicarrier code division multiplexing schemes. The HM for spreading over two symbols can be obtained from RRM using $\varphi = \pi/4$. A more general alternative lies on the use of CRM. CRM is a technique for achieving signal space diversity (SSD) in SISO and MIMO-OFDM/OFDMA systems and can be easily combined with turbo or LDPC codes in order to improve the system performance without a substantial reduction of the spectral efficiency.

The process of applying CRM is similar to RRM where a rotated supersymbol is obtained using

$$\mathbf{X} = \mathbf{A}_{M_{CRM}} \cdot \mathbf{S},\tag{4.83}$$

with S being an $M_{CRM} \times 1$ vector with a set of modulated symbols composing a supersymbol. Matrix $\mathbf{A}_{M_{CRM}}$ belongs to the family of the orthonormal (OCRM) or nonorthonormal (NCRM) complex matrices, which are defined as

$$(\text{OCRM})\mathbf{A}_{M_{CRM}} = \begin{cases} \begin{bmatrix} e^{j\varphi} & je^{-j\varphi} \\ -je^{j\varphi} & e^{-j\varphi} \end{bmatrix} \Big/ |\mathbf{A}_2|^{1/2}, M_{CRM} = 2 \\ |\mathbf{A}_2| = \det(A_2) = 2 \\ \begin{bmatrix} \mathbf{A}_{M_{CRM}/2} & \mathbf{A}_{M_{CRM}/2} \\ \mathbf{A}_{M_{CRM}/2} & -\mathbf{A}_{M_{CRM}/2} \end{bmatrix} \Big/ |\mathbf{A}_{M_{CRM}}|^{1/M_{CRM}}, M_{CRM} > 2 \end{cases} \tag{4.84}$$

and

$$(\text{NCRM})\mathbf{A}_{M_{CRM}} = \begin{cases} \begin{bmatrix} e^{j\varphi} & e^{-j\varphi} \\ -e^{-j\varphi} & e^{j\varphi} \end{bmatrix} \Big/ |\mathbf{A}_2|^{1/2}, M_{CRM} = 2 \\ |\mathbf{A}_2| = \det(A_2) = 2\cos(\varphi) \\ \begin{bmatrix} \mathbf{A}_{M_{CRM}/2} & \mathbf{A}_{M_{CRM}/2} \\ \mathbf{A}_{M_{CRM}/2} & -\mathbf{A}_{M_{CRM}/2} \end{bmatrix} \Big/ |\mathbf{A}_{M_{CRM}}|^{1/M_{CRM}}, M_{CRM} > 2 \end{cases}, \tag{4.85}$$

with $M_{CRM} = 2^n (n \geq 1)$, $|\mathbf{A}_{M_{CRM}}| = \det(\mathbf{A}_{M_{CRM}})$, and φ being the rotation angle [Correia 2002].

4.5.2 Transmitter for OFDM Schemes with CRM

CRM can be easily incorporated into OFDM systems. Figure 4.21 shows the block diagram of an OFDM transmitter with CRM and multiple transmitting antennas.

An information block is encoded, interleaved, and mapped according to the constellation symbols. A rotation matrix (RM) is then applied by grouping the symbols into M_{CRM}-tuples and multiplying them by rotation matrix $\mathbf{A}_{M_{CRM}}$. The resulting sequence is split into M_{tx} parallel streams, which are interleaved in the symbol interleaver. The objective of the symbol interleaver is to explore the characteristics of OFDM transmissions in severe time-dispersive environments whose channel frequency response can change significantly between different subcarriers. The interleaver ensures that samples of a supersymbol are mapped to distant subcarriers, thus taking advantage of the diversity in the frequency domain. Pilot symbols are inserted into the modulated symbols sequence, which is then converted into the time domain using an IDFT and transmitted as a conventional OFDM transmission (with the insertion of the CP, which is not explicitly shown).

4.5.3 Receivers for OFDM Schemes with CRM

Figure 4.22 represents the receiver block diagram for MIMO-OFDM transmissions with CRM. N_{rx} receiving antennas are assumed to be used. Although the

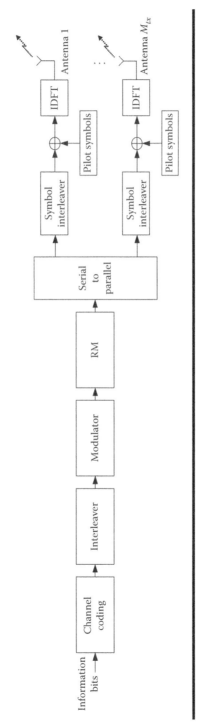

Figure 4.21   Transmitter block diagram for MIMO-OFDM transmissions using CRM.

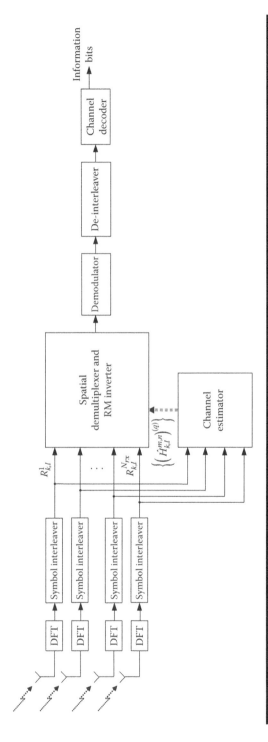

Figure 4.22 Receiver structure for MIMO-OFDM transmissions with CRM.

structure presented in Figure 4.22 corresponds to a noniterative receiver and assumes data multiplexed pilots, we could easily adopt an iterative receiver as well as implicit pilots from Section 4.3, since the only blocks that need to be modified and added are the 'spatial demultiplexer and RM inverter' block and the 'symbol interleaver' blocks.

According to Figure 4.22, the signal is sampled, the cyclic prefix removed, and the resulting signal is converted into the frequency domain with an appropriate size-N DFT operation. The sequence of symbols is then de-interleaved. If the cyclic prefix is longer than the overall channel impulse response, each received M_{CRM}-sized supersymbol can be expressed using matrix notation as

$$\mathbf{R} = \mathbf{H} \cdot \mathbf{X} + \mathbf{N}, \tag{4.86}$$

where $\mathbf{R}$ is a $(N_{rx.} \, M_{CRM}/M_{tx}) \times 1$ vector containing the samples of a supersymbol received in all the antennas and $\mathbf{H}$ is the frequency response channel matrix defined as a blockwise diagonal matrix

$$\mathbf{H} = \begin{bmatrix} \mathbf{H}_1 & & \mathbf{0} \\ & \ddots & \\ \mathbf{0} & & \mathbf{H}_{M_{CRM}/M_{tx}} \end{bmatrix}, \tag{4.87}$$

with

$$\mathbf{H}_k = \begin{bmatrix} H_k^{1,1} & \cdots & H_k^{1,M_{tx}} \\ \vdots & \ddots & \vdots \\ H_k^{N_{rx},1} & \cdots & H_k^{N_{rx},M_{tx}} \end{bmatrix}, \quad k = 1,\ldots,M_{CRM}/M_{tx}. \tag{4.88}$$

Index k represents a subcarrier position. It is important to note that due to the presence of the symbol interleaver, the different subcarriers denoted by index k may not be necessarily adjacent. To simplify the following explanations, we will assume that M_{CRM} is a multiple of the number of transmitting antennas M_{tx}. $\mathbf{N}$ is a $(N_{rx.} \, M_{CRM}/M_{tx}) \times 1$ vector containing AWGN samples.

The supersymbol's samples enter the spatial demultiplexer and CRM inverter block, which has the purpose of separating the streams transmitted simultaneously by the multiple antennas and invert the rotation applied at the transmitter.

In the following, some of the methods that were presented for traditional MIMO-OFDM systems in Sections 4.2.1 and 4.4 will be extended to MIMO-OFDM with CRM: the MMSE equalizer, the MLSO detector, and lattice-reduction-aided decoders.

The MMSE criterion can be applied to each individual subcarrier using

$$\hat{\mathbf{X}}_k = (\mathbf{H}_k)^H \cdot [\mathbf{H}_k(\mathbf{H}_k)^H + \sigma^2\mathbf{I}]^{-1}\mathbf{R}_k, \tag{4.89}$$

where $\hat{\mathbf{X}}_k$ is the $M_{tx} \times 1$ vector with the estimated subset of coordinates from the supersymbol mapped to subcarrier k, $\mathbf{R}_k$ is the $N_{rx} \times 1$ received signal vector in subcarrier k with one different receive antenna in each position and σ^2 is the noise variance. Using the rotated supersymbol estimates, $\hat{\mathbf{X}}_k$, the component symbol estimates are computed through

$$\hat{\mathbf{S}} = (\mathbf{A}_{M_{CRM}})^{-1} \cdot \hat{\mathbf{X}}. \tag{4.90}$$

In the case of using the MLSO criterion, each symbol estimate is computed as

$$\begin{aligned}
\hat{S}_l &= E\left[S_l|\mathbf{R}\right] \\
&= \sum_{s_i \in \Lambda} s_i \cdot P\left(S_l = s_i|\mathbf{R}\right) \\
&= \sum_{s_i \in \Lambda} s_i \cdot \frac{P\left(S_l = s_i\right)}{p(\mathbf{R})} p\left(\mathbf{R}|S_l = s_i\right),
\end{aligned} \tag{4.91}$$

with s_i representing a constellation symbol from the modulation alphabet Λ, $E[\cdot]$ denoting the expected value, $P(\cdot)$ denoting a discrete probability, and $p(\cdot)$ denoting a probability density function (PDF). Considering equiprobable symbols, we have $P(S_l = s_i) = 1/M$, where M is the constellation size. The PDF values required in Equation 4.91 can be computed as

$$p\left(\mathbf{R}|S_l = s_i\right) = \frac{1}{M^{M_{CRM}-1}} \sum_{S_l^{compl} \in \Lambda^{M_{CRM}-1}} p\left(\mathbf{R}|S_l = s_i, S_l^{compl}\right), \tag{4.92}$$

with

$$\begin{aligned}
p\left(\mathbf{R}|S_l = s_i, S_l^{compl}\right) &= \frac{1}{(2\pi\sigma^2)^{N_{rx} M_{CRM}/M_{tx}}} \\
&\quad \exp\left[\sum_{n=1}^{N_{rx} M_{CRM}/M_{tx}} -\frac{\left|R_n - \mathbf{H}(n,:) \cdot \mathbf{A}_{M_{CRM}} \cdot \mathbf{s}\right|^2}{2\sigma^2}\right],
\end{aligned} \tag{4.93}$$

where S_l^{compl} is a $(M_{CRM} - 1) \times 1$ vector representing a possible combination of symbols transmitted together with S_l in the same supersymbol, $\mathbf{s}$ is an $M_{CRM} \times 1$ vector comprising S_l^{compl} and s_i, R_n is the nth received sample in Equation 4.86, and $\mathbf{H}(n,:)$ is the nth line of channel matrix $\mathbf{H}$.

Finally, regarding the use of lattice-reduction-aided decoders, we start by rewriting Equation 4.86 as

$$\mathbf{Y} = \bar{\mathbf{H}}\mathbf{A}_{M_{CRM}}\mathbf{S} + \mathbf{N}, \tag{4.94}$$

where $\mathbf{Y}$ is the same received vector as $\mathbf{R}$ in Equation 4.86 and $\bar{\mathbf{H}}$ is the same frequency response channel matrix defined as a blockwise diagonal matrix in Equation 4.87 (we just changed the symbols in order to match the definitions used in Section 4.4). With the received vector represented as Equation 4.94, it is easy to conclude that all the expressions in Section 4.4, including the randomized lattice decoder algorithm (detailed in Chapter 2) and the corresponding expressions for computation of LLRs, can be directly applied. The only difference is the definition of $\mathbf{H}$, which becomes

$$\mathbf{H} = \alpha\bar{\mathbf{H}}\mathbf{A}_{M_{CRM}}. \tag{4.95}$$

Independent of the decoding method used, the resulting symbol estimates are serialized, demodulated, and de-interleaved before entering the channel decoder block, which produces the final estimate of the information sequence.

Although CRM is usually studied for usage with multiple transmitting antennas, the fast-varying channel frequency response of severe time-dispersive environments typical in OFDM systems can make the use of CRM very attractive even for single antenna transmissions. Figure 4.23 shows the BER performance of a SISO-OFDM transmission with QPSK modulation, LDPC codes with different coding rates, and two different CRM matrix sizes: $M_{CRM} = 2$ and $M_{CRM} = 16$. The angle used for the CRM matrices, which can affect the system performance as studied in [Seguro et al. 2011], was $\varphi = 30°$. The results we obtained using an UTRA LTE-Advanced Monte Carlo simulator for a 10 MHz bandwidth configuration whose corresponding parameters are the same as shown in Table 4.1 of Section 4.3.2. The channel impulse response is based on the typical urban (TypU) environment [3GPP 2011c] with Rayleigh fading assumed for the different paths. It is visible that the SSD gain provided by CRM increases with the size of the CRM matrix but is strongly dependent on the coding rate of the LDPC code. The SSD gain is higher for higher coding rates because of the lower coding gain of these codes (less channel bit redundancy). When the redundancy of the channel code increases, the SSD gain is not so noticeable due to the higher diversity gains that channel coding can offer compared to SSD.

Figure 4.24 illustrates the impact of the type of receiver on a MIMO-OFDM system with CRM. These curves were obtained for a 2×2 MIMO configuration combined with CRM16, 64-QAM modulation, and without coding. Curves for

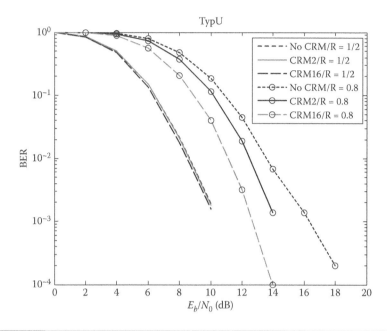

Figure 4.23 BLER performance of a SISO QPSK transmission with LDPC and CRM, TypU channel.

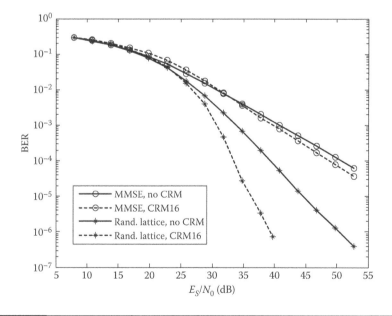

Figure 4.24 BER performance of an uncoded 2×2 MIMO 64-QAM transmission, with CRM16 and different receivers, VehA channel.

MMSE and randomized lattice decoder (applying MMSE with a candidate list size of $K = 25$) are shown. As a reference, performance curves of the receivers without CRM are also included. It is visible that, besides the large gap between the performances of the MMSE receiver and the randomized lattice decoder, which was already realized in Section 4.4.2, the performance gain achieved with the use of CRM is substantially larger than with the MMSE receiver.

Acknowledgments

This work was partially supported by FCT—Fundação para a Ciência e Tecnologia via projects ADIN—Advanced PHY/MAC Design for Infrastructure-less Networks PTDC/EEI-TEL/2990/2012, *LTE-Advanced enhancements using Femtocells* PTDC/EEA-TEL/120666/2010, and PEst-OE/EEI/LA0008/2013.

References

3GPP, Deployment aspects, TR 25.943 v10.0.0, April 2011.

3GPP, Evolved Universal Terrestrial Radio Access (E-UTRA); LTE physical layer; General description TS 36.201 v11.1.0, December 2012.

3GPP, Evolved Universal Terrestrial Radio Access (E-UTRA); Physical channels and modulation, TS 36.211 v11.3.0, June 2013.

3GPP, Evolved Universal Terrestrial Radio Access (E-UTRA); Multiplexing and channel coding, TS 36.212 v11.3.0, June 2013.

Alamouti, S. M. A simple transmit diversity technique for wireless communications, *IEEE J. Select. Areas Commun.*, 16(8), 1451–1458, 1998.

Babai, L. On Lovász' lattice reduction and the nearest lattice point problem, *Combinatorica*, 6(1), 1–13, 1986.

Bahl, L. R., J. Cocke, F. Jeinek, and J.Raviv. Optimal decoding of linear codes for minimizing symbol error rate, *IEEE Trans. Inform. Theory*, IT-20, 248–287, March 1974.

Benedetto S. and G. Montorsi, A soft-input soft-output APP module for iterative decoding of concatenated codes, *IEEE Commun. Lett.*, 1(1), 22–24, 1997.

Boutros J. and E. Viterbo, Signal space diversity: A power-and bandwidth-efficient diversity technique for the Rayleigh fading channel, *IEEE Trans. Inform. Theory*, 44(4), 1453–1467, 1998.

Cavers, J. K. An analysis of pilot symbol assisted modulation for Rayleigh fading channels, *IEEE Trans. Veh. Technol.*, 40(4), 686–693, 1991.

Coleri, S., M. Ergen, A. Puri, and A. Bahai, Channel estimation techniques based on pilot arrangement in OFDM systems, *IEEE Trans. Broadcast.*, 48(3), 223–229, 2002.

Correia, A. Optimised complex constellations for transmitter diversity, *Wireless Pers. Commun. J.*, 20(3), 267–284, 2002.

Deneire, L., B. Gyselinckx, and M. Engels, Training sequence versus cyclic prefix—A new look on single carrier communications, *IEEE Comm. Lett.*, 5, 292–294, 2001.

Divsalar D. and F. Pollara, Multiple turbo codes, *Proceedings of the IEEE Military Communications Conference*, San Siego, USA, pp. 279–285, November 1995.

Douillard, C., M. Jezequel, C. Berrou, A. Picart, P. Didier, and A. Glavieux, Iterative correction of intersymbol interference: Turbo-equalization, *Eur. Trans. Telecom.*, 6, 507–511, 1995.

ETSI, TR 101 112 v3.2.0, Selection procedures for the choice of radio transmission technologies of UMTS, Sophia Antipolis, France, 1998.

ETSI EN 302 307 V1.2.1: Digital Video Broadcasting (DVB); Second generation framing structure, channel coding and modulation systems for broadcasting, interactive services, news gathering and other broadband satellite applications (DVB-S2), 2009a.

ETSI EN 302 755 V1.1.1 Frame structure channel coding and modulation for a second generation digital terrestrial television broadcasting system (DVB-T2), 2009b.

Farhang-Boroujeny, B. Pilot-based channel identification: A proposal for semi-blind identification of communication channels, *Electron. Lett.*, 31(13), 1044–1046, 1995.

Foschini, G. J. Layered space-time architecture for wireless communication in a fading environment when using multiple antennas, *Bell Labs Tech. J.*, 1(2), 41–59, 1996.

Fossorier, M. P. C., F. Burkert, S. Lin, and J. Hagenauer, On the equivalence between SOVA and Max-Log-MAP decodings, *IEEE Commun. Lett.*, 2(5), May 1998.

Friedberg, S. H., A. J. Insel, and L. E. Spence. *Linear Algebra*, Prentice-Hall, USA, 2002.

Gallager, R. G. Low-density parity-check codes, *IRE Trans. Inform. Theory*, IT-8, 21–28, 1962.

Hagenauer J. and P. Hoher, A Viterbi algorithm with soft-decision outputs and its applications, *Proc. IEEE GLOBECOM'89*, Dallas, U.S., 1989.

Hanzo, L. L., Y. Akhtman, L. Wang, and M. Jiang, *MIMO-OFDM for LTE, WiFi and WiMAX: Coherent versus Non-Coherent and Cooperative Turbo Transceivers*, Wiley-IEEE Press, UK, 2010.

Heegard C. and S. B. Wicker, *Turbo Coding*, Kluwer Academic Publishers, USA, 1999.

Ho, C., B. Farhang-Boroujeny, and F. Chin, Added pilot semi-blind channel estimation scheme for OFDM in fading channels, *IEEE GLOBECOM'01*, San Antonio, TX, USA, November 2001.

Hoher, P., S. Kaiser, and P. Robertson, Pilot-symbol-aided channel estimation in time and frequency, *Proceedings of the IEEE Communication Theory Mini-Conference (CTMC)*, *IEEE GLOBECOM'97*, Phoenix, AZ, USA, pp. 90–96, 1997.

IEEE, IEEE Standard for local and metropolitan area networks—Part 16: Air Interface for Broadband Wireless Access Systems IEEE Std 802.16e-2005, February 2006.

IEEE, IEEE Standard for local and metropolitan area networks—Part 16: Air Interface for Broadband Wireless Access Systems IEEE Std 802.16 m-2011, May 2011a.

IEEE, IEEE Standard for Information technology—Telecommunications and information exchange between systems—Local and metropolitan area networks—specific requirements—Part 11: Wireless LAN Medium Access Control (MAC) and Physical Layer (PHY) Specifications. Amendment 5: Enhancements for Higher Throughput IEEE Std 802.11n-2009, October 2011b.

Kay, S. M. *Fundamentals of Statistical Signal Processing: Estimation Theory*. Prentice-Hall, Englewood Cliffs, NJ, 1993.

Kim, Y.-S., C.-J. Kim, G.-Y. Jeong, Y.-J. Bang, H.-K. Park, and S. S. Choi, New Rayleigh fading channel estimator based on PSAM channel sounding technique, *Proceedings of the IEEE International Conference on Communication*, Montreal, Canada, pp. 1518–1520, June 1997.

Klein, P. Finding the closest lattice vector when it's unusually close, *Proceedings of ACM-SIAM Symposium on Discrete Algorithms*, San Francisco, CA, USA, pp. 937–941, 2000.

Lenstra, A. K., H. W. Lenstra, and L. Lovász, Factoring polynomials with rational coefficients, *Mathematische Annalen*, 261(4), 515–534, 1982.

Liu, S., C. Ling, and D. Stehlé, Decoding by sampling: A randomized lattice algorithm for bounded distance decoding, *IEEE Trans. Inform. Theory*, 57(9), 5933–5945, 2011.

Orozco-Lugo, A., M. Lara, and D. McLernon, Channel estimation using implicit training, *IEEE Trans. Sig. Proc.*, 52(1), 240–254, 2004.

MacKay D. J. C. and R. M. Neal, Good codes based on very sparse matrices, in Boyd C. (Ed.), *Cryptography and Coding, 5th IMA Conference*, Springer, UK, pp. 100–111, 1995 (number 1025 in Lecture Notes in Computer Science).

MacKay D. J. C. and R. M. Neal, Near Shannon limit performance of low density parity check codes, *Electron. Lett.*, 33(6), 457–458, 1996.

Marques da Silva, M., A. Correia, R. Dinis, N. Souto, and J. Silva, *Transmission Techniques for Emergent Multicast and Broadcast Systems*, CRC Press, Taylor & Francis Group, Boca Raton, USA, 2010.

Moher, M. An iterative multiuser decoder for near-capacity communications, *IEEE Trans. Commun.*, 46(7), 870–880, 1998.

Rainish, D. Diversity transform for fading channels, *IEEE Trans. Commun.*, 44(12), 1653–1661, 1996.

Reed, M. C. Iterative multiuser detection for CDMA with FEC: Near single user performance, *IEEE Trans. Commun.*, 46(12), 1693–1699, 1998.

Richardson T. and R. Urbanke, The capacity of low-density parity-check codes under message-passing decoding, *IEEE Trans. Inform. Theory*, 47(2), 599–618, 2001a.

Richardson T. and R. Urbanke, Efficient encoding of low-density parity check codes, *IEEE Trans. Inform. Theory*, 47(2), 638–656, 2001b.

Robertson P., E. Villebrun, and P. Hoeher. A Comparison of Optimal and Sub-Optimal MAP Decoding Algorithms Operating in the Log Domain. *Proc. IEEE International Conference on Communications (ICC '95)*, vol. 2, pp. 1009-1013, Seattle, 1995.

Sanzi F. and J. Speidel, An adaptive two-dimensional channel estimator for wireless OFDM with application to mobile DVB-T, *IEEE Trans. Broadcast.*, 46, 128–133, 2000.

Seguro, J., P. Gomes, N. Souto, A. Correia, and R. Dinis, Rotation matrices for OFDM transmission, *Proceedings of EUROCON and CONFTELE 2011*, Lisbon, Portugal, April, 2011.

Souto, N., A. Correia, R. Dinis, J. C. Silva, and L. Abreu, Multiresolution MBMS transmissions for MIMO UTRA LTE systems, *Proceedings of the IEEE International Symposium on Broadband Multimedia Systems and Broadcasting*, Las Vegas, USA, March 31–April 2, 2008.

Tanner, R. M. A recursive approach to low complexity codes, *IEEE Trans. Inform. Theory*, 27(5), 533–547, 1981.

Valenti, M. C. Iterative channel estimation and decoding of pilot symbol assisted turbo codes over flat-fading channels, *IEEE J. Sel. Areas Commun.*, 19(9), 1697–1705, 2001.

Wang Z. and C. Ling, Derandomized sampling algorithm for lattice decoding, *IEEE Information Theory Workshop*, Lausanne, Switzerland, September 2012.

Wang X. and H. V. Poor, Iterative (turbo) soft interference cancellation and decoding for coded CDMA, *IEEE Trans. Commun.*, 47(7), 1046–1061, 1999.

Zhu, H., B. Farhang-Boroujeny, and C. Schlegel, Pilot embedding for joint channel estimation and data detection in MIMO communication systems, *IEEE Commun. Lett.*, 7(1), 30–32, 2003.

Chapter 5

MIMO Optimized for Single-Carrier Frequency-Domain Equalization

Rui Dinis, João Carlos Silva, and Nuno Souto

Contents

5.1 Introduction

Future wireless systems are required to support high quality of service at high data rates. Moreover, owing to power and bandwidth constraints, these systems are also required to operate with small transmit powers, especially at the mobile terminals (MTs), and to have high spectral efficiencies.

It is well known that, by employing multiple antennas at both the transmitter and the receiver, we can substantially increase the capacity of a given system [Foschini 1996]. To exploit this potential, several Bell Laboratory layered space–time (BLAST) architectures have been proposed for flat fading multiple-input, multiple-output (MIMO) channels [Foschini and Gans 1998; Wolniasky et al. 1998]. BLAST techniques were extended to frequency-selective channels through the use of time-domain MIMO decision feedback equalizers (MIMO-DFE) [Al-Dhahir and Sayed 2000; Lozano and Papadias 2002].

This concept can be extended to space-division multiple access (SDMA) techniques where we employ multiple antennas at the base station (BS) to increase the number of simultaneous users in a given cell, allowing a significant increase in the system spectral efficiency, while reducing the transmit power requirements for the MTs [Winters et al. 1992; Gitlin et al. 1994; Tidestav et al. 1999; Ariyavisitakul 2000; Abe and Matsumoto 2003; Roy and Falconer 2003; Sfar et al. 2003].

However, for the high data rates of broadband wireless systems, we can have severe time-dispersion effects associated with the multipath propagation. In this case, conventional time-domain equalization schemes are not practical, since the number of operations per symbol is proportional to the intersymbol interference (ISI) span. This can be more serious when conventional time-domain equalization methods are employed in high data rate MIMO systems.

Block transmission techniques, with appropriate cyclic extensions and employing frequency-domain equalization (FDE) techniques, have been shown to be suitable for high data rate transmission over severe time-dispersive channels, since the number of operations per symbol grows logarithmically with the block duration (and, therefore, the ISI span), due to the fast Fourier transform (FFT) implementation [Gusmão et al. 2000; Wang and Giannakis 2000; Falconer et al. 2002].

Orthogonal frequency division multiplexing (OFDM) schemes [Cimini 1985; Bingham 1990] are the most popular modulations based on this technique. Since an OFDM transmission over time-dispersive channels can be regarded as a parallel transmission over a set of nondispersive channels, one for each subcarrier, the extension of MIMO/BLAST techniques to OFDM schemes is straightforward, eventually with some preprocessing at the transmitter and/or employing adaptive loading schemes [Raleigh and Cioffi 1998; Raleigh and Jones 1999]. However, OFDM signals have high envelope fluctuations and a high peak-to-mean envelope power ratio (PMEPR) leading to amplification difficulties. For this reason, several techniques have been proposed for reducing the envelope fluctuations of OFDM signals [Jones and Wilkinson 1996; Müller et al. 1997; Dinis and Gusmão 2004]. However, these PMPER-reducing techniques require an increased signal-processing effort, especially at the transmitter side, and, possibly, some signal distortion when a nonlinear signal processing is employed. Moreover, even for the most sophisticated techniques, the transmitted signals still have PMEPRs higher than those for single-carrier (SC) signals based on similar constellations.

SC modulations, using block transmission techniques and FDE [Sari et al. 1994], are an alternative approach for broadband wireless systems. Similar to OFDM modulations, the data blocks are preceded by a cyclic prefix (CP), long enough to cope with the channel length. The received signal is transformed into the frequency domain, equalized in the frequency domain, and then transformed back to the time domain. The achievable performances as well as the overall implementation complexities are similar for SC schemes with FDE and OFDM schemes [Gusmão et al. 2000; Montezuma and Gusmão 2001a; Falconer et al. 2002]. However, the signal-processing effort is concentrated at the receiver for the SC case. This, combined with the lower envelope fluctuations of SC signals, makes them preferable for the uplink transmission (i.e., the transmission from the MT to the BS), while the OFDM schemes might be a better choice for the downlink transmission (i.e., the transmission from the BS to the MT). For this reason, a mixed OFDM plus SC mode air interface was proposed, employing an SC scheme with FDE in the uplink and an OFDM scheme in the downlink [Gusmão et al. 2000; Falconer et al. 2002].

Usually a linear FDE is employed at the receiver [Sari et al. 1994; Gusmão et al. 2000]. However, it is well known that nonlinear equalizers can significantly outperform linear equalizers [Proakis 1995]. For this reason, it is advantageous to design nonlinear equalizers for SC-FDE schemes [Benvenuto et al. 2010]. Among several different nonlinear equalizers, decision feedback equalizers (DFE) are especially popular due to its good performance/complexity trade-offs. A hybrid time–frequency SC-DFE was proposed employing a frequency-domain feedforward filter and a time-domain feedback filter [Benvenuto and Tomasin 2001]. The hybrid time–frequency-domain DFE can have better performance than a linear FDE but, as with conventional time-domain DFEs, it can suffer from error propagation, especially if the feedback filter has a large number of taps.

A promising iterative block DFE (IB-DFE) approach for SC transmission was proposed in [Benvenuto and Tomasin 2002] and extended to diversity scenarios [Dinis et al. 2003, 2004b] and spatial multiplexing schemes [Dinis et al. 2004; Kalbasi et al. 2004]. The IB-DFE schemes were also shown to allow excellent performance in many other scenarios, ranging from reduced cyclic prefix SC-FDE [Gusmão et al. 2007a, 2007b], code division multiple access (CDMA) scenarios [Dinis et al. 2004a; Silva and Dinis 2006a,b; Dinis et al. 2007a], network diversity schemes [Dinis et al.

2007b; Ganhão et al. 2011, 2012, 2013; Pereira et al. 2012, 2013a,b; Marques da Silva et al. 2012], and offset modulations [Luzio et al. 2012, 2013]. We can also define highly efficient techniques based on the IB-DFE concept for joint detection and estimation [Araújo and Dinis 2004; Lam et al. 2006; Dinis et al. 2008, 2010a; Pedrosa et al. 2010, 2012a,b; Silva et al. 2013] and to cope with severe nonlinear distortion effects [Silva and Dinis 2008, 2009; Dinis and Silva 2009; Dinis et al. 2009].

With IB-DFE schemes, both the feedforward and the feedback parts are implemented in the frequency domain. Since the feedback loop takes into account not just the hard decisions for each block but also the overall block reliability, the error propagation problem is significantly reduced. Consequently, the IB-DFE techniques offer much better performances than noniterative methods [Benvenuto and Tomasin 2002; Dinis et al. 2003]. In fact, IB-DFE techniques can be regarded as low-complexity turbo equalization schemes [Tüchler et al. 2003], since the feedback loop uses the equalizer outputs instead of the channel decoder outputs.

Earlier IB-DFE implementations considered hard decisions (weighted by the blockwise reliability) in the feedback loop. To improve the performance and to allow truly turbo FDE implementations, IB-DFE schemes with soft decisions were proposed [Benvenuto and Tomasin 2005; Gusmão et al. 2006, 2007a]. Usual IB-DFE implementations only consider QPSK constellations. However, larger constellations such as quadrature amplitude modulation (QAM) and phase shift keying (PSK) are often required when we want to increase the system's spectral efficiency. Furthermore, hierarchical constellations (which may be composed of nonuniformly spaced constellation points) are particularly interesting for broadcast/multicast systems since they are able to provide unequal bit error protection [Cover 1972; Jiang and Wilford 2005]. In this type of constellation, there are two or more classes of bits with different error protection, to which different streams of information can be mapped. Depending on the channel conditions, a given user can attempt to demodulate only the more protected bits or also the other bits that carry additional information. An application of these techniques is in the transmission of coded voice or video signals, where we can have different error protection associated with different resolutions [Ramchandran 1993; Jiang and Wilford 2005].

An important drawback of large constellations, in general, and nonuniform constellations, in particular, is that they are very sensitive to interference, namely, the residual ISI at the output of a practical equalizer that does not invert completely the channel effects (e.g., a linear equalizer optimized under the minimum mean squared error (MMSE) [Proakis 1995]). Therefore, we can expect significant performance improvements when we employ IB-DFE receivers with large constellations. However, there are some difficulties in the design of IB-DFE receivers for any constellation, namely, on the computation of the reliability of each block, as well as problems on the computation of the average symbol values conditioned on the FDE and/or the channel decoder output.

In this chapter, we consider MIMO SC-DFE schemes for different constellations. We start by characterizing the basic SC-FDE and IB-DFE schemes. Then we present a pragmatic approach for designing the receiver that employs a general method for the computation of the receiver parameters for any constellation. Our approach follows [Dinis et al. 2010b,c; Silva et al. 2012] and relies on an analytical characterization of the mapping rule where the constellation symbols are written as a linear function of the transmitted bits. This method is then employed to design turbo receivers implemented in the frequency domain for systems with both conventional constellations and multiresolution hierarchical constellations where we have different error protections. Finally, we describe MIMO receivers based on the IB-DFE concept.

5.2 SC-FDE Schemes

As already mentioned in Chapter 4, OFDM schemes are suitable for high data rate transmission over severely time-dispersive channels, since they have excellent performance with low-complexity receiver implementations. However, OFDM signals suffer from large envelope fluctuations, which lead to amplification difficulties. As an attractive alternative, SC-FDE schemes are also suitable for severely time-dispersive channels and the single-carrier nature of the transmitted signals means that they have much lower envelope fluctuations than OFDM signals based on the same constellation.

To understand SC-FDE schemes, let us consider an SC-based block transmission with N useful data symbols per block $\{s_n; n = 0, 1, \ldots, N-1\}$ resulting from the direct mapping of the original data into a selected signal constellation. As with OFDM schemes, a suitable CP is appended to the useful block, which corresponds to the repetition of the last N_G data symbols. The transmitted signal is then

$$s(t) = \sum_{n=-N_G}^{N-} s_n h_T(t - nT_S), \tag{5.1}$$

with T_S denoting the symbol duration, N_G denoting the number of samples at the cyclic prefix, and $h_T(t)$ denoting the adopted pulse shape. The transmitted symbols s_n belong to a given alphabet $\mathfrak{S}$ (i.e., a given constellation) with dimension $M = \#\mathfrak{S}$ and are selected according to the corresponding bits $\beta_n^{(m)}$, $m = 1, 2, \ldots,$ μ ($\mu = \log_2(M)$) that is, $s_n = f(b_n^{(1)}, b_n^{(2)}, \ldots, b_n^{(\mu)})$, with $b_n^{(m)} = 2\beta_n^{(m)} - 1$ (throughout this chapter, we assume that $\beta_n^{(m)}$ is the mth bit associated with the nth symbol and $b_n^{(m)}$ is the corresponding polar representation, i.e., $\beta_n^{(m)} = 0$ or 1 and $b_n^{(m)} = -1$ or $+1$, respectively). As with other cyclic-prefix-assisted block transmission schemes, it is assumed that the time-domain block is periodic, with period N, that is, $s_{-n}^{(m)} = s_{N-n}^{(m)}$.

If we discard the samples associated with the cyclic prefix at the receiver, then there is no interference between blocks, provided that the length of the cyclic prefix is higher than the length of the overall channel impulse response. Moreover, the linear convolution associated with the channel is equivalent to a cyclic convolution relatively to the N-length, useful part of the received block, $\{y_n; n = 0,1, \ldots, N-1\}$. This means that the corresponding frequency-domain block (i.e., the length-N DFT (discrete Fourier transform) of the block $\{y_n; n = 0,1, \ldots, N-1\}$) is $\{Y_k; k = 0,1, \ldots, N-1\}$, where

$$Y_k = S_k H_k + N_k, \tag{5.2}$$

with H_k denoting the channel frequency response for the kth subcarrier and N_k the corresponding channel noise. Clearly, the impact of a time-dispersive channel reduces to a scaling factor for each frequency.

The simplest FDE is the linear FDE depicted in Figure 5.1, where the frequency-domain signal at the FDE output is given by $\{\tilde{S}_k = F_k Y_k; k = 0,1, \ldots, N-1\}$ and the detection is based on the corresponding time-domain block $\{\tilde{s}_n; n = 0,1, \ldots, N-1\} = \text{IDFT } \{\tilde{S}_k = F_k Y_k; k = 0,1, \ldots, N-1\}$. If the FDE coefficients F_k are given by

$$F_k = \frac{1}{H_k}, \tag{5.3}$$

we eliminate completely the ISI at the FDE output (i.e., the FDE is optimized under the zero forcing (ZF) criterion). However, for a typical frequency-selective channel, we can have deep notches in the channel frequency response that lead to significant noise enhancement effects when the ZF criterion is employed. To minimize the combined effect of ISI and channel noise on the equalized samples, the FDE coefficients should be selected according to the MMSE criterion (minimum mean-squared error), which corresponds to using

$$F_k = \frac{H_k^*}{1/SNR + |H_k|^2}, \tag{5.4}$$

where the SNR (signal-to-noise ratio) is given by

$$SNR = \frac{E[|S_k|^2]}{E[|N_k|^2]}. \tag{5.5}$$

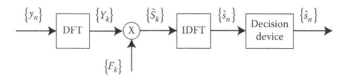

Figure 5.1 Basic structure of a linear FDE.

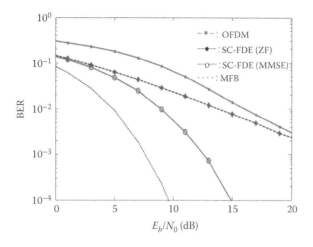

Figure 5.2 **BER performance for SC-FDE with QPSK constellations and a linear FDE optimized under the ZF or MMSE criterion, together with the corresponding MFB. For the sake of comparisons, we also include the uncoded BER performance for the corresponding OFDM schemes.**

Contrary to OFDM schemes where ZF and MMSE criterions yield the same performance, the SC-FDE performance is typically much better for the MMSE criterion [Gusmão et al. 2003]. An FDE optimized under the MMSE criterion does not attempt to fully invert the channel when we have a deep fade, reducing noise enhancement effects and allowing better performances. However, owing to the residual ISI, the performance of a linear FDE optimized under the MMSE criterion is still far from the matched filter bound (MFB), given by [Amaral et al. 2012]

$$P_{MFD} = E\left[Q\left(\sqrt{\frac{2E_b}{N_0} \sum_{k=0}^{N-1} \frac{|H_k|^2}{N}} \right) \right] \tag{5.6}$$

(see Figure 5.2).

5.3 IB-DFE Receivers

5.3.1 Basic IB-DFE Structure

Although the equalization can be accomplished employing a linear FDE, as explained in Section 5.2, the performance can be substantially improved with the

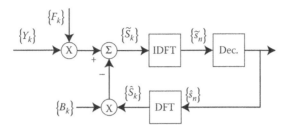

Figure 5.3   Basic structure of an IB-DFE.

use of an IB-DFE, whose structure is depicted in Figure 5.3. For a given iteration, the output samples are given by

$$\tilde{S}_k = F_k Y_k - B_k \hat{S}_k, \qquad (5.7)$$

with $\{F_k; k = 0,1, \ldots, N–1\}$ and $\{B_k; k = 0,1, \ldots, N–1\}$ denoting the feedforward and the feedback coefficients, respectively. $\{\hat{S}_k; k = 0,1,\ldots,N − 1\}$ is the DFT of the block $\{\hat{s}_n; n = 0,1,\ldots,N − 1\}$, with $\hat{s}_n$ denoting the "hard" estimate of s_n from the previous FDE iteration.

As shown in [Dinis et al. 2003],[*] the coefficients F_k and B_k that maximize the overall SNR in the samples $\tilde{S}_k$ are given by

$$F_k = \frac{\kappa H_k^*}{\alpha + (1 - \rho^2) |H_k|^2}; \qquad (5.8)$$

and

$$B_k = \rho(F_k H_k - 1), \qquad (5.9)$$

respectively, where

$$\alpha = E[|N_k|^2]/E[|S_k|^2] \qquad (5.10)$$

and κ is selected so as to ensure that

$$\sum_{k=0}^{N-1} F_k H_k / N = 1. \qquad (5.11)$$

---

[*] It should be noted that, contrary to [Dinis et al. 2003], we are considering a normalized feedforward filter.

The correlation coefficient ρ, which can be regarded as the blockwise reliability of the decisions used in the feedback loop (from the previous iteration), is given by

$$\rho = \frac{E[\hat{S}_k S_k^*]}{E[|S_k|^2]} = \frac{E[\hat{s}_n s_n^*]}{E[|s_n|^2]} \tag{5.12}$$

and can be estimated as described in [Dinis et al. 2003, 2007a] and [Silva et al. 2011].

It is important to note that although the IB-DFE just described is usually denoted as "IB-DFE with hard decisions," it would probably be more adequate to refer to it as "IB-DFE with blockwise soft decisions," as we will see in the following. In fact, Equation 5.7 could be written as

$$\tilde{S}_k = F_k Y_k - B_k' \overline{S}_k^{\text{Block}}, \tag{5.13}$$

with $\rho B_k' = B_k$ and $\overline{S}_k^{\text{Block}}$ denoting the average of the block of overall time-domain chips associated with a given iteration, given by $\overline{S}_k^{\text{Block}} = \rho \hat{S}_k$ (as mentioned earlier, ρ can be regarded as the blockwise reliability of the estimates $\{\hat{S}_k; n = 0, 1, \dots, M - 1\}$).

5.3.2 IB-DFE with Soft Decisions

In order to improve the performance, we can replace the "blockwise averages" used in the IB-DFE structure described earlier by "symbol averages," leading to what is usually denoted as "IB-DFE with soft decisions" [Benvenuto and Tomasin 2005; Gusmão et al. 2006, 2007a]. A simple way of achieving this is to replace the feedback input $\{\overline{S}_k^{\text{Block}}; k = 0, 1, \dots, N - 1$ by $\left\{\overline{S}_k^{\text{Symbol}} = \overline{S}_k; k = 0, 1, \dots, N - 1\right\} = \text{DFT}\left\{\overline{s}_n^{\text{Symbol}}; n = 0, 1, \dots, N - 1\right\}$, with $\overline{s}_n^{\text{Symbol}}$ denoting the average symbol values conditioned to the FDE output of the previous iteration $\tilde{s}_n$, with $\{\tilde{s}_n =; n = 0, 1, \dots, N - 1\}$ denoting the IDFT of the frequency-domain block $\{\tilde{S}_k; k = 0, 1, \dots, N - 1\}$. To simplify the notation, we will use $\overline{s}_n$ (and $\overline{S}_k$) instead of $\overline{s}_n^{\text{Symbol}}$ (and $\overline{S}_k^{\text{Symbol}}$) in the rest of the chapter.

For normalized QPSK constellations (i.e., $s_n = \pm 1 \pm j$) with Gray mapping, it is easy to show that [Gusmão et al. 2006, 2007a]

$$\overline{s}_n = \tanh\left(\frac{L_n^I}{2}\right) + j \tanh\left(\frac{L_n^Q}{2}\right) = \rho_n^I \hat{s}_n^I + j \rho_n^Q \hat{s}_n^Q, \tag{5.14}$$

with the log-likelihood ratios (LLR) of the "in-phase bit" and the "quadrature bit," associated with $s_n^I = Re\{s_n\}$ and $s_n^Q = Im\{s_n\}$, respectively, given by

$$L_n^I = 2\tilde{s}_n^I / \sigma^2 \tag{5.15}$$

and

$$L_n^Q = 2\tilde{s}_n^Q/\sigma^2, \tag{5.16}$$

respectively, where

$$\sigma^2 = \frac{1}{2}E[|s_n - \tilde{s}_n|^2] \approx \frac{1}{2N}\sum_{n=0}^{N-1}E[|\hat{s}_n - \tilde{s}_n|^2]. \tag{5.17}$$

The hard decisions $\hat{s}_n^I = \pm 1$ and $\hat{s}_n^Q = \pm 1$ are defined according to the signs of L_n^I and L_n^Q, respectively, and ρ_n^I and ρ_n^Q can be regarded as the reliabilities associated with the "in-phase" and "quadrature" bits of the nth symbol, given by

$$\rho_n^I = E[s_n^I \hat{s}_n^I]/E[|s_n^I|^2] = \tanh\left(|L_n^I|/2\right) \tag{5.18}$$

and

$$\rho_n^I = E[s_n^Q \hat{s}_n^Q]/E[|s_n^Q|^2] = \tanh\left(|L_n^Q|/2\right) \tag{5.19}$$

(for the first iteration, $\rho_n^I = \rho_n^Q = 0$ and $\bar{s}_n = 0$).

The feedforward coefficients are still obtained from Equation 5.8, with the blockwise reliability given by

$$\rho = \frac{1}{2N}\sum_{n=0}^{N-1}(\rho_n^I + \rho_n^Q). \tag{5.20}$$

Therefore, the receiver with "blockwise reliabilities" (hard decisions) and the receiver with "symbol reliabilities" (soft decisions) employ the same feedforward coefficients; however, in the first, the feedback loop uses the "hard decisions" on each data block, weighted by a common reliability factor, while in the second, the reliability factor changes from bit to bit. The receiver structure when we have soft decisions is depicted in Figure 5.4, which is closely related to the IB-DFE with hard decisions (Figure 5.3).

We can also define a frequency-domain turbo equalizer that employs the channel decoder outputs instead of the uncoded "soft decisions" in the feedback loop. The receiver structure is similar to the IB-DFE with soft decisions, but with a soft-in, soft-out (SISO) channel decoder employed in the feedback loop. The SISO block, which can be implemented as defined in [Vucetic and Yuan 2002], provides the LLRs of both the "information bits" and the "coded bits." The input of the SISO block is LLRs of the "coded bits" at the FDE. Once again, the feedforward coefficients are obtained from Equation 5.8, with the blockwise reliability given by Equation 5.20.

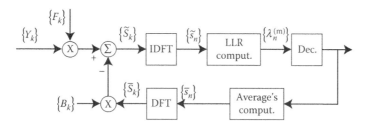

Figure 5.4 Regions associated with $\Psi_0^{(m)}$ and $\Psi_1^{(m)}(m = 1, 2, 3)$ for a uniform 8-PAM constellation with Gray mapping.

5.3.3 Multiresolution Systems

In a multiresolution system, we can have μ streams, each one associated with a different resolution and with a suitable error protection. Figure 5.5 illustrates the basic structure of a transmitter. According to this figure, the data stream associated with the mth resolution is encoded by a different channel encoder and the corresponding bits are interleaved, leading to the block $\{b_n^{(m)}, n = 0, 1, \ldots, N - 1\}$. The different blocks $\{b_n^{(m)}, n = 0, 1, \ldots, N - 1\}$, $m = 1, 2, \ldots, \mu$, are mapped onto the block of time-domain symbols $\{s_n, n = 0, 1, \ldots, N-1\}$ and the rest of the transmitter is similar to the transmitter for conventional constellations.[*]

The receiver can be the one depicted in Figure 5.6. Essentially, we have an IB-DFE receiver where the demapping block provides LLRs of the bits associated with each error protection level. The operations of detection/decoding and computation of average bit values are preformed separately for each resolution bit stream.

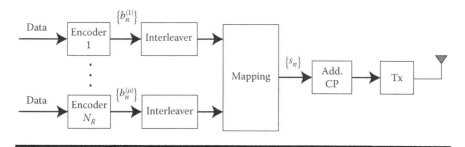

Figure 5.5 Transmitter structure for SC-FDE with multiresolution.

---

[*] Without loss of generality, we are assuming that the bit rate associated with each resolution is identical. The extension to the case where we have different bit rates for different resolutions is straightforward.

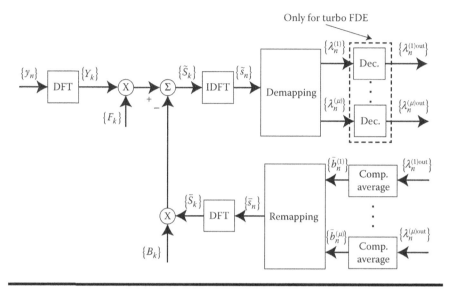

Figure 5.6 Iterative FDE receiver structure for SC-FDE with multiresolution (the dashed part corresponds to the SISO decoders and is only required for a turbo FDE).

5.3.4 Analytical Characterization of Mapping Rules

5.3.4.1 General Mapping

As shown in [Montezuma and Gusmão 2001a,b], it is possible to express the constellation symbols as a function of the corresponding bits using[*]

$$s_n = g_0 + g_1 b_n^{(1)} + g_2 b_n^{(2)} + g_3 b_n^{(1)} b_n^{(2)} + g_4 b_n^{(3)} + \cdots = \sum_{i=0}^{M-1} g_i \prod_{m=1}^{\mu} (b_n^{(m)})^{\gamma_{m,i}}, \quad (5.21)$$

for each $s_n \in \mathfrak{S}$, where $(\gamma_{\mu,i} \ \gamma_{\mu-1,i} \cdots \gamma_{2,i} \ \gamma_{1,i})$ is the binary representation of i. Since we have M constellation symbols in $\mathfrak{S}$ and M coefficients g_i, Equation 5.21 is a system of M equations that can be used to obtain the coefficients g_i, $i = 0,1,\ldots,\mu-1$. Writing Equation 5.21 in matrix format, we have

$$\mathbf{s} = \mathbf{W}\mathbf{g}, \quad (5.22)$$

---

[*] It should be noted that in this section s_n denotes the nth constellation point, but in the previous section, s_n denotes the nth transmitted symbol; the same applies to $b_n^{(m)}$ (or $\beta_n^{(m)}$) that here denotes the mth bit of the nth constellation point (instead of the mth bit of the nth transmitted symbol).

with

$$\mathbf{s} = [s_1 \ s_2 \ ... \ s_M]^T, \tag{5.23}$$

$$\mathbf{g} = [g_0 \ g_1 \ \cdots \ g_{\mu-1}]^T, \tag{5.24}$$

and W is a Hadamard matrix with dimensions $M \times M$. Clearly, the vector of constellation points s is the Hadamard transform of the vector of coefficients $\mathbf{g}$. Therefore, for a given constellation, we can obtain the corresponding coefficients g_i from the inverse Hadamard transform of the vector of constellation points.

5.3.5 Special Cases

In this section, we present some special cases where the mapping is particularly simple, avoiding the computation of Hadamard transforms.

5.3.5.1 PAM Constellations

For a uniform M-PAM constellation, we have $\mathfrak{S} = \{\pm1, \pm3, ..., \pm(M-1)\}$.

If we have a natural binary mapping, the only nonzero coefficients are g_1, g_2, g_4, ... , $g_{M/2}$ (i.e., the coefficients g_{2^i}, $i = 0,1, ... ,\mu-1$). Moreover, $g_{2^i} = 2^{\mu-i}$, which means that

$$s_n = g_1 b_n^{(1)} + g_2 b_n^{(2)} + g_4 b_n^{(3)} + \cdots = b_n^{(1)} + 2b_n^{(2)} + 4b_n^{(4)} + \cdots = \sum_{m=1}^{\mu} 2^{m-1} b_n^{(m)}. \tag{5.25}$$

For a Gray mapping, the only nonzero coefficients are the ones with binary representations $(0 \ ... \ 001)$, $(0 \ ... \ 011)$, ... , $(111 \ ... \ 1)$, that is, the coefficients $g_{2^i-1}, i = 0,1..., \mu-1$. Moreover, $g_{2^i-1} = 2^{\mu-i-1}$, which means that

$$s_n = 2^{\mu-1} b_n^{(1)} + 2^{\mu-2} b_n^{(1)} b_n^{(2)} + 2^{\mu-3} b_n^{(1)} b_n^{(2)} b_n^{(3)} + \cdots = \sum_{m=1}^{\mu} 2^{\mu-m} \prod_{m'=1}^{m} b_n^{(m')}. \tag{5.26}$$

For uniform 4-PAM constellations, we have

$$s_n = 2b_n^{(1)} + b_n^{(2)} \tag{5.27}$$

for a natural binary mapping and

$$s_n = 2b_n^{(1)} + b_n^{(1)} b_n^{(2)} \tag{5.28}$$

for a Gray mapping. For uniform 8-PAM constellations, we have

$$s_n = 4b_n^{(1)} + 2b_n^{(2)} + b_n^{(3)} \tag{5.29}$$

for a natural binary mapping and

$$s_n = 4b_n^{(1)} + 2b_n^{(1)}b_n^{(2)} + b_n^{(1)}b_n^{(2)}b_n^{(3)} \tag{5.30}$$

for a Gray mapping.

The same approach can be employed for nonuniform hierarchical constellations such as the ones adopted in multiresolution schemes [Jiang and Wilford 2005]. In fact, from Equations 5.25 and 5.26, an M-PAM constellation with either natural binary mapping or Gray mapping can be regarded as the sum of μ binary subconstellations, each one with twice the amplitude of the previous one. By reducing the amplitude of successive subconstellations, we obtain hierarchical constellations with different error protections.

5.3.5.2 QAM Constellations

The mapping of the bits on an M-QAM constellation is often performed independently to the in-phase and quadrature components, that is, half the bits are used to define the in-phase component (as in the previous case, for Gray mapping or natural binary mapping) and the other half is used to define the quadrature component. Therefore, an M-QAM constellation can be written as the sum of two $\sqrt{M}$-PAM constellations, one for the in-phase (real) component and the other for the quadrature (imaginary) component. For instance, for 16-QAM with Gray mapping, we have

$$s_n = 2b_n^{(1)} + b_n^{(1)}b_n^{(2)} + 2jb_n^{(3)} + jb_n^{(3)}b_n^{(4)} \tag{5.31}$$

and for 64-QAM with Gray mapping, we have

$$s_n = 4b_n^{(1)} + 2b_n^{(1)}b_n^{(2)} + b_n^{(1)}b_n^{(2)}b_n^{(3)} + 4jb_n^{(4)} + 2jb_n^{(4)}b_n^{(5)} + jb_n^{(4)}b_n^{(5)}b_n^{(6)}. \tag{5.32}$$

The extension to other mapping rules and/or nonuniform QAM constellations is straightforward.

5.3.5.3 M-PSK Constellations

Whereas the characterization of a BPSK constellation is trivial and QPSK constellations with Gray mapping are simply a special case within the aforementioned QAM class, characterizing analytically a given M-PSK constellation is in general complex[*] and we need to employ directly Equation 5.22.

---

[*] Naturally, we could define a PSK constellation as a complex exponential of a suitable PAM constellation, but this does not help us in the receiver design since the constellation symbols are not a linear function of the corresponding bits.

In the following, we present the analytical characterization of 8-PSK constellations (it is better to employ the Hadamard transform, as described earlier when we want to characterize larger PSK constellations). Let us assume that the constellation is defined by two points, s_1 and s_2, as well as their reflections in the real and imaginary axis, that is, $\mathfrak{S} = \{s_1, s_2, s_2^*, s_1^*, -s_1, -s_2, -s_2^*, -s_1^*\}$, corresponding to the "tri-bits" 010, 011, 111, 110, 000, 100, 101, 001, respectively (Gray mapping). Clearly, for a regular 8-PSK constellation, $s_1 = \exp(j3\pi/8)$ and $s_2 = \exp(j\pi/8)$. By using other values of s_1 and s_2, we can define 8-PSK constellations with two or three different error protections; we can also define some 8-APSK constellations (e.g., for $s_1 = \exp(j\pi/4)$ and $s_1 = 2\exp(j\pi/4)$).

If we define $\overline{s} = (s_1 + s_2)/2 = \overline{s}^R + j\overline{s}^I$ ($\overline{s}^R = Re\{\overline{s}\}$ and $\overline{s}^I = Im\{\overline{s}\}$) and $s_\Delta = s_1 - \overline{s} = s_\Delta^R + js_\Delta^I$ ($s_\Delta^R = Re\{s_\Delta\}$ and $s_\Delta^I = Im\{s_\Delta\}$), then the constellation point associated with the bits of $b_n^{(1)}$, $b_n^{(2)}$, and $b_n^{(3)}$ is

$$s_n = \overline{s}^R b_n^{(1)} + j\overline{s}^R b_n^{(2)} + s_\Delta^R b_n^{(1)} b_n^{(3)} + js_\Delta^I b_n^{(2)} b_n^{(3)}. \tag{5.33}$$

This means that $g_1 = \overline{s}^R$, $g_2 = j\overline{s}^R b_n^{(2)}$, $g_5 = s_\Delta^R$, and $g_6 = js_\Delta^I$, with the remaining being equal to zero.

5.3.6 Computation of Receiver Parameters

Taking into account the description provided in the previous sections, an IB-DFE receiver with soft decisions (as described in Section 5.3.2) has to carry the following constellation-dependent tasks (see Figure 5.7):

■ Demapping the time-domain samples at the output of the FDE, $\tilde{s}_n$, into the corresponding bits. This can be implemented by computing the log-likelihood ratios associated with each bit of each transmitted symbol.
■ Computation of the average symbol values conditioned to the FDE output of the previous iteration $\tilde{s}_n$, denoted by $\overline{s}_n$.
■ Computation of the blockwise reliability ρ, required for obtaining the feedforward coefficients (see Equation 5.8).

The log-likelihood ratio of the mth bit for the nth transmitted symbol is given by

$$\lambda_n^{(m)} = \log\left(\frac{Pr(\beta_n^{(m)} = 1 \mid \tilde{s}_n)}{Pr(\beta_n^{(m)} = 0 \mid \tilde{s}_n)}\right)$$

$$= \log\left(\frac{\sum_{s \in \Psi_1^{(m)}} \exp\left(-(\mid \tilde{s}_n - s \mid^2/2\sigma^2)\right)}{\sum_{s \in \Psi_0^{(m)}} \exp\left(-(\mid \tilde{s}_n - s \mid^2/2\sigma^2)\right)}\right), \tag{5.34}$$

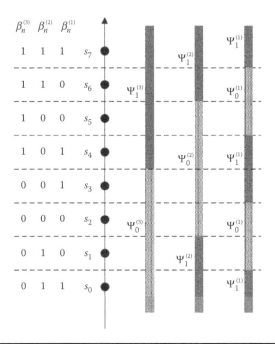

Figure 5.7   IB-DFE with soft decisions.

where $\Psi_1^{(m)}$ and $\Psi_0^{(m)}$ are the subsets of $\mathfrak{S}$ where $\beta_n^{(m)} = 1$ or 0, respectively (clearly, $\Psi_1^{(m)} \bigcup \Psi_0^{(m)} = \mathfrak{S}$ and $\Psi_1^{(m)} \bigcap \Psi_0^{(m)} = \varnothing$). As an example, Figure 5.4 shows the regions associated with $\Psi_0^{(m)}$ and $\Psi_1^{(m)}$ ($m = 1,2,3$) for a uniform 8-PAM constellation with Gray mapping.

For obtaining the average symbol values conditioned on the FDE output, $\bar{s}_n$, we need to obtain the average bit values conditioned on the FDE output, $\bar{b}_n^{(m)}$. These are related to the corresponding log-likelihood ratio as follows:

$$\bar{b}_n^{(m)} = \tanh\left(\frac{\lambda_n^{(m)}}{2}\right). \tag{5.35}$$

By taking advantage of the analytical characterization of the mapping rules Equation 5.21 (or, equivalently, the specific formulas of Section 5.3.5) and assuming uncorrelated bits (e.g., thanks to the adoption of suitable interleaving), we can write

$$\bar{s}_n = \sum_{i=0}^{M-1} g_i \prod_{m=1}^{\mu} \left(\bar{b}_n^{(m)}\right)^{\gamma_{m,i}}. \tag{5.36}$$

Finally, the reliability of the estimates to be used in the feedback loop is given by

$$\rho = \frac{E[\hat{s}_n s_n^*]}{E[|s_n|^2]} = \frac{\sum_{i=0}^{M-1} |g_i|^2 \prod_{m=1}^{\mu} \left(\rho_n^{(m)}\right)^{\gamma_{m,i}}}{\sum_{i=0}^{M-1} |g_i|^2}, \tag{5.37}$$

where $\rho_n^{(m)}$ is the reliability of the mth bit of the nth transmitted symbol, given by

$$\rho_n^{(m)} = \left|\bar{b}_n^{(m)}\right|. \tag{5.38}$$

For QPSK constellations, these results reduce to the ones presented in Section 5.2.2.

As an example, let us consider a uniform 4-PAM constellation with Gray mapping (i.e., the symbols are characterized by Equation 5.31). Figure 5.8 shows the LLR values of the different bits, $\lambda_n^{(m)}$, as a function of the output of the FDE, $\tilde{s}_n$, for different SNR values and Figure 5.9 shows the average value of each bit conditioned on the FDE output, $\bar{b}_n^{(m)}$ in the same condition. The regions where each bit is 0 or 1 are well defined when we have a high SNR, but for low SNRs, these regions are not so evident.

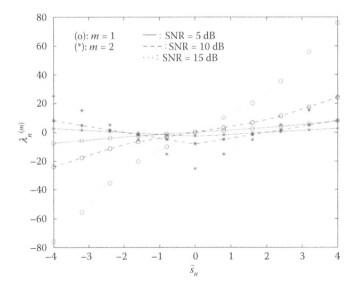

Figure 5.8 **Evolution of the LLR of the different bits, $\lambda_n^{(m)}$, for a uniform 4-PAM constellation with Gray mapping.**

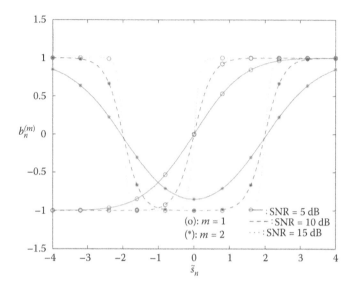

Figure 5.9 **Evolution of the average value of the different bits conditioned to the FDE output, $\bar{b}_n^{(m)}$, for a uniform 4-PAM constellation with Gray mapping.**

Figure 5.10 shows the average symbol value conditioned on the FDE output, $\bar{s}_n$, for the same settings of Figures 5.8 and 5.9. Once again, the four levels are clear for high SNR and the transition between levels becomes smoother as we reduce the SNR. The corresponding symbol reliability ρ_n is depicted in Figure 5.11. As expected, the reliability is lower between levels, becoming 0 for $\tilde{s}_n = 0$. For $\tilde{s}_n \approx 0$ or ±3, the reliability is close to 1, unless the SNR is very small.

5.3.7 Performance Results

In this section, some performance results concerning IB-DFE receivers with soft decisions for generalized constellations are presented. Blocks with $N = 256$ symbols plus an appropriate cyclic prefix are considered. The results assume a severely time-dispersive channel with perfect synchronization and channel estimation at the receiver.

Let us first consider a uniform 64-QAM constellation with Gray mapping based on two separate 8-PAM constellations characterized by $g_7/g_3 = g_3/g_1 = 0.5$. Figure 5.12 shows the BER performance for the IB-DFE receivers described in the previous sections. When compared with a conventional linear FDE, the performance improves significantly with the iterations, with a gain around 7 dB for BER $= 10^{-4}$ after four iterations.

Let us consider now a nonuniform 64-QAM constellation based on two 8-PAM constellations characterized by $g_7/g_3 = g_3/g_1 = 0.4$ (Gray mapping). These

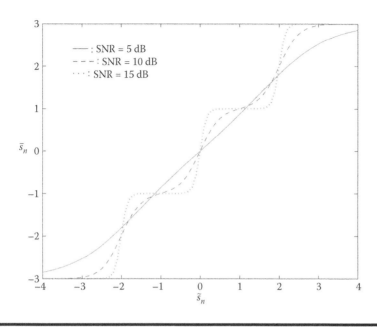

Figure 5.10 Evolution of $\bar{s}_n$ for a uniform 4-PAM constellation with Gray mapping.

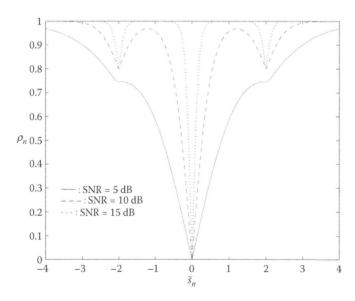

Figure 5.11 Evolution of ρ_n for a uniform 4-PAM constellation with Gray mapping.

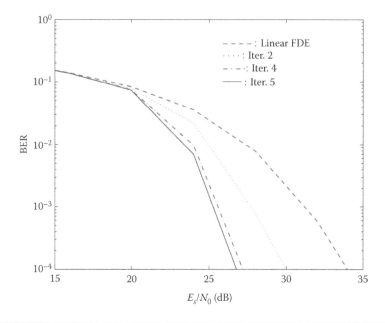

Figure 5.12   BER for a uniform 64-QAM with Gray mapping.

constellations allow bits with three different error protection levels, denoted LPB (least protected bits), IPB (intermediate protected bits), and MPB (most protected bits).

Figure 5.13 shows the uncoded BER performance for the different type of bits when we have a conventional IB-DFE receiver. These results are expressed as a function of E_s/N_0, with E_s denoting the average symbol energy and N_0 the noise power spectral density. Clearly, performance improves significantly with the number of iterations, outperforming significantly the linear FDE. This improvement is higher for LPB (more than 10 dB for BER = 10^{-4}), since they are more sensitive to the residual ISI that is inherent to a linear FDE optimized under the MMSE criterion.

Let us now consider the impact of channel coding on a conventional IB-DFE as well as on a turbo FDE (i.e., an IB-DFE where the channel decoder is involved in the feedback loop). We consider a rate-1/2 turbo code [Berrou et al. 1993] based on two identical recursive convolutional codes with two constituent codes characterized by $G(D) = [1(1 + D^2 + D^3)/(1 + D + D^3)]$ and interleaving depth corresponding to a single FFT block. We also assume three iterations of the turbo decoder for each iteration of the IB-FDE. Figure 5.14 shows the BER performance for the different types of bits. As expected, the turbo FDE outperforms the conventional IB-DFE (where the channel decoder is not involved in the feedback loop), with gains around 3 dB for BER = 10^{-4}.

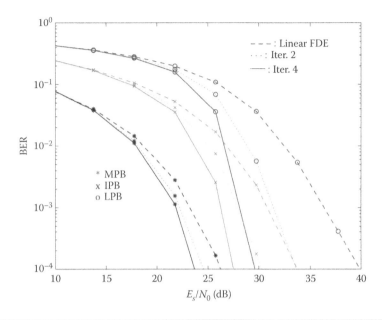

Figure 5.13 **BER for the different bits of a nonuniform 64-QAM modulation with Gray mapping.**

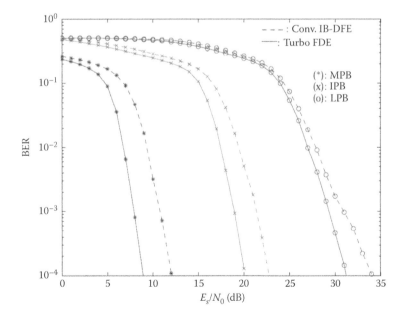

Figure 5.14 **Coded BER performance for the bits with different error protections.**

5.4 Iterative Frequency-Domain Detection for MIMO-SC

5.4.1 System Model

Let us now consider a MU-MIMO (multiuser MIMO) scenario where we have P MTs employing SC-FDE schemes and transmitting simultaneously to a BS equipped with R receive antennas, as depicted in Figure 5.15. For the sake of simplicity, we assume that each MT has a single transmit antenna, although we could easily extend it to the case where we have multiple antennas at one or several MTs. The time-domain block transmitted by the pth user is $\{s_{n,p}; n = 0,1, \ldots,N-1\}$, with $s_{n,p}$ denoting the nth data symbol of the pth user, which is selected from a given constellation (e.g., a QPSK (quadrature phase shift keying) constellation), under an appropriate mapping rule. A cyclic prefix, preceding each block, is used to avoid interblock interference and to make the linear convolution associated with the channel equivalent to a cyclic convolution with respect to the useful, length-N, part of the block. At the receiver, the cyclic prefix is discarded.

The time-domain block at the rth receive antenna is $\{y_n^{(r)}; n = 0,1, \ldots, N-1\}$. The corresponding frequency-domain block, obtained after an appropriate size-N DFT operation (discrete Fourier transform), is $\{Y_k^{(r)}; k = 0,1, \ldots, N-1\}$, where

$$Y_k^{(r)} = \sum_{p=1}^{P} S_{k,p} H_{k,p}^{(r)} + N_k^{(r)}, \tag{5.39}$$

with $H_{k,p}^{(r)}$ denoting the overall channel frequency response from the pth transmitting antenna to the rth receiver antenna, for the kth frequency, and $N_k^{(r)}$ denoting the corresponding channel noise. The block $\{S_{k,p}; k = 0,1,\ldots,M-1\}$ is the size-N DFT of the pth user's data block $\{s_{n,p}; n = 0,1,\ldots,N-1\}$.

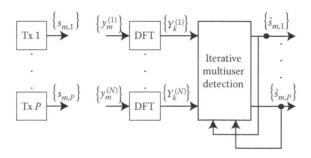

Figure 5.15   System model.

We consider a frequency-domain iterative multiuser detection that combines IB-DFE principles with interference cancelation. Each iteration consists of P detection stages, one for each user. When detecting a given user, the interference from previously detected users is canceled, as with conventional layered space–time (LST) receivers. However, in contrast to conventional LST receivers, we also cancel the residual ISI from the user that is being detected. Moreover, these interference and residual ISI cancelations take into account the reliability of each of the previously detected users. For a given iteration, the receiver structure for the detection of the pth user is illustrated in Figure 5.16. We have R frequency-domain feedforward filters (one for each receive antenna) and P frequency-domain feedback filters (one for each user). The feedforward filters are designed to minimize both the ISI and the multiuser interference that cannot be canceled by the feedback filters, due to decision errors in the previous detection steps. This structure can be regarded as an equalizer with multiuser interference suppression properties. After an IDFT operation, the corresponding time-domain outputs are passed through a hard-decision device so as to provide an estimate of the data block transmitted by the pth user.

We can employ two different iterative approaches for detecting the different users:

- A parallel interference cancelation (PIC) approach where we detect all users simultaneously at a given iteration, while removing interuser interference as well as residual ISI using data estimates from the previous iteration.
- A successive interference cancelation (SIC) approach where we cancel the interference from all users (using the most updated version of each user),

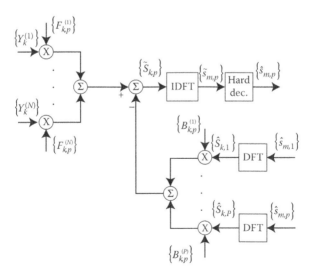

Figure 5.16   Detection of the *p*th user, for a given iteration.

as well as the residual ISI from the user that is being detected. In this case, the users should be detected according to their average received power, although our receiver is robust enough to cope with a wrong detection order (eventually requiring an additional iteration for convergence).

5.4.2 Computation of Receiver Coefficients

The frequency-domain samples associated with the pth user at the output of the equalizer/multiuser detector are given by

$$\tilde{S}_{k,p} = \sum_{r=1}^{R} F_{k,p}^{(r)} Y_k^{(r)} - \sum_{p'=1}^{P} B_{k,p}^{(p')} \hat{S}_{k,p'} = \sum_{r=1}^{R} F_{k,p}^{(r)} Y_k^{(r)} - B_{k,p}^{(p)} \hat{S}_{k,p} - \sum_{p' \neq p} B_{k,p}^{(p')} \hat{S}_{k,p'}, \quad (5.40)$$

where $F_{k,p}^{(r)} (k = 0,1,\ldots,N-1; r = 1,2,\ldots,R)$ denotes the feedforward coefficients and $B_{k,p}^{(p')} (k = 0,1,\ldots,N-1; p = 1,2,\ldots,P)$ denotes the feedback coefficients. The coefficients $\{B_{k,p}^{(p)}; k = 0,1,\ldots,N-1\}$ are used for residual ISI cancelation and the coefficients $\{B_{k,p}^{(p')}; k = 0,1,\ldots,N-1\} (p' \neq p)$ are used for interference cancelation. The block $\{\hat{S}_{k,p'}; k = 0,1,\ldots,M-1\}$ is the DFT of the block $\{\hat{s}_{n,p'}; n = 0,1,\ldots,N-1\}$, where the time-domain samples $\hat{s}_{n,p'}, n = 0,1,\ldots,N-1$ are the latest estimates for the p'th user transmitted symbols, that is, the hard decisions associated with the block of time-domain samples $\{\tilde{s}_{n,p'}; n = 0,1,\ldots,N-1\} = \text{IDFT}\{\tilde{S}_{k,p'}; k = 0,1,\ldots,N-1\}$. For the ith iteration of an SIC receiver, $\hat{s}_{n,p'}$ is associated with the ith iteration for $p' < p$ and with the $(i-1)$th iteration for $p' \geq p$ (in the first iteration, we do not have any information for $p' \geq p$ and $\hat{s}_{n,p'} = 0$); for the PIC receiver, $\hat{s}_{n,p'}$ is always associated with the previous iteration (for the first iteration, $\hat{s}_{n,p'} = 0$).

Owing to decision errors, we have $\hat{s}_{n,p} \neq s_{n,p}$ for some symbols. Consequently, $\hat{S}_{k,p} \neq S_{k,p}$. However, the frequency-domain estimates, $\hat{S}_{k,p}$, can be written as

$$\hat{S}_{k,p} = \rho_p S_{k,p} + \Delta_{k,p}, \quad (5.41)$$

where ρ_p denotes the correlation coefficient for the pth user, which can be obtained as described in the previous section, and $\Delta_{k,p}$ denotes a zero-mean error term. For the computation of the receiver coefficients, it is assumed that $E[\Delta_{k,p} S_{k',p}] \approx 0$, regardless of k and k'. Therefore,

$$E[|\Delta_{k,p}|^2] = (1 - \rho_p^2) N E_{S,p}. \quad (5.42)$$

By combining Equations 5.39, 5.40, and 5.41, we obtain

$$
\begin{aligned}
\tilde{S}_{k,p} &= \sum_{r=1}^{R} F_{k,p}^{(r)} \left(\sum_{p'=1}^{P} H_{k,p'}^{(r)} S_{k,p'} + N_k^{(r)} \right) - B_{k,p}^{(p)} \left(\rho_p S_{k,p} + \Delta_{k,p} \right) \\
&\quad - \sum_{p' \neq p} B_{k,p}^{(p')} \left(\rho_{p'} S_{k,p'} + \Delta_{k,p'} \right) \\
&= \gamma_p S_{k,p} + \left(\sum_{r=1}^{R} F_{k,p}^{(r)} H_{k,p}^{(r)} - \gamma_p - \rho_p B_{k,p}^{(p)} \right) S_{k,p} \\
&\quad + \sum_{p' \neq p} \left(\sum_{r=1}^{R} F_{k,p}^{(r)} H_{k,p'}^{(r)} - \rho_{p'} B_{k,p}^{(p')} \right) S_{k,p'} - \sum_{p'=1}^{P} B_{k,p}^{(p')} \Delta_{k,p'} + \sum_{r=1}^{R} F_{k,p}^{(n)} N_k^{(r)} \quad (5.43)
\end{aligned}
$$

with

$$
\gamma_p = \frac{1}{N} \sum_{k=0}^{N-1} \sum_{r=1}^{R} F_{k,p}^{(r)} H_{k,p}^{(r)} \qquad (5.44)
$$

denoting the average overall channel frequency response for the pth user, after combining the outputs of the R feedforward filters.

This means that $\tilde{S}_{k,p}$ has a "signal" component, $\gamma_p S_{k,p}$, and four "noise" components:

- $\left(\sum_{r=1}^{R} F_{k,p}^{(r)} H_{k,p}^{(r)} - \gamma_p - \rho_p B_{k,p}^{(p)} \right) S_{k,p}$ is the residual ISI.
- $\sum_{p' \neq p} \left(\sum_{r=1}^{R} F_{k,p}^{(r)} H_{k,p'}^{(r)} - \rho_{p'} B_{k,p}^{(p')} \right) S_{k,p'}$ is the residual multiuser interference.
- $\sum_{p'=1}^{P} B_{k,p}^{(p')} \Delta_{k,p'}$ accounts for the errors in $\hat{s}_{n,p'}$ ($P' = 1,2,\ldots,P$).
- $\sum_{r=1}^{R} F_{k,p}^{(r)} N_k^{(r)}$ comes from the channel noise.

The forward and backward coefficients, $\{F_{k,p}^{(r)}; k = 0,1,\ldots,N-1\}$, $r = 1,2,\ldots,R$, and $\{B_{k,p}^{(p')}; k = 0,1,\ldots,N-1\}$, $p' = 1,2,\ldots,P$, respectively, are chosen so as to maximize the "signal-to-noise plus interference ratio" (SNIR) for the pth user, defined as

$$
SNIR_p = \frac{|\gamma_p|^2 \, E_S}{E[|\varepsilon_{n,p}^{eq}|^2]}, \qquad (5.45)
$$

with $\varepsilon_{n,p}^{eq} = \tilde{s}_{n,p} - \gamma_p s_{n,p}$ denoting the overall noise. The SNIR can also be written as

$$
SNIR_p = \frac{1}{N} \sum_{k=0}^{N-1} SNIR_{k,p}^F, \qquad (5.46)
$$

where $SNIR^F_{k,p}$ denotes the SNIR associated with the corresponding frequency-domain samples, defined as

$$SNIR^F_{k,p} = \frac{|\gamma_p|^2 \, ME_S}{E[|\varepsilon^{Eq}_{k,p}|^2]}, \qquad (5.47)$$

with the block $\{\varepsilon^{Eq}_{k,p}; k = 0,1,\ldots,N-1\}$ denoting the DFT of the block $\{\varepsilon^{eq}_{n,p}; n = 0,1,\ldots,N-1\}$. Clearly, the maximization of the SNIR in the time-domain samples $\tilde{s}_{n,p}$ is equivalent to the maximization of the SNIR in the corresponding frequency-domain samples, $SNIR^F_{k,p}$.

It is simple to verify that

$$E\left[\left|\varepsilon^{Eq}_{k,p}\right|^2\right] = \left|\sum_{r=1}^{R} F^{(r)}_{k,p} H^{(r)}_{k,p} - \gamma_p - \rho_p B^{(p)}_{k,p}\right|^2 NE_S + \sum_{p'\neq p}\left|\sum_{r=1}^{R} F^{(r)}_{k,p} H^{(r)}_{k,p'} - \rho_{p'} B^{(p')}_{k,p}\right|^2 NE_S$$

$$+ \sum_{p'=1}^{P}\left|B^{(p')}_{k,p}\right|^2 (1 - \rho^2_{p'})NE_S + \sum_{r=1}^{R}\left|F^{(r)}_{k,p}\right|^2 2N\sigma^2_N, \qquad (5.48)$$

with $\sigma^2_N = (1/2N)E[|N^{(r)}_k|^2]$ denoting the variance of both the in-phase and quadrature components of the channel noise at the input of each receive antenna.

The optimum feedforward and feedback coefficients are obtained by maximizing $SNIR^F_{k,p}$, which is equivalent to minimizing $E[|\varepsilon^{Eq}_{k,p}|^2]$ when $\gamma_p = 1$. This can be accomplished by solving the following set of $R + P$ equations for each frequency:

$$\frac{\partial E\left[\left|\varepsilon^{Eq}_{k,p}\right|^2\right]}{\partial F^{(r)}_{k,p}} = 2NE_S H^{(r)*}_{k,p}\left(\sum_{r'=1}^{R} F^{(r')}_{k,p} H^{(r')}_{k,p} - \gamma_p - \rho_p B^{(p)}_{k,p}\right)$$

$$+ 2NE_S \sum_{p'\neq p} H^{(r)*}_{k,p'}\left(\sum_{r'=1}^{R} F^{(r')}_{k,p} H^{(r')}_{k,p'} - \rho_{p'} B^{(p')}_{k,p}\right)$$

$$+ 4N\sigma^2_N F^{(e)}_{k,p} = 0, \quad r = 1,2,\ldots,R, \qquad (5.49)$$

$$\frac{\partial E\left[\left|\varepsilon^{Eq}_{k,p}\right|^2\right]^{(p)}}{\partial B_{k,p}} = -2NE_S\rho_p\left(\sum_{r'=1}^{R} F^{(r')}_{k,p} H^{(r')}_{k,p} - \gamma_p - \rho_p B^{(p)}_{k,p}\right)$$

$$+ 2NE_S(1 - \rho^2_p)B^{(p)}_{k,p} = 0 \qquad (5.50)$$

and

$$
\frac{\partial E\left[\left|\varepsilon_{k,p'}^{Eq}\right|^2\right]}{\partial B_{k,p}^{(p')}} = -2NE_S\rho_{p'}\left(\sum_{r'=1}^{R}F_{k,p}^{(n')}H_{k,p'}^{(r')} - \rho_{p'}B_{k,p}^{(p')}\right)
$$
$$
+ 2NE_S(1 - \rho_{p'}^2)B_{k,p}^{(p')} = 0, \quad p' \neq p. \tag{5.51}
$$

These equations can be rewritten in the form

$$
H_{k,p}^{(r)*}\left(\sum_{r'=1}^{R}F_{k,p}^{(n')}H_{k,p}^{(r')} - \gamma_p - \rho_p B_{k,p}^{(p)}\right) + \sum_{p'\neq p}H_{k,p'}^{(r)*}\left(\sum_{r'=1}^{R}F_{k,p}^{(n')}H_{k,p'}^{(r')} - \rho_{p'}B_{k,p}^{(p')}\right)
$$
$$
+ \frac{F_{k,p}^{(n)}}{SNR_p} = 0, \quad r = 1,2,\ldots,R \tag{5.52}
$$

and

$$
\rho_{p'}\left(\sum_{r'=1}^{R}F_{k,p}^{(r')}H_{k,p'}^{(r')} - \gamma_p \delta_{p,p'} - \rho_{p'}B_{k,p}^{(p')}\right) = (1 - \rho_{p'}^2)B_{k,p}^{(p')}, \quad p' \neq p, \tag{5.53}
$$

where $SNR_p = (E_{S,p}/2\sigma_N^2)$.

From Equation 5.9, the optimum values of $B_{k,p}^{(p')}$ are

$$
B_{k,p}^{(p')} = \rho_{p'}\left(\sum_{r'=1}^{R}F_{k,p}^{(r',i)}H_{k,p}^{(r')} - \delta_{p,p'}\gamma_p\right), \tag{5.54}
$$

with $\delta_{p,p'} = 1$ for $p = p'$ and 0 otherwise.

This leads to the set of R equations

$$
(1 - \rho_p^2)H_{k,p}^{(r)*}\sum_{r'=1}^{R}F_{k,p}^{(r')}H_{k,p}^{(r')} + \sum_{p'\neq p}(1 - \rho_p^2)H_{k,p'}^{(r)*}\sum_{r'=1}^{R}F_{k,p}^{(r')}H_{k,p'}^{(r')} + \frac{F_{k,p}^{(r)}}{SNR_p}
$$
$$
= \kappa_p(1 - \rho_p^2)H_{k,p}^{(r)*}, \quad r = 1,2,\ldots,R, \tag{5.55}
$$

where κ_p is selected to ensure that we have a normalized FDE output with $\gamma_p = 1$. These feedforward coefficients can be used for obtaining the feedback coefficients $B_{k,p}^{(p')}$ and $B_{k,p}^{(p')}, p' \neq p$, respectively.

It can be shown that the solution of this system of equations can be written in the form

$$F_{k,p}^{(r)} = \sum_{p'=1}^{P} H_{k,p'}^{(r)*} C_{k,p}^{(p')},$$

(5.56)

with the set of coefficients $\{C_k^{(p')}; p' = 1, 2, \ldots, P\}$ satisfying the set of P equations

$$\sum_{p''=1}^{P} C_{k,p}^{(p'')}\left((1 - \rho_{p'}^2)\sum_{r'=1}^{R} H_{k,p''}^{(r')*} H_{k,p'}^{(r')} + \frac{1}{SNR_p}\delta_{p',p''}\right) = \delta_{p,p'}, \quad p' = 1, 2, \ldots, P.$$

(5.57)

The computation of the feedforward coefficients from Equation 5.57 is simpler than the direct computation from Equation 5.56, especially when $P < R$.

For the special case where $P = 1$ (and $p = 1$), it can be shown that

$$F_{k,1}^{(r)} = \frac{SNR_1 \cdot H_{k,1}^{(r)*}}{1 + SNR_1(1 - \rho^2)\sum_{r'=1}^{N}| H_{k,1}^{(r')} |^2}, \quad r = 1, 2, \ldots, R,$$

(5.58)

which corresponds to the feedforward coefficients of an IB-DFE with R-branch space diversity [Dinis et al. 2003]. For the first iteration

$$F_{k,1}^{(r)} = \frac{SNR_1 \cdot H_{k,1}^{(r)*}}{1 + SNR_1\sum_{r'=1}^{R}| H_{k,1}^{(r')} |^2}, \quad r = 1, 2, \ldots, R,$$

(5.59)

corresponding to a linear FDE with R-branch space diversity [Gusmão et al. 2003].

It should be noted that, for the first iteration ($i = 0$), we do not have any information about $S_{k,p'}$ for $p' \leq p$, for the SIC receiver, or all $S_{k,p'}$, for the PIC receiver. Therefore, the corresponding correlation coefficients are zero, leading to $B_{k,p}^{(p')} = 0$. After the first iteration, and if the residual BER (bit error rate) is not too high, $\hat{s}_{n,p'} = s_{n,p'}$ for most of the data symbols, and $\hat{S}_{n,p'} \approx S_{k,p'}$; this means that we can use the feedback coefficients to eliminate a significant part of the residual ISI, as well as the residual multiuser interference.

5.4.3 Soft Decisions

The previous IB-DFE receiver considered hard decisions in the feedback loop (or symbolwise soft decisions). The receiver can be easily modified for soft decisions by following the approach described in the previous section. Essentially, the FDE output is given by

$$\tilde{S}_{k,p} = \sum_{r=1}^{R} F_{k,p}^{(r)} Y_k^{(r)} - \sum_{p'=1}^{P} B_{k,p}^{(p')} \overline{S}_{k,p'}, \tag{5.60}$$

where the average transmitted frequency-domain block conditioned to the FDE output $\overline{S}_{k,p'}$ can be computed as described in the previous section. It can be easily shown that the optimum feedback coefficients are still given by Equation 5.52 and the feedback values are given by

$$B_{k,p}^{(p')} = \sum_{r'=1}^{R} F_{k,p}^{(r',i)} H_{k,p}^{(r')} - \delta_{p,p'}. \tag{5.61}$$

5.4.4 Complexity Issues

Both SIC and PIC receivers require R size-N DFT operations, one for each receiver antenna, and a pair of DFT/IDFT operations (with size N) for the detection of each user at each iteration. For the equalization/interference cancelation, we need NRP multiplications for the first iteration of a PIC receiver and $NRP + P(P–1)N/2$ multiplications for the first iteration of an SIC receiver. For the remaining iterations, we need $(R + P)PN$ multiplications for both SIC and PIC receivers.

The most complex part of the algorithm is the computation of the feedforward coefficients, since we need to solve R systems of P equations, for each iteration and each user. Naturally, for slow-varying channels, this operation is not required for all blocks.

5.4.5 Performance Results

In this section, we consider the use of the proposed receiver in an SDMA system where each user has one transmit antenna and the base station has R receive antennas. The data block consists of $N = 256$ QPSK data symbols, plus an appropriate cyclic prefix. We consider an uncoded scenario, with square-root raised-cosine filtering with zero roll-off and a severely time-dispersive channel with rich multipath propagation. We have perfect synchronization and channel estimation conditions.

Figure 5.17 shows the BER performances for different users and different iterations, when we have $P = 4$ users with the same average receive power and $N = 4$ receive antennas at the BS. An SIC receiver is assumed. For the sake of comparisons, we also include the MFB performance.

From this figure, we can observe that, for the first iteration, the users have very different performances: more than 6 dB from user 1 to user 4, at BER = 10^{-4}. This

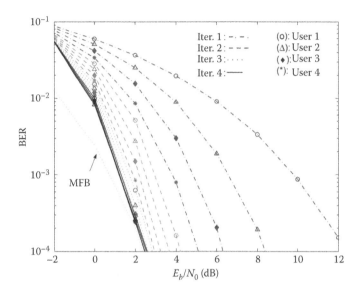

Figure 5.17   BER for the different users and the different iterations, as well as the corresponding MFB, for $R = P = 4$.

difference decreases as we increase the number of iterations, with all users having almost the same performance after three iterations. Moreover, the resulting performance is very close to the MFB after four iterations. This shows that the proposed receiver is able to eliminate a significant part of the ISI and multiuser interference.

Figure 5.18 shows the average BER performances (averaged over all the users) for SIC and PIC receivers for the different iterations. After the first iteration, the performance of the PIC receiver is almost 2 dB worse than the performance of the SIC receiver. After iteration 3, this gap reduces to less than 1 dB and after iteration 4 to 0.2 dB. Once again, the BER performances after four iterations are very close to the corresponding MFB for both structures. It should be mentioned that, for the PIC receiver, all users have the same average BER.

Figure 5.19 shows the average BER performance after four iterations, for different values of R and P. Once again, the BER performances after four iterations are close to the corresponding MFBs, regardless of R and P.

Let us consider now a scenario where the received powers are not the same for all users (e.g., due to a wrong power control and/or different uncoded BER requirements for the different users). We consider $P = 4$ users and an SIC receiver with $R = 4$ receive antennas. The average received power for users 1 and 2 is 6 dB larger than the received powers for users 3 and 4. For a given iteration, the receiver detects first the high-power and then the low-power users. Figure 5.20 shows the BER performances for the different users. Once again, the iterative detection procedure allows significant performance gains and after four iterations, the BER

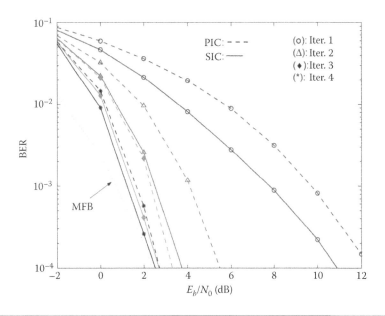

Figure 5.18 Average BER of an SIC receiver or a PIC receiver, as well as the corresponding MFB, for $R = P = 4$.

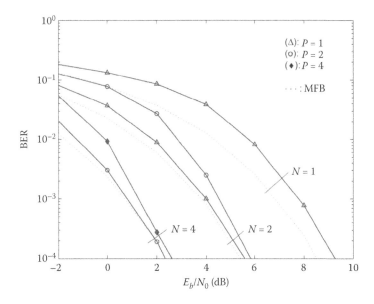

Figure 5.19 Average BER performance for different values of N and P, after four iterations.

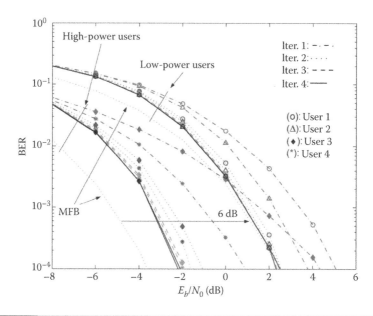

Figure 5.20 **BER performance for low-power and high-power users, as a function of the E_b/N_0 of the low-power users.**

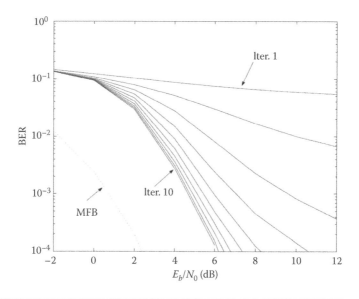

Figure 5.21 **Average BER performance for $R = 4$ and $P = 6$.**

performances are similar for users with the same average power. Clearly, the performance of low-power users asymptotically approaches the MFB when we increase the number of iterations. However, for the high-power users, we are still 2 dB from the MFB at BER = 10^{-4}. This somewhat unexpected behavior can be explained from the fact that the BER is much lower for the high-power users, allowing an almost perfect interference cancelation of its effects on low-power users and performances close to the MFB. The lower BERs for the low-power users preclude an appropriate interference cancelation on the high-power users.

Figure 5.21 refers to the situation where $R = 4$ and $P = 6$, that is, an overloaded scenario where the number of users is larger than the number of receive antennas at the BS. In this case, a perfect multiuser separation is not possible since we have $R = 4$ degrees of freedom to separate $P = 6$ users. However, the iterative receiver structure presented here has an acceptable performance, with significant interference cancelation, although, even after 10 iterations, the achievable performance is still about 4 dB from the MFB.

Acknowledgment

This work was supported by the FCT (Fundação para a Ciência e Tecnologia) via project PEst-OE/EEI/LA0008/2013.

References

Abe, T. and T. Matsumoto, Space–time turbo equalization in frequency-selective MIMO channels, *IEEE Trans. Veh. Tech.*, 52(3), 469–475, 2003.

Al-Dhahir, N. and A. H. Sayed, The finite-length multi-input multi-output MMSE-DFE, *IEEE Trans. Signal Process.*, 48, 2921–2936, 2000.

Amaral, F., R. Dinis, P. Montezuma, and N. Souto, Approaching the MFB with block transmission techniques, *Eur. Trans. Telecommun.*, 23(1), 76–86, 2012.

Araújo, T. and R. Dinis, Iterative equalization and carrier synchronization for single carrier transmission over severe time-dispersive channels, in *Proceedings of IEEE GLOBECOM'04*, Dallas, USA, November–December 2004.

Ariyavisitakul, S. Turbo space–time processing to improve wireless channel capacity, in *Proceedings of IEEE ICC'02*, 3, 1238–1242, 2000.

Benvenuto, N., R. Dinis, D. Falconer, and S. Tomasin, Single carrier modulation with non linear frequency domain equalization: An idea whose time has come—again, *Proc. IEEE*, 98(1), 69–96, 2010.

Benvenuto, N. and S. Tomasin, On the comparison between OFDM and single carrier with DFE using a frequency domain feedforward filter, *IEEE Trans. Commun.*, 50(6), 947–955, 2002.

Benvenuto, N. and S. Tomasin, Block iterative DFE for single carrier modulation, *IEE Electron. Lett.*, 38(19), 1144–1145, 2002.

Benvenuto, N. and S. Tomasin, Iterative design and detection of a DFE in the frequency domain, *IEEE Trans. Commun.*, 53(11), 1867–1875, 2005.

Berrou, C., A. Glavieux, and P. Thitimajshima, Near Shannon limit error correcting coding and decoding: Turbo-codes, in *Proceedings of IEEE ICC'93*, Geneva, Switzerland, Vol. 2, pp. 1064–1070, May 1993.

Bingham, J. Multicarrier modulation for data transmission: An idea whose time has come, *IEEE Commun. Mag.*, 28(5), 5–14, 1990.

Cimini Jr., L. Analysis and simulation of a digital mobile channel using orthogonal frequency division multiplexing, *IEEE Trans. Commun.*, 33(7), 665–675, 1985.

Cover, T. Broadcast channels, *IEEE Trans. Inform. Theory*, IT-18, 2–14, 1972.

Dinis, R., T. Araújo, P. Pedrosa, and F. Nunes, Joint turbo equalization and carrier synchronization for SC-FDE schemes, *Eur. Trans. Telecommun.*, 21(2), 131–141, 2010a.

Dinis, R., P. Carvalho, L. Bernardo, R. Oliveira, M. Pereira, and P. Pinto, Frequency-domain multipacket detection: A high throughput technique for SC-FDE systems, *IEEE Trans. Wireless Commun.*, 8(7), 3798–3807, 2009.

Dinis, R., D. Falconer, C. Lam, and M. Sabbaghian, A multiple access scheme for the uplink of broadband wireless systems, in *Proceedings of IEEE GLOBECOM'04*, Dallas, USA, December 2004a.

Dinis, R. and A. Gusmão, A class of nonlinear signal processing schemes for bandwidth-efficient OFDM transmission with low envelope fluctuation, *IEEE Trans. Commun.*, 52(11), 2009–2018, 2004.

Dinis, R., A. Gusmão, and N. Esteves, On broadband block transmission over strongly frequency-selective fading channels, in *Proceedings of Wireless 2003*, Calgary, Canada, 261–269, July 2003.

Dinis, R., A. Gusmão, and N. Esteves, Iterative block-DFE techniques for single-carrier-based broadband communications with transmit/receive space diversity, in *Proceedings of IEEE ISWCS'04*, Port Louis, Mauritius, September 2004b.

Dinis, R., C. Lam, and D. Falconer, Joint frequency-domain equalization and channel estimation using superimposed pilots, in *Proceedings of IEEE WCNC'08*, Las Vegas, USA, March 2008.

Dinis, R., R. Kalbasi, D. Falconer, and A. Banihashemi, Iterative layered space–time receivers for single-carrier transmission over severe time-dispersive channels, *IEEE Commun. Lett.*, 8(9), 579–581, 2004.

Dinis, R. and P. Silva, Iterative detection of multicode DS-CDMA signals with strong nonlinear distortion effects, *IEEE Trans. Veh. Technol.*, 58(8), 4169–4181, 2009.

Dinis, R., P. Silva, and T. Araújo, Turbo equalization with cancelation of nonlinear distortion for CP-assisted and zero-padded MC-CDM schemes, *IEEE Trans. Commun.*, 57(8), 2185–2189, 2009.

Dinis, R., P. Silva, and Gusmão, IB-DFE receivers with space diversity for CP-assisted DS-CDMA and MC-CDMA systems, *Eur. Trans. Telecommun.*, 18(7), 791–802, 2007a.

Dinis, R., J. Silva, N. Souto, and P. Montezuma, On the design of frequency-domain equalizers for OQPSK modulations, in *Proceedings of 33rd IEEE Sarnoff Symposium'2010*, Princeton, USA, April 2010b.

Dinis, R., J. Silva, N. Souto, and P. Montezuma, On the design of turbo equalizers for SC-FDE schemes with different error protections, in *Proceedings of IEEE VTC'2010 (Fall)*, Ottawa, Canada, September 2010c.

Falconer, D., S. Ariyavisitakul, A. Benyamin-Seyar, and B. Eidson, Frequency domain equalization for single-carrier broadband wireless systems, *IEEE Commun. Mag.*, 40(4), 58–66, 2002.

Foschini, G. Layered-space–time architecture for wireless communication in a fading environment when using multi-element antennas, *Bell Labs Tech. J.*, 41–59, Autumn 1996.

Foschini, G. J. and M. J. Gans, On limits of wireless communications in fading enviroments when using multiple antennas, *Wireless Pers. Commun.*, 6, 315–335, 1998.

Ganhão, F., M. Pereira, L. Bernardo, R. Dinis, R. Oliveira, and P. Pinto, Performance of hybrid ARQ for NDMA access schemes with uniform average power control, *J. Commun.*, 6(9), 691–699, 2011.

Ganhão, F., R. Dinis, L. Bernardo, and R. Oliveira, Analytical BER and PER performance of frequency-domain diversity combining, multipacket detection and hybrid schemes, *IEEE Trans. Commun.*, 60(8), 2353–2362, 2012.

Ganhão, F., M. Pereira, L. Bernardo, R. Dinis, R. Oliveira, and P. Pinto, Performance analysis of an hybrid ARQ adaptation of NDMA schemes, *IEEE Trans. Commun.*, 61(8), 3304–3317, 2013.

Gitlin, R., J. Salz, and J. Winters, The impact of antenna diversity on the capacity of wireless communication systems, *IEEE Trans. Commun.*, 42(2/3/4), 1740–1751, 1994.

Gusmão, A., R. Dinis, J. Conceição, and N. Esteves, Comparison of two modulation choices for broadband wireless communications, in *Proceedings of IEEE VTC'00 (Spring)*, 2, 1300–1305, 2000.

Gusmão, A., R. Dinis, and N. Esteves, On frequency-domain equalization and diversity combining for broadband wireless communications, *IEEE Trans. Commun.*, 51(7), 1029–1033, 2003.

Gusmão, A., P. Torres, R. Dinis, and N. Esteves, A class of iterative FDE techniques for reduced-CP SC-based block transmission, in *Proceedings of the International Symposium on Turbo Codes*, Munich, Germany, April 2006.

Gusmão, A., P. Torres, R. Dinis, and N. Esteves, A turbo FDE technique for reduced-CP SC-based block transmission systems, *IEEE Trans. Commun.*, 55(1), 16–20, 2007a.

Gusmão, A., P. Torres, R. Dinis, and N. Esteves, A reduced-CP approach to SC/FDE block transmission for broadband wireless communications, *IEEE Trans. Commun.*, 55(4), 801–809, 2007b.

Jiang, H. and P. Wilford, A hierarchical modulation for upgrading digital, *IEEE Trans. Broadcast.*, 51(2), 223–229, 2005.

Jones, A. and T. Wilkinson, Combined coding for error control and increased robustness to system nonlinearities in OFDM, in *Proceedings of IEEE VTC'96*, 2, 904–908, 1996.

Kalbasi, R., R. Dinis, D. Falconer, and A. Banihashemi, Hybrid time-frequency layered space–time receivers for severe time-dispersive channels, in *Proceedings of IEEE SPAWC'04*, Lisbon, Portugal, July 2004.

Lam, C., D. Falconer, F. Danilo-Lemoine, and R. Dinis, Channel estimation for SC-FDE systems using frequency domain multiplexed pilots, in *Proceedings of IEEE VTC'06 (Fall)*, Montreal, Canada, September 2006.

Lozano, A. and C. Papadias, Layered space–time receiver for frequency-selective wireless channels, *IEEE Trans. Commun.*, 50(1), 65–73, 2002.

Luzio, J., R. Dinis, and P. Montezuma, SC-FDE for offset modulations: An efficient transmission technique for broadband wireless systems, *IEEE Trans. Commun.*, 60(7), 1851–1861, 2012.

Luzio, M., R. Dinis, and P. Montezuma, Pragmatic frequency domain equalization for single carrier with offset modulations, *IEEE Trans. Wireless Commun.*, 12(9), 4496–4505, 2013.

Montezuma, P. and A. Gusmão, A pragmatic coded modulation choice for future broadband wireless communications, in *Proceedings of IEEE VTC'01 (Spring)*, 2, 1324–1328, 2001a.

Montezuma, P. and A. Gusmão, On analytically described trellis-coded modulation schemes, in *Proceedings of ISCTA'01*, Ambleside, UK, July 2001b.

Müller, S., R. Bäuml, R. Fischer, and J. Huber, OFDM with reduced peak-to-average power ratio by multiple signal representation, *Ann. Telecommun.*, 52, 58–67, 1997.

Pedrosa, P., R. Dinis, and F. Nunes, Iterative frequency domain equalization and carrier synchronization for multi-resolution constellations, *IEEE Trans. Broadcast.*, 56(4), 551–557, 2010.

Pedrosa, P., R. Dinis, and F. Nunes, Analytical performance evaluation of a class of receivers with joint equalization and residual CFO estimation, *Trans. Emerg. Telecommun. Technol.*, 23(8), 749–763, December 2012a.

Pedrosa, P., R. Dinis, F. Nunes, and J. Bioucas-Dias, Phase drift estimation and symbol detection in digital communications: A stochastic recursive filtering approach, *IEEE Commun. Lett.*, 16(6), 854–857, 2012b.

Pereira, M., L. Bernardo, R. Dinis, R. Oliveira, P. Montezuma, and P. Pinto, Performance of diversity combining ARQ error control in a TDMA SC-FDE system, *IEEE Trans. Commun.*, 60(3), 735–746, 2012.

Pereira, M., L. Bernardo, R. Dinis, R. Oliveira, and P. Pinto, Frequency-domain cross-layer diversity techniques: An efficient way of coping with lost packets in broadband wireless systems, *IEEE Wireless Commun. Mag.*, 20(5), 100–108, 2013a.

Pereira, M., L. Bernardo, R. Dinis, R. Oliveira, and P. Pinto, On the use of frequency-domain cross-layer diversity techniques to cope with lost packets, *Phys. Commun.*, Elsevier Physical Communication, 8, 56–68, 2013b.

Proakis, J. *Digital Communications*, McGraw-Hill, New York, USA, 1995.

Raleigh, G. and J. Cioffi, Spatio-temporal coding for wireless communication, *IEEE Trans. Commun.*, 46(3), 357–366, 1998.

Raleigh, G. and V. Jones, Multivariate modulation and coding for wireless communication, *IEEE J. Select. Areas Commun.*, 17(5), 851–866, 1999.

Ramchandran, K., A. Ortega, K. Uz, and M. Vetterli, Multi-resolution broadcast for digital HDTV using joint source/channel coding, *IEEE J. Select. Areas Commun.*, 11, 6–23, 1993.

Roy, S. and D. Falconer, Optimum infinite-length MMSE multi-user decision-feedback space–time processing in broadband cellular radio, *Wireless Pers. Commun.*, 27, 1–32, 2003.

Sari, H., G. Karam, and I. Jeanclaude, An analysis of orthogonal frequency-division multiplexing for mobile radio applications, in *Proceedings of IEEE VTC'94*, 3, 1635–1639, 1994.

Sfar, S., R. Murch, and K. Letaief, Layered space–time multiuser detection over wireless uplink systems, *IEEE Trans. Wireless Commun.*, 2(4), 653–668, 2003.

Silva, P. and R. Dinis, Frequency-domain multiuser detection for CP-assisted DS-CDMA signals, in *Proceedings of IEEE VTC'06 (Spring)*, Melbourne, Australia, May 2006a.

Silva, P. and R. Dinis, Multiuser detection for the uplink of prefix-assisted DS-CDMA systems employing multiple transmit and receive antennas, in *Proceedings of IEEE VTC'06 (Fall)*, Montreal, Canada, September 2006b.

Silva, P. and R. Dinis, A turbo SDMA receiver for strongly nonlinearly distorted MC-CDMA signals, *Can. J Elect. Comput. Eng.*, 33(1), 39–44, 2008.

Silva, P. and R. Dinis, Joint turbo equalization and multiuser detection of MC-CDMA signals with strongly nonlinear transmitters, *IEEE Trans. Veh. Technol.*, 58(5), 2288–2298, 2009.

Silva, F., R. Dinis, and P. Montezuma, Estimation of the feedback reliability for IB-DFE receivers, *ISRN Commun. Networking J.*, 2011, 2011.

Silva, J., R. Dinis, N. Souto, and P. Montezuma, Single-carrier frequency domain equalisation with hierarchical constellations: An efficient transmission technique for broadcast and multicast systems, *IET Commun.*, 6(13), 2065–2073, 2012.

Silva, F., R. Dinis, and P. Montezuma, Channel estimation and equalization for asynchronous single frequency networks, *IEEE Trans. Broadcast.*, 2013.

Silva, M., R. Dinis, and P. Montezuma, Iterative frequency-domain packet combining techniques for UWB systems with strong interference levels, *Wireless Pers. Commun.*, 9(9), 2065–2073, 2012.

Tidestav, C., M. Sternad, and A. Ahlèn, Reuse within a cell—Interference rejection or multiuser detection?, *IEEE Trans. Commun.*, 47(10), 1511–1522, 1999.

Tüchler, M., R. Koetter, and A. Singer, Turbo equalization: Principles and new results, *IEEE Trans. Commun.*, 50(5), 754–767, 2002.

Vucetic, B. and J. Yuan, *Turbo Codes: Principles and Applications*, Kluwer Academic publishers, London, 2002.

Wang, Z. and G. Giannakis, Wireless multicarrier communications—Where Fourier meets Shannon, *IEEE Signal Proc. Mag.*, 17(3), 29–48, 2000.

Winters, J., J. Salz, and R. Gitlin, The capacity of wireless communication systems can be substantially increased by the use of antenna diversity, in *Proceedings of ICUPC'92*, Dallas, USA, pp. 02.01/1–02.01/5, September–October 1992.

Wolniasky, P. W., G. J. Foschini, G. D. Golden, and R. A. Valenzuela, V-BLAST: An architecture for realizing very high rates over rich-scattering wireless channel, in *Proceedings of ISSSE'98*, Pisa, Italy, pp. 295–300, October 1998.

Chapter 6

MIMO Optimized for W-CDMA

João Carlos Silva, Nuno Souto, and Rui Dinis

Contents

In this chapter, we will study the use of linear equalizers for wideband code division multiple access (W-CDMA) using multiple-input multiple-output (MIMO) systems. The W-CDMA concept will be introduced and the minimum mean square error (MMSE) alongside the zero-forcing (ZF) methods will be studied and analyzed.

The well-known RAKE receiver is not sufficient to cope with the interference of fully loaded systems [Brunner et al. 1999]. One of the main problems consists in the uncanceled multipath components pertaining to the same message; since the ratio of used channels per spreading factor (SF) is close to one, the cross-correlation between sequences is very significant, and left uncanceled. By employing an MPIC (multipath interference canceler), substantial performance gains were observed, although still very far from the single user bound for fully loaded scenarios. Since the MPIC has its operation stemmed from the RAKE, it is only able to cancel part of the inter-symbolic interference (ISI) (the RAKE leaves a significant noise component attached). However, for higher-order modulations than quadrature phase shift keying (QPSK) and/or MIMO systems, where interference from other transmitting antennas (ICI—interchannel interference) significantly increases the multiple access interference (MAI) and ISI, a more powerful receiver needs to be employed.

Equalization-based receivers try to take into account all effects that the symbols are subject to in the transmission chain, namely, the joint compensation of MAI, ICI, and ISI. In this chapter, linear equalizers were employed, namely, MMSE and ZF.

6.1 W-CDMA Fundamentals

The fundamentals of W-CDMA are addressed in this section. To begin with, the CDMA concept is analyzed and compared to the alternative multiple access techniques, such as time division multiple access (TDMA) and frequency division multiple access (FDMA). The direct-sequence (DS) spread-spectrum (SS) technique is also compared to other methods, including the time hopping (TH) and frequency hopping (FH) methods. Narrowband and wideband channels are also discussed, with the multipath effect being explained in the latter case.

After defining the W-CDMA model, the conventional RAKE receiver is described, and the concept of maximal ratio combining (MRC) is analyzed. In order to obtain better performance results by canceling a substantial part of ISI/MPI (intersymbolic/multipath interference), the standard MPIC is also discussed.

6.1.1 CDMA Concept

Multiple access (MA) communications refer to a communication system that allows more than one user to transmit over one physical channel resource at the same time. This does not imply that the multiple signals are necessarily overlapping in time; users only need to feel that they are accessing the channel all the time (for instance, user 1 may be transmitting and receiving short bursts of 100 μs every 1 ms, but the 900 μs gap is imperceptible and of no consequence, as long as all the information he needs to transmit is accommodated within his periodic 100 μs bursts—this is the principle of TDMA, explained later on).

MA is distinct from multiplexing, though the two concepts are very similar. MA is reserved for dynamic systems of channel sharing in which users occupy part of the channel only if they need to transmit; multiplexing refers to a fixed allocation of channel resources such as FM radio broadcasting, and the splitting of a given bandwidth into two segments for uplink and downlink communications (FDD—frequency division duplexing).

The MA techniques can be divided into two classes:

- Fixed assignment—a central controller (base station) assigns channels to users, who request channels through a common reservation protocol when they need to establish a connection.
- Random access—users in the system transmit whenever they need to, based on a common protocol. If two or more data packets are sent at the same time, collisions occur and the lost packets must be retransmitted.

6.1.2 Fixed Assignment Multiple Access Techniques

Three of the most popular forms of fixed assignment MA are

- TDMA—In TDMA, users must be time-synchronized, and are assigned time slots within a frame in which they can transmit. The entire frequency bandwidth is used by each user, but there is no interference because time slots are nonoverlapping.

■ FDMA—In FDMA, users transmit all the time, but each one is allowed only one segment of the total system bandwidth. Like TDMA, there is no interference from other users in FDMA, as long as the bandwidth segments of each user are separated by a frequency guard wide enough to prevent adjacent-band interference. A filter at the receiver is employed to use solely the data contained in the wanted frequency segment.

■ CDMA—In CDMA, users transmit over all the time over the same frequency, being separated on the basis of their different symbol-pulse waveforms (or codes). CDMA requires that the bandwidth occupied by each user be several times that of the data bandwidth; hence CDMA is possible only with SS modulation.

Both TDMA and FDMA signals can be made noninterfering or orthogonal. In practice, this requires the insertion of guard intervals and guard bands between users, respectively, to account for nonidealities such as synchronization error and delay spread caused by the channel. CDMA signals can be made orthogonal only when the users are perfectly synchronized in time, and the channel has no delay spread.

For the uplink, the first requirement is impossible to meet; the users are not synchronized with each other, but even if they were, since they have different channels, it would be impossible to have the received signals orthogonal to each other. Also, in wireless channels, the delay spread is usually significant—especially in W-CDMA, characterized by frequency-selective fading and different delay spreads. Therefore, CDMA is necessarily a nonorthogonal form of MA and suffers from MAI.

There are, however, several advantages to the usage of CDMA, namely

■ Capacity (measured as bits/second/Hz/user). In fact, since all of the bandwidth is used for all the time, the systems' potential is effectively exploited [Viterbi 1995].
■ Robustness to frequency-selective fading (multipath diversity helps reduce the channel fading effect—this is also common to TDMA).
■ Asynchronous nature of the system (contrary to TDMA).
■ Universal frequency reuse (eliminating the need for cell planning in FDMA).
■ Linear performance degradation with loading. There is no hard limit on the number of users allowed—more users can be admitted as long as the quality of service (QoS) for the existing users remains adequate.
■ No periodic on–off switching (huge problem in hospitals for TDMA equipment, due to their regular on–off switching), and less peak power requirement on the amplifiers.

Comparing the three fixed assignment MAs, FDMA is clearly the least attractive scheme. Between CDMA and TDMA, it could seem that CDMA would be the obvious choice, although practical implementations of CDMA and TDMA exhibit little advantages over each other. The European 3G standard, universal mobile telecommunication system (UMTS), has both a CDMA and TDMA standard, whereas the American standard, CDMA2000, has just CDMA.

6.1.3 Spread-Spectrum Communications

SS signals are characterized for having a bandwidth, W, much greater than the information rate (in bps) [Proakis 2001]. Therefore, if the information bit rate is R, the bandwidth expansion factor, usually called processing gain or spreading factor (SF—the main designation adopted in UMTS and this chapter), is defined as

$$SF = \frac{W}{R} \gg 1 \qquad (6.1)$$

The concept of SS was introduced by the United States for military communications after World War II, primarily because of its antijamming capabilities. At the time, an enemy would attempt to jam your communications by sweeping a powerful narrowband signal (just a tone, for instance) over all likely frequencies. However, with SS, a matched-filter (MF) detector is just another SS modulator (the narrowband jamming signal appears at the MF output as a low-level wideband signal, but the wideband desired signal appears as a high-power narrowband signal). After lowpass filtering, the desired signal is recovered, while the jamming signal is severely attenuated by the filtering. The power level of the jamming signal is reduced by the SF of the system. The greater the value of SF, the more robust the system is against narrowband jamming, and frequency-selective fading. A very good introduction to SS is given in [Proakis and Salehi 2002]. SS communications can be employed using different types of techniques; the most common techniques are portrayed in Figure 6.1:

- DS—The information symbols are encoded by a high-rate channel code (usually between 1/2 and 1/3), after which they are spreaded (with a low-rate code, $1/SF$, normally using orthogonal spreading sequences in order to distinguish between users/physical channels), and then modulated in combination with a pseudorandom sequence (commonly referred to as scrambling, with

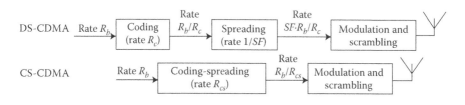

Figure 6.1   DS-CDMA and CS-CDMA schemes.

unitary rate and generally used to distinguish between transmitters). In the receiver, the received signal is descrambled, demodulated, despreaded, and, finally, decoded.

■ Code spread (CS)—An alternative to DS is to perform all the spreading of the signal using a low-rate code (equivalent to R_c/SF, $R_c < 1$) and then adding the scrambling code to the signal. In [Viterbi 1995], it is shown that this technique can achieve a greater performance level than the DS technique (if optimal decoding is employed by the receiver).

■ FH—In this case, the available bandwidth is divided into several contiguous subband frequency slots. A pseudorandom sequence is used for selecting the frequency slot for transmission in each signaling interval.

■ TH—In this method, a time interval is divided into several time slots and the coded information symbols are transmitted in a time slot selected according to a pseudorandom sequence. The coded symbols are transmitted in the selected time slot as blocks of one or more code words.

■ Hybrid techniques—DS, CS, FH, and TH can be combined to obtain other types of SS signals. For example, a system can use a combination of DS and FH, where the transmitted signal is associated with two code sequences. One of the sequences is multiplied by the signal to be transmitted while the second is used to select the frequency slot for transmission in each signaling interval.

6.1.4 Narrowband versus Wideband CDMA

In CDMA communications, several users share the same channel bandwidth for transmitting information simultaneously. Assuming that all the users employ the same channel encoder, the transmitted signals use the same frequency band and may be distinguished from one another by using a different pseudorandom (DS-CDMA)/coded pseudorandom (CS-CDMA) sequence for each transmitted signal.

CDMA systems can be considered narrowband or wideband depending on the mobile propagation conditions. If the transmitted signal bandwidth is lower than the coherence bandwidth of the channel, there will be only one macroscopic distinguishable received replica of the signal. In this case, the system is narrowband CDMA. If the transmitted signal bandwidth is greater than the coherence bandwidth of the channel, then it will be possible to resolve several multipath components resulting in a W-CDMA system.

Formally, the transmitted CDMA signal $x_T(t)$ can be described as

$$x_{T,\text{modulated}}(t) = Re\{x_T(t)\}. \tag{6.2}$$

where

$$x_T(t) = \sum_{k=1}^{K} \sum_{i=0}^{\infty} b'_{k,i} \, c_{k,i} \, p(t - iT_c) \tag{6.3}$$

is the baseband representation of the signal transmitted by the user k. $\{b'_{k,i}\}$ is a sequence of complex symbols, where each consecutive set of SF symbols is the repetition of the same modulated symbol (SF is given by $SF = T_b/T_c$). $T_b = 1/R_b$ is the bit period and T_c is the chip period. The original sequence of unrepeated symbols is $\{b_{k,i}\}$, and the function $p(t)$ represents a pulse having a limited bandwidth. $\{c_{k,i}\}$ is the sequence of values of the combined spreading, scrambling, and modulation code (usually periodic) for user k, being composed of two component sequences c_{I_i} and c_{Q_i} (with amplitudes varying between -1 and $+1$) for the in-phase and the in-quadrature components, respectively. To keep the absolute value of the complex code equal to 1, the sequence is constructed as (QPSK is considered)

$$c_i = \frac{\sqrt{2}}{2}[c_{I_i} + jc_{Q_i}]. \tag{6.4}$$

There is a significant difference between narrowband and wideband CDMA, namely, in the channel effect and the appearance of MPI—multipath interference. A brief discussion on both concepts is given below.

6.1.5 Narrowband CDMA

In a narrowband CDMA system, admitting that there are K users transmitting simultaneously, the decision variable for symbol $\{b_{0,i}\}$ of user 0 obtained in the receiver (Figure 6.2) is

$$
\begin{aligned}
y_{0,i} = \hat{b}_{0,i} = {} & \frac{1}{SF} \sum_{m=0}^{SF-1} b'_{0,i\cdot SF+m} \cdot c_{0,i\cdot SF+m} \cdot c^*_{0,i\cdot SF+m} \cdot \alpha_{0,i\cdot SF+m} \cdot \alpha^*_{0,i\cdot SF+m} \\
& + \frac{1}{SF} \sum_{m=0}^{SF-1} \left(\sum_{k=1}^{K-1} b'_{k,i\cdot SF+m} \cdot c_{k,i\cdot SF+m} \cdot \alpha_{k,i\cdot SF+m} \right) \cdot c^*_{0,i\cdot SF+m} \cdot \alpha^*_{0,i\cdot SF+m} \\
& + \frac{1}{SF} \sum_{m=0}^{SF-1} (n_{i\cdot SF+m}) \cdot c^*_{0,i\cdot SF+m} \cdot \alpha^*_{0,i\cdot SF+m},
\end{aligned}
\tag{6.5}
$$

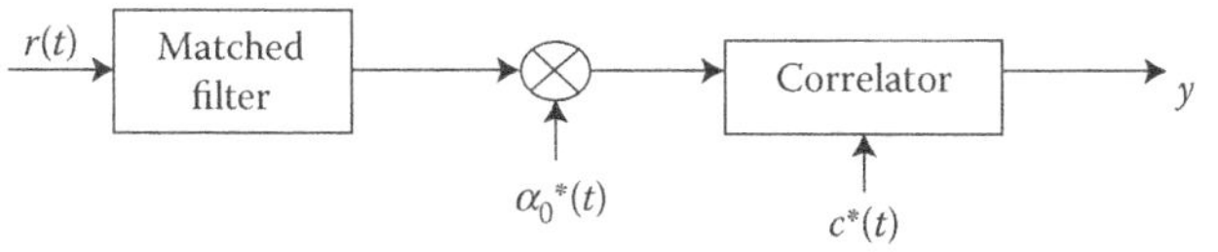

Figure 6.2 DS-CDMA receiver scheme.

where $\alpha_{u,i}^{*}$ is the complex conjugate of the channel coefficient of user k. To simplify the notation, it was considered that there were no delays between the received signals of all the users. Since $c_i \cdot c_i^{*} = 1$, the expression simplifies to

$$
\begin{aligned}
y_{0,i} &= \frac{1}{SF} \sum_{m=0}^{SF-1} b_{0,i \cdot SF+m}^{'} \cdot \left|\alpha_{0,i \cdot SF+m}\right|^2 + \frac{1}{SF} \sum_{m=0}^{SF-1} \left(\sum_{k=1}^{K-1} b_{k,i \cdot SF+m}^{'} \cdot c_{k,i \cdot SF+m} \cdot \alpha_{k,i \cdot SF+m} \right) \\
&\quad \times c_{0,i \cdot SF+m}^{*} \cdot \alpha_{0,i \cdot SF+m}^{*} + \frac{1}{SF} \sum_{m=0}^{SF-1} (n_{i \cdot SF+m}) \cdot c_{0,i \cdot SF+m}^{*} \cdot \alpha_{0,i \cdot SF+m}^{*} \\
&= \frac{1}{SF} b_{0,i \cdot SF+m}^{'} \sum_{m=0}^{SF-1} \left|\alpha_{0,i \cdot SF+m}\right|^2 + \frac{1}{SF} \sum_{m=0}^{SF-1} \left(\sum_{k=1}^{K-1} b_{k,i \cdot SF+m}^{'} \cdot c_{k,i \cdot SF+m} \cdot \alpha_{k,i \cdot SF+m} \right) \\
&\quad \times c_{0,i \cdot SF+m}^{*} \cdot \alpha_{0,i \cdot SF+m}^{*} + \frac{1}{SF} \sum_{m=0}^{SF-1} (n_{i \cdot SF+m}) \cdot c_{0,i \cdot SF+m}^{*} \cdot \alpha_{0,i \cdot SF+m}^{*}.
\end{aligned}
$$

$$(6.6)$$

In this expression, the second term represents the interference component, I_i, from the other users

$$
\begin{aligned}
\tilde{I}_i &= \frac{1}{SF} \sum_{m=0}^{SF-1} \left(\sum_{k=1}^{K-1} b_{k,i \cdot SF+m}^{'} \cdot c_{k,i \cdot SF+m} \cdot \alpha_{k,i \cdot SF+m} \right) \cdot c_{0,i \cdot SF+m}^{*} \cdot \alpha_{0,i \cdot SF+m}^{*} \\
&= \frac{1}{SF} \sum_{k=1}^{K-1} \left(\sum_{m=0}^{SF-1} b_{k,i \cdot SF+m}^{'} \cdot c_{k,i \cdot SF+m} \cdot \alpha_{k,i \cdot SF+m} \cdot c_{0,i \cdot SF+m}^{*} \cdot \alpha_{0,i \cdot SF+m}^{*} \right) \\
&= \frac{1}{SF} \sum_{k=1}^{K-1} b_{k,i \cdot SF+m} \left(\sum_{m=0}^{SF-1} c_{k,i \cdot SF+m} \cdot c_{0,i \cdot SF+m}^{*} \cdot \alpha_{0,i \cdot SF+m}^{*} \cdot \alpha_{k,i \cdot SF+m} \right).
\end{aligned}
$$

$$(6.7)$$

Admitting that the channel is approximately constant during SF chip period results in

$$
I_i = \frac{1}{SF} \sum_{k=1}^{K-1} b_{k,i \cdot SF+m} \cdot \alpha_{0,i \cdot SF}^{*} \cdot \alpha_{k,i \cdot SF} \left(\sum_{m=0}^{SF-1} c_{k,i \cdot SF+m} \cdot c_{0,i \cdot SF+m}^{*} \right).
$$

$$(6.8)$$

Therefore, to minimize the interference in the received signal, the cross-correlation of the spreading sequences, given by

$$
\sum_{m=0}^{SF-1} c_{k,i \cdot SF+m} \cdot c_{0,i \cdot SF+m}^{*},
$$

$$(6.9)$$

should have an absolute value as low as possible. Ideally, this value should be equal to zero so that no interference appears in the decision variable but that is only possible when using orthogonal spreading sequences and the users' received signals are all synchronized (like the downlink connection of a mobile cellular radio communication system). In the uplink, since there is no synchronization and each user has a different channel, usually pseudo-noise (PN) sequences are used, which exhibit small cross-correlation values between misaligned sequences. Figure 6.2 represents the DS-CDMA receiver scheme.

6.1.6 Wideband CDMA

In W-CDMA systems, replicas arrive at the receiver during a certain (continuous) period of time, due to temporal dispersion. In order to facilitate the handling of the channel, the tapped delay line (TDL) channel model was created, where the replicas are modeled as being discrete (intervals of T_c), being dubbed as multipaths. The multipath replicas carry the information about the transmitted signal and suffer fading that is uncorrelated between them. So, when a replica is severely attenuated due to the fading, there is the possibility that the others are received in good conditions and can be used to recover the transmitted signal. Owing to the good autocorrelation properties of the spreading codes usually used in CDMA systems, it is possible to distinguish and extract the strongest replicas present in the received signal.

After the transmission, the signal passes through the channel. The channel impulse response $h_c(\tau,t)$ can be described using the previously referred TDL model, where the multipaths are considered discrete and their number is $L(t)$:

$$h_c(\tau,t) = \sum_{l=1}^{L(t)} \alpha_l(\tau_l(t),t) \cdot \delta(\tau - \tau_l(t)). \tag{6.10}$$

This model represents the response of the channel at instant t to an impulse applied at $t - \tau$. This formulation does not correspond to the usual impulse response formulation for time-variant linear systems $h(t,t_0')$ that represents the response at time t to an impulse applied at time t_0. These two formulations are related through $h_c(\tau,t) = h(t,t_0)$ with $\tau = t - t_0$. Admitting a channel with a fixed number of discrete multipath components, L, and respective delays τ_l, results in the following channel impulse response:

$$h_c(\tau,t) = \sum_{l=1}^{L} \alpha_l(t) \cdot \delta(\tau - \tau_l) \tag{6.11}$$

The signal after the channel is given by

$$x_{\text{channel}}(t) = h_c(\tau,t) \otimes x_T(t). \tag{6.12}$$

The symbol $\otimes$ in the last expression represents a superposition integral for time-variant linear systems:

$$h_c(\tau,t) \otimes x_T(t) = \int_{-\infty}^{\infty} h_c(\tau,t) x_T(t - \tau) d\tau. \tag{6.13}$$

In the case of time-invariant systems, it is reduced to a convolution integral (with the symbol $*$). At the receiver, the signal $y_R(t)$ is expressed by

$$y_R(t) = h_c(\tau,t) * y_T(t) + I(t) + n(t) = \sum_{l=1}^{L} \alpha_l(t) \cdot x_T(t - \tau_l) + I(t) + n(t), \tag{6.14}$$

where $I(t)$ represents the overall MAI, while $n(t)$ is Gaussian noise.

The replicas at the receiver can be combined using a RAKE receiver, as shown in Figure 6.3, thus obtaining diversity. According to the RAKE scheme, the received signal is first filtered using a matched filter. Then, each of the L branches of the receiver, also called "fingers," is used to detect separately one of the L strongest replicas. The L estimates of the signal obtained by the "fingers" are then added to compute the final estimate of the transmitted signal. MRC is assumed to be used, which means that each of the L estimates is weighted using the complex conjugate of the corresponding channel coefficient $\alpha_i^*(t)$. This technique results in the highest mitigation of fading effects.

For correct operation, the RAKE receiver must be able to identify the strongest multipath components present in the received signal and also estimate their relative delays, amplitudes, and phases. This can be accomplished by transmitting periodic preambles or pilot codes.

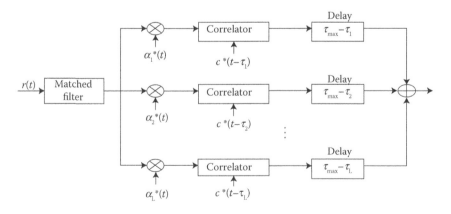

Figure 6.3   RAKE receiver scheme.

Usually, only one sample per chip is used after the filter, in order to reduce complexity and simulation time. These samples are obtained at the optimum sampling instants according to the Nyquist pulse-shaping criterion. If the multipath components (taps) are not aligned with the sampling time (real-life scenario), then the work done in [Silva et al. 2003] should be taken into account, in order to continue working with only one sample per chip.

In each "finger" of the RAKE, the samples suffer an advance τ_l identical to those specified by an environment model and are then multiplied by the samples of the complex conjugate of the correspondent channel coefficient α_l so that the original signal phase is recovered. For theoretical, perfect synchronization is admitted so the delay values, τ_l, are directly used in the RAKE. After the multiplication by the conjugates of the channel coefficients, the signal (with components in-phase and in quadrature since it can be QPSK or M-QAM modulated) enters the correlator where it is multiplied by the complex conjugate of the spreading code $c(t)$ synchronized with the respective replica. The result is then integrated for each bit period, thus achieving the despreading of the signal. The resulting symbols are then demodulated to obtain "soft" estimates of the transmitted symbols. These symbols' estimates are then summed with the respective estimates of the other "fingers." Note that this addition is weighted due to the previous multiplication of the estimates by the respective complex conjugate of the channel coefficients.

In the RAKE receiver, the signal is sampled and the resulting sequence is multiplied in each "finger" by the samples of the complex conjugate of the respective channel coefficient. Then, it is multiplied by the complex conjugate of the spreading code $\{c_i^*\}$ with a delay equal to the delay of the replica being extracted by the "finger," and each set of SF symbols is integrated, thus resulting in a despreaded sequence. The estimates of symbol sequences obtained in each "finger" are all time aligned and summed, thus obtaining a combined estimate.

The decision variable associated with each despreaded QPSK symbol i obtained in "finger" l is given by

$$
\begin{aligned}
y_i^l &= \frac{1}{SF} \sum_{m=0}^{SF-1} b'_{i \cdot SF+m} \cdot c_{i \cdot SF+m} \cdot c_{i \cdot SF+m}^* \cdot \alpha_{l,i \cdot SF+m} \cdot \alpha_{l,i \cdot SF+m}^* \\
&+ \frac{1}{SF} \sum_{m=0}^{SF-1} (n_{l,i \cdot SF+m} + I_{l,i \cdot SF+m} + Z_{l,i \cdot SF+m}) \cdot c_{i \cdot SF+m}^* \cdot \alpha_{l,i \cdot SF+m}^* \\
&= \frac{1}{SF} \sum_{m=0}^{SF-1} b'_{i \cdot SF+m} \left| \alpha_{l,i \cdot SF+m} \right|^2 + \frac{1}{SF} \sum_{m=0}^{SF-1} (n_{l,i \cdot SF+m} + I_{l,i \cdot SF+m} + Z_{l,i \cdot SF+m}) \\
&\quad \cdot c_{i \cdot SF+m}^* \cdot \alpha_{l,i \cdot SF+m}^*,
\end{aligned}
\tag{6.15}
$$

where $|c_i| = 1$. $\alpha_{l,i \cdot SF+m}$, $I_{l,i \cdot SF+m}$, and $n_{l,i \cdot SF+m}$ represent the samples of the channel coefficients, MAI overall and noise, respectively, at time instants

$t = (i \cdot SF + m) \cdot T_c - \tau_l$. $Z_{l,i \cdot SF+m}$ corresponds to the samples of the interference created by the replicas of the transmitted signal

$$Z_{l,i \cdot SF+m} = \sum_{\substack{j=1 \\ j \neq l}}^{L} \alpha_j (t = (i \cdot SF + m)T_c - \tau_l) \cdot x_T (t = (i \cdot SF + m)T_c - \tau_l). \quad (6.16)$$

Note that after the receiver filter and before the sampling, the noise power is

$$\sigma_n^2 = \int_{-\infty}^{\infty} G_w(f) |H_R(f)|^2 \, df, \quad (6.17)$$

where $G_w(f)$ is the spectral density function, which for additive white Gaussian noise (AWGN) is $G_w = N_0/2$. $H_R(f)$ is the frequency response of the receiver filter. So, in the case of AWGN and square-root raised cosine receiver filter (which outputs the same noise power of an ideal lowpass filter with bandwidth $1/(2T_c)$) results in

$$\sigma_n^2 = \frac{N_0}{2} \int_{-\infty}^{\infty} |H_R(f)|^2 \, df = \frac{N_0}{2} \cdot 2 \cdot \frac{1}{2T_c} = \frac{N_0}{2} \cdot \frac{1}{T_c}. \quad (6.18)$$

After summing the decision variables of each finger at the output of the RAKE, we get

$$y_i = \frac{1}{SF} \sum_{l=1}^{L} \sum_{m=0}^{SF-1} b'_{i \cdot SF+m} \left| \alpha_{l,i \cdot SF+m} \right|^2 + \frac{1}{SF} \sum_{l=1}^{L} \sum_{m=0}^{SF-1} (n_{l,i \cdot SF+m} + I_{l,i \cdot SF+m} + Z_{l,i \cdot SF+m})$$

$$\cdot c^*_{i \cdot SF+m} \cdot \alpha^*_{l,i \cdot SF+m}$$

$$= b_i \frac{1}{SF} \sum_{l=1}^{L} \sum_{m=0}^{SF-1} \left| \alpha_{l,i \cdot SF+m} \right|^2 + \frac{1}{SF} \sum_{l=1}^{L} \sum_{m=0}^{SF-1} (n_{l,i \cdot SF+m} + I_{l,i \cdot SF+m} + Z_{l,i \cdot SF+m})$$

$$\cdot c^*_{i \cdot SF+m} \cdot \alpha^*_{l,i \cdot SF+m}$$

$$= b_i \frac{1}{SF} \sum_{l=1}^{L} \sum_{m=0}^{SF-1} \left| \alpha_{l,i \cdot SF+m} \right|^2 + \frac{1}{SF} \sum_{l=1}^{L} \sum_{m=0}^{SF-1} (I_{l,i \cdot SF+m} + Z_{l,i \cdot SF+m}) \cdot c^*_{i \cdot SF+m} \cdot \alpha^*_{l,i \cdot SF+m}$$

$$+ \frac{1}{SF} \sum_{l=1}^{L} \sum_{m=0}^{SF-1} n_{l,i \cdot SF+m} \cdot c^*_{i \cdot SF+m} \cdot \alpha^*_{l,i \cdot SF+m}$$

$$= \sum \text{information} + \sum \text{interference} + \sum \text{noise}.$$

$$(6.19)$$

In this expression, the first term represents the transmitted symbol that must be decoded at the receiver. The second term represents the interference from other users and multipath components that degrade the decoding of the transmitted symbols and the third term is noise.

This type of signal combination is denominated by "MRC" and results in the maximization of the received signal-to-noise ratio (SNR). In fact, for the general case of a received signal y obtained by the combination of L replicas of the transmitted signal suffering fading $\alpha_k(t)$ that are uncorrelated between them, we have

$$y = \sum_{k=1}^{L} G_k(\alpha_k b + n_k) = \sum_{k=1}^{L} G_k \alpha_k b + \sum_{k=1}^{L} \alpha_k n_k. \tag{6.20}$$

$G_k(t)$ are complex weights that are multiplied by each received signal replica at the receiver. The received SNR is (assuming a simple binary phase shift keying—BPSK modulation)

$$\frac{S}{N} = \frac{E_b}{N_0} \frac{\left| \sum_{k=1}^{L} G_k \alpha_k \right|^2}{\left| \sum_{k=1}^{L} G_k \right|^2}, \tag{6.21}$$

with E_b denoting the average energy per transmitted bit. The weights $G_k(t)$ that maximize S/N can be obtained through the Cauchy–Schwarz inequality:

$$\left| \sum_{k=1}^{L} a_k b_k^* \right| \leq \left(\sum_{k=1}^{L} |a_k|^2 \right) \left(\sum_{k=1}^{L} |b_k|^2 \right), \tag{6.22}$$

in which the equality is verified when $a_k = c \cdot b_k$, for any c.

Applying this theorem to the SNR expression results in

$$\frac{S}{N} \leq \frac{E_b}{N_0} \frac{\sum_{k=1}^{L} |G_k|^2 \sum_{k=1}^{L} |\alpha_k|^2}{\left| \sum_{k=1}^{L} G_k \right|^2} \tag{6.23}$$

for which the equality $G_k = \alpha_k$ results in an SNR that is equal to the sum of the SNR of the individual "fingers," that is

$$\frac{S}{N} = \frac{E_b}{N_0} \sum_{k=1}^{L} |\alpha_k|^2. \tag{6.24}$$

The average SNR is

$$\frac{S}{N}_{\text{avg}} = \frac{E_b}{N_0} \sum_{k=1}^{L} E\{|\alpha_k(t)|^2\}, \tag{6.25}$$

where $E\{|\alpha_k(t)|^2\}$ is the average value of the channel coefficient $|\alpha_k(t)|^2$.

When using QPSK modulation, for example, for the downlink transmission of the UMTS, we have:

$$T_c = 2 \cdot T_b \cdot R \tag{6.26}$$

(the 2 factor is due to the modulation being QPSK; $\log_2 M$ for M-QAM modulations),

$$(E_c)_{I|Q} = \frac{1}{2} \cdot E_c = \frac{1}{2} \cdot A^2 \cdot T_c = \frac{1}{2} \cdot A^2 \cdot 2 \cdot T_b \cdot R = E_b \cdot R \tag{6.27}$$

(E_c is the average energy by complex chip),

$$\left(\frac{S}{N}\right)_{I|Q} = \frac{(E_c)_{I|Q}/T_c}{2 \cdot B \cdot N_0/2} = \frac{(E_c)_{I|Q}/T_c}{r_c/2 \cdot N_0} = \frac{2 \cdot T_c \cdot (E_c)_{I|Q}/T_c}{N_0} = \frac{2 \cdot (E_c)_{I|Q}}{N_0}$$

$$= \frac{2E_{bR}}{N_0} a \tag{6.28}$$

or, equivalently,

$$N_{I|Q} = \frac{S_{I|Q}}{2 \cdot R \cdot E_b/N_0}. \tag{6.29}$$

In these expressions, E_c and T_c correspond to the average complex chip energy and chip period, respectively. R is the ratio between the total number of information bits over the total number of transmitted chips, thus including the code rate and the spreading factor.

In the uplink connection of the UMTS, since each bit is transmitted simultaneously in the "I" and "Q" branches, we have:

$$T_c = T_b \cdot R, \tag{6.30}$$

$$(E_c)_{I|Q} = \frac{1}{2} \cdot E_c = \frac{1}{2} \cdot A^2 \cdot T_c = \frac{1}{2} \cdot A^2 \cdot T_b \cdot R = \frac{1}{2} \cdot E_b \cdot R, \tag{6.31}$$

$$\left(\frac{S}{N}\right)_{I|Q} = \frac{2 \cdot (E_c)_{I|Q}}{N_0} = \frac{E_b \cdot R}{N_0}, \tag{6.32}$$

$$\left(\frac{S}{N}\right)_{I|Q} = R \cdot \frac{E_b}{N_0} \Leftrightarrow N_{I|Q} = \frac{S_{I|Q}}{R \cdot E_b/N_0}. \tag{6.33}$$

Note that after demodulation, the expression for the SNR is the same of a regular polar transmission, that is, the SNR doubles

$$\left(\frac{S}{N}\right) = 2 \cdot R \cdot \frac{E_b}{N_0} \Leftrightarrow N = \frac{S}{2 \cdot R \cdot E_b/N_0}. \tag{6.34}$$

In the case of the UMTS HSDPA connection using 16-QAM modulation, we obtain:

$$T_c = 4 \cdot T_b \cdot R, \tag{6.35}$$

$$(E_c)_{I|Q} = \frac{1}{2} \cdot E_c = \frac{1}{2} \cdot \overline{A^2} \cdot T_c = \frac{1}{2} \cdot \overline{A^2} \cdot 4 \cdot T_b \cdot R = 2 \cdot E_b \cdot R, \tag{6.36}$$

$$\left(\frac{S}{N}\right)_{I|Q} = \frac{2 \cdot (E_c)_{I|Q}}{N_0} = \frac{4 \cdot E_b \cdot R}{N_0}, \tag{6.37}$$

$$\left(\frac{S}{N}\right)_{I|Q} = 4 \cdot R \cdot \frac{E_b}{N_0} \Leftrightarrow N_{I|Q} = \frac{S_{I|Q}}{4 \cdot R \cdot E_b/N_0}. \tag{6.38}$$

In the case of a RAKE receiver, the expression to compute the necessary noise power to achieve certain average bit energy to noise spectral density value is different:

$$\frac{S}{N_{avg}} = \frac{E_b}{N_0} \sum_{k=1}^{L} \left(\left|\tilde{W}_k(t)\right|^2\right)_{avg} \Leftrightarrow \frac{E_b}{N_0} = \frac{S}{N_{avg}} \cdot \frac{1}{\sum_{k=1}^{L} \left(\left|\tilde{W}_k(t)\right|^2\right)_{avg}}. \tag{6.39}$$

For the downlink connection, the noise power becomes

$$N_{I|Q} = \frac{S_{I|Q}}{2 \cdot R \cdot E_b/N_0} = \frac{S_{I|Q}}{2 \cdot R \cdot S/N_{avg} \cdot 1/\sum_{k=1}^{L} \left(\left|\tilde{W}_k(t)\right|^2\right)_{avg}}. \tag{6.40}$$

Care must be taken when speaking of either receive or transmit E_b/N_0, since both values are different. The receive E_b/N_0 is related to the transmit E_b/N_0 by a diversity order factor (DOF), which essentially accounts for all types of diversity in the transmission, assuming that there is no free-space power loss and normalizing the most powerful tap to one (wideband system).

Defining tx_{div} as the transmit diversity (number of transmit antennas), rx_{div} as the receive diversity (number of receive antennas), and mp_{div} as the channel diversity, that is, the diversity provided by the channel multipaths, given by

$$mp_{div} = \frac{\sum P\left(\sum_i \alpha_i\right)}{P(\alpha_{main})}, \qquad (6.41)$$

the DOF is heuristically given by

$$DOF = tx_{div} \cdot rx_{div} \cdot mp_{div}. \qquad (6.42)$$

Some DOF values are presented in Table 6.1. The multipath diversity of the main UMTS channels is presented in Table 6.1a. The transmit diversity may be accounted for in terms of number of antennas and multipath components, weighed by their transmit power levels. For illustration purposes, the transmit diversity of three transmit diversity arrangements is presented in Table 6.1b.

For the receive diversity, it suffices to consider the total number of receive antennas for most number of cases. As an example, for the SIMO (single input, multiple output) case of Figure 6.4, where a two-tap channel is assumed, a DOF equal to 4 would be obtained:

Table 6.1 Multipath Diversity for the Main UMTS Channels (a) and Different Transmit Diversity Schemes (b)

(a)			
Channel			
Pedestrain A	Pedestrain B	Indoor A	Vehicular A
1.05	2.46	1.1	2.02

(b)			
Transmit Diversity			
1 antenna	2 antennas equal power	2 antennas 3 dB difference	
1	2	1.5	

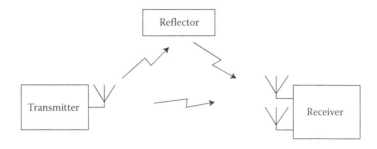

Figure 6.4 SIMO case with two receivers and two equal-channel taps.

$$DOF = tx_{\mathrm{div}} \cdot rx_{\mathrm{div}} \cdot mp_{\mathrm{div}} = 1 \cdot 2 \cdot 2 = 4. \tag{6.43}$$

The relationship between the transmit and receive E_b/N_0 is simply

$$\frac{E_b}{N_{0\,RX}} = DOF \frac{E_b}{N_{0TX}}. \tag{6.44}$$

Differently from most of the other chapters, in this chapter, the received E_b/N_0 is used in all simulation results, unless stated otherwise. MIMO simulation results will reveal comparable system performance for similar values of E_b/N_0, unaffected by the order of the receive diversity. Another advantage of using the receive E_b/N_0 value is the possibility of extrapolating coding results from the performance curves, since the decoder's performance depends on the received E_b/N_0.

6.2 MMSE System Matrices for a MIMO DS-CDMA System

The design of the system matrices for the MF and the ZF/MMSE schemes are presented in this section (note that the MF is also discussed since it is an integral part of both ZF and MMSE schemes). A MIMO arrangement is assumed, in which the data streams for each user are either split into different streams for each transmit antenna to increase the bit rate/capacity, or replicated (with interleaving) for each antenna, to increase the transmit diversity, reducing the necessary transmit power for nominal operations.

The transmitted signal associated with the kth spreading code and the txth transmit antenna is given by

$$x_{tx}^{(k)}(t) = A_{tx}^{(k)} \sum_{n=1}^{N \cdot SF} b_{\lfloor n/SF \rfloor, tx}^{(k)} \, c_n^{(k)} f_T(t - nT_C) = A_{tx}^{(k)} \sum_n \beta_{n,tx}^{(k)} f_T(t - nT_C), \tag{6.45}$$

where N is the number of data symbols to be transmitted by each antenna, $A_{tx}^{(k)} = \sqrt{E_k}$ (admitting $f_T(0) = 1$, $E\left[|b^{(k)}|^2\right] = 1$ and $E\left[|c^{(k)}|^2\right] = 1$), with E_k denoting the symbol energy, SF is the spreading factor, K is the total number of spreading codes per antenna/physical channels (or users, if each user uses only one physical channel), and $R_C = 1/T_C$ is the chip rate. The nth chip associated with the kth user and the txth antenna is $\beta_{n,tx}^{(k)} = b_{\lfloor n/SF \rfloor,tx}^{(k)} c_n^{(k)}$. The data symbols to be transmitted at the txth transmit antenna are $b_{n,tx}^{(k)}$, $n = 1,2,\ldots, N$, and the combined spreading and scrambling signature is $c_n^{(k)}$, $n = 1,2,\ldots, N \cdot SF$. $f_T(t)$ is the adopted pulse shape filter (a square-root raised cosine filtering is assumed).

At the receiver, we have N_{RX} antennas. The signal for any given receive antenna rx, prior to the reception filter, is given by

$$y_{rx}(t) = \sum_{k=1}^{K} \sum_{n=1}^{N \cdot SF} \sum_{tx=1}^{N_{TX}} \left(x_{n,tx}^{(k)}(t) * h_{n,rx,tx}^{(k)}(t)\right) + n(t), \tag{6.46}$$

with $n(t)$ denoting the channel noise, assumed Gaussian, with zero mean and the variance of the real and imaginary components denoted by σ^2. The channel impulse response between the transmit antenna tx and the receive antenna rx is

$$h_{n,rx,tx}^{(k)}(t) = \sum_{l=1}^{L} \alpha_{n,rx,tx,l}^{(k)} \delta(t - \tau_l^{(k)}), \tag{6.47}$$

with L denoting the total number of the channel's multipath components, $\alpha_{n,tx,rx,l}^{(k)}$ denoting the complex attenuation (fading) factor for the lth path, and $\tau_l^{(k)}$ is the propagation delay associated with the lth path (for the sake of simplicity, it is assumed that this delay is constant; the generalization to other cases is straightforward). The received signal associated with each antenna is submitted to a reception filter, with impulse response $f_R(t)$, which is assumed to be matched to $f_T(t)$ (i.e., $f_R(t) = f_T^*(-t)$), leading to the signal

$$r_{rx}(t) = \sum_{tx=1}^{N_{Tx}} \sum_{k=1}^{K} A_{tx}^{(k)} \sum_{l=1}^{L} \sum_{n=1}^{N \cdot SF} b_{\lfloor n/K \rfloor,tx}^{(k)} c_n^{(k)} h_{n,rx,tx,l}^{(k)} p(t - nT_C - \tau_l^{(k)}), \tag{6.48}$$

where $p(t) = f_T(t) * f_R(t)$ (for the Nyquist pulses considered in this work, $p(t) = f_T(t) * f_R(t) = f_T(t) * f_T^*(-t)$ is such that $p(nT_C) = 0$ for integer $n \neq 0$).

This signal is sampled at the chip rate and the corresponding samples can be written as

$$r_{rx,n} \triangleq r_{rx}(nT_C + \tau_0) = \sum_{tx=1}^{N_{Tx}} \sum_{k=1}^{K} A_{tx}^{(k)} \sum_{l=1}^{L} \sum_{n'} \beta_{n',tx}^{(k)} h_{n',rx,tx,l}^{(k)} p((n-n')T_C - (\tau_l^{(k)} - \tau_0^{(k)}))$$

$$= \sum_{tx=1}^{N_{Tx}} \sum_{k=1}^{K} A_{tx}^{(k)} \sum_{l=1}^{L} \sum_{n''} \beta_{n-n'',tx}^{(k)} h_{n-n'',rx,tx,l}^{(k)} p(n'' T_C - (\tau_l^{(k)} - \tau_0^{(k)})), \qquad (6.49)$$

with

$$\beta_{n,tx}^{(k)} = b_{\lfloor n/K \rfloor, tx}^{(k)} c_n^{(k)}. \qquad (6.50)$$

It should be noted that $n \in \mathbb{Z}$, but in practical terms, it is enough to consider just a few components in the vicinity of $\tau_l^{(k)} - \tau_0^{(k)}$. The above formulas are valid for both the downlink and uplink transmissions. In the case of the uplink transmission, special care must be taken, namely, in the introduction of different delays per user (that can be accounted for in the symbols from different transmit antennas).

6.2.1 Main System Matrices

Using matrix algebra, the received vector is as follows:

$$\mathbf{r}_v = \mathbf{SCAb} + \mathbf{n} \qquad (6.51)$$

where **S**, **C**, and **A** are the spreading, channel, and amplitude matrices, respectively, built in such a way that the expression in Equation 6.49 is reproduced. The receive vector $\mathbf{r}_v$ encompasses the messages for all receive antennas, such that

$$\mathbf{r}_v = \begin{bmatrix} \mathbf{r}_{rx=1}(t) \\ \mathbf{r}_{rx=2}(t) \\ \vdots \\ \mathbf{r}_{rx=N_{RX}}(t) \end{bmatrix}$$

Note that the channel matrix encompasses not only the channel coefficients but also the filter's coefficients. The spreading matrix accounts for the spreading and scrambling codes, as well as for the delays between users and channel replicas. The channel matrix accounts for the fading coefficients for all links between each transmit and receive antenna. For simplicity, the spreading matrix for the downlink will be described, and the assumption that the spreading and scrambling codes are the same for all the transmit antennas will be made, having a direct effect on both the spreading and the channel matrix. The other structures remain the same, for both the downlink and uplink transmissions.

6.2.1.1 Downlink with Equal Scrambling for All Transmit Antennas—S and C Matrices

The downlink spreading matrix $\mathbf{S}$ has dimensions $(SF \cdot N \cdot N_{RX} + \psi_{MAX} \cdot N_{RX}) \times (K \cdot L \cdot N \cdot N_{RX})$ (ψ_{max} is the maximum delay of the channel's impulse response, normalized to the number of chips, $\psi_{max} = \lceil \tau_{max}/T_c \rceil$, where T_c is the chip period), and it is composed of submatrices $\mathbf{S}_{RX}$ in its diagonal for each receive antenna $\mathbf{S} = \text{diag}(\mathbf{S}_{RX=1},\dots,\mathbf{S}_{RX=N_{RX}})$. Each of these submatrices has dimensions $(SF \cdot N + \psi_{max}) \times (K \cdot L \cdot N)$, and they are further composed of smaller matrices $\mathbf{S}_n^L$, one for each bit position, with size $(\mathbf{SF} + \psi_{MAX}) \times (K \cdot L)$. The $\mathbf{S}_{RX}$ matrix structure is made of $\mathbf{S}_{RX}=[\mathbf{S}_{\varepsilon,1},\dots,\mathbf{S}_{\varepsilon,N}]$ with

$$\mathbf{S}_{\varepsilon,n} = \begin{bmatrix} \mathbf{0}_{(SF\cdot(n-1))\times(K\cdot L)} \\ S_n^L \\ \mathbf{0}_{(SF\cdot(N-n))\times(K\cdot L)} \end{bmatrix}.$$

The $\mathbf{S}_n^L$ matrices are made of $K \cdot L$ columns $\mathbf{S}_n^L=[\mathbf{S}_{\text{col}(k=1,l=1),n},\dots,\mathbf{S}_{\text{col}(k=K,l=L),n}]$. Each of these columns is composed of

$$s_{\text{col}(kl),n} = \Big[\mathbf{0}_{(1\times\text{delay}(l))}, c_n(k)_{1\times SF}, \mathbf{0}_{(1\times(\psi_{MAX}-\text{delay}(l)))} \Big]^T,$$

where $c_n(k)$ is the combined spreading and scrambling for the bit n of user k.

These $\mathbf{S}^L$ matrices are either all alike, if no long scrambling code is used, or different, if the scrambling sequence is longer than the SF. The $\mathbf{S}^L$ matrices represent the combined spreading and scrambling sequences, conjugated with the channel delays. The shifted spreading vectors for the multipath components are all equal to the original sequence of the specific user

$$\mathbf{S}_n^L = \begin{bmatrix} \mathbf{S}_{1,1,1,n} & & \cdots & & \cdots & \mathbf{S}_{K,1,1,n} & & \\ \vdots & \ddots & \mathbf{S}_{1,1,L,n} & \cdots & & \vdots & \ddots & \mathbf{S}_{K,1,L,n} \\ \mathbf{S}_{1,SF,1,n} & & \vdots & \cdots & \mathbf{S}_{K,SF,1,n} & & \\ & \ddots & \mathbf{S}_{1,SF,L,n} & \cdots & & \cdots & \ddots & \mathbf{S}_{K,SF,L,n} \end{bmatrix}.$$

Note that, to correctly model the multipath interference between symbols, there is an overlap between the $\mathbf{S}^L$ matrices, of ψ_{max}. As opposed to the SISO multipath case, the matrix is not trimmed for the last multipath components.

The channel matrix $\mathbf{C}$ is a $(K \cdot L \cdot N \cdot N_{RX}) \times (K \cdot N_{TX} \cdot N)$ matrix, and it is composed of N_{RX} submatrices, each one for a receive antenna

$$
\mathbf{C} = \begin{bmatrix} \mathbf{C}_{RX=1} \\ \vdots \\ \mathbf{C}_{RX=N_{RX}} \end{bmatrix}.
$$

Each $\mathbf{C}_{RX}$ matrix is composed of N matrices $\mathbf{C}^{KT}$ alongside its diagonals.

$$
\mathbf{C}_{RX} = \begin{bmatrix} \mathbf{C}_{1,RX}^{KT} & & \\ & \ddots & \\ & & \mathbf{C}_{N,RX}^{KT} \end{bmatrix}.
$$

Each $\mathbf{C}_{n,RX}^{KT}$ matrix is $(K \cdot L) \times (K \cdot N_{TX})$ and represents the fading coefficients for the current symbol of each path, user, transmit antenna, and receive antenna. The matrix structure is made up of further smaller matrices alongside the diagonal of $\mathbf{C}_{n,RX}^{KT}$, $\mathbf{C}_{n,RX}^{KT} = \mathrm{diag}\left(\mathbf{C}_{n,RX,K=1}^{T},\ldots,\mathbf{C}_{n,RX,K=K}^{T}\right)$ with $\mathbf{C}^{T}$ of dimensions $L \times N_{TX}$, representing the combination of fading coefficients and filters' coefficients for the user's multipath and *tx*th antenna component.

By defining

$$
c_{n,rx,tx,l}^{(k)} \triangleq \sum_{n'} \alpha_{n-n',rx,tx,l}^{(k)} \, p(n'T_C - (\tau_l^{(k)} - \tau_0^{(k)})), \tag{6.52}
$$

we have

$$
C_{n,RX}^{KT} = \begin{bmatrix} c_{1,RX,1,1} & \cdots & c_{1,RX,N_{TX},1} & & & \\ \vdots & & \vdots & & & \\ c_{1,RX,1,L} & \cdots & c_{1,RX,N_{TX},L} & & & \\ & & & \ddots & & \\ & & & & c_{K,RX,1,1} & \cdots & c_{K,RX,N_{TX},1} \\ & & & & \vdots & & \vdots \\ & & & & c_{K,RX,1,L} & \cdots & c_{K,RX,N_{TX},L} \end{bmatrix}
$$

The $\mathbf{A}$ matrix is a diagonal of dimension $(K \cdot N_{TX} \cdot N)$ and represents the amplitude of each user per transmission antenna and symbol, $\mathbf{A} = \mathrm{diag}\,(\mathbf{A}_{1,1,1},\ldots,\mathbf{A}_{N_{TX},1,1},\ldots, \mathbf{A}_{N_{TX},K,1},\ldots,\mathbf{A}_{N_{TX},K,N})$.

6.2.1.2 Different Scrambling for All Transmit Antennas and Uplink Modifications—S and C Matrices

The previous $\mathbf{S}$ and $\mathbf{C}$ matrices assume that all antennas use the same spreading and scrambling code. However, when not operating under full-loading conditions, best results are obtained when different scrambling sequences are used at each transmit antenna, in order to increase diversity.

Therefore, the main changes to the matrices would be the $\mathbf{S}_n^L$ submatrix for the $\mathbf{S}$ matrix

$$
\mathbf{S}_n^L =
\begin{bmatrix}
\mathbf{S}_{1,1,1,1,n} & \cdots & \mathbf{S}_{1,1,1,N_{TX},n} & & & & \\
\vdots & & \vdots & \ddots & \mathbf{S}_{1,1,L,1,n} & \cdots & \mathbf{S}_{1,1,L,N_{TX},n} \\
\mathbf{S}_{1,SF,1,1,n} & \cdots & \mathbf{S}_{1,SF,1,N_{TX},n} & & \vdots & & \vdots \\
& & & \ddots & \mathbf{S}_{1,SF,L,1,n} & \cdots & \mathbf{S}_{1,SF,L,N_{TX},n} \\
\cdots & \mathbf{S}_{K,1,1,1,n} & \cdots & \mathbf{S}_{K,1,1,N_{TX},n} & & & \\
\cdots & \vdots & & \vdots & \ddots & \mathbf{S}_{K,1,L,1,n} & \cdots & \mathbf{S}_{K,1,L,N_{TX},n} \\
\cdots & \mathbf{S}_{K,SF,1,1,n} & \cdots & \mathbf{S}_{K,SF,1,N_{TX},n} & & \vdots & \\
\cdots & & & & \ddots & \mathbf{S}_{K,SF,L,1,n} & \cdots & \mathbf{S}_{K,SF,L,N_{TX},n}
\end{bmatrix},
$$

and the $\mathbf{C}^{KT}$ submatrix for the $\mathbf{C}$ matrix

$$
\mathbf{C}^{KT} =
\begin{bmatrix}
c_{1,1,1} & & & & & & & \\
& \ddots & & & & & & \\
& & c_{N_{TX},1,1} & & & & & \\
\vdots & \vdots & \vdots & & & & & \\
c_{1,L,1} & & & & & & & \\
& \ddots & & & & & & \\
& & c_{N_{TX},L,1} & & & & & \\
& & & \ddots & & & & \\
& & & & c_{1,1,K} & & & \\
& & & & & \ddots & & \\
& & & & & & c_{N_{TX},1,K} & \\
& & & & \vdots & \vdots & \vdots & \\
& & & & c_{1,L,K} & & & \\
& & & & & \ddots & & \\
& & & & & & c_{N_{TX},L,K}
\end{bmatrix}.
$$

For the uplink transmission, the **S** matrix should also portray the delays between different users (in this case, different transmit antennas). The $\mathbf{S}_n^L$ should thus be adjusted, with the columns being shifted either upwards or downwards, depending on the offsets between users. Note that the size (number of lines) of the $\mathbf{S}_n^L$ matrix can be increased, to account for the delays between users. Although the final **S** matrix can have a bigger number of lines than for the downlink case, its overall structure remains the same.

6.2.1.3 Design of the Remaining Structures

The resulting matrix from the **SCA** operation (henceforth known as **SCA** matrix) is depicted in Figure 6.5. It is an $N_{RX} \cdot (N \cdot SF + \psi_{max}) \times N_{TX} \cdot K \cdot N$ matrix, and it is the reference matrix for the decoding algorithms. Note that the SCA matrix is sparse in nature.

The resulting **SCA** matrix will have the same size as before; only the number of operations increases while constructing the **SCA** matrix (the **SC** multiplication has an increase in complexity equal to the number of transmit antennas), since values from different antennas must be treated differently.

Vector **b** represents the information symbols. It has length $(K \cdot N_{TX} \cdot N)$, and it has the following structure:

$$\mathbf{b} = \left[\mathbf{b}_{1,1,1}, \ldots, \mathbf{b}_{N_{TX},1,1}, \ldots, \mathbf{b}_{1,K,1}, \ldots, \mathbf{b}_{N_{TX},K,1}, \ldots, \mathbf{b}_{N_{TX},K,N} \right]^T. \qquad (6.53)$$

Note that the bits of each transmit antenna are grouped together in the first level, and the bits of other interferer in the second level. This is to guarantee that the resulting matrix to be inverted has all its nonzero values as close to the diagonal as possible. Also note that there is usually a higher correlation between bits from different antennas using the same spreading code than between bits with different spreading codes.

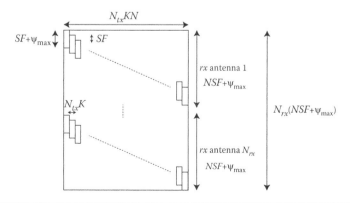

Figure 6.5   **Layout of the SCA matrix.**

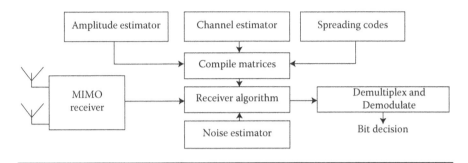

Figure 6.6 MIMO receiver for W-CDMA schemes.

Finally, the **n** vector is a $(N \cdot SF \cdot N_{RX} + N_{RX} \cdot \psi_{max})$ vector with noise components to be added to the received vector r_v, which is partitioned by N_{RX} antennas

$$\mathbf{r}_v = \begin{bmatrix} \mathbf{r}_{1,1,1}, \ldots, \mathbf{r}_{1,SF,1}, \ldots, \mathbf{r}_{N,1,1}, \ldots, \mathbf{r}_{N,SF+\psi_{max},1}, \ldots, \mathbf{r}_{N,1,N_{RX}}, \ldots, \mathbf{r}_{N,SF+\psi_{max},N_{RX}} \end{bmatrix}^T \quad (6.54)$$

(the delay ψ_{max} is only used for the final bit, though its effects are present throughout $\mathbf{r}_v$). Figure 6.6 illustrates the main blocks from which the receiver is compiled.

If transmit diversity is applied, the **b** vector should be arranged accordingly, with replicas (after multiplexing) for each antenna.

6.2.2 Receiver Schemes Using the System Matrices

Equalization-based receivers such as the MMSE and ZF take into account all effects that the symbols are subject to in the transmission chain, namely, the MAI, ISI, and the channel effect. Using as basis the (unnormalized) matched filter output (obtained by applying the Hermitian to the **SCA** matrix and multiplying by the received vector)

$$\mathbf{y}_{MF} = (\mathbf{SCA})^H \mathbf{r}_v \quad (6.55)$$

and defining **R** as

$$\mathbf{R} = \mathbf{A} \cdot \mathbf{C}^H \cdot \mathbf{S}^H \cdot \mathbf{S} \cdot \mathbf{C} \cdot \mathbf{A} \quad (6.56)$$

the equalization matrix $(\mathbf{E}_M)$ for the MF and ZF can be written as

$$\mathbf{E}_{M,MFZF} = \mathbf{R}, \quad (6.57)$$

where the normalized MF estimate is given by

$$\mathbf{y}_{NMF} = \mathrm{diag}\left(\mathbf{E}_{M,MFZF}^{-1}\right)\mathbf{y}_{MF}, \tag{6.58}$$

and the ZF estimate as

$$\mathbf{y}_{ZF} = \mathbf{E}_{M,MFZF}^{-1}\mathbf{y}_{MF}, \tag{6.59}$$

which is simply applying the inverse of all effects the message was subject to. Once again, to prevent an ill-conditioned matrix for inversion (the $\mathbf{E}_M$ might become ill-conditioned when the system is fully loaded [Divsalar et al. 1998], depending on the cross-correlations between the users' signature sequences), a small value (e.g., 1e^{-6}) should be added to all elements in the main diagonal of the E_M. In order to avoid round-off problems, the E_M should be rounded at a value above the minimum machine precision.

The MMSE estimate aims to minimize $E(|\mathbf{b} - \hat{\mathbf{b}}|^2)$. From [Kay 1993], the $\mathbf{E}_M$ includes the estimated noise power σ^2, and is represented by

$$\mathbf{E}_{M,MMSE} = \mathbf{R} + \sigma^2\mathbf{I}. \tag{6.60}$$

The MMSE estimate is thus

$$\mathbf{y}_{MMSE} = \hat{\mathbf{b}} = \mathbf{E}_{M,MMSE}^{-1}\mathbf{y}_{MF}. \tag{6.61}$$

Both the ZF- and MMSE-based receivers are seldom used due to their perceived complexity, especially for wideband MIMO systems (with frequency-selective fading channels). Owing to the multipath-causing ISI, the whole information block is usually decoded at once (although there are some decoding variants in which the block is divided into smaller blocks [Shoumin and Zhi 2004; Silva et al. 2005], requiring some overlapping between symbols, in order to provide the best results), requiring the use of a significant amount of memory and computing power for the algebraic operations.

However, if the sparseness of the matrices is taken into account, only a fraction of the memory and computing power is required. As can be inferred from the previously described matrices, all system matrices are sparse in nature and consist of submatrices that are sparse themselves.

The most troublesome matrix to deal with is the $\mathbf{E}_M$, due to its inversion (more precisely, the resolution of the equation system leading to the final estimate). Fortunately, the $\mathbf{E}_M$ is also sparse, being possible to handle it with simplicity. For instance, considering a maximum-loading simulation case using 16-QAM modulation, $SF = 16$ (i.e., 16 physical channels), $L = 2$ multipaths (with the second multipath with a 1 chip delay, resembling the Indoor A or Pedestrian A channel), a MIMO of 2-transmit/2-receive antenna system, and a block size of 1024 bits ($N = 256$ symbols, with 4 bits per symbol) per each physical channel

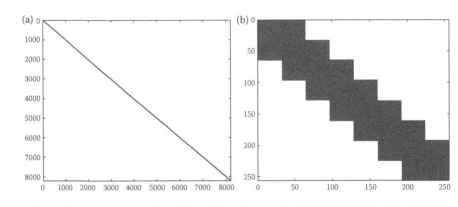

Figure 6.7 **Equalization matrix for a two-tap channel, $K = 16$, $N_{TX} = 2$ (a), and diagonal close-up for the same case (b).**

of each transmit antenna, the matrix's diagonal width (MDW), which in this case is $(K \cdot L \cdot N_{TX}) \cdot 3/2 = 96$, is roughly 1.2% of the matrix's width (MW) of $(K = 16) \cdot (N_{TX} = 2) \cdot (N = 256) = 8192$, MDW = 1.2% MW.

Another aspect of the E_M is that it is Hermitian positive definite, and thus it can be decomposed using the Cholesky decomposition. Since it is a banded matrix (with all elements concentrated on its diagonal), there is no Cholesky fill-in since the band is dense (cases with small chip delays (Figure 6.7)), and thus it presents itself as if the sparse reverse Cuthill–McKee ordering algorithm [Liu et al. 1981] had been applied to it (a good preordering for lower upper factorization of matrices (LU) or Cholesky factorization of matrices that come from long, skinny problems). The next section introduces a detailed description of the EM, and enhanced algorithms to solve it.

6.3 Enhanced Algorithms for Solving the Equalization Matrix

In this section, the complexity of the solution to the LMMSE and the ZF receiver's equations is studied. It will be shown that the equation can be solved with optimized Gauss, Cholesky, or block-Fourier algorithms, and thus saved some operations when using the normal Gauss method, which has a complexity of d^3, d being the number of lines/columns of the matrix to be inverted. Some of these solutions are very computationally efficient and thus allow for the use of the LMMSE in fully loaded MIMO systems.

6.3.1 Matrix Reordering

Matrix reordering is used in order to simplify solving the MMSE equation. While in the original version, the **SCA** matrix is devised in such a way as to make the received vector partitioned per receive antenna in order to make the system matrices more perceivable, the reordering of the structure of the **SCA** matrix is done solely to simplify the processing.

Replacing $\mathbf{T} = \text{order}(\mathbf{SCA})$; $\mathbf{N} = \sigma^2\mathbf{I}$; $\hat{\mathbf{d}} = \mathbf{y}_{MMSE}$ and $\mathbf{e} = \text{order}(\mathbf{r}_V)$ in the MMSE equation, a simpler version is obtained:

$$\hat{\mathbf{d}} = (\mathbf{T}^H\mathbf{T} + \mathbf{N})^{-1}\mathbf{T}^H\mathbf{e}, \qquad (6.62)$$

where order($\mathbf{x}$) represents a line reordering of vector or matrix $\mathbf{x}$, where the lines of each antenna are intercalated with the purpose of making a more compact and almost block-circulant matrix

$$\text{order}(\mathbf{x}) = \begin{bmatrix} \mathbf{x}_{TX=1}(1,:) \\ \mathbf{x}_{TX=2}(1,:) \\ \vdots \\ \mathbf{x}_{TX=N_{TX}}(1,:) \\ \vdots \\ \mathbf{x}_{TX=1}(N,:) \\ \mathbf{x}_{TX=2}(N,:) \\ \vdots \\ \mathbf{x}_{TX=N_{TX}}(N,:) \end{bmatrix} \qquad (6.63)$$

Figure 6.8 shows the reordering result for a two-antenna matrix.

For high SNRs, Equation 6.62 becomes the ZF detector equation:

$$\hat{\mathbf{d}} = (\mathbf{T}^H\mathbf{T})^{-1}\mathbf{T}^H\mathbf{e}. \qquad (6.64)$$

Since usually $\tau_{MAX} \leq SF$, the $\mathbf{T}^H\mathbf{T}$ product results in a square matrix with the structure presented in Figure 6.9 with $a = 2\ KN_{TX}$ and $n = KN_{TX}N$. It can be shown that $\mathbf{T}^H\mathbf{T}$ is a positive-definite Hermitian matrix.

Earlier works [Machauer et al. 2001; Vollmer et al. 2001] dealt only with the ZF detector equation for constant-channel situations. Here, the validity of those algorithms for unsteady channels situations will be evaluated. New algorithms for unsteady channel situations will be proposed and some optimizations will also be presented in pseudo-code form. Finally, all the algorithms will be adapted to the LMMSE detector.

6.3.2 Standard Algorithms for the Exact Solution

Equation 6.64 can be written as an $\mathbf{Ax} = \mathbf{b}$ system with $\mathbf{A}$ being a positive-definite Hermitian matrix, where $A = \mathbf{T}^H\mathbf{T}$, $x = \hat{\mathbf{d}}$, and $\mathbf{b} = \mathbf{T}^H\mathbf{e}$. It can be solved for $\mathbf{x}$ using

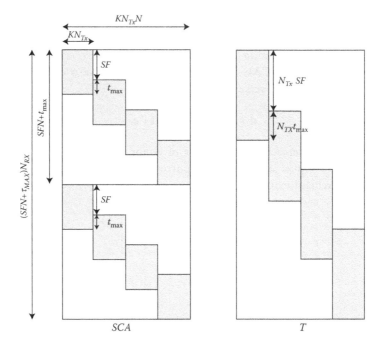

Figure 6.8 Line reordering sample of *SCA* matrix for $N_{TX} = 2$.

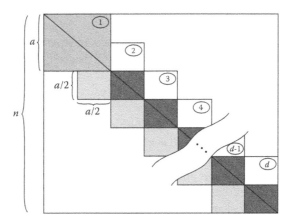

Figure 6.9 Typical correlation matrix of the E_M.

the Gauss elimination or the Cholesky algorithm. Since the $\mathbf{Ax} = \mathbf{b}$ system needs to be solved for a particular $\mathbf{b}$ vector only, there is no need to invert the $\mathbf{A}$ matrix. The Gauss elimination can be used to transform the $\mathbf{Ax} = \mathbf{b}$ system in a $\mathbf{Ux} = \mathbf{b}'$ where $\mathbf{U}$ is an upper triangular matrix and then $\mathbf{x}$ can be obtained by direct substitution. The Cholesky method is a little bit more complex: first $\mathbf{A}$ is factorized in $\mathbf{A} = \mathbf{U}^H \mathbf{U}$

by the Cholesky algorithm, then $\mathbf{U}^H\mathbf{U}\mathbf{x} = \mathbf{b}$ can be decomposed in $\mathbf{U}^H\mathbf{c} = \mathbf{b}$ and $\mathbf{U}\mathbf{x} = \mathbf{c}$; these two systems can be solved by direct substitution. The Cholesky algorithm can save almost half of the floating point operations needed in the Gauss elimination because it takes advantage of the symmetry of the $\mathbf{A}$ matrix, but the Gauss elimination is less complex and requires no square roots to be calculated.

Table 6.2 shows the number of floating point operations required by both methods. The additions are separated into real and complex ($R+$ and $C+$, respectively). The extra operations wasted by the Gauss algorithm are partially recovered in the substitution phase, where the Cholesky method requires the solution of two triangular systems and hence twice the operations of the Gauss algorithm. The number of operations required in this phase is also included in Table 6.2. The "Order" column presents the highest power of the total number of operations considering each multiplication–addition pair as a single operation.

6.3.2.1 Optimizations

A generic positive-definite Hermitian matrix that is nonzero only in equally overlapped squares centered along the diagonal is represented in Figure 6.10.

Table 6.2 Number of Floating Operations Needed to Solve the $Ax = b$ System with Standard Methods

	$\div$	$\times$	$C+$	$R+$	$\sqrt{}$	$Subs\div$	$Subs$ $\times/C+$	$Order$
Gauss	$\dfrac{n^2 - n}{2}$	$\dfrac{n^3 - n}{3}$	$\dfrac{n^3 - n}{3}$	0	0	n	$\dfrac{n^2 - n}{2}$	$\dfrac{n^3}{3}$
Cholesky	$\dfrac{n^2 - n}{2}$	$\dfrac{n^3 - n}{6}$	$\dfrac{n^3 - 3n^2 + 2n}{6}$	$\dfrac{n^2 + n}{2}$	n	$2n$	$n^2 - n$	$\dfrac{n^3}{6}$

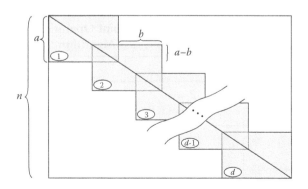

Figure 6.10   Generalized EM.

The number of overlapped squares is $d = (n - a/b) + 1$. The Gauss algorithm can be optimized for this type of matrix by eliminating the operations involving zero elements. The idea is presented in Figure 6.11.

First, the standard Gauss algorithm is applied to the *r*1 square submatrix. There is no need to change the *r*4 rectangle. The next step is the elimination of the *r*2 block using the last $a-b$ pivots of *r1* (the pivots are the diagonal elements after the elimination phase). During this phase, *r*3 is updated. Finally, the standard algorithm is applied to *r*3. This process is repeated until all blocks are updated. During this process, as each line is updated, the correspondent element of vector **b** is simultaneously updated. Note that the matrix diagonal is fully contained in the diagonal squares. Table 6.3 presents the number of floating point operations required in each of the three described phases.

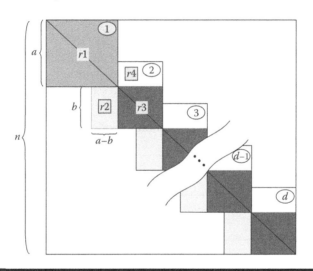

Figure 6.11   Optimized Gauss algorithm for the EM.

Table 6.3   Number of Floating Point Operations Needed for the Optimized Gauss Algorithm

Phase	÷	×/C+
*r*1	$\dfrac{a^2 - a}{2}$	$\dfrac{a^3 - a}{3}$
*r*2	$b(a-b)$	$\dfrac{3a^3 - 6a^2}{16}$
*r*3	$\dfrac{b^2 - b}{2}$	$\dfrac{b^3 - b}{3}$

The total number of operations can be calculated from

$$N_{Gauss\,Opt} = N_{Gauss\,r1} + (d-1)\left(N_{Gauss\,r2} + N_{Gauss\,r3}\right), \tag{6.65}$$

leading to

$$N^+_{Gauss\,Opt} = n\left(a - \frac{b}{2} - \frac{1}{2}\right) - \frac{a(a-b)}{2}, \tag{6.66}$$

and

$$
\begin{aligned}
N^\times_{Gauss\,Opt} &= N^{C+}_{Gauss\,Opt} \\
&= n\frac{9a^3 - 18a^2 + 16b\left(b^2 - 1\right)}{48b} - a\frac{9a^3 - 2a^2\left(8b + 9\right) + 16b^3}{48b}.
\end{aligned} \tag{6.67}
$$

A similar adaptation can be developed for the column-Cholesky factorization algorithm (link node), portrayed in Figure 6.12 (note that similar results could be achieved for the line version of that algorithm). The algorithm is optimized according to the specific matrix structure of the $\mathbf{E}_M$ in Figure 6.13.

In this case, only the upper triangle has to be accessed. First, the standard column-Cholesky algorithm is applied to the $r1$ triangle. In a second step, the rectangle $r2$ is calculated accessing only the elements of $r2$ and $r1$. In the next step, the triangle $r3$ is computed using only elements of $r2$ and $r3$. The last two steps are repeated for all remaining blocks, using only elements of the last and current block. As in the optimized Gauss algorithm, the rectangular blocks do not contain the diagonal. Table 6.4 presents the number of floating point operations of the three phases described.

$$
\begin{aligned}
&\textit{for } j = 1, n \\
&\quad \textit{for } k = 1, j - 1 \\
&\quad\quad \textit{for } i = j, n \\
&\quad\quad\quad a_{ij} = a_{ij} - a_{ik} \cdot a_{jk} \\
&\quad a_{jj} = \sqrt{a_{jj}} \\
&\quad \textit{for } k = j + 1, n \\
&\quad\quad a_{kj} = a_{kj}/a_{jj}
\end{aligned}
$$

Figure 6.12   Column-oriented Cholesky factorization.

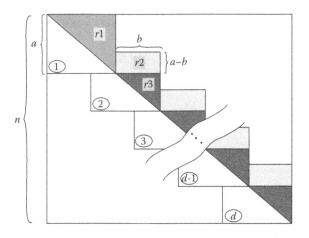

Figure 6.13   Optimized Cholesky algorithm for the EM.

Table 6.4   Number of Floating Point Operations Needed for the Optimized Cholesky Algorithm

Phase	÷	×	C+	R+	$\sqrt{}$
r1	$\dfrac{a^2 - a}{2}$	$\dfrac{a^3 - a}{6}$	$\dfrac{a^3 - 3a^2 + 2a}{6}$	$\dfrac{a^2 + a}{2}$	a
r2	$b(a - b)$	$\dfrac{b}{2}\left[(a - b)^2 - a + b\right]$	$\dfrac{b}{2}\left[(a - b)^2 - a + b\right]$	0	0
r3	$\dfrac{b^2 - b}{2}$	$\dfrac{a}{2}(b^2 - b) - \dfrac{b}{6}(2b^2 - 3b + 1)$	$\dfrac{a}{2}(b^2 - b) - \dfrac{b}{3}(b^2 - 1)$	$ab - \dfrac{b^2}{2} + \dfrac{b}{2}$	b

The total number of operations can be calculated from

$$N_{Chol\,Opt} = N_{Chol\,r1} + (d - 1)\left(N_{Chol\,r2} + N_{Chol\,r3}\right), \tag{6.68}$$

leading to

$$N^+_{Chol\,Opt} = n\left(a - \frac{b}{2} - \frac{1}{2}\right) - \frac{a(a - b)}{2}, \tag{6.69}$$

$$N^\times_{Chol\,Opt} = n\left(\frac{a^2}{2} - \frac{a}{2}(b + 2) + \frac{b^2}{6} + b - \frac{1}{6}\right) - a\left(\frac{a^2}{3} - \frac{a}{2}(b + 2) + b\left(\frac{b}{6} + 1\right)\right), \tag{6.70}$$

$$N_{Chol\,Opt}^{C+} = n\left(\frac{a^2}{2} - \frac{a}{2}(b+2) + \frac{b^2}{6} + \frac{b}{2} + \frac{1}{3} \right) - a\left(\frac{a^2}{3} - \frac{a}{2}(b+1) + b\left(\frac{b}{6} + \frac{1}{2} \right) \right),$$

$$(6.71)$$

$$N_{Chol\,Opt}^{R+} = n\left(a - \frac{b}{2} + \frac{1}{2} \right) + \frac{a}{2}(b-a),$$

$$(6.72)$$

$$N_{Chol\,Opt}^{\sqrt{}} = n.$$

$$(6.73)$$

Both Gauss and Cholesky methods need final substitution phases. These substitutions can also be optimized since the resulting matrices have a structure similar to the original **A** matrix but with nonzero elements only above (or below) the diagonal, as shown in Figure 6.14.

The solution of a system **Ax** = **b** with **A** having a structure similar to the structure presented in Figure 6.14 requires one division for each line of the matrix and one pair multiplication–addition for each nonzero element.

Since there are $d((a^2/2) - a) - (d-1)((a-b)^2/2 - (a-b))$ nonzero elements, the needed number of floating point operations can be written as:

$$N_{Subs\,Opt}^{+} = n$$

$$(6.74)$$

$$N_{Subs\,Opt}^{\times} = N_{Subs\,Opt}^{C+} = n\left(a - \frac{b}{2} - 1 \right) - \frac{a(a-b)}{2}.$$

$$(6.75)$$

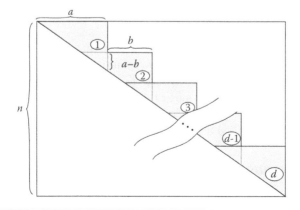

Figure 6.14 Resulting matrix structure after Gauss elimination/Cholesky factorization.

We are interested in a special type of block-diagonal positive-definite Hermitian matrices, with

$$b = a/2, \tag{6.76}$$

as the one represented in Figure 6.9. Rewriting the above equations for this special case and keeping only the n-dependent terms ($n \gg a$) results in Table 6.5. Figure 6.15 presents the optimized version of the Cholesky algorithm. A block version of the optimized Cholesky algorithm was developed to take advantage of the basic linear algebra subprograms (BLAS). In this version, element-by-element and vector-by-vector operations were transformed in matrix operations that can be more easily optimized by sublayer software or even be executed in dedicated hardware. Figure 6.16 presents such an approach.

The optimized Cholesky algorithm can save almost 30% of the number of operations required for the optimized Gauss algorithm, despite its increased complexity and need of square root operations.

6.3.3 Partial Cholesky Approximation

The Cholesky decomposition of block-Toeplitz matrices is an upper (or lower) matrix approximately block-Toeplitz with the same block size as the original matrix. This means that the **U** matrix can be approximated by calculating only the first L block-rows and assuming that the remaining block-rows are identical to the last calculated block-row.

Looking at **U** in Figure 6.17, only the dark shaded part is computed. The last computed block (marked as L) is then repeated until the full matrix is completed. This approximation is very effective when the channel is constant. Figure 6.18 shows the maximum relative error to the solution without any approximation, for each approximation level (L), that is, the number of calculated blocks, for different speeds in a Pedestrian A channel with 1 antenna.

As can be seen for a constant channel, calculating only the first one or two blocks allows approximations in the system solution with relative error below 10^{-4} or 10^{-9}. This can be used to greatly reduce the number of operations necessary to solve the system. If only two blocks are calculated, the number of operations can be reduced approximately by a factor of a/n. When the channel changes, even slowly, this approach cannot be used due to the high errors.

6.3.4 Partitioning

Partitioning the block-diagonal system $\mathbf{Ax} = \mathbf{b}$ would be very useful to reduce the number of floating point operations needed to solve the system (if no overlap is used; i.e., in good channel conditions) and could also permit the introduction of parallelism in algorithms that are intrinsically sequential, like the algorithms

Table 6.5 Number of Floating Point Operations Needed When $b = a/2$

	÷	×	C+	R+		Subs÷	Subsx	Subs C+
Gauss	$\dfrac{n^2-n}{2}$	$\dfrac{n^3-n}{3}$	$\dfrac{n^3-n}{3}$	0	0	n	$\dfrac{n^2-n}{2}$	$\dfrac{n^2-n}{2}$
Cholesky	$\dfrac{n^2-n}{2}$	$\dfrac{n^3-n}{6}$	$\dfrac{n^3-3n^2+2n}{6}$	$\dfrac{n^2+n}{2}$	n	$2n$	n^2-n	n^2-n
Gauss	$n\left(\dfrac{3a}{4}-\dfrac{1}{2}\right)$	$\dfrac{n(11a^2-18a-8)}{24}$	$\dfrac{n(11a^2-18a-8)}{24}$	0	0	n	$n\left(\dfrac{3}{4}a-1\right)$	$n\left(\dfrac{3}{4}a-1\right)$
Cholesky	$n\left(\dfrac{3}{4}a-\dfrac{1}{2}\right)$	$\dfrac{n}{24}(7a^2-18a+8)$	$n\left(\dfrac{7}{24}a^2-\dfrac{a}{2}-\dfrac{1}{6}\right)$	$n\left(\dfrac{3}{4}a+\dfrac{1}{2}\right)$	n	$2n$	$n\left(\dfrac{3}{2}a-2\right)$	$n\left(\dfrac{3}{2}a-2\right)$

```
lMin = 1;
cMin = 1;
cMax = a;

while (1)
    for j=cMin:cMax
        s=0;
        for k=cMin:j-1
            M(k,j) = (M(k,j)-dot(M(lMin:k-1,k),M(lMin:k-1,j)))/M(k,k);
        end
        for k=lMin:j-1
            s = s+real(M(k,j)*conj(M(k,j)));
        end
        s = real(M(j,j))-s;
        M(j,j)= sqrt(s);
    end

    if(cMax >= n) break; end

    cMax = cMax+b;
    cMin = cMax-b+1;
    lMin = lMin+b;

    for j=cMin:cMax
        for k=lMin:cMin-1
            M(k,j) = (M(k,j)-dot(M(lMin:k-1,k),M(lMin:k-1,j)))/M(k,k);
        end
    end
end
```

Figure 6.15   Matlab code for the optimized Cholesky algorithm for the EM.

```
for j=1:NB:n-NB

    JNB = j+NB;
    JNB1 = JNB-1;
    J2NB1 = j+2*NB-1;

    M(j:JNB1 , j:JNB1) = ccholesky(M(j:JNB1, j:JNB1), NB);

    M(j;JNB1, JNB:J2NB1) = M(j:JNB1, j:JNB1)' \M(j:JNB1, JNB:J2NB1|;

    M(JNB:J2NB1, JNB:J2NB1) =
            M(JNB:J2NB1, JNB:J2NB1) -M(j:JNB1, JNB:J2NB1)' * M(j:JNB1, JNB:J2NB1);
end

j = j+NB;
JNB1 = j+NB-1;
M(j:JNB1, j:JNB1) = ccholesky(M(j:JNB1,j:JNB1), NB);
```

Figure 6.16   Matlab code for the block version in the optimized Cholesky algorithm.

presented in the previous sections. In this section, different partitioning approaches will be discussed.

Since **A** is block-diagonal and has generally decreasing values as we get farther from its diagonal, it is expected that it can be divided into smaller matrices that produce smaller systems whose combined solutions would approximate the solution of the original system.

Figure 6.19 presents a sample solution of a system divided into two (Figure 6.19a), with the division sketched in Figure 6.20, and four (Figure 6.19b) subsystems. Note

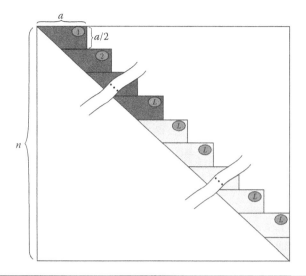

Figure 6.17 Structure of the *U* matrix for the partial Cholesky approximation.

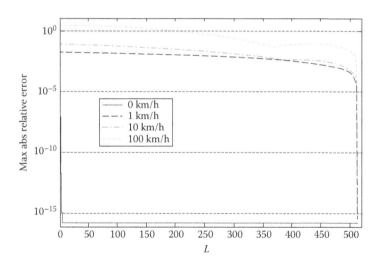

Figure 6.18 Relative error to the ZF solution when using the partial Cholesky approximation, for the Pedestrian A channel with $k = 1$.

that there are two $m \times m$ blocks completely ignored at the middle of the **A** matrix. Surprisingly, the maximum relative error is only 12% of the exact solution. Figure 6.19b shows the same data when the **A** matrix is divided into four slices. The maximum error level is approximately the same as in the previous case, but now there are three high error regions.

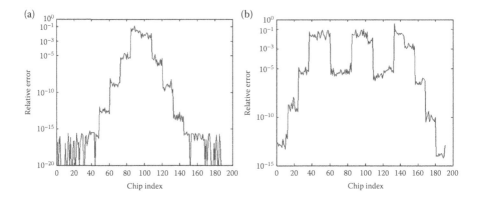

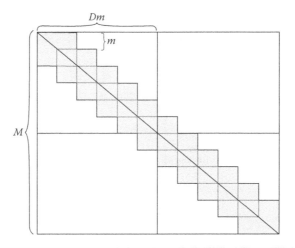

Figure 6.19 Relative error to the ZF solution for the Pedestrian A channel, obtained solving a (nonoverlap) partitioned system when $M = 192$, $m = 12$, for the cases of $D = 8$ ($192/(8 \times 12) = 2$ blocks) and $v = 10$ km/h (a), $D = 4$ ($192/(4 \times 12) = 4$ blocks) and $v = 10$ km/h (b).

Figure 6.20 Partitioning without overlapping scheme.

Although the error obtained with the last partitioning method is not extremely high and appears only around the division lines, much better results can be obtained if overlapping partitions are considered.

This proceeding is sketched in Figure 6.21. Each slice overlaps the last in $2 \times lap$ blocks (where *lap* is the number of blocks that are discarded from each overlapping side of each computed slice). Note that the last slice can be smaller. From each slice $(D - 2lap)m$ values are obtained for the solution vector **x**, except in the case of the first and last slices, where $(D - lap)m$ and $M - ((L - 1)D + lap)m$ values are obtained, respectively.

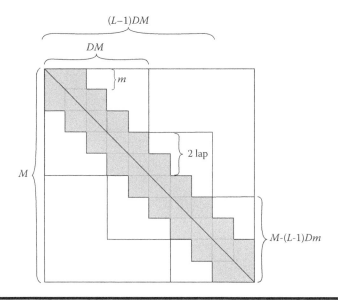

Figure 6.21 Partitioning with overlapping scheme.

As seen in Figure 6.19, the error level rises at the beginning and end of each slice, so the overlapping method should discard these values. In each interaction, $lap \times m$ values are discarded from the beginning and $lap \times m$ values from the end, except in the first end last slice, where only the last $lap \times m$ and first $lap \times m$ values are discarded, respectively. The algorithm is presented in detail in Figure 6.22. An alternative to have a final slice with size smaller than the early ones is to extend the **b** vector with zeros until $M/m - 2lap$ becomes a multiple of $D - 2lap$.

Several simulations were run for different channel changing speeds (using the Pedestrian A channel). Similar results were obtained for all matrices tested. Sample results are presented in Table 6.6 and Figure 6.23, for a $v = 10$ km/h condition. All the values presented should be multiplied by the corresponding column factor to obtain the maximum error of the partition algorithm relative to the exact solution of the original $\mathbf{Ax} = \mathbf{b}$ system. The column for $lap = 0$ corresponds to the lap-less situation.

Table 6.6 shows that the maximum error level depends almost exclusively from the overlapping level, so the proper overlapping can be easily selected just by knowing the maximum error allowed in the real system. The number of blocks D processed by each thread can be selected from the total number of threads L that can be executed simultaneously by the hardware, using the relation

$$L = \left\lceil \frac{M/m - 2lap}{D - 2lap} \right\rceil, \tag{6.77}$$

where $\lceil x \rceil$ represents the smallest integer greater than or equal to x.

```
x = zeros(M, 1);

NB = M/m;
inc = D-2*lap;
L = ceil((NB-2*lap)/inc);

BL = B(1:D*m);
SL = S(1:D*m, 1:D*m);
xL = SL\BL;
x(1:(D-lap)*m) = xL(1:(D-lap)*m);

start = inc;

for l=2:L-1
    BL = B(start*m+1:(start+D)*m);
    SL = S(start*m+1:(start+D)*m, start*m+1:(start+D)*m);
    xL = SL\BL;
    x((start+lap)*m+1:(start+D-lap)*m) = xL(lap*m+1:(D-lap)*m);
    start = start+inc;
end

BL = B(start*m +1 : M);
SL = S(start*m +1 : M, start*m + 1: M);
xL = SL\BL;
x((start + lap)*m + 1 : M) = xL(lap*m +1: lap*m + M-(start + lap)*m);
```

Figure 6.22   Pseudo-code for partitioning with overlapping algorithm.

6.3.5 Solving the Equalization Matrix with the Block-Fourier Algorithm

The block-Fourier algorithm is introduced in this section. The concept of circulant and block-circulant matrices is boarded, in order to explain the Fourier algorithm in detail.

6.3.5.1 Diagonalizing Circulant Matrices

A circulant matrix is a square Toeplitz matrix with each column being a rotated version of the column to the left of it:

$$
C = \begin{bmatrix}
c_1 & c_n & c_{n-1} & \cdots & c_2 \\
c_2 & c_1 & c_n & \cdots & c_3 \\
c_3 & c_2 & c_1 & \cdots & c_4 \\
\vdots & \vdots & \vdots & \ddots & \vdots \\
c_n & c_{n-1} & c_3 & \cdots & c_1
\end{bmatrix}.
\tag{6.78}
$$

The interesting property of circulant matrices is that its eigenvectors matrix is equal to the orthonormal discrete Fourier transform (DFT) matrix F_n' of corresponding dimension n. F_n' can be written as

$$
\mathbf{F}_n' = \sqrt{n} \cdot \mathbf{F}_n,
\tag{6.79}
$$

Table 6.6 Maximum Relative Error for the Partitioning Algorithm; $v = 10$ km/h

		Lap							
		0	1	2	3	4	5	6	7
Factor		*1.E + 00*	*1.E–04*	*1.E–07*	*1.E–11*	*1.E–15*	*1.E–15*	*1.E–15*	*1.E–15*
D	1	0.506							
	2	0.459							
	3	0.152	0.911						
	4	0.221	0.494						
	5	0.127	0.494	0.090					
	6	0.082	0.494	0.123					
	7	0.459	0.105	0.123	0.247				
	8	0.221	0.494	0.021	0.063				
	9	0.083	0.640	0.042	0.038	0.898			
	10	0.099	0.194	0.123	0.063	0.637			
	11	0.079	0.190	0.037	0.222	0.817	0.505		
	12	0.033	0.082	0.020	0.038	0.646	0.623		
	13	0.159	0.059	0.015	0.036	0.642	0.552	0.416	
	14	0.459	0.494	0.018	0.047	0.572	0.432	0.504	
	15	0.127	0.845	0.090	0.033	0.420	0.284	0.252	0.378
	16		0.009	0.005	0.066	0.479	0.232	0.432	0.504

where F_n is the nonorthonormal DFT matrix:

$$
\mathbf{F}_n = \begin{bmatrix}
1 & 1 & 1 & \cdots & 1 \\
1 & \omega & \omega^2 & \cdots & \omega^{n-1} \\
1 & \omega^2 & \omega^4 & \cdots & \omega^{2(n-1)} \\
\vdots & \vdots & \vdots & \ddots & \vdots \\
1 & \omega^{n-1} & \omega^{2(n-1)} & \cdots & \omega^{(n-1)(n-1)}
\end{bmatrix}
\tag{6.80}
$$

with $\omega = e^{-j2\pi/n}$; $j = \sqrt{-1}$.

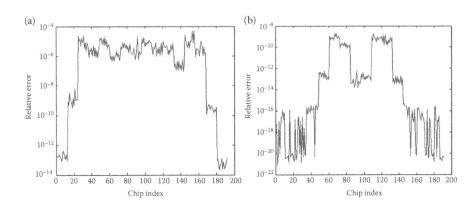

Figure 6.23 Relative error to the ZF solution for the Pedestrian A channel, obtained solving an overlapped partitioned system, for the conditions of (a) $D = 4$, $lap = 1$, $v = 10$ km/h and (b) $D = 8$, $lap = 2$, $v = 10$ km/h.

Using this property, a circulant matrix $\mathbf{C}$ can be decomposed in

$$\mathbf{C} = \mathbf{F}_n'^{-1} \boldsymbol{\Lambda} \mathbf{F}_n' = \mathbf{F}^{-1} \boldsymbol{\Lambda} \mathbf{F}, \tag{6.81}$$

where $\boldsymbol{\Lambda}$ is a diagonal matrix that contains the eigenvalues of C. The $\boldsymbol{\Lambda}$ matrix can be easily computed from

$$\boldsymbol{\Lambda} = \mathrm{diag}(\mathbf{F}\mathbf{C}(:,1)) \tag{6.82}$$

where $\mathrm{diag}(\mathbf{x})$ represents the diagonal matrix whose diagonal elements are taken from the $\mathbf{x}$ vector. Substituting $\mathbf{C}$ by Equation 6.81 in the linear system

$$\mathbf{C}\mathbf{x} = \mathbf{b} \tag{6.83}$$

and solving for $\mathbf{x}$ results in

$$\mathbf{x} = \mathbf{F}^{-1} \boldsymbol{\Lambda}^{-1} \mathbf{F}\mathbf{b} \tag{6.84}$$

Equation 6.84 can be computed efficiently with three discrete Fourier transforms and the inversion of the diagonal matrix $\boldsymbol{\Lambda}$. The complete solution would require $3n^2$ complex multiplication–addition pairs for the three DFT matrix/vector multiplications; n complex divisions to invert $\boldsymbol{\Lambda}$; and n complex multiplications to multiply $\boldsymbol{\Lambda}^{-1}$ by $\mathbf{F}\mathbf{b}$.

Using the Cooley and Tukey fast Fourier transform algorithm [Vollmer et al. 2001], the complex multiplication–addition operation pair needed to compute a size n DFT is

$$\mathbf{N}_{FFT}^{C\oplus\otimes} = \frac{n\log_2 n}{2} \tag{6.85}$$

This means that Equation 6.84 can be computed spending only $n\left(\frac{3}{2}\log_2 n + 2\right)$ complex floating point operations (considering each multiplication–addition pair as one operation). The memory requirements to solve the system using such algorithm are also very modest. It is only necessary to keep in memory two size n vectors: the **b** vector and the first column of C. All operations can be made in-place as the solution vector **x** replaces **b**. Further economy can be achieved if **C** is sparse (band or block–diagonal, for example).

6.3.5.2 Application to Block-Circulant Matrices

A block-circulant matrix can be visualized as a circulant matrix where each element is a matrix instead of a scalar value. Consider a block-circulant matrix $\mathbf{C}_{(PQ)}$ composed of $N \times N$ blocks of size $P \times Q$. If $\mathbf{C}_{(PQ)} \in \mathbb{C}^{NP \times NQ}$ is block-circulant, it must satisfy

$$\mathbf{C}_{(PQ)i,j} = \mathbf{C}_{(PQ)\tilde{i},\tilde{j}}, \tag{6.86}$$

with

$$\begin{aligned}
\tilde{i} &= ((i + P - 1)\bmod NP) + 1 \\
\tilde{j} &= ((j + Q - 1)\bmod NQ) + 1.
\end{aligned} \tag{6.87}$$

This means that each element of $\mathbf{C}_{(PQ)}$ is repeated P rows below and Q columns to the right of it. Indices that exceed the NP lines or NQ columns wrap around to the first lines and first columns, respectively. From now on, only square block matrices with $N \times N$ blocks will be considered, so the N index will be omitted for simplicity. The block dimensions of the matrix will be represented in subscript between curve brackets, and in the case of square blocks matrices, only a dimension is represented.

To deal with block-circulant matrices, the block-Fourier transformation needs to be introduced. The block-Fourier matrix is defined as [Joy and Tavsanoglu 1995]

$$\mathbf{F}_{(K)} = \mathbf{F}_N \odot \mathbf{I}_K, \tag{6.88}$$

where $\mathbf{F}_N$ is a nonorthonormal DFT matrix of dimension N as defined in Equation 6.79, $\mathbf{I}_K$ is the K size identity matrix, and $\odot$ denotes the Kronecker product.

$$\mathbf{Z} = \mathbf{X} \odot \mathbf{Y} = \begin{bmatrix} x_{11}\mathbf{Y} & x_{12}\mathbf{Y} & \cdots & x_{1N}\mathbf{Y} \\ x_{21}\mathbf{Y} & x_{22}\mathbf{Y} & \cdots & x_{2N}\mathbf{Y} \\ \vdots & \vdots & \ddots & \vdots \\ x_{M1}\mathbf{Y} & x_{M2}\mathbf{Y} & \cdots & x_{MN}\mathbf{Y} \end{bmatrix}, \tag{6.89}$$

with $\mathbf{X} \in \mathbb{C}^{M \times N}$; $\mathbf{Y} \in \mathbb{C}^{P \times Q}$; $\mathbf{Z} \in \mathbb{C}^{MP \times NQ}$. Similar to the last section, a block-circulant matrix can also be decomposed using block-Fourier transforms:

$$\mathbf{C}_{(PQ)} = \mathbf{F}_{(P)}^{-1} \mathbf{\Lambda}_{(PQ)} \mathbf{F}_{(Q)}, \tag{6.90}$$

where $\mathbf{\Lambda}_{(PQ)}$ is a block-diagonal matrix computed from

$$\mathbf{\Lambda}_{(PQ)} = \mathrm{diag}_{(PQ)}(\mathbf{F}_{(Q)}\mathbf{C}_{(PQ)}(:,1:Q)), \tag{6.91}$$

with $\mathrm{diag}_{(PQ)}(\mathbf{x})$ representing the block-diagonal matrix whose block elements are the $P \times Q$-sized blocks of x. Similar to the circulant systems, a block-circulant system can also be efficiently solved using the block-Fourier decomposition. The block-circulant system

$$\mathbf{C}_{(P)}\mathbf{x} = \mathbf{b}, \tag{6.92}$$

with $\mathbf{C}_{(P)} \in \mathbb{C}^{NP \times NP}$; $\mathbf{x} \in \mathbb{C}^{NP}$; $\mathbf{b} \in \mathbb{C}^{NP}$. It can be solved with

$$\mathbf{x} = \mathbf{F}_{(P)}^{-1} \mathbf{\Lambda}_{(P)}^{-1} \mathbf{F}_{(P)} \mathbf{b}. \tag{6.93}$$

If the blocks are not square, the Moore–Penrose [Ben-Israel and Greville 1977] pseudoinverse concept can be used. Consider a block-circulant matrix $\mathbf{C}_{(PQ)} \in \mathbb{C}^{NP \times NQ}$, with $P \times Q$-sized blocks.

$$\mathbf{C}_{(PQ)}\mathbf{x} = \mathbf{b}, \quad \mathbf{C}_{(PQ)} \in \mathbb{C}^{NP \times NQ}; \quad \mathbf{x} \in \mathbb{C}^{NQ}; \quad \mathbf{b} \in \mathbb{C}^{NP} Q \leq P. \tag{6.94}$$

The system (6.94) can be solved using the Moore–Penrose pseudoinverse of the complex matrix $\mathbf{C}_{(PQ)}$

$$\mathbf{x} = \left(\mathbf{C}_{(PQ)}^{H}\mathbf{C}_{(PQ)}\right)^{-1} \mathbf{C}_{(PQ)}^{H}\mathbf{b}, \tag{6.95}$$

if $\mathbf{C}_{(PQ)}^{H}\mathbf{C}_{(PQ)}$ is invertible. This solution is the least squares solution (ZF) to the system (6.94), as previously shown. Applying the block-Fourier decomposition of Equation 6.90 to Equation 6.95 results in

$$\mathbf{x} = \left[\left(\mathbf{F}_{(P)}^{-1}\mathbf{\Lambda}_{(PQ)}\mathbf{F}_{(Q)}\right)^{H}\mathbf{F}_{(P)}^{-1}\mathbf{\Lambda}_{(PQ)}\mathbf{F}_{(Q)}\right]^{-1}\cdot\left(\mathbf{F}_{(P)}^{-1}\mathbf{\Lambda}_{(PQ)}\mathbf{F}_{(Q)}\right)^{H}\mathbf{b}, \qquad (6.96)$$

with $\mathbf{\Lambda}_{(PQ)}$ defined as in Equation 6.91. Equation 6.96 can be simplified, considering that $\mathbf{F}_{(K)}^{H} = \mathbf{F}_{(K)}^{-1}$:

$$\mathbf{x} = \mathbf{F}_{(Q)}^{-1}\left[\mathbf{\Lambda}_{(PQ)}^{H}\mathbf{\Lambda}_{(PQ)}\right]^{-1}\mathbf{\Lambda}_{(PQ)}^{H}\mathbf{F}_{(P)}\mathbf{b}. \qquad (6.97)$$

This solution can be computed with only three block-Fourier transforms, the inversion of $\mathbf{\Lambda}_{(PQ)}^{H}\mathbf{\Lambda}_{(PQ)}$ and the multiplication of two block-diagonal matrices by a column vector. The multiplication

$$\left[\mathbf{\Lambda}_{(PQ)}^{H}\mathbf{\Lambda}_{(PQ)}\right]^{-1}\cdot\mathbf{\Lambda}_{(PQ)}^{H}\cdot\left(\mathbf{F}_{(P)}\mathbf{b}\right), \qquad (6.98)$$

must be performed from right to left to minimize the number of operations required, since a matrix–vector multiplication is faster than a matrix–matrix one. Regarding that

$$\left[\mathbf{\Lambda}_{(PQ)}^{H}\mathbf{\Lambda}_{(PQ)}\right]^{-1} \in \mathbb{C}^{NQ\times NQ}$$
$$\mathbf{\Lambda}_{(PQ)}^{H}\in \mathbb{C}^{NQ\times NP} \qquad (6.99)$$
$$\mathbf{F}_{(P)}\mathbf{b} \in \mathbb{C}^{NP},$$

the multiplication (6.98) requires $(NP + NQ)NQ$ pairs of complex multiplications/additions. From the definition of block-Fourier transform in Equation 6.88, it can be shown that given

$$\mathbf{y} = \mathbf{F}_{(K)}\mathbf{x}, \qquad (6.100)$$

with

$$\mathbf{x},\mathbf{y} \in \mathbb{C}^{NK};\mathbf{F}_{(K)} \in \mathbb{C}^{NK\times NK},$$

one has

$$\mathbf{y}\left(i : K : (N-1)K + i\right) = \mathbf{F}_{N}\mathbf{x}\left(i : K : (N-1)K + i\right), \qquad (6.101)$$

with $1 \le i \le K$, where F_N represents a nonorthonormal DFT matrix of dimension N as defined in Equation 6.80. This simply means that each block-Fourier transform of block size K can be decomposed in K Fourier transforms of size N. Furthermore, each Fourier transform can be executed independently of each other and thus the advantage of parallel hardware implementations can be taken. Taking this into consideration, recalling Equation 6.85 and considering that are needed two block-Fourier transforms of block-size Q and one of block-size P to compute Equation 6.97, the number of operations required for that three Fourier transforms is

$$N^{C \oplus \otimes}_{3 \, block - FFT} = (2Q + P) \frac{N \log_2 N}{2}. \tag{6.102}$$

There are two options to calculate $[\Lambda^H_{(PQ)} \Lambda_{(PQ)}]^{-1}$:

■ Start by using the already-calculated $\Lambda_{(PQ)}$ defined in Equation 6.91; then compute $\Sigma_{(QQ)} = \Lambda^H_{(PQ)} \Lambda_{(PQ)}$ in the frequency domain; and finally invert the resultant block-diagonal matrix.

■ Compute $S = C^H_{(PQ)} C_{(PQ)}$ in the time domain; apply the Fourier transform $\Sigma_{(QQ)} = \text{diag}_{(QQ)}(F_{(Q)} S(:,1 : Q))$; and then invert $\Sigma_{(QQ)}$.

In both cases, the inversion of the $\Sigma_{(QQ)}$ block-diagonal matrix is needed. Since $\Lambda_{(PQ)}$ always has to be computed for the solution of Equation 6.97, the second case needs an extra block-fast Fourier transform (FFT) to calculate $\Sigma_{(QQ)}$ but because this extra block-FFT only uses the first Q columns of S there is no need to compute the remainescence. The calculation of the first Q columns of $C^H_{(PQ)} C_{(PQ)}$ is equivalent to $C^H_{(PQ)} C_{(PQ)}(:,1 : Q)$ and requires $N^2 PQ^2$ operations, while the calculation of $\Lambda^H_{(PQ)} \Lambda_{(PQ)}$ requires only NPQ^2 complex operations because $\Lambda_{(PQ)}$ is block-diagonal. Table 6.7 compares the number of operations required in both methods.

In general, the first method is less complex, but the number of operations needed to compute the first Q columns of $C^H_{(PQ)} C_{(PQ)}$ can be reduced if C is sparse (band or block-diagonal, for example). In that case, method 2 can become advantageous for some matrices. $\Sigma_{(QQ)}$ is block-square matrix with $N \times N$ blocks of size $Q \times Q$.

Since $\Sigma_{(QQ)}$ is also block-diagonal, it can be inverted by inverting each one of the N $Q \times Q$ block submatrices separately. Because $\Sigma_{(QQ)}$ is Hermitian positive-definite, each block matrix is also Hermitian positive-definite and therefore can be inverted using the Cholesky decomposition. Let Σ_k be the kth block of the diagonal of $\Sigma_{(QQ)}$. It can be decomposed using the Cholesky algorithm into

$$\Sigma_k = R^H_k R_k, \quad 1 \le k \le N, \tag{6.103}$$

where each R_k is an upper triangular matrix. Each Σ_k can then be easily inverted by inverting the corresponding R_k:

Table 6.7 Number of Operations Required to Compute $\left[\Lambda_{(PQ)}^H \Lambda_{(PQ)}\right]^{-1}$ Using Two Distinct Methods

Method 1	Number of Operations
$\Lambda_{(PQ)} = \mathrm{diag}_{(PQ)}(\mathbf{F}_{(P)}\mathbf{C}(:,1{:}Q))$	0 (previously calculated)
$\Sigma_{(QQ)} = \Lambda_{(PQ)}^H \Lambda_{(PQ)}$	NPQ^2
Method 2	Number of Operations
$\mathbf{S} = \mathbf{C}_{(PQ)}^H \mathbf{C}_{(PQ)}(1^{\text{st}}\ Q\ \text{columns})$	N^2PQ^2
$\Sigma_{(QQ)} = \mathrm{diag}_{(QQ)}(\mathbf{F}_{(Q)}\mathbf{S}(:,1{:}Q))$	$\dfrac{NQ^2 \log_2 N}{2}$

$$\Sigma_k^{-1} = \left(\mathbf{R}_k^H \mathbf{R}_k\right)^{-1} = \mathbf{R}_k^{-1}\left(\mathbf{R}_k^{-1}\right)^H. \tag{6.104}$$

At this point, the straightforward way would be compute each Σ_k^{-1} by multiplying each pair $\mathbf{R}_k^{-1}(\mathbf{R}_k^{-1})^H$ independently. This would require N multiplications of $Q \times Q$ matrix pairs, which absorb NQ^3 complex scalar multiplications.

A better approach is to compose $\mathbf{R}_{(QQ)}^{-1}$ as

$$\mathbf{R}_{(QQ)}^{-1}\left((k-1)Q + 1 : kQ,(k-1)Q + 1 : kQ\right) = \mathbf{R}_k^{-1}, \quad 1 \le k \le N. \tag{6.105}$$

In other words, $\mathbf{R}_{(QQ)}^{-1}$ is a block-diagonal matrix composed of the N $\mathbf{R}_k^{-1}$ matrices sequentially disposed along its diagonal. $\mathbf{R}_{(QQ)}$ can be viewed as the Cholesky decomposition of $\Sigma_{(QQ)}$:

$$\Sigma_{(QQ)} = \mathbf{R}_{(QQ)}^H \mathbf{R}_{(QQ)}. \tag{6.106}$$

Therefore, $\Sigma_{(QQ)}$ can be inverted as

$$\Sigma_{(QQ)}^{-1} = \mathbf{R}_{(QQ)}^{-1}\left(\mathbf{R}_{(QQ)}^{-1}\right)^H. \tag{6.107}$$

Some operations can be saved by first multiplying $(\mathbf{R}_{(QQ)}^{-1})^{H}$ by the already-calculated vector obtained from $\mathbf{\Lambda}_{(PQ)}^{H}(\mathbf{F}_{(P)}\mathbf{b})$. Next, $\mathbf{R}_{(QQ)}^{-1}$ is multiplied by the resulting vector in a similar fashion as is done in Equation 6.98. In the multiplication of $\mathbf{R}_{(QQ)}^{-1}$ by an NQ size vector, only NQ^2 operations are spent, in opposition to the $N^2 Q^3$ operations necessary to compute all $N \mathbf{\Sigma}_{k}^{-1}$ blocks of $\mathbf{\Sigma}_{(QQ)}^{-1}$ as in Equation 6.104.

In fact, it was ignored that each $\mathbf{R}_{k}^{-1}$ is a triangular matrix. This means that multiplying by its transpose conjugate requires only $(6Q^3 + 9Q^2 + 3N)/(16)$ operations. In the same way, $\mathbf{R}_{(QQ)}^{-1}$ is block-diagonal with triangular blocks. Its multiplication by a column vector requires $(Q^2 + Q)/(2)$ per block. This means that the first method requires $N \cdot (6Q^3 + 9Q^2 + 3N)/(16)$ while the second consumes only $N \cdot (Q^2 + Q)/(2)$ operations. Therefore, the best solution is to compose the $\mathbf{R}_{(QQ)}^{-1}$ matrix and ban the matrix–matrix multiplications, as deduced initially. Rewriting Equation 6.97, we obtain

$$\mathbf{x} = \mathbf{F}_{(Q)}^{-1}\mathbf{R}_{(QQ)}^{-1}\left(\mathbf{R}_{(QQ)}^{-1}\right)^{H}\mathbf{\Lambda}_{(PQ)}^{H}\mathbf{F}_{(P)}\mathbf{b}. \tag{6.108}$$

To minimize the number of operations, Equation 6.108 must be computed from right to left, avoiding matrix–matrix multiplications. Finally, the number of operations involved in the inversion of each $\mathbf{R}_{k}^{-1}$ block must be calculated. The inversion of an $\mathbf{R}_{k}^{-1}$ block involves a Cholesky factorization and a triangular matrix inversion. Table 6.8 presents the number of floating point operations necessary to compute the Cholesky factor of an $n \times n$ complex matrix and required to invert an $n \times n$ triangular complex matrix. The order column presents the highest n power of the total operation number, considering each multiplication–addition pair as a single operation.

Considering the data of Table 6.8, it is easy to see that the number of operations needed to compute $\mathbf{R}_{(QQ)}^{-1}$ is $N \cdot (Q^3 + 9Q^2 + 2Q)/(6)$, approximately. To sum it up, the shortest-length least squares solution of Equation 6.94 can be computed with the algorithms of Figure 6.24, depending on the calculation method used for $\mathbf{\Sigma}_{(QQ)}$. The complexity of the algorithms of Figure 6.24 is given in Table 6.9.

Table 6.8 Number of Operations Required by Triangular Matrix Inversion and Cholesky Factorization of $n \times n$ Complex Matrices

	$\mathbb{C}+$	$\mathbb{C}\times$	$\mathbb{C}+$	$R+$	$\mathbb{R}\sqrt{}$	Order
Cholesky decomposition	$\dfrac{n^2 - n}{2}$	$\dfrac{n^3 - n}{3}$	$\dfrac{n^3 - 3n^2 + 2n}{6}$	$\dfrac{n^2 + n}{6}$	n	$\dfrac{n^3}{6}$
Triangular matrix inversion	$\dfrac{n^2 - n}{2}$	$\dfrac{n^2 - n}{2}$		—	—	$\dfrac{n^3}{6}$

(a)

$$\Lambda_{(PQ)} \leftarrow \mathrm{diag}_{(PQ)}(\mathbf{F}_{(P)}\mathbf{C}_{(PQ)}(:,1:Q))$$

$$\Sigma_{(QQ)} = \Lambda_{(PQ)}^{H}\Lambda_{(PQ)}$$

for $k = 1$ *to* N

$\qquad \Sigma_k \leftarrow \Sigma_{(QQ)}((k-1)Q + 1{:}kQ,(k-1)Q + 1{:}kQ)$

$\qquad \mathbf{R}_k \leftarrow \mathrm{chol}(\Sigma_k)$

$\qquad \mathbf{R}_k^{-1} \leftarrow \mathrm{TriangInv}(\mathbf{R}_k)$

$\qquad \mathbf{R}_{(QQ)}^{-1}((k-1)Q + 1{:}kQ,(k-1)Q + 1{:}kQ)$

$$\mathbf{x} \leftarrow \mathbf{F}_{(Q)}^{-1}\left(\mathbf{R}_{(QQ)}^{-1}\left(\left(\mathbf{R}_{(QQ)}^{-1}\right)^{H}\left(\Lambda_{(PQ)}^{H}\left(\mathbf{F}_{(P)}\mathbf{b}\right)\right)\right)\right)$$

(b)

$$\Lambda_{(PQ)} \leftarrow \mathrm{diag}_{(PQ)}(\mathbf{F}_{(P)}\mathbf{C}_{(PQ)}(:,1:Q))$$

$$\mathbf{S}_{(:,1:Q)} = \mathbf{C}_{(PQ)}^{H}\mathbf{C}_{(PQ)}(:,1:Q)$$

$$\Sigma_{(QQ)} = \mathrm{diag}_{(QQ)}(\mathbf{F}_{(Q)}\mathbf{S}(:,1:Q))$$

for $k = 1$ *to* N

$\qquad \Sigma_k \leftarrow \Sigma_{(QQ)}((k-1)Q + 1{:}kQ,(k-1)Q + 1{:}kQ)$

$\qquad \mathbf{R}_k \leftarrow \mathrm{chol}(\Sigma_k)$

$\qquad \mathbf{R}_k^{-1} \leftarrow \mathrm{TriangInv}(\mathbf{R}_k)$

$\qquad \mathbf{R}_{(QQ)}^{-1}((k-1)Q + 1{:}kQ,(k-1)Q + 1{:}kQ) = \mathbf{R}_k^{-1}$

$$\mathbf{x} \leftarrow \mathbf{F}_{(Q)}^{-1}\left(\mathbf{R}_{(QQ)}^{-1}\left(\left(\mathbf{R}_{(QQ)}^{-1}\right)^{H}\left(\Lambda_{(PQ)}^{H}\left(\mathbf{F}_{(P)}b\right)\right)\right)\right)$$

Figure 6.24 Pseudo-codes of block-Fourier algorithms to compute the shortest-length least square solution of a block-circulant system, for (a) method 1 and (b) method 2.

Considering expression (6.101) and keeping in memory only the nonzero parts of the matrices, algorithms of Figure 6.24a and b can be simplified as presented in Figures 6.25 and 6.26, respectively.

Owing to finite precision round errors, the $\mathbf{R}_k$ matrices may not be Hermitian positive-definite even if $\mathbf{C}_{(PQ)}^{H}\mathbf{C}_{(PQ)}$ is Hermitian positive-definite. This can be

Table 6.9 Number of Operations Required for the Algorithms of Figure 6.24 (Direct Correspondence between (a) and (b))

Step	Rep	Number of Operations
Method 1		
$\mathbf{F}_{(P)}\,\mathbf{C}_{(PQ)}(:,1:Q)$	1	$\dfrac{PQN\log_2 N}{2}$
$\mathbf{\Lambda}^H_{(PQ)}\mathbf{\Lambda}_{(PQ)}$	1	NPQ^2
$\text{chol}(\Sigma_k)$	N	$\dfrac{Q^3 + 3Q^2 + 2Q}{6}$
$\text{TriangInv}(\mathbf{R}_k)$	N	Q^2
$\mathbf{F}_{(P)}\mathbf{b}$	1	$\dfrac{PN\log_2 N}{2}$
$\mathbf{\Lambda}^H_{(PQ)}\cdot(\dots)$	1	NQP^2
$(\mathbf{R}^{-1}_{(QQ)})^k\cdot(\dots)$	1	$N\dfrac{Q^2+Q}{2}$
$\mathbf{R}^{-1}_{(QQ)}\cdot(\dots)$	1	$N\dfrac{Q^2+Q}{2}$
$\mathbf{F}^{-1}_{(Q)}\cdot(\dots)$	1	$\dfrac{QN\log_2 N}{2}$
$\text{Total} = N\left[\dfrac{Q^3 + 15Q^2 + 8Q}{6} + PQ^2 + QP^2 + (Q + QP + P)\dfrac{\log_2 N}{2}\right]$		
Method 2		
$\mathbf{F}_{(P)}\mathbf{C}_{(PQ)}(:,1:Q)$	1	$\dfrac{PQN\log_2 N}{2}$
$\mathbf{C}^H_{(PQ)}\mathbf{C}_{(PQ)}(:,1:Q)$	1	N^2PQ^2
$\mathbf{F}_{(Q)}\mathbf{S}(:,1:Q)$	1	$\dfrac{Q^2N\log_2 N}{2}$

Table 6.9 (Continued) Number of Operations Required for the Algorithms of Figure 6.24 (Direct Correspondence between (a) and (b))

chol(Σ_k)	N	$\dfrac{Q^3 + 3Q^2 + 2Q}{6}$
TriangInv(R_k)	N	Q^2
$\mathbf{F}_{(P)}\mathbf{b}$	1	$\dfrac{PN\log_2 N}{2}$
$\mathbf{\Lambda}^H_{(PQ)} \cdot (...)$	1	NQP^2
$(\mathbf{R}^{-1}_{(QQ)})^k \cdot (...)$	1	$N\dfrac{Q^2 + Q}{2}$
$\mathbf{R}^{-1}_{(QQ)} \cdot (...)$	1	$N\dfrac{Q^2 + Q}{2}$
$\mathbf{F}^{-1}_{(Q)} \cdot (...)$	1	$\dfrac{QN\log_2 N}{2}$
Total $= N\left[\dfrac{Q^3 + 15Q^2 + 8Q}{6} + NPQ^2 + QP^2 + (Q^2 + Q + QP + P)\dfrac{\log_2 N}{2}\right]$		

corrected simply by removing the imaginary part of the diagonal elements and zeroing all other elements that have complex modulus below some threshold value, before applying the Cholesky factorization. This new simplified versions require the same number of floating point operations as derived before because null element operations were not considered from the beginning.

6.3.5.3 Application to the Zero-Forcing Detector

The solution of the ZF detector is [Machauer et al. 2001]

$$\mathbf{d} = \left(\mathbf{T}^H\mathbf{T}\right)^{-1}\mathbf{T}^H\mathbf{e} \tag{6.109}$$

This is the shortest-length least squares solution of

$$\mathbf{e} = \mathbf{T}\mathbf{d} \tag{6.110}$$

for $k = 1$ to P

$$\begin{cases} \mathbf{C}_f\left(k{:}P{:}(N-1)P+k, 1{:}Q\right) \leftarrow \text{FFT}\left(\mathbf{C}_{(PQ)}(k{:}P{:}(N-1)P+k, 1{:}Q)\right) \\ \mathbf{b}_f\left(k{:}P{:}(N-1)P+k, 1\right) \leftarrow \text{FFT}\left(\mathbf{b}(k{:}P{:}(N-1)P+k)\right) \end{cases}$$

for $k = 1$ to N

$$\begin{cases} \mathbf{\Sigma}_K \leftarrow \left(\mathbf{C}_f\left((k-1)P+1{:}kP, 1{:}Q\right)\right)^H \mathbf{C}_f\left((k-1)P+1{:}kP, 1{:}Q\right) \\ \mathbf{R}_k \leftarrow \text{chol}(\mathbf{\Sigma}_K) \\ \mathbf{R}_k^{-1} \leftarrow \text{TriangInv}(\mathbf{R}_k) \\ \mathbf{x}_f\left((k-1)Q+1{:}kQ, 1\right) \leftarrow \mathbf{R}_k^{-1}\left((\mathbf{R}_k^{-1})^H\left((\mathbf{C}_f((k-1)P+1{:}kP, 1{:}Q))\right)^H \\ \qquad \times \mathbf{b}_f\left((k-1)P+1{:}kP\right)\right) \end{cases}$$

for $k = 1$ to Q

$$\mathbf{x}\left(k{:}Q{:}(N-1)Q+k, 1\right) \leftarrow \text{IFFT}\left(\mathbf{x}_f\left(k{:}Q{:}(N-1)Q+k\right)\right)$$

Figure 6.25   Pseudo-code for the optimized block-Fourier algorithm for the E_M—method 1.

where $\mathbf{T}$ is not square, in general. Figure 6.27 represents different structures of $\mathbf{T}$ matrices. Figure 6.27a represents the normal structure of $\mathbf{T}$. $\mathbf{T}$ is a block matrix with V_n blocks disposed along its diagonal.

All N blocks are equal in a constant-channel condition. Even if the channel varies slowly, it can be a reasonable approximation to consider all V_n block equal as will be investigated later. In a constant-channel condition, it is easy to extend matrix $\mathbf{T}$ to become block-circulant, simply by adding extra block-columns to it, as shown in Figure 6.27b. The elements below the last V_a block wrap around the top of the columns.

The number of extra columns needed to make $\mathbf{T}$ block-circulant is

$$E = N - n = \lceil H/P \rceil - 1. \tag{6.111}$$

The resulting matrix is block-circulant with $N \times N$ blocks of $P \times Q$ size. This new matrix will be represented as $\mathbf{T}_{(PQ)}$. Now, Equation 6.109 can be transformed in

$$
\begin{aligned}
&\textit{for } k = 1 \text{ to } P\\
&\begin{cases}
\mathbf{C}_f\left(k{:}P{:}(N-1)P+k, 1{:}Q\right) \leftarrow \text{FFT}\left(\mathbf{C}_{(PQ)}(k{:}P{:}(N-1)P+k, 1{:}Q)\right)\\
\mathbf{b}_f\left(k{:}P{:}(N-1)P+k, 1\right) \leftarrow \text{FFT}\left(b(k{:}P{:}(N-1)P+k)\right)
\end{cases}
\end{aligned}
$$

$$
\mathbf{S} \leftarrow \mathbf{C}_{(PQ)}^H \mathbf{C}_{(PQ)}\left(:,1{:}Q\right)
$$

$$
\begin{aligned}
&\textit{for } k = 1 \text{ to } Q\\
&\mathbf{S}_f\left(k{:}Q(N-1)Q+k,1{:}Q\right) \leftarrow \text{FFT}\left(\mathbf{S}\left(k{:}Q{:}(N-1)Q+K,\ 1{:}Q\right)\right)
\end{aligned}
$$

$$
\begin{aligned}
&\textit{for } k = 1 \text{ to } N\\
&\begin{cases}
\boldsymbol{\Sigma}_K \leftarrow \mathbf{S}_f\left((k-1)Q+1{:}kQ, 1{:}Q\right)\\
\mathbf{R}_k \leftarrow \text{chol}\left(\sum_K\right)\\
\mathbf{R}_k^{-1} \leftarrow \text{TriangInv}(\mathbf{R}_k)\\
\mathbf{x}_f\left((k-1)Q+1{:}kQ, 1\right) \leftarrow \mathbf{R}_k^{-1}\left((\mathbf{R}_k^{-1})^H\left((\mathbf{C}_f((k-1)P+1{:}kP, 1{:}Q))\right)^H\right.\\
\qquad \left. \times\, \mathbf{b}_f\left((k-1)P+1{:}kP\right)\right)
\end{cases}
\end{aligned}
$$

$$
\begin{aligned}
&\textit{for } k = 1 \text{ to } Q\\
&\mathbf{x}\left(k{:}Q{:}(N-1)Q+k,1\right) \leftarrow \text{IFFT}\left(\mathbf{x}_f\left(k{:}Q{:}(N-1)Q+k\right)\right)
\end{aligned}
$$

Figure 6.26 Pseudo-code for the optimized block-Fourier algorithm for the E_M—method 2.

$$
\hat{\mathbf{d}}^{\approx} = \left(\mathbf{T}_{(PQ)}^H \mathbf{T}_{(PQ)}\right)^{-1} \mathbf{T}_{(PQ)}^H \mathbf{e}',\ \mathbf{T}_{(PQ)} \in \mathbb{C}^{NP \times NQ};\ \hat{\mathbf{d}} \in \mathbb{C}^{NQ};\ \mathbf{e}' \in \mathbb{C}^{NP}, \quad (6.112)
$$

and solved with the Fourier method, as done for Equation 6.94 in the last section. The $\mathbf{e}'$ vector can be obtained from $\mathbf{e}$ by padding at its end with $(N-n+1)P-H$ zeros, and $\hat{\mathbf{d}}$ can be extracted from the first nQ elements of $\hat{\mathbf{d}}$.

There are two approximations in the transformation of $\mathbf{T}$ in $\mathbf{T}_{(PQ)}$:

- All V_n blocks were made equal.
- Extra columns/lines were added to the matrix.

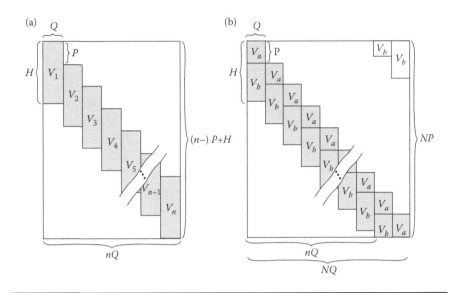

Figure 6.27   Block structure of the *T* matrix (a) and structure of the extended *T* matrix for a constant-channel condition (b).

The error level introduced by each approximation will be determined next. Figure 6.28 shows the relative error introduced in the solution of Equation 6.109 by considering that the V_n blocks are all equal to the first block V_1. This error was determined comparing the solution of Equation 6.109 using the original **T** matrix with the solution of the same system using the block-constant version of **T**.

The **e** vector was obtained from Equation 6.110 using a random **d** vector:

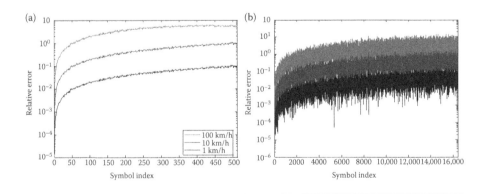

Figure 6.28   Relative error to the ZF solution for a constant-channel approximation using the first block and the Pedestrian A channel for (a) *k* = 1, SISO and (b) *k* = SF, MIMO 2 × 2.

$$d(i) = \frac{\sqrt{2}}{2}\left(\pm 1 \pm j\right), \quad 1 \le i \le nQ. \tag{6.113}$$

As it can be seen, the higher the speed, the higher the error introduced and the quickest the divergence from the correct value as we get farther from the initial block position. The approximation is acceptable for the first blocks, but never for the whole $\hat{\mathbf{d}}$ vector, even if low speeds are considered. For high speeds, the relative error level rises quickly above 100%.

In Figure 6.28, all the V_n blocks were made equal to the first block V_1. A better approximation would be expected if a middle block was used. Let us use the middle block of $\mathbf{T}$ if n is even or the left-middle block if n is odd:

$$V = V_{(\lfloor n/2 \rfloor)} \tag{6.114}$$

Using this method, better approximations in a wider central range can be attained, as shown in Figure 6.29.

Nevertheless, the approximation is very poor for the first and last elements of the $\hat{\mathbf{d}}$ vector. The last figures make it obvious that the methods presented in [Vollmer et al. 2001] or [Machauer et al. 2001], although valid for constant-channel condition, are not very useful if the channel changes, even if low speeds are considered.

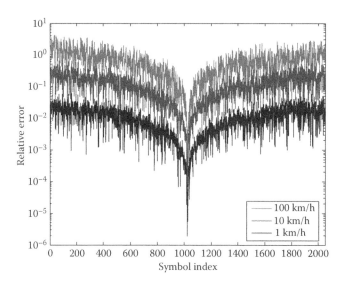

Figure 6.29 Relative error to the ZF solution for constant-channel approximation, using the middle block for the Pedestrian A channel, $k = 1$ and MIMO 4×4.

Constant-channel conditions will be considered for the remainder of this section. First, the error level introduced by the addition of extra columns/lines to the **T** matrix, transforming it in the block-circulant matrix $\mathbf{T}_{(PQ)}$, will be determined, as was previously explained, alongside the error of the block-Fourier algorithm. Later, we will return to the more general case of the varying channel.

6.3.5.4 Constant-Channel Conditions

Table 6.10 shows the error level introduced in the determination of vector $\hat{\mathbf{d}}$ by each phase of the detection process in 12 constant-channel situations. Each value is the maximum complex modulus of the relative difference between the estimated $\hat{\mathbf{d}}$ and the real $\mathbf{d}$ obtained in 100 runs with distinct random d vectors.

The correlated values described in Table 6.10 were taken from [3GPP 2003]. A squared matrix disposition was considered for the 4×4 case. The "Estimation" column indicates the maximum error of the estimation if Equation 6.109 is solved directly. The "Circulant" column shows the maximum error introduced by the extension of the **T** matrix to become $\mathbf{T}_{(PQ)}$ relative to the estimated solution. The "Fourier1" and "Fourier2" columns show the maximum error introduced by the use of the Fourier algorithms 1 and 2 compared with the circulant system solution.

Table 6.10 Maximum Absolute Error to the ZF Solution Introduced by Each Phase of the Zero-Forcing Detector

				Maximum Absolute Error			
	Antennas	Load	Correlated	Estimation	Circulant	Fourier1	Fourier2
Pedestrain A	1×1	Min	—	4.E–16	5.E–16	2.E–15	3.E–15
		Max	—	9.E–14	1.E–13	1.E–13	1.E–13
	2×2	Min	—	8.E–15	6.E–15	1.E–14	1.E–14
		Max	—	4.E–14	5.E–14	5.E–14	4.E–14
		Min	Yes	1.E–14	1.E–14	2.E–14	2.E–14
	4×4	Min	—	4.E–14	2.E–14	7.E–14	5.E–14
		Min	Yes	5.E–14	3.E–14	8.E–14	7.E–14
Vehicular A	1×1	Min	—	7.E–16	8.E–16	3.E–15	3.E–15
		Max	—	1.E–14	3.E–14	3.E–14	3.E–14
	2×2	Min	—	2.E–14	7.E–15	3.E–14	2.E–14
		Max	—	1.E–12	2.E–12	2.E–12	2.E–12
	4×4	Min	—	5.E–14	3.E–14	8.E–14	7.E–14

Since an *SF* of 16 is considered, the minimum load situation corresponds to 0 interferers while the maximum load situation corresponds to 15 interferers (with each user having only one physical channel).

As can be seen, the errors are very low for all the tested matrices and are only slightly above the floating point precision used. Furthermore, the **T** matrix extension is not the main error cause. Figure 6.30 shows the error at the end of the block-Fourier algorithm relative to the original **d** vector for a sample case.

It can be seen that the error level is constant along the entire symbol vector and no beginning or end high error levels appear, since no multipath interference from adjacent blocks is being considered. Excluding the midamble from the detection and splitting the process into two independent detections, each one involving only data symbols, some errors can be introduced at the beginning of the second data chunk. This can be corrected by including some symbols of the midamble in the second detection process.

Figures 6.31 and 6.32 show the used algorithms, and Table 6.11 their complexity. The algorithm of Figure 6.31 is similar to that of Figure 6.25 but the algorithm of Figure 6.32 makes use of the structure of *T* to save some operations. This is done using only the nonzero elements of *T* and regarding that

$$(\mathbf{A}^H \mathbf{B})^H = \mathbf{B}^H \mathbf{A}. \tag{6.115}$$

This saves the computation of the last *E* blocks of $\mathbf{S}(:, 1:Q)$.

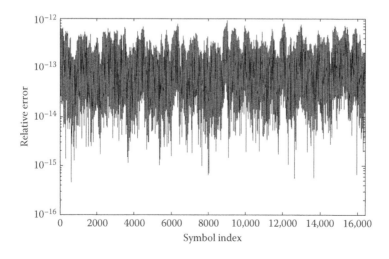

Figure 6.30 Relative error to the ZF solution introduced by the block-Fourier algorithm (sample case).

for $k = 1$ to P

$$\begin{cases} \mathbf{T}_f\big(k{:}P{:}(N-1)P+k, 1{:}Q\big) \leftarrow \mathrm{FFT}\big(\mathbf{T}(k{:}P{:}(N-1)P+k, 1{:}Q)\big) \\ \mathbf{e}_f\big(k{:}P{:}(N-1)P+k, 1\big) \leftarrow \mathrm{FFT}\big(\mathbf{e}(k{:}P{:}(N-1)P+k)\big) \end{cases}$$

for $k = 1$ to N

$$\begin{cases} \mathbf{\Sigma}_K \leftarrow \big(\mathbf{T}_f\big((k-1)P+1{:}kP, 1{:}Q\big)\big)^H \mathbf{T}_f\big((k-1)P+1{:}kP, 1{:}Q\big) \\ \mathbf{R}_k \leftarrow \mathrm{chol}\big(\mathbf{\Sigma}_K\big) \\ \mathbf{R}_k^{-1} \leftarrow \mathrm{TriangInv}(\mathbf{R}_k) \\ \mathbf{d}_f\big((k-1)Q+1{:}kQ, 1\big) \leftarrow \mathbf{R}_k^{-1}\big((\mathbf{R}_k^{-1})^H\big((\mathbf{T}_f((k-1)P+1{:}kP, 1{:}Q))\big)^H \\ \quad \times \mathbf{e}_f\big((k-1)P+1{:}kP\big)\big) \end{cases}$$

for $k = 1$ to Q

$$\mathbf{d}\big(k{:}Q{:}(N-1)Q+k, 1\big) \leftarrow \mathrm{IFFT}\big(df\big(k{:}Q{:}(N-1)Q+k\big)\big)$$

Figure 6.31 Pseudo-code of the block-Fourier algorithm.

Comparing the two methods, it can be seen that method 2 is advantageous if

$$NPQ^2 > NQ^2 \frac{\log_2 N}{2} + Q^2\left(H + HE - P\frac{E^2 + E}{2}\right). \tag{6.116}$$

Since $N \gg Q$ and $N \gg H \geq P$, the last inequality usually resumes to

$$P > \frac{\log_2 N}{2}. \tag{6.117}$$

Since P is the number of antennas multiplied by the spreading factor, it is easily greater than the right side of inequality (6.117) and the block-Fourier method 2 is in general advantageous.

6.3.5.5 Block-Fourier Algorithm with Partitioning

The algorithms proposed in last section reveal already many parallel paths that could be exploited for parallel processing in adequate hardware. Nevertheless, the

for $k = 1$ *to* P

$$\begin{cases} \mathbf{T}_f(k{:}P{:}(N-1)P+k,1{:}Q) \leftarrow \text{FFT}\,(\mathbf{T}(k{:}P{:}(N-1)P+k,1{:}Q)) \\ \mathbf{e}_f\,(k{:}P{:}(N-1)P+k,1) \leftarrow \text{FFT}\,(\mathbf{e}(k{:}P{:}(N-1)P+k)) \end{cases}$$

$E \leftarrow \lceil H/P \rceil - 1$

$\mathbf{S}(1{:}Q,1{:}Q) \leftarrow (\mathbf{T}(1{:}H,1{:}Q))^H\,\mathbf{T}(1{:}H,1{:}Q)$

for $k = 1$ *to* P

$$\begin{cases} \mathbf{S}_k \leftarrow (\mathbf{T}(1{:}H-kp,1{:}Q))^H\,\mathbf{T}(kP+1{:}H,1{:}Q) \\ \mathbf{S}(kQ+1{:}(k+1)Q,\,1{:}Q) \leftarrow \mathbf{S}_k \\ \mathbf{S}((N-k)Q+1{:}(N-k+1)Q,1{:}Q) \leftarrow \mathbf{S}_k^H \end{cases}$$

for $k = 1$ *to* Q

$Sf(k{:}Q{:}(N-1)Q+k,1{:}Q) \leftarrow \text{FFT}\,(\mathbf{S}(k{:}Q{:}(N-1)Q+k,1{:}Q)$

for $k = 1$ *to* N

$$\begin{cases} \mathbf{\Sigma}_k \leftarrow \mathbf{S}_f((k-1)Q+1{:}kQ,\,1{:}Q)) \\ \mathbf{R}_k \leftarrow \text{chol}(\mathbf{\Sigma}_k) \\ \mathbf{R}_k^{-1} \leftarrow \text{TriangInv}(\mathbf{R}_k) \\ \mathbf{d}_f\big((k-1)Q+1{:}kQ,1\big) \leftarrow \mathbf{R}_k^{-1}\big((\mathbf{R}_k^{-1})^H\big((\mathbf{T}_f((k-1)P+1{:}kP,\,1{:}Q))^H\big) \\ \qquad \times \mathbf{e}_f\big((k-1)P+1{:}kP\big)\big) \end{cases}$$

for $k = 1$ *to* Q

$\mathbf{d}(k{:}Q(N-1)Q+k,1) \leftarrow \text{IFFT}(\mathbf{d}_f(k{:}Q(N-1)Q+k,1))$

Figure 6.32   Pseudo-code for the optimized block-Fourier algorithm.

algorithm remains globally sequential since it only determines the estimated $\hat{d}^{\approx}$ vector at the end.

Figure 6.33 illustrates an approach to split the extended ZF Equation 6.112 in smaller systems. Figure 6.33 represents $\mathbf{S}_Q = \mathbf{T}_{(PQ)}^H \mathbf{T}_{(PQ)}$ and the estimated vector $\hat{\mathbf{d}}^{\approx}$. The idea is to split the $\mathbf{S}_Q$ matrix in smaller ones. This can be a reasonable approximation because the $\mathbf{S}_Q$ has the greater values concentrated around the diagonal and decreasing modulus as the values are further from the diagonal. This means that each element of vector $\hat{\mathbf{d}}$ depends mainly on the same index value of vector $\mathbf{T}_{(PQ)}^H \mathbf{e}'$ and it depends increasingly less on the values as we get farther from that same index value.

Since each partition just approximates well the middle values, the $\hat{\mathbf{d}}$ values of the beginning and end of each partition can be discarded. The first l^- and last l^+

Table 6.11 Number of Operations Required by the Block-Fourier Algorithm

Step	Rep	Number of Operations
Block-Fourier Algorithm in Figure 6.31		
$\mathbf{T}_f$	P	$\dfrac{QN\log_2 N}{2}$
$\mathbf{e}_f$	P	$\dfrac{N\log_2 N}{2}$
Σ_k	N	PQ^2
$\mathbf{R}_k$	N	$\dfrac{Q^3 + 3Q^2 + 2Q}{6}$
$\mathbf{R}_1^{-1}$	N	Q^2
$\mathbf{d}_f$	N	$Q^2 + Q + QP^2$
$\mathbf{d}$	Q	$\dfrac{N\log_2 N}{2}$
$\text{Total} = N\left[\dfrac{Q^3 + 15Q^2 + 8Q}{6} + PQ^2 + QP^2 + (Q + QP + P)\dfrac{\log_2 N}{2}\right]$		
Block-Fourier Algorithm in Figure 6.32		
$\mathbf{T}_f$	P	$\dfrac{QN\log_2 N}{2}$
$\mathbf{e}_f$	P	$\dfrac{N\log_2 N}{2}$
$\mathbf{S}_{(IQ,\,IQ)}$	1	HQ^2
$\mathbf{S}_k$	1	$Q^2\left(HE - P\dfrac{E^2 + E}{2}\right)$
$\mathbf{S}_f$	Q	$\dfrac{QN\log_2 N}{2}$

Table 6.11 (Continued) Number of Operations Required by the Block-Fourier Algorithm

Step	Rep	Number of Operations
$\mathbf{R}_k$	N	$\dfrac{Q^3 + 3Q^2 + 2Q}{6}$
$\mathbf{R}_1^{-1}$	N	Q^2
$\mathbf{d}_f$	N	$Q^2 + Q + QP^2$
$\mathbf{d}$	Q	$\dfrac{N\log_2 N}{2}$

$$\text{Total} = N\left[\frac{Q^3 + 15Q^2 + 8Q}{6} + QP^2 + (Q^2 + Q + QP + P)\frac{\log_2 N}{2}\right.$$
$$\left. + Q^2\left(HE - P\frac{E^2 + E}{2}\right)\right)$$

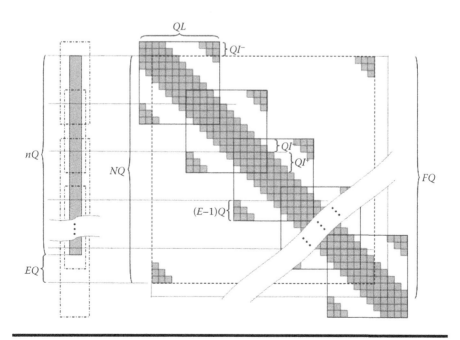

Figure 6.33 Partitioning for the block-Fourier algorithm for $\hat{\mathbf{d}}$ and S_Q.

elements of each $\hat{\mathbf{d}}$ partition are thus discarded. Note that since the block-Fourier algorithm will be applied at each partition, and because each partition has to be approximated as a block-circulant matrix, high error will also appear in the first elements of the first partition and in the last elements of the last partition. This would not happen if each partition would be solved with an exact method like Gauss or Cholesky. This is the reason why those elements are also discarded in Figure 6.33. Figures 6.34 and 6.35 represent such algorithms, based on the two versions presented earlier. Table 6.12 analyzes their computational requirements.

The overlapping length must be carefully selected according to the precision required. The bigger the overlapping, the more expensive the computation will be. In the derivation of the algorithms, a prelap l^- and postlap l^+ were defined in a similar way as done in [Machauer et al. 2001], but in the simulations, it will always be used $l^- = l^+$ since there is no advantage of defining dissimilar overlapping lengths; in fact, the best results are obtained with equal overlapping as could be easily guessed from symmetry considerations.

The partition length can be selected according to the number of intended partitions and the overlapping selected, and this is usually determined by the hardware structure.

6.3.5.6 Solving the E_M for the Unsteady Channel

As can be seen in the next section, the block-Fourier algorithms presented so far work well just under constant-channel conditions or with very slow changing speeds. Even at $v = 1$ km/h, significant errors arise. The standard block-Fourier algorithm cannot be adapted for unsteady-channel conditions because it just works for block-circulant matrices, but the partitioned version can be easily adapted just by using in each partition a different block of the original $\mathbf{T}$ matrix. If the middle block in each partition of the original $\mathbf{T}$ matrix is used to construct each extended approximate $\mathbf{T}_{(PQ)}$, the block-Fourier algorithm can be used for each partition as done in the last section, and $\hat{\mathbf{d}}$ obtained from the middle elements of each partition computation.

Figures 6.36 and 6.37 represent the two versions of such an algorithm. Table 6.13 analyzes the respective computational requirements.

The overlapping length must be selected according to the precision required as for the constant channel. The bigger the overlapping, the better the approximation but more expensive will the computation be.

The partition length now must also be selected according to the precision required: bigger partitions will yield faster results with small precisions, whereas smaller partitions will require too much computational power. Partition and overlapping length can also be determined by the hardware structure available, if some kind of parallel processing is available in an already-developed hardware platform. Also, very small partitions or very long overlapping can become incompatible, making the required error level just not attainable for high-speed channels.

function $\mathbf{d} = BFPC1(\mathbf{T}(1{:}LP,1{:}Q),e,n,Q,P,H,L,l^-,l^+)$

$N = n + \lceil H/P \rceil - 1$

$l = L - l^- - l^+$

$F = \lceil N/l \rceil$

$E \leftarrow \lceil H/P \rceil - 1$

$\mathbf{e}' = \left[\mathbf{0}_{(Pl^-)}; \mathbf{e}; \mathbf{0}((Fl + l + -n + 1)P - H) \right]$

for $k = 1$ *to* P

$\mathbf{T}_f(k{:}P(L-1)P + k, 1{:}Q) \leftarrow \text{FFT}(\mathbf{T}(k{:}P{:}(L-1)P + k, 1{:}Q))$

for $k = 1$ *to* L

$\begin{cases} \mathbf{\Sigma}_k \leftarrow (\mathbf{T}_f((k-1)P + 1{:}kPk, 1{:}Q))^H \mathbf{T}_f((k-1)P + 1{:}kPk, 1{:}Q) \\ \mathbf{R}_k \leftarrow \text{chol}(\mathbf{\Sigma}_k) \\ \mathbf{R}_k^{-1} \leftarrow \text{TrianInv}(\mathbf{R}_k) \end{cases}$

for $f = 1$ *to* F

$\begin{cases} \textit{for } k = 1 \text{ to } P \\ \mathbf{e}_f(k{:}P{:}(L-1)P + k, 1) \leftarrow \text{FFT}\left(\mathbf{e}'((f-1)lP + k : P : (fl + l^- + l^+ - 1)P + k)\right) \\ \textit{for } k = 1 \text{ to } L \\ \mathbf{d}_f((k-1)Q + 1{:}kQ, 1) \leftarrow \mathbf{R}_k^{-1}\left((\mathbf{R}_k^{-1})^H\left((\mathbf{T}_f((k-1)P + 1{:}kP, 1{:}Q))^H\right)\right. \\ \qquad \left. \times \mathbf{e}_f((k-1)P + 1{:}kP)\right) \\ \textit{for } k = 1 \text{ to } Q \\ \mathbf{d}_l(k{:}Q(L-1)Q + k, 1) \leftarrow \text{IFFT}\left(\mathbf{d}_f(k{:}Q(L-1)Q + k)\right) \\ \mathbf{d}'((f-1)lQ + 1{:}flQ) \leftarrow \mathbf{d}_l\left(l^-Q + 1{:}(l^- + l)Q\right) \end{cases}$

$\mathbf{d} = \mathbf{d}'_{(1{:}nQ)}$

Figure 6.34   Pseudo-code for the block-Fourier partitioned algorithm for constant channels—method 1.

function $\mathbf{d} = BFPC2O\Big(T\big(1{:}LP, 1{:}Q\big), e, n, Q, P, H, L, l^-, l^+\Big)$

$N = n + \big[H/P\big] - 1$
$l = L - l^- - l^+$
$F = \big[N/l\big]$
$E \leftarrow \big[H/P\big] - 1$

$$\mathbf{e}' = \left[\mathbf{0}_{(Pl^-)}; \mathbf{e}; \mathbf{0}_{\big((Fl + l^+ - n + 1)P - H\big)}\right]$$

for $k = 1$ *to* P
$\mathbf{T}_f\big(k{:}P{:}(L-1)P + k, 1{:}Q\big) \leftarrow \text{FFT}\Big(\mathbf{T}\big(k{:}P{:}(L-1)P + k, 1{:}Q\big)\Big)$
$\mathbf{S}(1{:}Q, 1{:}Q) \leftarrow \big(\mathbf{T}\big(1{:}H, 1{:}Q\big)\big)^H \mathbf{T}\big(1{:}H, 1{:}Q\big)$
for $k = 1$ *to* E
$\left|\begin{array}{l} \mathbf{S}_k \leftarrow \big(\mathbf{T}\big(1{:}H - kP, 1{:}Q\big)\big)^H \mathbf{T}\big(kP + 1{:}H, 1{:}Q\big) \\ \mathbf{S}\big(kQ + 1{:}(k+1)Q, 1{:}Q\big) \leftarrow \mathbf{S}_k \\ \mathbf{S}\big((L-k)Q + 1{:}(L-k+1)Q, 1{:}Q\big) \leftarrow \mathbf{S}_k^H \end{array}\right.$
for $k = 1$ *to* Q
$\mathbf{S}_f\big(k{:}Q(L-1)Q + k, 1{:}Q\big) \leftarrow \text{FFT}\Big(\mathbf{S}\big(k{:}Q(L-1)Q + k, 1{:}Q\big)\Big)$
for $k = 1$ *to* L
$\left|\begin{array}{l} \boldsymbol{\Sigma}_k \leftarrow \mathbf{S}_f\big((k-1)Q + 1{:}kQ, 1{:}Q\big) \\ \mathbf{R}_k \leftarrow \text{chol}(\boldsymbol{\Sigma}_k) \\ \mathbf{R}_k^{-1} \leftarrow \text{TriangInv}(\mathbf{R}_k) \end{array}\right.$
for $k = 1$ *to* F
$\left|\begin{array}{l} \textit{for } k = 1 \textit{ to } P \\ \mathbf{e}_f\big(k{:}P{:}(L-1)P + k, 1\big) \leftarrow \text{FFT}\Big(\mathbf{e}'\big((f-1)lP + k : P : (fl + l^- + l^+ - 1)P + k\big)\Big) \\ \textit{for } k = 1 \textit{ to } L \\ \mathbf{d}_f\big((k-1)Q + 1{:}kQ, 1\big) \leftarrow \mathbf{R}_k^{-1}\Big(\big(\mathbf{R}_k^{-1}\big)^H \big(\big(\mathbf{T}_f\big((k-1)P + 1{:}kP, 1{:}Q\big)\big)^H\big) \\ \qquad \times \mathbf{e}_f\big((k-1)P + 1{:}kP\big)\Big) \\ \textit{for } k = 1 \textit{ to } Q \\ \mathbf{d}_l\big(k{:}Q(L-1)Q + k, 1\big) \leftarrow \text{IFFT}\Big(\mathbf{d}_f\big(k{:}Q(L-1)Q + k\big)\Big) \\ \mathbf{d}'\big((f-1)lQ + 1{:}flQ\big) \leftarrow \mathbf{d}_l\big(l^-Q + 1{:}(l^- + l)Q\big) \end{array}\right.$
$\mathbf{d} = \mathbf{d}'(1{:}nQ)$

Figure 6.35 **Pseudo-code for the block-Fourier partitioned algorithm for constant channels—method 2.**

Table 6.12   Number of Operations Required by the Partitioned Block-Fourier Algorithm

Step	Rep	Number of Operations
Method 1		
$\mathbf{T}_f$	P	$\dfrac{QL\log_2 L}{2}$
$\boldsymbol{\Sigma}_k$	L	PQ^2
$\mathbf{R}_k$	L	$\dfrac{Q^3 + 3Q^2 + 2Q}{6}$
$\mathbf{R}_1^{-1}$	L	Q^2
$\mathbf{e}_f$	FP	$\dfrac{L\log_2 L}{2}$
$\mathbf{d}_f$	FL	$Q^2 + Q + QP^2$
$\mathbf{d}_2$	FQ	$\dfrac{L\log_2 L}{2}$
$\text{Total} = L\left[\dfrac{Q^3 + 9Q^2 + 2Q}{6} + F(Q^2 + Q + QP^2) + PQ^2 + Q^2 + (QF + QP + FP)\dfrac{\log_2 L}{2} \right]$		
Method 2		
$\mathbf{T}_f$	P	$\dfrac{QL\log_2 L}{2}$
$\mathbf{S}_{(IQ,\,IQ)}$	1	HQ^2
$\mathbf{S}_k$	1	$Q^2\left(HE - P\dfrac{E^2 + E}{2} \right)$
$\mathbf{S}_f$	Q	$\dfrac{QL\log_2 L}{2}$
$\mathbf{R}_k$	L	$\dfrac{Q^3 + 3Q^2 + 2Q}{6}$

continued

Table 6.12   (Continued) Number of Operations Required by the Partitioned Block-Fourier Algorithm

Step	Rep	Number of Operations
$\mathbf{R}_1^{-1}$	L	Q^2
$\mathbf{e}_f$	FP	$\dfrac{L\log_2 L}{2}$
$\mathbf{d}_f$	FL	$Q^2 + Q + QP^2$
$\mathbf{d}_2$	FQ	$\dfrac{L\log_2 L}{2}$
Total $= L\left[\dfrac{Q^3 + 9Q^2 + 2Q}{6} + F(Q^2 + Q + QP^2) + (Q^2 + QF + QP + FP)\dfrac{\log_2 L}{2}\right]$ $+ Q^2\left(H + HE - P\dfrac{E^2 + E}{2}\right)$		

6.3.5.7 *Application to the LMMSE Detector*

The equation of the LMMSE detector is

$$\hat{\mathbf{d}} = \left(\mathbf{T}^H\mathbf{T} + \sigma^2\mathbf{I}_{nQ}\right)^{-1}\mathbf{T}^H\mathbf{e}. \qquad (6.118)$$

As done for the ZF detector, a block-circulant version of the MMSE detector equation can be obtained just by adding some extra columns and lines to matrix **T**. In this case, the noise matrix $\mathbf{N} = \sigma^2\,\mathbf{I}_{nQ}$ also has to be extended. This can be done just by adding E (defined in Equation 6.111) lines/columns to it, obtaining the new noise matrix

$$\mathbf{N}_{NQ} = \sigma^2\mathbf{I}_{NQ}. \qquad (6.119)$$

The block-circulant version of the MMSE detector becomes

$$\hat{\mathbf{d}} = \left(\mathbf{T}_{(PQ)}^H\mathbf{T}_{(PQ)} + \mathbf{N}_{NQ}\right)^{-1}\mathbf{T}_{(PQ)}^H\mathbf{e}',\ \mathbf{T}_{(PQ)} \in \mathbb{C}^{NP\times NQ};\ \mathbf{N}_{NQ} \in \mathbb{R}^{NQ\times NQ};$$
$$\hat{\mathbf{d}} \in \mathbb{C}^{NQ};\ \mathbf{e}' \in \mathbb{C}^{NP}$$

$$(6.120)$$

$$function\ \mathbf{d} = BFP1\left(T, \mathbf{e}, n, Q, P, H, L, l^-, l^+\right)$$

$$N = n + \left[H/P\right] - 1$$
$$l = L - l^- - l^+$$
$$F = \left[N/l\right]$$
$$E \leftarrow \left[H/P\right] - 1$$
$$m \leftarrow \left[l/2\right]$$

$$\mathbf{e}' = \left[\mathbf{0}_{(Pl^-)}; \mathbf{e}; \mathbf{0}_{\left((Fl + l^+ - n + 1)P - H\right)}\right]$$

for $f = 1$ to F

$\quad$ *for* $k = 1$ to P

$\qquad \mathbf{T}_f\left(k{:}P{:}(L-1)P + k, 1{:}Q\right) \leftarrow \mathrm{FFT}\left(\mathbf{T}(mP + k{:}P{:}(m + L - 1)P\right.$
$\qquad \left.+k, mQ + 1{:}(m+1)Q)\right)$

$\qquad \mathbf{e}_f\left(k{:}P{:}(L-1)P + k, 1\right) \leftarrow \mathrm{FFT}\left(\mathbf{e}'((f-1)lP + k{:}P{:}(fl + l^- + l^+ - 1)P + k)\right)$

$\quad$ *for* $k = 1$ to L

$\qquad \boldsymbol{\Sigma}_k \leftarrow \left(\mathbf{T}_f\left((k-1)P + 1{:}kP, 1{:}Q\right)\right)^H \mathbf{T}_f\left((k-1)P + 1{:}kP, 1{:}Q\right) + \sigma^2 \mathbf{I}_Q$

$\qquad \mathbf{R}_k \leftarrow \mathrm{chol}(\boldsymbol{\Sigma}_k)$

$\qquad \mathbf{R}_k^{-1} \leftarrow \mathrm{TriangInv}(\mathbf{R}_k)$

$\qquad \mathbf{d}_f\left((k-1)Q + 1{:}kQ, 1\right) \leftarrow \mathbf{R}_k^{-1}\left((\mathbf{R}_k^{-1})^H \left((\mathbf{T}_f((k-1)P + 1{:}kP, 1{:}Q))^H\right)\right.$
$\qquad \left. \times \mathbf{e}_f((k-1)P + 1{:}kP)\right)$

$\quad$ *for* $k = 1$ to Q

$\qquad \mathbf{d}_l\left(k{:}Q(L-1)Q + k, 1\right) \leftarrow \mathrm{IFFT}\left(\mathbf{d}_f\left(k{:}Q(L-1)Q + k\right)\right)$

$\qquad \mathbf{d}'\left((f-1)lQ + 1{:}flQ\right) \leftarrow \mathbf{d}_l\left(l^-Q + 1{:}(l^- + l)Q\right)$

$\qquad m \leftarrow \min(m + l, n - 1)$

$$\mathbf{d} = \mathbf{d}'(1{:}nQ)$$

Figure 6.36   Pseudo-code for the block-Fourier partitioned algorithm for unsteady channels, using method 1.

$function\ \mathbf{d} = BFP2O\big(T,e,n,Q,P,H,L,l^-,l^+\big)$

$N = n + \lceil H/P \rceil - 1$

$l = L - l^- - l^+$

$F = \lceil N/l \rceil$

$E \leftarrow \lceil H/P \rceil - 1$

$m \leftarrow \lceil l/2 \rceil$

$\mathbf{e}' = \Big[\mathbf{0}_{(Pl^-)};\mathbf{e};\mathbf{0}_{((Fl+l^+-n+1)P-H)} \Big]$

$for\ f = 1\ to\ F$

$\quad\Big[$ $for\ k = 1\ to\ P$

$\qquad\mathbf{T}_f\big(k{:}P{:}(L-1)P+k,1{:}Q\big) \leftarrow \mathrm{FFT}\big(\mathbf{T}(mP+k{:}P{:}(m+L-1)P+k,\ mQ+1{:}(m+1)Q)\big)$

$\qquad\mathbf{e}_f\big(k{:}P{:}(L-1)P+k,1\big) \leftarrow \mathrm{FFT}\big(\mathbf{e}'((f-1)lP+k{:}P{:}(fl+l^-+l^+-1)P+k)\big)$

$\quad\mathbf{S}(1{:}Q,1{:}Q) \leftarrow \big(\mathbf{T}\big(mP+1{:}mP+H,\ mQ+1{:}(m+1)Q\big)\big)^H\,\mathbf{T}$

$\qquad\times\big(mP+1{:}mP+H,\ mQ+1{:}(m+1)Q\big)$

$\quad for\ k = 1\ to\ E$

$\qquad\Big[$ $\mathbf{S}_k \leftarrow \big(\mathbf{T}\big(mP+1{:}(m-k)P+H,\ mQ+1{:}(m+1)Q\big)\big)^H\,\mathbf{T}$

$\qquad\times\big((m+k)P+1{:}mP+H,\ mQ+1{:}(m+1)Q\big)$

$\qquad\mathbf{S}\big(kQ+1{:}(k+1)Q,1{:}Q\big) \leftarrow \mathbf{S}_k$

$\qquad\mathbf{S}\big((L-k)Q+1{:}(L-k+1)Q,1{:}Q\big) \leftarrow \mathbf{S}_k^H$

$\quad for\ k = 1\ to\ Q$

$\quad\mathbf{S}_f\big(k{:}Q(L-1)Q+k,1{:}Q\big) \leftarrow \mathrm{FFT}\big(\mathbf{S}\big(k{:}Q(L-1)Q+k,1{:}Q\big)\big)$

$\quad for\ k = 1\ to\ L$

$\qquad\Big[$ $\mathbf{\Sigma}_k \leftarrow \mathbf{S}_f\big((k-1)Q+1{:}kQ,1{:}Q\big) + \sigma^2 \mathbf{I}_Q$

$\qquad\mathbf{R}_k \leftarrow \mathrm{chol}(\mathbf{\Sigma}_k)$

$\qquad\mathbf{R}_k^{-1} \leftarrow \mathrm{TriangInv}(\mathbf{R}_k)$

$\qquad\mathbf{d}_f\big((k-1)Q+1{:}kQ,1\big) \leftarrow \mathbf{R}_k^{-1}\Big(\big(R_k^{-1}\big)^H\Big(\big(Tf\big((k-1)P+1{:}kP,1{:}Q\big)\big)^H\,\mathbf{e}_f\big((k-1)P+1{:}kP\big)\Big)\Big)$

$\quad for\ k = 1\ to\ Q$

$\quad\mathbf{d}_l\big(k{:}Q(L-1)Q+k,1\big) \leftarrow \mathrm{IFFT}\big(\mathbf{d}_f\big(k{:}Q(L-1)Q+k\big)\big)$

$\quad\mathbf{d}'\big((f-1)lQ+1{:}flQ\big) \leftarrow \mathbf{d}\big(l^-Q+1{:}(l^-+l)Q\big)$

$\quad m \leftarrow \min(m+l,\ n-1)$

$\mathbf{d} = \mathbf{d}'(1{:}nQ)$

Figure 6.37 Pseudo-code for the block-Fourier partitioned algorithm for unsteady channels, using method 2.

Table 6.13   Number of Operations Required by the Algorithms in (a) Figure 6.36 and (b) Figure 6.37

Step	Rep	Number of Operations
Method 1		
$\mathbf{T}_f$	FP	$\dfrac{QL\log_2 L}{2}$
$\mathbf{e}_f$	FP	$\dfrac{L\log_2 L}{2}$
$\mathbf{\Sigma}_k$	FL	PQ^2
$\mathbf{R}_k$	FL	$\dfrac{Q^3 + 3Q^2 + 2Q}{6}$
$\mathbf{R}_1^{-1}$	FL	Q^2
$\mathbf{d}_f$	FL	$Q^2 + Q + QP^2$
$\mathbf{d}$	FQ	$\dfrac{L\log_2 L}{2}$
$\text{Total} = FL\left[\dfrac{Q^3 + 15Q^2 + 8Q}{6} + PQ^2 + QP^2 + (Q + QP + P)\dfrac{\log_2 L}{2}\right]$		
Method 2		
$\mathbf{T}_f$	FP	$\dfrac{QL\log_2 L}{2}$
$\mathbf{e}_f$	FP	$\dfrac{L\log_2 L}{2}$
$\mathbf{S}_{(IQ,\,IQ)}$	F	HQ^2
$\mathbf{S}_k$	F	$Q^2\left(HE - P\dfrac{E^2 + E}{2}\right)$
$\mathbf{S}_f$	FQ	$\dfrac{QL\log_2 L}{2}$

continued

Table 6.13 (Continued) Number of Operations Required by the Algorithms in (a) Figure 6.36 and (b) Figure 6.37

(b)

Step	Rep	Number of Operations
R_k	FL	$\dfrac{Q^3 + 3Q^2 + 2Q}{6}$
R_k^{-1}	FL	Q^2
d_f	FL	$Q^2 + Q + QP^2$
d_2	FQ	$\dfrac{L\log_2 L}{2}$

$$\text{Total} = FL\left[\frac{Q^3 + 15Q^2 + 8Q}{6} + QP^2 + (Q^2 + Q + QP + P)\frac{\log_2 L}{2}\right]$$
$$+ FQ^2\left(H + HE - P\frac{E^2 + E}{2}\right)$$

As before, the block-circulant matrices can be decomposed using the block-Fourier transform as

$$\mathbf{T}_{(PQ)} = \mathbf{F}_{(P)}^{-1}\mathbf{\Lambda}_{(PQ)}\mathbf{F}_{(Q)} \tag{6.121}$$

$$\mathbf{N}_{NQ} = \mathbf{F}_{(Q)}^{-1}\mathbf{\Lambda}_N\mathbf{F}_{(Q)}, \tag{6.122}$$

where

$$\mathbf{\Lambda}_{(PQ)} = \text{diag}_{(PQ)}(\mathbf{F}_{(Q)}\mathbf{T}_{(PQ)}(:,1:Q)) \tag{6.123}$$

$$\mathbf{\Lambda}_N = \text{diag}_{(QQ)}(\mathbf{F}_{(Q)}\mathbf{N}_{NQ}(:,1:Q)), \ \mathbf{\Lambda}_N \in \mathbb{C}^{NQ \times NQ}. \tag{6.124}$$

Replacing Equations 6.121 and 6.122 in Equation 6.120 results in

$$\hat{\mathbf{d}} = \left[\left(\mathbf{F}_{(P)}^{-1}\mathbf{\Lambda}_{(PQ)}\mathbf{F}_{(Q)}\right)^H \mathbf{F}_{(P)}^{-1}\mathbf{\Lambda}_{(PQ)}\mathbf{F}_{(Q)} + \mathbf{F}_{(Q)}^{-1}\mathbf{\Lambda}_N\mathbf{F}_{(Q)}\right]^{-1} \cdot \left(\mathbf{F}_{(P)}^{-1}\mathbf{\Lambda}_{(PQ)}\mathbf{F}_{(Q)}\right)^H \mathbf{e}$$

$$\tag{6.125}$$

Considering that $\mathbf{F}_{(K)}^H = \mathbf{F}_{(K)}^{-1}$, Equation 6.125 can be simplified to

$$\hat{\mathbf{d}} = \mathbf{F}_{(Q)}^{-1}\left(\mathbf{\Lambda}_{(PQ)}^H\mathbf{\Lambda}_{(PQ)} + \mathbf{\Lambda}_N\right)^{-1}\cdot\mathbf{\Lambda}_{(PQ)}^H\mathbf{F}_{(P)}\mathbf{e} \tag{6.126}$$

Since $\mathbf{N}_{NQ}$ is a diagonal matrix, and from the definition of the block-Fourier matrix (6.88), it can be shown that

$$\text{diag}_{(QQ)}(\mathbf{F}_{(Q)}\mathbf{N}_{NQ}(:,1:Q)) = \mathbf{N}_{NQ} \tag{6.127}$$

This is true only when the $\mathbf{N}_{NQ}$ matrix is a block-diagonal matrix composed of N equal blocks of size Q.

Now, Equation 6.126 can be rewritten as

$$\hat{\mathbf{d}} = \mathbf{F}_{(Q)}^{-1}\left(\mathbf{\Lambda}_{(PQ)}^H\mathbf{\Lambda}_{(PQ)} + \sigma^2\mathbf{I}_{NQ}\right)^{-1}\cdot\mathbf{\Lambda}_{(PQ)}^H\mathbf{F}_{(P)}\mathbf{e} \tag{6.128}$$

This equation can be used in the derivation of new algorithms based on the work already done for the ZF detector. The extension is very obvious since $\sigma^2\mathbf{I}_{NQ}$ is a real positive diagonal matrix, which added to $\mathbf{\Lambda}_{(PQ)}^H\mathbf{\Lambda}_{(PQ)}$ results in a block-diagonal matrix whose blocks remain positive-definite Hermitian. This means that they can be inverted with the Cholesky algorithm as done for the ZF detector. The MMSE algorithms require more NQ real floating point additions in the nonpartitioned case, more LQ real floating point additions in the constant-channel partitioned case, and more FLQ real floating point additions in the partitioned unsteady channel case. The resulting algorithms and respective computational requirements are presented from Figures 6.38 through 6.42.

6.3.5.8 Zero-Forcing Results

One hundred different **T** matrices were created for a Pedestrian A without interferers. The simulator created 100 random data vectors and used each of them with each of the **T** matrices to create the corresponding **e** vectors from $e = Td$. Then all the algorithms were applied in turn to estimate **d** from each **e** vector. Finally, the estimated and original **d** vectors were compared and the wrong bits count up. The large number of bits used was necessary to reach valid statistical results. This procedure was repeated for different velocities and antennas configuration as presented in Table 6.14.

Tables 6.15 and 6.16 present the BER results for the different algorithms. Note that, for the stationary case, no errors would be obtained. Simulations for the $v = 0$ km/h (stationary situation) were also run, but the results were omitted from the tables since there were no errors with any of the algorithms. The "Est"

for $k = 1$ *to* P

$$\begin{cases} \mathbf{T}_f\left(k{:}P{:}(N-1)P + k, 1{:}Q\right) \leftarrow \text{FFT}\left(T\left(k{:}P{:}(N-1)P + k, 1{:}Q\right)\right) \\ \mathbf{e}_f\left(k{:}P{:}(N-1)P + k, 1\right) \leftarrow \text{FFT}\left(e\left(k{:}P{:}(N-1)P + k\right)\right) \end{cases}$$

$E \leftarrow \lceil H/P \rceil - 1$

$$\mathbf{S}(1{:}Q, 1{:}Q) \leftarrow \left(\mathbf{T}\left(1{:}H, 1{:}Q\right)\right)^H \mathbf{T}\left(1{:}H, 1{:}Q\right)$$

for $k = 1$ *to* E

$$\begin{cases} \mathbf{S}_k \leftarrow \left(\mathbf{T}\left(1{:}H - kP, 1{:}Q\right)\right)^H \mathbf{T}\left(kP + 1{:}H, 1{:}Q\right) \\ \mathbf{S}\left(kQ + 1{:}(k+1)Q, 1{:}Q\right) \leftarrow \mathbf{S}_k \\ \mathbf{S}\left((N-k)Q + 1{:}(N-k+1)Q, 1{:}Q\right) \leftarrow \mathbf{S}_k^H \end{cases}$$

for $k = 1$ *to* Q

$$\mathbf{S}_f\left(k{:}Q(N-1)Q + k, 1{:}Q\right) \leftarrow \text{FFT}\left(\mathbf{S}\left(k{:}Q(N-1)Q + k, 1{:}Q\right)\right)$$

for $k = 1$ *to* N

$$\begin{cases} \mathbf{\Sigma}_k \leftarrow \mathbf{S}_f\left((k-1)Q + 1{:}kQ, 1{:}Q\right) + \sigma^2 \mathbf{I}_Q \\ \mathbf{R}_k \leftarrow \text{chol}(\mathbf{\Sigma}_k) \\ \mathbf{R}_k^{-1} \leftarrow \text{TriangInv}(\mathbf{R}_k) \\ \mathbf{d}_f\left((k-1)Q + 1{:}kQ, 1\right) \leftarrow \mathbf{R}_k^{-1}\left((\mathbf{R}_k^{-1})^H \left((\mathbf{T}_f((k-1)P + 1{:}kP, 1{:}Q))^H\right)\right. \\ \quad \left. \times \mathbf{e}_f((k-1)P + 1{:}kP)\right) \end{cases}$$

for $k = 1$ *to* Q

$$\mathbf{d}\left(k{:}Q(N-1)Q + k, 1\right) \leftarrow \text{IFFT}\left(\mathbf{d}_f\left(k{:}Q(N-1)Q + k\right)\right)$$

Figure 6.38 MMSE block-Fourier partitioned algorithm for steady channels—method 1.

algorithm represents the "exact" solution of the ZF equation, that is, its solution with an algorithm that does not include extra approximations further than the numeric floating point precision of the simulator. It can represent the Gauss or Cholesky algorithms as well as their optimized versions. As expected, the "Est" algorithm always estimated the correct data vector, since no noise was considered in the simulation.

The "Fourier1" and "Fourier2" algorithms correspond to the two versions of the unpartitioned Fourier algorithm described earlier. The "Fourier1m" and "Fourier2m" are the same algorithms but using the middle block of the **T** matrix,

$$\text{function } \mathbf{d} = BFPC2On\left(\mathbf{T}\left(1:LP,1:Q\right),\mathbf{e},n,Q,P,H,L,l^{-},l^{+},\sigma^{2}\right)$$

$$N = n + \left\lceil H/P\right\rceil - 1$$
$$l = L - l^{-} - l^{+}$$
$$F = \left\lceil N/l\right\rceil$$
$$E \leftarrow \left\lceil H/P\right\rceil - 1$$

$$\mathbf{e}' = \left[\mathbf{0}_{(Pl^{-})};\mathbf{e};\mathbf{0}_{((Fl+l^{+}-n+1)P-H)}\right]$$
$$\text{for } k = 1 \text{ to } P$$
$$\mathbf{T}_{f}\left(k{:}P{:}(L-1)P+k,1{:}Q\right) \leftarrow \text{FFT}\left(\mathbf{T}\left(k{:}P{:}(L-1)P+k,1{:}Q\right)\right)$$
$$\mathbf{S}(1{:}Q,1{:}Q) \leftarrow \left(\mathbf{T}\left(1{:}H,1{:}Q\right)\right)^{H}\mathbf{T}\left(1{:}H,1{:}Q\right)$$
$$\text{for } k = 1 \text{ to } E$$
$$\left|\begin{array}{l}\mathbf{S}_{k} \leftarrow \left(\mathbf{T}\left(1{:}H-kP,1{:}Q\right)\right)^{H}\mathbf{T}\left(kP+1{:}H,1{:}Q\right)\\ \mathbf{S}\left(kQ+1{:}(k+1)Q,1{:}Q\right) \leftarrow \mathbf{S}_{k}\\ \mathbf{S}\left((L-k)Q+1{:}(L-k+1)Q,1{:}Q\right) \leftarrow \mathbf{S}_{k}^{H}\end{array}\right.$$
$$\text{for } k = 1 \text{ to } Q$$
$$\mathbf{S}_{f}\left(k{:}Q(L-1)Q+k,1{:}Q\right) \leftarrow \text{FFT}\left(\mathbf{S}\left(k{:}Q(L-1)Q+k,1{:}Q\right)\right)$$
$$\text{for } k = 1 \text{ to } L$$
$$\left|\begin{array}{l}\mathbf{\Sigma}_{k} \leftarrow \mathbf{S}_{f}\left((k-1)Q+1{:}kQ,1{:}Q\right) + \sigma^{2}\mathbf{I}_{Q}\\ \mathbf{R}_{k} \leftarrow \text{chol}(\mathbf{\Sigma}_{k})\\ \mathbf{R}_{k}^{-1} \leftarrow \text{TriangInv}(\mathbf{R}_{k})\end{array}\right.$$
$$\text{for } k = 1 \text{ to } F$$
$$\left|\begin{array}{l}\text{for } k = 1 \text{ to } P\\ \mathbf{e}_{f}\left(k{:}P{:}(L-1)P+k,1\right) \leftarrow \text{FFT}\left(\mathbf{e}'((f-1)lP+k:P:(fl+l^{-}+l^{+}-1)P+k)\right)\\ \text{for } k = 1 \text{ to } L\\ \mathbf{d}_{f}\left((k-1)Q+1{:}kQ,1\right) \leftarrow \mathbf{R}_{k}^{-1}\left(\left(\mathbf{R}_{k}^{-1}\right)^{H}\left(\left(\mathbf{T}_{f}\left((k-1)P+1{:}kP,1{:}Q\right)\right)^{H}\right)\right.\\ \qquad \times \mathbf{e}_{f}\left((k-1)P+1{:}kP\right)\right)\\ \text{for } k = 1 \text{ to } Q\\ \mathbf{d}_{l}\left(k{:}Q(L-1)Q+k,1\right) \leftarrow \text{IFFT}\left(\mathbf{d}_{f}\left(k{:}Q(L-1)Q+k\right)\right)\\ \mathbf{d}'\left((f-1)lQ+1{:}flQ\right) \leftarrow \mathbf{d}_{l}\left(l^{-}Q+1{:}(l^{-}+l)Q\right)\end{array}\right.$$
$$\mathbf{d} = \mathbf{d}'_{(1{:}nQ)}$$

Figure 6.39 **MMSE block-Fourier partitioned algorithm for steady channels—method 2.**

function $\mathbf{d} = BFP1n(\mathbf{T}, \mathbf{e}, n, Q, P, H, L, l^-, l^+, \sigma^2)$

$N = n + \left[H/P\right] - 1$

$l = L - l^- - l^+$

$F = \left[N/l\right]$

$E \leftarrow \left[H/P\right] - 1$

$m \leftarrow \left[l/2\right]$

$e' = \left[\mathbf{0}_{(Pl^-)}; \mathbf{e}; \mathbf{0}_{\left((Fl + l^+ - n + 1)P - H\right)}\right]$

for $f = 1$ *to* F

   for $k = 1$ *to* P

     $\mathbf{T}_f\left(k{:}P{:}(L-1)P + k, 1{:}Q\right) \leftarrow \text{FFT}\left(\mathbf{T}(mP + k{:}P{:}(m + L - 1)P + k, \, mQ + 1{:}(m+1)Q)\right)$

     $\mathbf{e}_f\left(k{:}P{:}(L-1)P + k, 1\right) \leftarrow \text{FFT}\left(e'((f-1)lP + k{:}P{:}(fl + l^- + l^+ - 1)P + k)\right)$

   for $k = 1$ *to* L

     $\boldsymbol{\Sigma}_k \leftarrow \left(\mathbf{T}_f\left((k-1)P + 1{:}kP, 1{:}Q\right)\right)^H \mathbf{T}_f\left((k-1)P + 1{:}kP, 1{:}Q\right) + \sigma^2\mathbf{I}_Q$

     $\mathbf{R}_k \leftarrow \text{chol}(\boldsymbol{\Sigma}_k)$

     $\mathbf{R}_k^{-1} \leftarrow \text{TriangInv}(\mathbf{R}_k)$

     $\mathbf{d}_f\left((k-1)Q + 1{:}kQ, 1\right) \leftarrow \mathbf{R}_k^{-1}\left(\left(\mathbf{R}_k^{-1}\right)^H\left(\left(\mathbf{T}_f\left((k-1)P + 1{:}kP, 1{:}Q\right)\right)^H\right)\right)$

     $\text{xef}\left((k-1)P + 1{:}kP\right)\big)\big)$

   for $k = 1$ *to* Q

   $\mathbf{d}_l\left(k{:}Q(L-1)Q + k, 1\right) \leftarrow \text{IFFT}\left(\mathbf{d}_f\left(k{:}Q(L-1)Q + k\right)\right)$

   $\mathbf{d}'((f-1)lQ + 1{:}flQ) \leftarrow \mathbf{d}_l\left(l^-Q + 1{:}(l^- + l)Q\right)$

   $m \leftarrow \min(m + l, n - 1)$

$\mathbf{d} = \mathbf{d}'(1{:}nQ)$

Figure 6.40   MMSE block-Fourier partitioned algorithm for unsteady channels—method 1.

---

while "Fourier1" and "Fourier2" use the first block. As can be seen in the tables, both versions (1 and 2) have identical results, since they are numerically equivalent. Also, as expected, the use of the middle block gives the best results except for low speeds, where both methods are equivalent.

"FourierP2c" refers to the partitioned Fourier algorithm for constant channels based in the second version of the basic Fourier algorithm, while "FourierP2v" is the corresponding unsteady channel version. As expected, "FourierP2v" always gives better results, except for very low speeds, where "FourierP2c" can give equivalent results with less floating point operations. The numbers after "FourierP2c" or

function $\mathbf{d} = BFP2On\left(\mathbf{T}, \mathbf{e}, n, Q, P, H, L, l^-, l^+, \sigma^2\right)$

$N = n + \lceil H/P \rceil - 1$

$l = L - l^- - l^+$

$F = \lceil N/l \rceil$

$E \leftarrow \lceil H/P \rceil - 1$

$m \leftarrow \lceil l/2 \rceil$

$\mathbf{e}' = \left[\mathbf{0}_{(Pl^-)}; \mathbf{e}; \mathbf{0}_{((Fl+l^+-n+1)P-H)}\right]$

for $f = 1$ *to* F

   for $k = 1$ *to* P

      $\mathbf{T}_f\left(k{:}P{:}(L-1)P+k, 1{:}Q\right) \leftarrow \mathrm{FFT}\left(\mathbf{T}(mP+k{:}P{:}(m+L-1)P+k, mQ+1{:}(m+1)Q)\right)$

      $\mathbf{e}_f\left(k{:}P{:}(L-1)P+k, 1\right) \leftarrow \mathrm{FFT}\left(\mathbf{e}'((f-1)lP+k{:}P{:}(fl+l^-+l^+-1)P+k)\right)$

   $\mathbf{S}(1{:}Q, 1{:}Q) \leftarrow \left(\mathbf{T}\big(mP+1{:}mP+H, mQ+1{:}(m+1)Q\big)\right)^H \mathbf{T}$

      $\times\big(mP+1{:}mP+H, mQ+1{:}(m+1)Q\big)$

   for $k = 1$ *to* E

      $\mathbf{S}_k \leftarrow \left(\mathbf{T}\big(mP+1{:}(m-k)P+H, mQ+1{:}(m+1)Q\big)\right)^H \mathbf{T}$

      $\times\big((m+k)P+1{:}mP+H, mQ+1{:}(m+1)Q\big)$

      $\mathbf{S}\big(kQ+1{:}(k+1)Q, 1{:}Q\big) \leftarrow \mathbf{S}_k$

      $\mathbf{S}\big((L-k)Q+1{:}(L-k+1)Q, 1{:}Q\big) \leftarrow \mathbf{S}_k^H$

   for $k = 1$ *to* Q

   $\mathbf{S}_f\left(k{:}Q(L-1)Q+k, 1{:}Q\right) \leftarrow \mathrm{FFT}\left(\mathbf{S}\big(k{:}Q(L-1)Q+k, 1{:}Q\big)\right)$

   for $k = 1$ *to* L

      $\boldsymbol{\Sigma}_k \leftarrow \mathbf{S}_f\big((k-1)Q+1{:}kQ, 1{:}Q\big) + \sigma^2\mathbf{I}_Q$

      $\mathbf{R}_k \leftarrow \mathrm{chol}(\boldsymbol{\Sigma}_k)$

      $\mathbf{R}_k^{-1} \leftarrow \mathrm{TriangInv}(\mathbf{R}_k)$

      $\mathbf{d}_f\big((k-1)Q+1{:}kQ, 1\big) \leftarrow \mathbf{R}_k^{-1}\left[\left(\mathbf{R}_k^{-1}\right)^H\left(\left(\mathbf{T}_f\big((k-1)P+1{:}kP, 1{:}Q\big)\right)^H\right)\right.$

      $\left.\mathbf{e}_f\big((k-1)P+1{:}kP\big)\right)$

   for $k = 1$ *to* P

   $\mathbf{d}_l\big(k{:}Q(L-1)Q+k, 1\big) \leftarrow \mathrm{IFFT}\big(\mathbf{d}_f\big(k{:}Q(L-1)Q+k\big)\big)$

   $\mathbf{d}'\big((f-1)lQ+1{:}flQ\big) \leftarrow \mathbf{d}\big(l^-Q+1{:}(l^-+l)Q\big)$

   $m \leftarrow \min(m+l, n-1)$

$\mathbf{d} = \mathbf{d}'(1{:}nQ)$

Figure 6.41 **MMSE block-Fourier partitioned algorithm for unsteady channels—
method 2.**

$N\left[\dfrac{Q^3 + 15Q^2 + 8Q}{6} + QP^2 + (Q^2 + Q + QP + P)\dfrac{\log_2 N}{2}\right]$ $+ Q^2\left(H + HE - P\dfrac{E^2 + E}{2}\right)$	(a)
$L\left[\dfrac{Q^3 + 9Q^2 + 8Q}{6} + F(Q^2 + Q + QP^2) + (Q^2 + QF + QP + FP)\dfrac{\log_2 L}{2}\right]$ $Q^2\left(H + HE - P\dfrac{E^2 + E}{2}\right)$	(b)
$FL\left[\dfrac{Q^3 + 15Q^2 + 14Q}{6} + PQ^2 + QP^2 + (Q + QP + P)\dfrac{\log_2 L}{2}\right]$	(c)
$FL\left[\dfrac{Q^3 + 15Q^2 + 14Q}{6} + QP^2 + (Q^2 + Q + QP + P)\dfrac{\log_2 L}{2}\right]$ $+ FQ^2\left(H + HE - P\dfrac{E^2 + E}{2}\right)$	(d)

Figure 6.42   Number of operations required by the algorithm of (a) Figure 6.38, (b) Figure 6.39, (c) Figure 6.40, and (d) Figure 6.41.

Table 6.14   Simulated Conditions for the ZF Detector

Channel	Pedestrain A
Interferers	0
Number of antennas	1; 2; 3; 4
Velocity	0; 1; 10; 100 (km/h)

"FourierP2v" indicate the number of blocks used in each partition and the prelap and postlap blocks number. For example, "FourierP2v_008_002_002" refers to an eight-block partition with two blocks prelap and two blocks postlap algorithm. From Tables 6.15 and 6.16, it can be concluded that it is advantageous to reduce the size of partitions, especially for high speeds (as expected, since in each partition, the channel is approximated as constant). It is also easy to see that, for a particular partition size, better results are attained as larger laps are used. For low speeds, the size of the partitions does not have a very strong influence on the correctness of the estimation, but greater lap sizes are also advantageous as noticed for high speeds. The conjunction of these two factors makes it hard to find the best algorithm for high speeds, since small partitions cannot have large overlaps. In a similar way,

Table 6.15 BER Performance for the ZF Using 1 and 2 Antennas

Interferers	0				0		
Channel Model Environment	*Pedestrian A*				*Pedestrian A*		
Antennas	1				2		
Number of bits (Mbits)	10.24				20.48		
Velocity (km/h)	1	10	100		1	10	100
Est	0.0E + 0	0.0E + 0	0.0E + 0		0.0E + 0	0.0E + 0	0.0E + 0
Fourierl	5.6E–4	9.8E–3	2.7E–1		0.0E + 0	4.5E–3	2.6E–1
Fourierlm	1.7E–4	7.5E–4	9.5E–2		0.0E + 0	6.2E–4	1.2E–1
Fourier2	5.6E–4	9.8E–3	2.7E–1		0.0E + 0	4.5E–3	2.6E–1
Fourier2 m	1.7E–4	7.5E–4	9.5E–2		0.0E + 0	6.2E–4	1.2E–1
FourierP2c_128_000_000	1.8E–4	7.6E–4	9.5E–2		0.0E + 0	6.4E–4	1.2E–1
FourierP2c_128_002_002	1.7E–4	7.5E–4	9.5E–2		0.0E + 0	6.2E–4	1.2E–1
FourierP2c_128_004_004	1.7E–4	7.5E–4	9.5E–2		0.0E + 0	6.2E–4	1.2E–1
FourierP2c_128_008_008	1.7E–4	7.5E–4	9.5E–2		0.0E + 0	6.2E–4	1.2E–1
FourierP2c_128_016_016	1.7E–4	7.5E–4	9.5E–2		0.0E + 0	6.2E–4	1.2E–1

continued

Table 6.15 (Continued) BER Performance for the ZF Using 1 and 2 Antennas

Interferers	0			0		
Channel Model Environment	Pedestrian A			Pedestrian A		
Antennas	1			2		
FourierP2c_128_032_032	1.7E–4	7.5E–4	9.5E–2	0.0E + 0	6.2E–4	1.2E–1
FourierP2c_064_000_000	1.9E–4	7.7E–4	9.5E–2	0.0E + 0	6.7E–4	1.2E–1
FourierP2c_064_002_002	1.7E–4	7.5E–4	9.5E–2	0.0E + 0	6.2E–4	1.2E–1
FourierP2c_064_004_004	1.7E–4	7.5E–4	9.5E–2	0.0E + 0	6.2E–4	1.2E–1
FourierP2c_064_008_008	1.7E–4	7.5E–4	9.5E–2	0.0E + 0	6.2E–4	1.2E–1
FourierP2c_064_016_016	1.7E–4	7.5E–4	9.5E–2	0.0E + 0	6.2E–4	1.2E–1
FourierP2c_032_000_000	2.1E–4	8.0E–4	9.5E–2	0.0E + 0	7.1E–4	1.2E–1
FourierP2c_032_002_002	1.7E–4	7.5E–4	9.5E–2	0.0E + 0	6.2E–4	1.2E–1
FourierP2c_032_004_004	1.7E–4	7.5E–4	9.5E–2	0.0E + 0	6.2E–4	1.2E–1
FourierP2c_032_008_008	1.7E–4	7.5E–4	9.5E–2	0.0E + 0	6.2E–4	1.2E–1
FourierP2c_016_000_000	2.4E–4	8.5E–4	9.5E–2	0.0E + 0	8.2E–4	1.2E–1
FourierP2c_016_002_002	1.8E–4	7.5E–4	9.5E–2	0.0E + 0	6.4E–4	1.2E–1
FourierP2c_016_004_000	1.7E–4	7.5E–4	9.5E–2	0.0E + 0	6.2E–4	1.2E–1
FourierP2c_008_000_002	3.1E–4	9.5E–4	9.5E–2	0.0E + 0	1.1E–3	1.2E–1

FourierP2c_008_002_002	2.0E-4	7.5E-4	9.5E-2	0.0E + 0	7.3E-4	1.2E-1
FourierP2c_004_000_002	5.4E-4	1.2E-3	9.5E-2	0.0E + 0	1.7E-3	1.2E-1
FourierP2c_004_001_001	3.6E-4	7.9E-4	9.5E-2	0.0E + 0	1.1E-3	1.2E-1
FourierP2v_128_000_000	7.6E-5	1.8E-4	6.2E-3	0.0E + 0	1.5E-5	1.0E-2
FourierP2v_128_002_002	6.5E-5	1.7E-4	6.0E-3	0.0E + 0	2.2E-6	8.8E-3
FourierP2v_128_004_004	5.9E-5	1.8E-4	6.1E-3	0.0E + 0	3.2E-6	8.3E-3
FourierP2v_128_008_008	5.0E-5	2.2E-4	4.8E-3	0.0E + 0	6.8E-6	6.3E-3
FourierP2v_128_016_016	3.1E-5	1.5E-4	2.9E-3	0.0E + 0	1.5E-5	4.3E-3
FourierP2v_128_032_032	1.9E-5	7.2E-5	9.3E-4	0.0E + 0	9.8E-7	1.7E-3
FourierP2v_064_000_000	2.2E-5	8.4E-5	9.6E-4	0.0E + 0	2.3E-5	1.7E-3
FourierP2v_064_002_002	3.0E-5	4.0E-5	8.3E-4	0.0E + 0	3.8E-6	1.5E-3
FourierP2v_064_004_004	3.8E-5	3.2E-5	7.7E-4	0.0E + 0	8.7E-6	1.3E-3
FourierP2v_064_008_008	5.5E-5	6.4E-5	5.4E-4	0.0E + 0	1.2E-5	9.9E-4
FourierP2v_064_016_016	1.9E-5	5.9E-5	1.2E-4	0.0E + 0	9.8E-7	5.9E-4
FourierP2v_032_000_000	2.2E-5	6.6E-5	1.7E-4	0.0E + 0	4.5E-5	6.6E-4
FourierP2v_032_002_002	1.5E-5	3.2E-5	1.2E-4	0.0E + 0	2.5E-6	5.0E-4
FourierP2v_032_004_004	0.0E + 0	2.9E-5	8.8E-5	0.0E + 0	4.0E-6	4.0E-4

continued

Table 6.15 (Continued) BER Performance for the ZF Using 1 and 2 Antennas

Interferers	0			0		
Channel Model Environment	Pedestrian A			Pedestrian A		
Antennas	1			2		
FourierP2v_032_008_008	1.9E–5	2.4E–5	8.9E–5	0.0E+0	3.5E–6	1.2E–4
FourierP2v_016_000_000	2.3E–5	3.8E–5	1.4E–4	0.0E+0	8.9E–5	2.7E–4
FourierP2v_016_002_002	0.0E+0	1.4E–5	6.1E–5	0.0E+0	1.0E–6	2.1E–4
FourierP2v_016_004_000	0.0E+0	1.1E–5	2.2E–5	0.0E+0	4.9E–8	2.0E–4
FourierP2v_008_000_002	2.2E–6	1.6E–5	7.8E–5	0.0E+0	1.6E–4	3.7E–4
FourierP2v_008_002_002	0.0E+0	2.3E–6	1.3E–5	0.0E+0	3.4E–7	1.5E–4
FourierP2v_004_000_002	2.3E–6	6.4E–6	9.1E–5	0.0E+0	2.7E–4	4.1E–4
FourierP2v_004_001_001	0.0E+0	0.0E+0	2.1E–5	0.0E+0	3.9E–7	1.8E–4

Table 6.16 BER Performance for the ZF Using Three and Four Antennas

Interferers	0			0		
Channel Model Environment	*Pedestrian A*			*Pedestrian A*		
Antennas	3			4		
Number of bits (Mbits)	30.78			40.96		
Velocity (km/h)	1	10	100	1	10	100
Est	0.0E+0	0.0E+0	0.0E+0	0.0E+0	0.0E+0	0.0E+0
Fourierl	0.0E+0	1.4E−3	2.9E−1	0.0E+0	6.9E−3	2.9E−1
Fourierl m	0.0E+0	1.2E−4	1.5E−1	0.0E+0	9.4E−4	1.6E−1
Fourier2	0.0E+0	1.4E−3	2.9E−1	0.0E+0	6.9E−3	2.9E−1
Fourier2 m	0.0E+0	1.2E−4	1.5E−1	0.0E+0	9.4E−4	1.6E−1
FourierP2c_128_000_000	0.0E+0	1.2E−4	1.5E−1	0.0E+0	9.6E−4	1.6E−1
FourierP2c_128_002_002	0.0E+0	1.2E−4	1.5E−1	0.0E+0	9.4E−4	1.6E−1
FourierP2c_128_004_004	0.0E+0	1.2E−4	1.5E−1	0.0E+0	9.4E−4	1.6E−1
FourierP2c_128_008_008	0.0E+0	1.2E−4	1.5E−1	0.0E+0	9.4E−4	1.6E−1
FourierP2c_128_016_016	0.0E+0	1.2E−4	1.5E−1	0.0E+0	9.4E−4	1.6E−1

continued

Table 6.16 (Continued) BER Performance for the ZF Using Three and Four Antennas

Interferers	0			0		
Channel Model Environment	Pedestrian A			Pedestrian A		
Antennas	3			4		
Number of bits (Mbits)	30.78			40.96		
FourierP2c_128_032_032	0.0E+0	1.2E-4	1.5E-1	0.0E+0	9.4E-4	1.6E-1
FourierP2c_064_000_000	0.0E+0	1.3E-4	1.5E-1	0.0E+0	9.8E-4	1.6E-1
FourierP2c_064_002_002	0.0E+0	1.2E-4	1.5E-1	0.0E+0	9.4E-4	1.6E-1
FourierP2c_064_004_004	0.0E+0	1.2E-4	1.5E-1	0.0E+0	9.4E-4	1.6E-1
FourierP2c_064_008_008	0.0E+0	1.2E-4	1.5E-1	0.0E+0	9.4E-4	1.6E-1
FourierP2c_064_016_016	0.0E+0	1.2E-4	1.5E-1	0.0E+0	9.4E-4	1.6E-1
FourierP2c_032_000_000	0.0E+0	1.5E-4	1.5E-1	0.0E+0	1.0E-3	1.6E-1
FourierP2c_032_002_002	0.0E+0	1.2E-4	1.5E-1	0.0E+0	9.4E-4	1.6E-1
FourierP2c_032_004_004	0.0E+0	1.2E-4	1.5E-1	0.0E+0	9.4E-4	1.6E-1
FourierP2c_032_008_008	0.0E+0	1.2E-4	1.5E-1	0.0E+0	9.4E-4	1.6E-1
FourierP2c_016_000_000	0.0E+0	1.9E-4	1.5E-1	4.9E-8	1.1E-3	1.6E-1
FourierP2c_016_002_002	0.0E+0	1.2E-4	1.5E-1	0.0E+0	9.5E-4	1.6E-1
FourierP2c_016_004_000	0.0E+0	1.2E-4	1.5E-1	0.0E+0	9.4E-4	1.6E-1

FourierP2c_008_000_002	0.0E+0	2.7E-4	1.5E-1	0.0E+0	1.3E-3	1.6E-1
FourierP2c_008_002_002	0.0E+0	1.2E-4	1.5E-1	0.0E+0	1.0E-3	1.6E-1
FourierP2c_004_000_002	0.0E+0	4.4E-4	1.5E-1	0.0E+0	1.8E-3	1.6E-1
FourierP2c_004_001_001	0.0E+0	1.8E-4	1.5E-1	0.0E+0	1.3E-3	1.6E-1
FourierP2v_128_000_000	0.0E+0	5.8E-6	1.1E-2	0.0E+0	8.1E-5	1.3E-2
FourierP2v_128_002_002	0.0E+0	0.0E+0	1.1E-2	0.0E+0	8.8E-5	1.2E-2
FourierP2v_128_004_004	0.0E+0	9.7E-8	1.0E-2	0.0E+0	8.8E-5	1.0E-2
FourierP2v_128_008_008	0.0E+0	1.6E-7	7.9E-3	0.0E+0	6.9E-5	9.0E-3
FourierP2v_128_016_016	0.0E+0	0.0E+0	5.4E-3	0.0E+0	5.0E-5	5.4E-3
FourierP2v_128_032_032	0.0E+0	0.0E+0	1.7E-3	0.0E+0	2.9E-5	1.6E-3
FourierP2v_064_000_000	0.0E+0	9.1E-6	1.8E-3	0.0E+0	4.3E-5	1.7E-3
FourierP2v_064_002_002	0.0E+0	9.7E-8	1.1E-3	0.0E+0	2.3E-5	1.4E-3
FourierP2v_064_004_004	0.0E+0	0.0E+0	9.5E-4	0.0E+0	2.5E-5	1.1E-3
FourierP2v_064_008_008	0.0E+0	0.0E+0	6.5E-4	0.0E+0	2.2E-5	7.0E-4
FourierP2v_064_016_016	0.0E+0	0.0E+0	2.6E-4	0.0E+0	1.2E-5	2.3E-4
FourierP2v_032_000_000	0.0E+0	1.9E-5	3.3E-4	0.0E+0	3.7E-5	3.2E-4

continued

Table 6.16 (Continued) BER Performance for the ZF Using Three and Four Antennas

Interferers	0		0	
Channel Model Environment	*Pedestrian A*		*Pedestrian A*	
Antennas	3		4	
Number of bits (Mbits)	30.78		40.96	
FourierP2v_032_002_002	0.0E + 0	1.9E–4	1.0E–5	1.4E–4
FourierP2v_032_004_004	0.0E + 0	1.0E–4	1.1E–5	1.1E–4
FourierP2v_032_008_008	0.0E + 0	1.1E–4	6.5E–6	7.6E–5
FourierP2v_016_000_000	3.9E–5	2.0E–4	4.3E–5	1.9E–4
FourierP2v_016_002_002	3.2E–8	9.4E–5	5.3E–6	7.8E–5
FourierP2v_016_004_000	0.0E + 0	5.2E–5	2.7E–6	4.8E–5
FourierP2v_008_000_002	7.7E–5	1.8E–4	6.5E–5	1.8E–4
FourierP2v_008_002_002	0.0E + 0	7.0E–5	1.5E–6	6.1E–5
FourierP2v_004_000_002	1.3E–4	2.5E–4	9.8E–5	2.6E–4
FourierP2v_004_001_001	6.2E–7	9.5E–5	2.4E–8	7.9E–5

for low speeds, large overlaps imply very large partition and a compromise has to be made in each situation. Nevertheless, it is clear that for high speeds, the most important factor is the size of the partitions, while for low speeds, the overlap size is the key factor.

6.3.5.9 LMMSE Results

A Pedestrian A channel with $v = 100$ km/h, $k = 16$, and a MIMO 2×2 arrangement was considered, and the performance results for the BER and BLER are portrayed in Figure 6.43. The "standard" method in the figures refers to estimation obtained by solving the LMMSE equation with the standard Gauss algorithm. The legend "EstPr_xx_yy_yy" represents the results of the application of an "exact" partitioned algorithm (Gauss or Cholesky, for example), with partitions of size xx and prelaps and postlaps of size yy.

As expected, the constant-channel approximations are useless for a $v = 100$ km/h situation. This is evident from the two block flourier simulations (curves above 10^{-1} for the BER, and of 1 for the BLER) for constant channels run. The partitioned algorithms revealed to be very effective, since their solutions are only distinguishable from the standard Gauss algorithm solution for high E_b/N_0 values. Note also that the Fourier-partitioned algorithms for variable channels are only slightly worse than the corresponding partitioned "Est" algorithms.

The "FourierP2vr_04_01_01" and the "EstPr_32_04_04" methods provide almost identical results as the full Gauss method, while the "FourierP2vr_32_04_04" and the "EstPr_32_00_00" provide worse results only for high E_b/N_0 values.

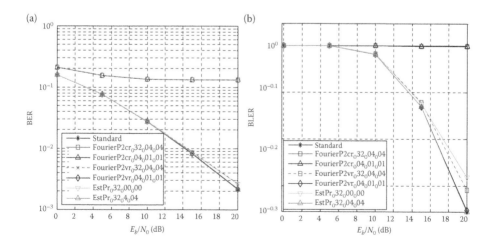

Figure 6.43 BER (a) and BLER (b) performance results for MMSE, under the Pedestrian A channel with $v = 100$ km/h, $k = 16$, and MIMO 2×2.

6.3.6 Final Remarks

In this section, optimized versions of the Gauss and Cholesky algorithms were presented, for usage for the solution of a ZF or LMMSE detector in MIMO/BLAST systems. The optimizations were based simply on the removal of the unnecessary operations regarding the structure of the involved matrices. However, the simulation of an MMSE receiver for a real-life MIMO setting (e.g., a standard-sized block for a specific channel quality indicator (CQI) in the UMTS) would be very complicated to perform without these optimizations.

The block-Fourier algorithms presented in [Vollmer et al. 2001] and [Machauer et al. 2001] for the ZF algorithm under constant-channel conditions were also tested under unsteady channel situations, having revealed to be useless in conditions of medium or high speeds. New versions of those algorithms, capable of dealing with detection in unsteady channels with speeds until $v = 100$ km/h, were derived and tested. These new algorithms were based on the partitioned block-Fourier algorithms of earlier works [Vollmer et al. 2001; Machauer et al. 2001], but extra steps were added to take into consideration the channel change from partition to partition. Inside each partition, the channel is considered constant. The new algorithms, although more computationally expensive than the original block-Fourier ones, are not as expensive as the Gauss or Cholesky ones (even if optimized versions are considered).

The best algorithm must be selected according to the channel conditions; for almost constant channels, constant block-Fourier algorithms could be used with good results, while for high speeds, the new block-Fourier algorithms proposed must be used, preferably with small-sized partitions. An optimized Gauss or Cholesky approach could also be employed if sufficient computational power becomes available, with the advantage of being a velocity-independent solution. Also, the benefit of parallel processing that can be exploited with the block-Fourier algorithms can be extended to the Gauss and Cholesky methods (at least partially) with the introduction of partitioning.

6.4 Performance Results of the Equalization Techniques

Both the MF and the ZF/MMSE schemes were simulated. Since the ZF/MMSE results are based on the equalization of the MF results, it is interesting to compare the three schemes. Simulations were run considering block sizes of 1024 bits, and the reference UMTS channels. A MIMO setting with an equal number of transmitters and receivers was implemented. An *SF* of 16 was used (hence 16 physical channels per transmit antenna). Minimum (usage of 1 physical channel per transmit antenna) and maximum (usage of 16 physical channels per transmit antenna) loading conditions were considered.

6.4.1 MF Results

The MF results (Figure 6.44) are important for the MMSE receiver, since it is a crucial part of the algorithm. It can be seen that, for the minimum loading case (Figure 6.44a), results of Vehicular A are best and of Pedestrian A are worse, due to multipath diversity. The diagonal of the $\mathbf{E}_M$ was used for normalization of $\mathbf{y}_{MF}$, instead of using just the estimated channel coefficients as is usually done for standard RAKE receivers. The extra information from the $\mathbf{E}_M$ allows minimizing the correlation effect, and thus the multipath diversity can be exploited for higher-order modulations when there is little interference, contrary to the normal RAKE.

Owing to interference from other antennas, and the fact that the simple MF algorithm does not perform any type of interference canceling nor equalization, the lowest MIMO orders provide the best results. For the fully loaded case (Figure 6.44b), results for Pedestrian A are better than the Vehicular A channel, due to the high amount of multipath interference. The lowest MIMO orders still provide the best results, due to the reduced interference.

The uplink transmission of the MF receiver (Figure 6.45) yields worse results, due to the increased interference caused by the correlation of the spreading codes, which are no longer orthogonal. Note that the error floor is now over 10%, making it virtually impossible to operate at full loading with this scheme.

6.4.2 MMSE and ZF Results

Figure 6.46 portrays some BER results for the MMSE receivers. As expected, the best results were obtained for the minimum loading case (Figure 6.46a) of the highest MIMO orders (highest receive diversity). Since the Vehicular A channel has greater number of taps, best results are obtained for this channel (note that perfect channel estimation is assumed). Indoor A is the second-best, since it has a

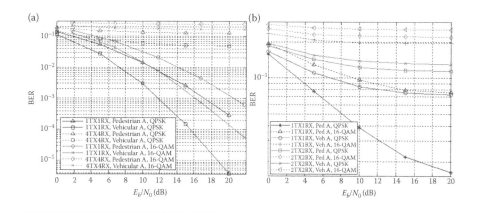

Figure 6.44 DL BER performance for MF, using (a) $k = 1$ and (b) $k = $ SF.

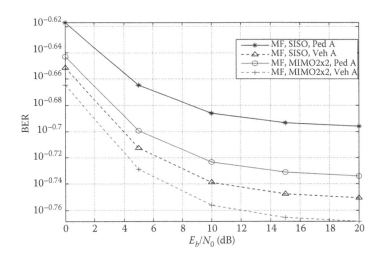

Figure 6.45   UL BER performance for MF, using *k* = SF.

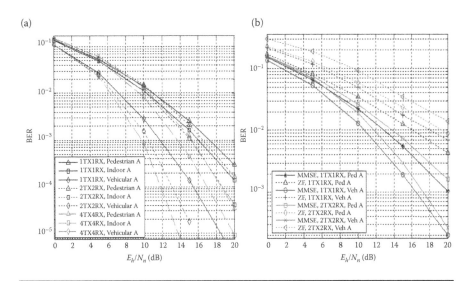

Figure 6.46   BER performance for (a) MMSE using *k* = 1 and SF = 16 and (b) MMSE and ZF, with *k* = SF.

second tap of greater power than the pedestrian A channel, which is predominantly a one-tap channel.

For the fully loaded system (Figures 6.46b and Figure 6.47), it can be seen that the situation is quite different, with the lowest MIMO orders yielding the best results, due to the reduction of interference. Thus, for both modulations, the

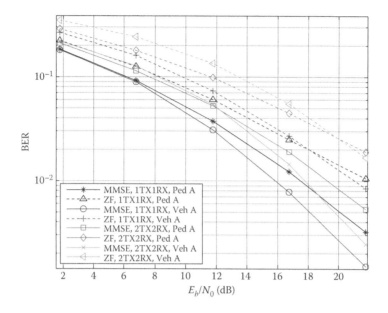

Figure 6.47   BER performance for MMSE and ZF, using 16-QAM and k = SF.

best and worst channel's performance are still Vehicular A and Pedestrian A, for high values of E_b/N_0. For the 16-QAM case with low values of E_b/N_0, the channel's performance order is modified, with the worst being Vehicular A. This is due to the nature of the 16-QAM modulation, which is dependent on the symbol's amplitude, being highly influenced by high levels of multipath interference. Note also that the performance of both Pedestrian A and Indoor A are much closer for the 16-QAM case, due to the modulation's inaptness to exploit multipath when compared to the QPSK modulation.

The MMSE results are much better than for the MF alone, due to the equalization. The only performance curves from both receivers that are closer to each other are for the case with no interferers and SISO, where interference is minimal, caused only by the channel's wideband effect (ISI).

Regarding the ZF results, it can be seen that these are significantly worse that those of the MMSE algorithm, with differences of performance over 10 dB. Noise estimation plays a determining role in the MMSE algorithm, especially in fully loaded systems, where the interference is high.

Some uplink results for both the MMSE and ZF are presented in Figure 6.48. The performance is also worse than for the downlink, due to the added interference between spreading sequences. The ZF results are disastrous (above 10%), augmenting the importance of the estimation of the noise component for the MMSE.

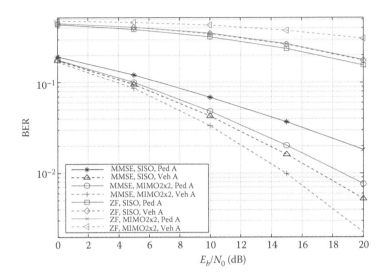

Figure 6.48   UL BER performance for MMSE and ZF, using *k* = SF.

Acknowledgment

This work was supported by the FCT (Fundação para a Ciência e Tecnologia) via project PEst-OE/EEI/LA0008/2013.

References

3GPP Technical Report 25.996 v6.1.0 (2003-09).

Ben-Israel, A. and T. Greville, *Generalized Inverses: Theory and Applications*. New York: Wiley, 1977.

Brunner, C., M. Haardt, and J.A. Nossek, Onspace-time RAKE receiver structures for WCDMA, signals, systems, and computers. *Conference Record of the Thirty-Third Asilomar Conference, Volume 2*, 1999, pp. 1546–1551, doi: 10.1109/ACSSC.1999.832008.

Cavers, J.K. An analysis of pilot symbol assisted modulation for Rayleigh fading channels, *IEEE Transactions on Vehicular Technology*, 40, 686–693, November 1991.

Divsalar, D., M. Simon, and D. Raphaeli, Improved parallel interference cancellation in CDMA, *IEEE Transactions on Communications*, 46, 258–268, February 1998.

Joy, M.P. and V. Tavsanoglu, Circulant matrices and the stability of ring CNNs, *ISCAS 95, IEEE International Symposium on Circuits and Systems*, 1, 501–504, 28 April–3 May 1995.

Kay, S. *Fundamentals of Statistical Signal Processing: Estimation Theory*, Englewood Cliffs, NJ: Prentice-Hall, 1993.

Machauer, R., M. Iurascu, and F. Jondral, FFT speed multiuser detection for high rate data mode in UTRA-FDD, *in Proc. of the IEEE Vehicular Technology Conference*, 2001.

Proakis, J.G. *Digital Communications*, New York: McGraw-Hill, 4th ed., 2001.

Proakis, J.G. and M. Salehi, *Communication Systems Engineering*, Harlow, UK: Prentice Hall, 2nd ed., 2002 (section 10.3).

Shoumin, L. and T. Zhi, Near-optimum soft decision equalization for frequency selective MIMO channels, *IEEE Transactions on Signal Processing*, 52(3), 721–733, March 2004.

Silva, J.C., N. Souto, A. Rodrigues, A. Correia, F. Cercas, and R. Dinis, A L-MMSE DS-CDMA detector for MIMO/BLAST systems with frequency selective fading, *Proceedings of the IEEE IST Mobile & Wireless Communications Summit*, Dresden, Germany, June 19–23, 2005.

Silva, J.C., N. Souto, A. Rodrigues, F. Cercas, and A. Correia, Conversion of reference tapped delay line channel models to discrete time channel models, *Proceedings of the IEEE Vehicular Technology Conference—VTC 2003 Fall*, Orlando, USA, October 6–9, 2003.

Viterbi, J. *CDMA: Principles of Spread Spectrum Communication*, Reading, MA: Addison-Wesley, 1995.

Vollmer, M., M. Haardt, and J. Gotze, Comparative study of joint-detection techniques for TD-CDMA based mobile radio systems, *IEEE Journal on Selected Areas in Communications*, 19(8), August 2001.

Chapter 7

Ultrawideband Systems and MIMO

Vit Sipal, Ben Allen, David J. Edwards, and Wasim Q. Malik

Contents

Ultrawideband (UWB) technology uses extremely wide bandwidth signals with low-power spectral density. This approach changes the wireless communications paradigm fundamentally because the large signal bandwidth leads to the emergence of some interesting and unique channel properties [Allen et al. 2006, Sipal et al. 2012d]. There has been an enormous interest in the technology during the last 20 years because UWB, and specifically UWB multiple-input and multiple-output (MIMO), enables very high data-rate wireless communications, wireless asset localization, and microwave medical imaging, with properties and capabilities unmatched by conventional wireless technologies. The objective of the chapter is to introduce UWB MIMO technology and to provide an overview of its major potential and constraints for these three fields.

7.1 UWB for Wireless Communication: Introduction

As the name suggests, UWB systems are wireless systems with a very large bandwidth. Based on the current generally accepted definition and industry specifications, a wireless system can be described as UWB if it uses an absolute bandwidth larger than 500 MHz, or a relative bandwidth larger than 20% [Allen et al. 2006].

In this section, we discuss the motivation for UWB wireless systems. This is followed by a brief overview of the history of UWB systems and their outlook for the future. Finally, the main modulation schemes and the worldwide regulatory framework for UWB regulation are discussed.

7.1.1 Motivation for UWB

The history of wireless communication started with Hertz's experimental transmission of short "UWB-like" impulses in the 1880s [Allen et al. 2006]. The practical applications of these experiments emerged in 1895 when Marconi demonstrated the transmission of information over a distance of a mile for nonline of sight conditions [Falciasecca et al. 2009]. Since then, the advances in wireless communications technology have been rapid. For wireless engineers, this presents many interesting challenges, of which two key challenges are the following:

- Currently, the radio spectrum suitable for wireless transmission is almost fully allocated; therefore further expansion of the bandwidth for wireless services is limited [Goldsmith et al. 2009].
- Despite all the progress we have made in the past 100 years, the main force behind the increase of data rate in a single link is the increase of bandwidth.

UWB addresses both issues. In terms of the allocated spectrum, UWB uses the same spectrum that is being used by the primary users. However, due to the strict regulations on the effective isotropic-radiated power (EIRP) spectral density, for example [Sipal et al. 2012d], the wideband radiation causes only negligible interference with the primary users. For the primary user, the UWB radiation appears to be white noise, and due to its low power UWB does not significantly deteriorate the performance of the primary user [Allen et al. 2006]. For the UWB user, the primary user represents an interference source which, due to its narrowband nature, corrupts only a small part of the UWB signal so that coexistence is possible [Allen et al. 2006].

In terms of the second statement about bandwidth being the main force for data-rate increase, we use the example of the IEEE802.11 systems and discuss the main forces behind the increase in its capacity in the evolution of the system.

In the IEEE802.11a standard enacted in 1999, the single stream data rate was 54 Mbps and multiple users were accommodated by frequency division multiplex (IEEE802.11). With the IEEE802.11ac standard of 2013, a single stream data rate in the IEEE802.11ac systems is now capable of offering up to 866.7 Mbps (IEEE802.11). In other words, the data rate per single stream increased by a factor of 16 over 14 years.

The contributors toward this development are: bandwidth and MIMO; reduction of the guard interval; increase of the modulation level; and increase of the code rate. The contribution of these to the overall factor of 16 is illustrated in Figure 7.1.

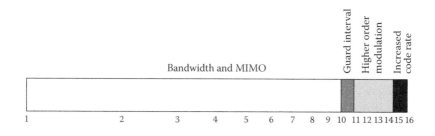

Figure 7.1 Contributions toward the single-user data-rate increase between IEEE802.11a and IEEE802.11ac.

Note that bandwidth and MIMO are the main contributors to the increase in capacity by a factor of 9.9. Multiuser MIMO-enabled users share the same frequency band. As a result, the bandwidth of the orthogonal frequency-division multiplexing (OFDM) symbol is increased from 20 to 160 MHz. This alone gives an increase by a factor of 8. Furthermore, the relative number of pilot and guard subcarriers has been reduced. As a result, the number of data subcarriers in IEEE802.11ac is 476 compared to 48 in IEEE802.11a.

The reduction of the guard interval from 800 to 400 ns means that the symbol repetition rate, and hence the data rate, has increased by a factor of 1.1. The increase of modulation order from 64-QAM to 256-QAM represents a significant challenge for analogue-to-digital conversion but contributes only by a factor of $4/3 = 1.33$. Finally, the increase of code rate from 3/4 to 5/6 has provided a data rate increase by factor of 1.11.

The numbers in Figure 7.1 confirm the importance of bandwidth and MIMO. Multiple-user MIMO enables the reuse of spectral resources by multiple users but one still requires bandwidth to provide data rate in the single data link. In the logarithmic sense, the bandwidth in IEEE802.11ac is closer to UWB (factor 3.125) than to its predecessor IEEE802.11a (factor of 8). This underlines the importance of studying UWB channels and UWB MIMO. On the one hand, the large bandwidth represents a challenge. On the other, the bandwidth offers a higher temporal resolution which enables a better understanding of the underlying physical principles in a wireless channel.

7.1.2 History and Current UWB Standards

The pioneering experiments of Hertz and Marconi employed electric sparks and are often considered as the first UWB experiments [Allen et al. 2006]. Wireless technology, however, soon realized the advantage of heterodyne narrowband transmission and the art of wireless using extremely short pulses was forgotten until the 1960s. In the 1960s, this interest was revived by a US military project mainly due to an interest in radar applications, as radar resolution increases with bandwidth [Allen et al. 2006]. Commercial interest in UWB technology for communications began in the 1990s. However, the real trigger for the intensive research efforts had been marked by the Part 15 ruling of the US FCC that allowed unlicensed UWB radiation subject to strict EIRP requirements [Allen et al. 2006].

Standardization efforts started in 2002, and are sometimes considered to be one of the reasons for the premature commercial failure of the technology [Sipal et al. 2012d]. The high data-rate UWB standard was deliberated upon by the IEEE 802.15.3a task group. However, two opposing industrial alliances, WiMedia alliance and UWB forum, were unable to find a compromise and as a consequence the standardization efforts within the IEEE were stopped in early 2006 [Heidari 2008, Sipal et al. 2012d].

The UWB forum was dismantled soon after Motorola left it in 2006 [Heidari 2008]. The WiMedia alliance brought their proposal forward to the European Association for Standardizing Information and Communication Systems (ECMA), and their multiband OFDM scheme was standardized by the ECMA-368 standard. The ECMA-368 standard foresees data rates ranging from 53.3 to 480 Mbps using OFDM symbols with 128 subcarriers and 528 MHz bandwidth. More details of the WiMedia's ECMA standard can be found in the standard itself (ECMA 368) or in [Heidari 2008].

The dismantling of the IEEE 802.15.3a task group has, however, not been the end of UWB standardization within the IEEE. Two low-data-rate standards have since been approved, namely, the IEEE802.15.4a for personal area networks (IEEE802.15.4) and IEEE802.15.6 for body area networks (IEEE802.15.6).

These standards are attractive because they offer the possibility to implement communication systems with properties desirable for wireless sensor networks. The systems have low-power consumption, they are robust against the impact of frequency selective fading, their inherently low range combined with time hopping enables dense spatial reuse of the frequency resources, and in the case of IEEE802.15.4 they also provide ranging capabilities required by location-aware wireless sensor networks.

The IEEE802.15.4a for personal area networks relies on the combination of burst-position modulation (BPM) and binary-phase-shift-keying (BPSK) (IEEE802.15.4). The symbol structure is presented in Figure 7.2. Each symbol transmits two information bits. The first bit is defined by the position of the burst of pulses in the symbol—in the first or in the second half of the symbol as

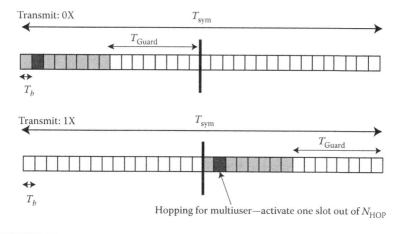

Figure 7.2 **The symbol structure in the 802.15.4a UWB standard. The light shade of gray is used to show the PPM modulation. To enable spectrum sharing, time hopping is used and transmitted only in one time slot T_b (dark gray) is used for bit transmission.**

shown in Figure 7.2. The second data bit is coded by the polarity of the pulse burst. Additionally, the pulse burst can be positioned in one of N_{HOP} bins (N_{HOP} is integer representing the number of bins/slots for time-hopping sequence) as the standard foresees time hopping to enable spectrum sharing among multiple users. The number of chips (pulses per burst) and the number of bins for hopping are defined by the standard and vary between different data rates (IEEE802.15.4). The mandatory bandwidth of the pulses is 499.2 MHz, but optionally can also reach 1081.6, 1331.2, and 1354.97 MHz (IEEE802.15.4).

The IEEE805.15.4a UWB standard foresees nonmandatory ranging capability which relies on time-of-arrival estimation (IEEE802.15.4). The range is determined using the roundtrip duration of a signal between two nodes, and the known duration of the processing at the distant transceiver (IEEE802.15.4). The ranging precision as defined by the standard can achieve resolution of up to 1 cm (IEEE802.15.4).

The IEEE802.15.6 system for body-area-networks is a low complexity low data-rate standard providing data rates from 0.3948 to 12.636 Mbps (IEEE802.15.6). The mandatory modulation is impulse radio (IR) using on–off-keying (OOK) and low complexity energy detection. Optional IR with differential BPSK/QPSK or frequency modulation UWB can be implemented (IEEE802.15.6).

In OOK IR UWB, the data bits are modulated on a symbol as presented in Figure 7.3. For the transmission of "0," the pulse is transmitted in the first half of the symbol; for the transmission of "1," the pulse is transmitted in the second half of the symbol, as shown in Figure 7.3. Within each half of the symbol, there are 16 bins with duration T_W, which are used for multiuser access by time hopping. The duration of the pulse T_W can be equal to 2, 4, 8, 16, 32, or 64 ns. The increase of the pulse duration reduces the data rate by a factor of 2, but it improves the sensitivity

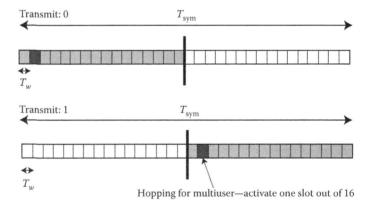

Figure 7.3 The symbol structure in IEEE802.15.6. The light gray color shows the information carried in the first or the second half of the symbol. Each half consists of 16 time slots for time-hopping is employed and only one time slot (dark gray) per symbol is actively used for communication.

of the receiver by 3 dB. The pulse bandwidth is 499.2 MHz (IEEE802.15.6). The standard does not define the pulse waveform, but chirp pulses or chaotic waveforms to create pulses long in the time domain yet satisfying the UWB criteria in the frequency domain are suggested (IEEE802.15.6).

7.1.3 UWB Regulations Worldwide

The standards described in Section 7.1.2 define the framework for communications, but the transmitters are also required to comply with the local regulations on the EIRP spectral density. The corresponding regulations in Europe, the United States, Korea, and Japan are presented in Figure 7.4. As can be seen, the maximum emission limit in all regions is the EIRP spectral density of −41.3 dBm/MHz. More details on the regulations can be found, for example, in [Sipal et al. 2012d].

These regulatory limits have significant impact on the design and operation of MIMO in UWB systems because all UWB regulations are EIRP spectral density bound. It means that UWB MIMO systems must ensure the maximum transmitted power spectral density in any direction does not exceed the limit. In contrast, for the IEEE802.11, the radiation is power bound. Hence, as long as the total gain of transmission does not exceed 6 dBi, the power spectral density radiated in a selected direction can be increased [Vithanage et al. 2009].

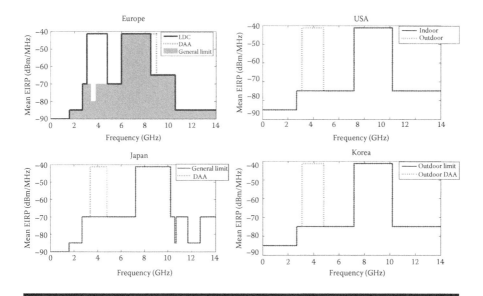

Figure 7.4   Spectral emmission masks for UWB region in Europe, the United States, Japan, and Korea. In bands denoted as DAA, radiation is permissible only for transmitters equipped with detect-and-avoid capability. In bands denoted as LDC, only low-duty-cycle operation is allowed.

In summary, the understanding of the regulations for UWB systems is important as these regulations set a different framework. In some cases, UWB systems, therefore, have to resort to different MIMO optimization strategies than IEEE802.11, making UWB system design and implementation rather specific [Vithanage et al. 2009].

7.2 UWB Wireless SISO Channel

The UWB wireless channel differs significantly from the narrowband channel because its bandwidth enables the emergence of some unique and interesting channel properties. These properties represent the main opportunity for the technology, but at the same time they can also be seen as its main limitation. Hence, the UWB wireless channel is introduced here.

In this section, the path gain, channel impulse response in SISO UWB wireless channel, and the robustness of UWB wireless channel to fading are discussed. The main objective of this section is to demystify the UWB wireless channels.

7.2.1 Path Gain in the UWB Wireless Channel

Path gain is the most important component in the link budget. It enables designers to perform initial range and feasibility estimates. Here, only the general concept of path gain is introduced and readers interested in more details on path gain are referred, for example, to [Molisch et al. 2004] and [de la Roche et al. 2013].

Path gain is the mean expected attenuation a signal experiences during propagation over range r in a particular environment, and is defined as [de la Roche et al. 2013]

$$PG(r) = \mathbf{E}\left\{ \int_{f_c - B/2}^{f_c + B/2} |H(f,r)|^2 df \right\} \tag{7.1}$$

where $E\{\cdot\}$ is the expected value excluding fading (see Section 7.2.5); $H(f, r)$ is the channel transfer function for range r; B is the bandwidth; and f_c is the center frequency.

The reader should note the inverse relationship between path gain and path loss. As a result, path loss increases with range, whereas path gain is reduced.

Path gain is expressed in decibel (dB) scale as

$$PG_{dB}(r) = PG_0 - 10n \log_{10}\left(\frac{r}{r_0}\right) \tag{7.2}$$

where PG_0 is the mean path gain for reference range r_0 (for UWB channels r_0 is typically selected as 1 m [de la Roche et al. 2013]); n is the path-gain coefficient (in free space $n = 2$, for line-of-sight condition in an indoor environment its value is lower since the path gain is increased by the energy carried by the multipaths [de la Roche et al. 2013]).

In contrast with narrowband channels, the UWB channel occupies a wide range of frequencies and $H(f, r)$ depends strongly on frequency. Thus, path gain depends on the bandwidth and center frequency of the channel [Molisch et al. 2004].

Practical systems introduced in Section 7.1.2 have typical bandwidths of 500 MHz (with exception of IEEE802.15.4 which considers bandwidths of 500–1355 MHz). Therefore, only the impact of center frequency has to be considered and variations within the symbol bandwidth can be neglected [de la Roche et al. 2013].

7.2.2 Channel Impulse Response of the SISO UWB Wireless Channel

In principle, the impulse response of a UWB channel does not differ from that of a narrowband channel and can be expressed as the sum of consecutive multipath components (rays). For simulations that explore system performance, it is important to know what the typical statistical properties of the channel impulse response are, that is, the typical delays of consecutive rays, the typical number of rays, and so on [de la Roche et al. 2013].

It is important to have a model that describes the channel accurately but remains understandable and simple. Unfortunately, most models reported in the literature, including the IEEE802.15.4a channel model for UWB, are rather complex and error prone. It can be shown that due to a systematic error in measurement processing, an error was embedded in the IEEE802.15.4a channel model and the number of rays was overestimated by factor of 4 or 5 [Sipal et al. 2010].

7.2.3 Systematic Errors in Wireless Channel Measurements

Owing to the vast bandwidth, UWB channel measurements are often performed in the frequency domain by vector network analyzers (VNA). These measurements provide the measured channel transfer function $H_M(f)$. The application of inverse Fourier transform on $H_M(f)$ to obtain the channel impulse response introduces artifacts which we discuss next.

As noted above, the actual channel impulse response, $h_C(t)$, without the impact of the antennas, is a real function and can be described as a sum of N multipath components:

$$h_C(t) = \sum_{i=1}^{N} \alpha_i * \delta(t - \tau_i) \tag{7.3}$$

where τ_i is the delay of the ith multipath component; $\delta(t)$ is the Dirac delta function; $\alpha_i(t)$ is the real amplitude of the ith multipath component; and $*$ denotes convolution.

Since practical communication systems (including practical UWB systems) are bandpass systems, the baseband descriptors are often preferred [Molisch et al. 2004]. In the baseband equivalent, the channel impulse response $h_{CB}(t)$ is a complex function and the relationship between $h_C(t)$ and $h_{CB}(t)$ is known to be [Ohm and Lueke 2006]:

$$h_C(t) = \Re\{h_{CB}(t) \cdot \exp(j2\pi f_C t)\} \tag{7.4}$$

where f_C is the center frequency of the system and $\Re\{\cdot\}$ takes the real component of $\{\cdot\}$.

Using Equation 7.3, it is possible to express the measured channel transfer function $H_M(f)$ as

$$H_M(f) = \left[\sum_{i=1}^{N} \alpha_i \cdot \exp(-j2\pi f \tau_i) \cdot H_{A,i}(f)\right] \cdot \text{rect}\left(\frac{f - f_C}{B}\right) \tag{7.5}$$

where B is the bandwidth of the measurement around the center frequency f_C; $\text{rect}(x) = \begin{cases} 1 & \text{for } x < 0.5 \\ 0 & \text{otherwise} \end{cases}$; and $H_{A,i}(f)$ is the combined channel transfer function of the transmit and receive antenna.

$H_{A,i}(f)$ is specific to each multipath component to reflect the fact that the radiation properties of antennas are dependent on the angle-of-departure and angle-of-arrival (AoD and AoA), that is, the gain of the antenna is different in different directions. $H_{A,i}(f)$ also includes the variation of antenna gain as a function of frequency.

Equation 7.5 represents the data obtained from a measurement with a VNA. The inverse Fourier transform of Equation 7.5 yields

$$h_M(t) = \sum_{i=1}^{N} \alpha_i \cdot \left\{h_{A,i}(t - \tau_i) * \left[\text{sinc}(Bt)\cos(2\pi f_C t)\right]\right\} \tag{7.6}$$

where $\text{sinc}(x) = \sin(\pi x)/\pi x$.

Let us compare Equation 7.3 with Equation 7.6. In Equation 7.3, each ray has a simple form $\delta(t - \tau_i)$. In the measured impulse response in Equation 7.6, each ray is represented by $h_{A,i}(t - \tau_i) * [\text{sinc}(Bt)\cos(2\pi f_C t)]$ which includes the impact of antenna impulse responses $h_{A,i}(t)$ and the impact of the measurement with limited bandwidth around the center frequency [sinc(Bt) cos($2\pi f_C t$)]. This is a systematic error which has an impact on the observation of the number of rays.

First, the [sinc(Bt) cos($2\pi f_C t$)] term produces multiple peaks which are not independent multipath components. The impact of cos($2\pi f_C t$) can be removed by transformation into equivalent baseband, but even then the sinc(Bt) term remains and its artifacts can be erroneously interpreted as independent multipaths. The first two sidelobes in sinc(Bt) are spaced $1.5/B$ around the main lobe attenuated by 13.3 dB. Further, sidelobes are then spaced by $1/B$. It is possible to reduce the impact of the sinc-function by applying different windowing functions, but at the cost of filtering out some information about the channel.

Second, even if the impact of the sinc function is eliminated by advanced signal processing, this is not always true about the impact of manifestation of the antenna impulse responses $h_{A,i}(t)$, which is a convolution of the impulse response of the transmit antenna and the impulse response of the receiving antenna. Figure 7.5 shows a simulated antenna impulse response $h_{A,i}(t)$ for transmission between two example antennas in free space, that is, the impulse response corresponds to a single ray. Thus, Figure 7.5 shows the difficulty of telling apart the manifestation of antenna impulse response $h_{A,i}(t)$ and independent rays.

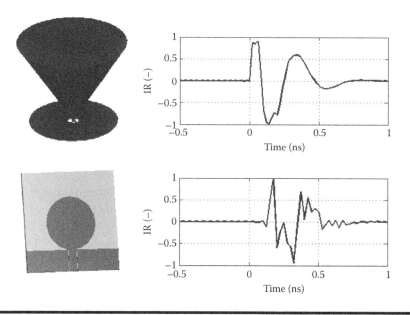

Figure 7.5 **Manifestation of antenna impulse responses for free-space transmission between two discone antennas (top) and between two circular monopole antennas facing broadside (bottom).**

The removal of $h_{A,i}(t)$ from Equation 7.6 is possible, as shown in [Sipal et al. 2010] where the known antenna impulse response was deconvolved from a measured channel to leave only the sinc function present. However, the deconvolution requires perfect knowledge of $h_{A,i}(t)$ for each ray, which is only possible when the AoD and AoA for each ray are known. This is not the case for a general channel measurement [Malik et al. 2008].

To conclude, this section shows that the traditional channel measurements are subject to systematic error which can result in overestimation of the number of independent rays in the channel. With this knowledge, it is possible to correct and simplify many channel models as is shown in the subsequent section simplifying the IEEE802.15.4a channel model.

7.2.4 Removal of Systematic Measurement Error from Channel Models

The IEEE802.15.4 channel model is the most widely used channel model for UWB system modeling. The channel impulse response, according to the IEEE802.15.4a, is defined for the equivalent baseband channel, that is, it is a complex variable, and it defines that the rays arrive in N_K clusters, each of which consists of up to N_L rays [Molisch et al. 2004]:

$$h_C(t) = \sum_{k=1}^{N_K} \sum_{l=1}^{N_L} \alpha_{k,l} e^{j\theta_{k,l}} \delta(t - T_k - \tau_{k,l}) \tag{7.7}$$

where T_k is the cluster delay; $\tau_{k,l}$ is the delay within the cluster; $\theta_{k,l}$ is the phase of the ray.

The intercluster delays $T_k - T_{k-1}$ are a Poisson process [Molisch et al. 2004]. The intracluster delays $\tau_{k,l} - \tau_{k,l-1}$ are the mixture of two Poisson processes. However, when this mixture of Poisson processes is critically studied under the consideration of the systematic error from the preceding section, it becomes apparent that only one of the Poisson processes represents the actual ray, whereas the second one is caused by the systematic error in evaluation of the measured results.

For instance, in the office environment, the two Poisson processes are reported to have mean ray arrival times 0.337 and 5.26 ns [Molisch et al. 2004] and the process with mean ray arrival time of 0.337 ns dominates with relative occurrence of 98% [Molisch et al. 2004]. Based on the previous section, it is reasonable to argue that the IEEE802.15.4a channel model overestimates the number of paths [Sipal et al. 2010].

Thus, it is possible to conclude that the mean arrival time of a ray in a cluster is a single Poisson process. In the office environment, the mean ray arrival time is 5.26 ns. For other environments, it is the Poisson process with the longer mean ray arrival time from the two offered by the IEEE802.15.4a channel model. This significantly simplifies the channel model as the number of independent rays is significantly reduced [Sipal et al. 2010] and it enables us to reuse the results of the IEEE802.15.4a channel model.

7.2.5 Multipath Fading in UWB Wireless Channels

Fading in the wireless channels is caused by multipath superposition at the receiver. The superposition of multiple signal copies can be either constructive or destructive. As a result, the received signal strength can vary significantly from values expected based on path gain. Similarly, even a small displacement of the transmitter or receiver can convert constructive superposition into destructive interference and vice versa [de la Roche et al. 2013].

UWB systems are often quoted to be robust against multipath fading. This statement is generally true but robustness to multipath fading also depends on the architecture of the UWB system [de la Roche et al. 2013]. Thus, this section discusses frequency-selective fading in more detail.

The channel transfer function $H(f)$ is obtained as the Fourier transform of Equation 7.3:

$$H(f) = \sum_{i=1}^{N} \alpha_i \exp(-j2\pi f \tau_i) \tag{7.8}$$

The received channel energy for a channel with center frequency f_c and bandwidth B can be determined as follows:

$$\int_{f_c-B/2}^{f_c+B/2} |H(f)|^2 \, df = \int_{f_c-B/2}^{f_c+B/2} \left| \sum_{k=1}^{N} \sum_{i=1}^{N} \alpha_i \alpha_k \exp(-j2\pi f \tau_i) \exp(j2\pi f \tau_k) \right|^2 df \tag{7.9}$$

Some basic algebra results in

$$\int_{f_c-B/2}^{f_c+B/2} |H(f)|^2 \, df = \int_{f_c-B/2}^{f_c+B/2} \sum_{i=1}^{N} \alpha_i^2 \, df + \int_{f_c-B/2}^{f_c+B/2} \sum_{\substack{k=1 \\ k \neq i}}^{N} \sum_{i=1}^{N} \alpha_i \alpha_k \exp(-j2\pi f \tau_i) \exp(j2\pi f \tau_k) df$$

$$\int_{f_c-B/2}^{f_c+B/2} |H(f)|^2 \, df = B \sum_{i=1}^{N} \alpha_i^2 + \left[\sum_{\substack{k=1 \\ k \neq i}}^{N} \sum_{i=1}^{N} \alpha_i \alpha_k \cos[2\pi f(\tau_k - \tau_i)] \right]_{f_c-B/2}^{f_c+B/2}$$

$$\int_{f_c-B/2}^{f_c+B/2} |H(f)|^2 \, df = B \sum_{i=1}^{N} \alpha_i^2 + \sum_{\substack{k=1 \\ k \neq i}}^{N} \sum_{i=1}^{N} \frac{1}{\pi(\tau_k - \tau_i)} \alpha_i \alpha_k \sin[\pi B(\tau_k - \tau_i)] \cos[2\pi f_c(\tau_k - \tau_i)]$$

$$\int_{f_c-B/2}^{f_c+B/2} |H(f)|^2 \, df = B\left\{ \sum_{i=1}^{N} \alpha_i^2 + \sum_{\substack{k=1 \\ k\neq i}}^{N} \sum_{i=1}^{N} \alpha_i \alpha_k \mathrm{sinc}[B(\tau_k - \tau_i)]\cos[2\pi f_c(\tau_k - \tau_i)] \right\} \quad (7.10)$$

where $\mathrm{sinc}(x) = \dfrac{\sin(\pi x)}{\pi x}$.

Equation 7.10 can be used to explain fading in narrowband and UWB chan-nels. The single-sum is the energy carried by the multipath and represents the mean path gain. The double-sum represents the fading and the superposition of the mul-tipath components in the time domain. Depending on the delay between mul-tipath components as well as the amplitudes of α_i, the components in the double sum either increase or reduce the channel energy.

Owing to the sinc term, the contribution of the second sum in Equation 7.10 is reduced with bandwidth. For narrowband channels, $\mathrm{sinc}[B(\tau_k - \tau_i)] \approx 1$, and the variation of path gain can be significant because even for a small displacement, the terms $\cos[2\pi f_c(\tau_k - \tau_i)]$ can change the amplitude significantly. However, for a large bandwidth B the second term is weighted by $\mathrm{sinc}[B(\tau_k - \tau_i)] \approx 0$. As a result, the channel energy from Equation 7.1 does not change for small displacements. This is illustrated in Figures 7.6 and 7.7.

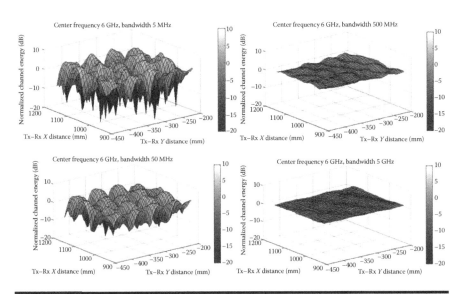

Figure 7.6 Example of the spatial variation of the channel energy for various band-width B in Equation 7.10, measurements obtained for a laboratory environment as described in Sipal (2012). (Adapted from V. Sipal, Impact of the wireless channel on the performance of ultrawideband communication systems, PhD thesis, University of Oxford, Dec. 2012.)

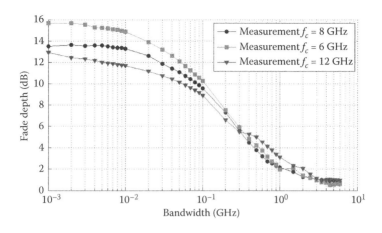

Figure 7.7 Fade depth scaling for channel from Figure 7.6 for various center frequencies. Fade depth is defined as three times the standard deviation of the normalized channel energy in dB.

Figure 7.6 presents the measured channel energy (normalized to a local mean) evaluated according to Equation 7.10 for bandwidths 5, 50, 500, and 5000 MHz. All measurements were performed for the same 1600 channels (grid 40 × 40) in a laboratory environment. The details about the measurement can be found in [Sipal et al. 2011]. In terms of fading, Figure 7.6 shows the variation of the channel energy for a small spatial displacement and it also illustrates the increasing impact of the sinc term as the bandwidth in Equation 7.10 is increased.

Figure 7.7 focuses in more detail on the impact of the scaling of the sinc term with bandwidth. It analyzes the measurement from Figure 7.6 for bandwidths from 1 MHz to 6 GHz. For each bandwidth, the normalized energy for each data point of the 40 × 40 grid is calculated. Then, the standard deviation of these energies is determined and used to calculate the fade depth, which is defined as three times the standard deviation of the spatial variation of the channel energy [Malik et al. 2007, Allen et al. 2008, Sipal et al. 2011]. Figure 7.7 shows how this measure scales with increasing bandwidth. It shows that for bandwidths beyond 1 GHz the channel energy variation in an indoor channel is negligible.

On the basis of Figures 7.6 and 7.7, it is apparent that systems with larger bandwidth are less prone to the impact of fading. However, this statement should be used with caution as the impact of fading depends on the modulation and other factors.

For IR systems such as those described by IEEE802.15.4 and IEEE802.15.6 standard, the impact of fading is negligible and fade margin does not need to be considered in link budget. In terms of MIMO, the fact that the energy of the received signal is independent of small spatial displacement means that IR UWB systems cannot benefit from spatial diversity systems, however, they still can benefit from spatial filtering techniques.

For OFDM systems, the mean energy of the received signal is independent of the small spatial displacement, but this does not apply to individual subcarriers. In fact, the energy variation among individual subcarriers can be up to 20 dB [Sipal et al. 2012b]. As a result, even though the mean received signal power is 10 dB above the receiver sensitivity, a significant number of weak subcarriers may have negative SNR [Sipal et al. 2012b]. These subcarriers often cause a significant increase of the overall BER and lead to failure of the communication for UWB OFDM systems with fixed modulation [Sipal et al. 2013b].

This property, however, means that OFDM systems can benefit from both spatial filtering and spatial diversity schemes. Both spatial filtering and spatial diversity in UWB OFDM systems are performed at the subcarrier level. Thus, there is little conceptual difference between a UWB OFDM system and a narrowband OFDM/single-carrier system.

7.3 MIMO UWB Wireless Channel

The preceding section introduced the channel impulse response model for a UWB channel. In this section, it is shown how the model can be extended to describe MIMO channels.

First, let us assume that the observations are performed in the far field of the transmit antenna. Then, the rays in the antenna impulse response are manifestations of plane waves and each ray manifests itself at all antenna elements of a MIMO system.

This assumption that the channel impulse response in adjacent channels contains the same rays is confirmed by Figures 7.8 and 7.9. Data for Figures 7.8 and 7.9 were obtained in a rigorous channel measurement campaign reported in [Sipal et al. 2010, 2011]. The transmitter had a fixed position and the receiver was positioned in a grid of 40 × 40 points spaced by 6 mm. The measurement was performed in a laboratory environment across frequencies 3–20 GHz.

Figure 7.8 shows the channel impulse response of five adjacent channels. The channel impulse response was determined for the real channel (i.e., no baseband transformation) and therefore contains the artifacts discussed and described in Section 7.2.2 in Equation 7.6. It is apparent that the same rays with the same artifacts can be observed in all five channels. The gray dashed lines show that the delay of the rays is changing linearly with linear displacement. Figure 7.9 presents a time snapshot of the entire grid of 1600 channels to show the first two rays from Figure 7.8. It is apparent that if the measurement artifacts were removed, a single wave would be propagating across the plotted grid. It is also apparent that the wave is spherical (due to range of 1.1 m) but for small displacements, that is,

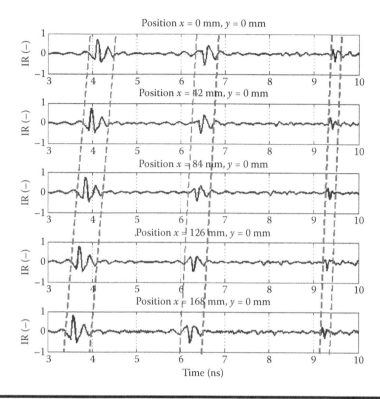

Figure 7.8 Channel impulse response for five adjacent channels spaced by 42 mm each, showing that the same rays, with a linear shift in delays, can be identified in each channel. (Reprinted from V. Sipal, Impact of the wireless channel on the performance of ultrawideband communication systems, PhD thesis, University of Oxford, Dec. 2012.)

smaller than 100 mm, the error due to plane wave approximation is acceptably small.

To conclude, Figures 7.8 and 7.9 confirm the hypothesis that for a small displacement matching rays can be identified in the impulse responses of the wireless channels and that it can be assumed that many of these rays are plane waves.

Thus, the channel impulse response of a MIMO channel with N_R receiver antennas and N_T transmitter antennas can be mathematically formulated as follows [Malik et al. 2005]:

$$h_{C,m,n}(t) = \sum_{i=1}^{N} \alpha_i \delta(t - \tau_i - \tau_{i,m,n}) \qquad (7.11)$$

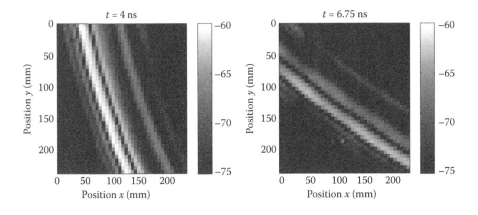

Figure 7.9 Snapshots of 40 × 40 wireless channels showing the plane wave. (Reprinted from V. Sipal, Impact of the wireless channel on the performance of ultrawideband communication systems, PhD thesis, University of Oxford, Dec. 2012.)

where $h_{C,m,n}(t)$ is the channel impulse response between antennas m and n and $\tau_{i,m,n}$ are the delays incurred due to the small displacement of the antennas.

Under the plane wave assumption, $\tau_{i,m,n}$ can be expressed using simple path delay geometry as a function of antenna spacing and AoA/AoD when the geometry and orientation of the transmitter and receiver arrays are known.

As mentioned earlier, these equations are valid only for a small displacement. In most cases for an indoor environment and range larger than 1 m, displacements below 10 cm are usually sufficiently small so that the path gain of individual rays does not change significantly, that is, change in α_i due to change in path gain, shadowing, or antenna gain (due to changed AoA/AoD) can be neglected. For ranges below 1 m, the limit of what counts as a small displacement is smaller because the same absolute displacement results in larger relative change of path loss, that is, the α_i coefficient change faster with displacement.

The overall MIMO channel impulse response comprises the individual channels from Equation 7.11. The specific results, however, depend on the MIMO operation applied to the signals received by individual antenna elements.

7.4 UWB Antenna Arrays

Antenna arrays are a fundamental component of most MIMO systems. Owing to the available bandwidth, UWB antenna arrays possess some specific properties not present in conventional antenna arrays [Kaiser et al. 2009]. For practical applications, however, caution is required because as with fading in UWB channels, many statements about UWB arrays are not general. They are applicable to certain array geometry and certain signal types with specific bandwidth. These limitations will be discussed in more detail here.

7.4.1 Difference between Narrowband and UWB Array, Beam Patterns

There are two basic differences between a UWB and a narrowband antenna array behavior—the duration of the signals processed by the array and the bandwidth of the signal.

In narrowband systems, signal bandwidths are on the order of up to a few MHz. For instance, the IEEE802.11 standard specifies the subcarrier bandwidth of 0.3125 MHz (IEEE802.11). The duration of such a signal multiplied by the speed of propagation is significantly larger than the physical dimension of typical arrays. Antenna arrays are thus described by array factors [Balanis 2005]. The array factor description assumes a signal at a single frequency being radiated/received by each array element. To describe the array reception/radiation in different directions, the signals are summed in the complex domain considering the phase shifts caused by geometric path differences [Balanis 2005].

For UWB arrays, the signal can be extremely short, that is, less than 1 ns. Hence, it is possible that for some antenna geometries and azimuth angles, the received/ transmitted signal is a pulse train. In other words, the signals do not superpose in the time domain. Additionally, due to the large bandwidth, a single phase shift as a descriptor for the geometrical path differences between rays received/transmitted by different antenna elements cannot be used.

As a result, time-domain descriptors for UWB antenna arrays, referred to as beam patterns, have been introduced, for example, in [Kaiser et al. 2009]. Consider Figure 7.10 showing a uniform linear array with N elements spaced by a distance d.

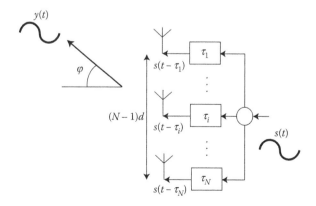

Figure 7.10 **UWB antenna array transmitting signal $s(t)$ with limited duration.**

The transmitted signal $s(t)$ is delayed by delays τ_i which can be adjusted to control the direction of the main beam.

The transmitted signal $y(t,\varphi)$ in direction φ in the far-field can then be described as

$$y(t,\varphi) = \sum_{i=1}^{N} s\left(t - (i-1)\frac{d}{c}\sin\varphi - \tau_i\right) \tag{7.12}$$

For far-field beam steering in direction φ_0, that is, $\tau_{i+1} = \tau_i + d/c\,\sin\varphi_0$, it is possible to rewrite Equation 7.12 as follows [Malik et al. 2006]:

$$y(t,\varphi) = \sum_{i=1}^{N} s\left(t - (i-1)\frac{d}{c}(\sin\varphi - \sin\varphi_0)\right) \tag{7.13}$$

The beam patterns then characterize the properties $y(t,\varphi)$ as a function of azimuth angle. These properties can be: the maximum power; the total radiated energy; or the peak signal amplitude [Kaiser et al. 2009]. Here, the total radiated energy criterion is investigated in more detail to show the relationship between narrowband and UWB arrays. We define beam pattern $\varepsilon(\varphi, \varphi_0)$ as follows:

$$\varepsilon(\varphi,\varphi_0) = \frac{\int_{-\infty}^{\infty} |y(t,\varphi)|^2 \, dt}{\int_{-\infty}^{\infty} |s(t)|^2 \, dt} \tag{7.14}$$

Using Equation 7.13 and some basic algebra, it is possible to express $\varepsilon(\varphi, \varphi_0)$ as

$$\varepsilon(\varphi,\varphi_0) = \frac{\sum_{i=0}^{N-1}\sum_{j=0}^{N-1} R_{ss}((i-j)\Delta\tau)}{R_{ss}(0)} \tag{7.15}$$

where $R_{ss}(\tau) = \int_{-\infty}^{\infty} s(t)s^*(t-\tau)\,dt$ is the autocorrelation function with * denoting complex conjugation, and $\Delta\tau\, d/c\,(\sin\varphi - \sin\varphi_0)$.

Considering that autocorrelation is a Hermitian function ($R_{ss}(\tau) = R_{ss}^*(-\tau)$), it is possible to further simplify Equation 7.15 into

$$\varepsilon(\varphi,\varphi_0) = N + \frac{\sum_{i=1}^{N-1}(N-i)\Re\{R_{ss}(i\Delta\tau)\}}{R_{ss}(0)} \tag{7.16}$$

Assuming that $s(t)$ is a bandpass signal $s(t) = b(t)\cos(2\pi f_c t)$ with pulse envelope $b(t)$. This is a realistic assumption, because the IEEE802.15.4 and IEEE802.15.6 standards indeed use bandpass signals. The final expression for $\varepsilon(\varphi, \varphi_0)$ can be obtained as

$$\varepsilon(\varphi,\varphi_0) = N + \frac{\sum_{i=1}^{N-1}(N-i)\Re\{R_{bb}(i\Delta\tau)\}\cos(2\pi f_c \Delta\tau)}{R_{bb}(0)} \tag{7.17}$$

Equation 7.17 can be used to explain the behavior of narrowband and UWB arrays. For narrowband antennas, the geometric path delay in the array is negligible compared to the duration of the signal. Thus, it is possible to use the approximation $R_{bb}(i\Delta\tau) \approx R_{bb}(0)$. As a result, the beam pattern is defined only by the phase shifts in the term $\cos(2\pi f_c \Delta\tau)$. Indeed, the result from Equation 7.17 for this case equals to the square of the classical array factor.

For UWB signals, specifically for the impulse system, the properties of the array also depend on the shape of the envelope signal $b(t)$. As bandwidth increases, the impact of the cos term in Equation 7.17 is reduced. This dependency is typically manifested as smoothing of nulls in the beam pattern as shown in Figure 7.11.

From Equation 7.17, it is apparent that for large bandwidth and azimuth angles far from the main beam, the beam pattern assumes a constant level $10\log_{10}N$, which is due to the fact that for these angles it is possible to approximate $R_{bb}(i\Delta\tau) = 0$ for all $i \geq 1$. This is observed for the sparse array and bandwidth of 4 GHz in Figure 7.11.

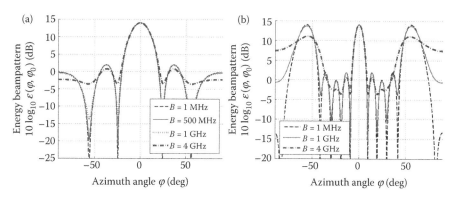

Figure 7.11 **Beampatterns for six element UWB antenna array transmitting signal Gaussian pulse $b(t)$ with various bandwidths modulated at center frequency of 7.25 GHz. Elements spaced by (a) 2 cm and (b) 5 cm (enabling the existence of grating lobes). (Reprinted from V. Sipal, Impact of the wireless channel on the performance of ultrawideband communication systems, PhD thesis, University of Oxford, Dec. 2012.)**

7.4.2 Impact of Beam Patterns on Practical Systems

From Equation 7.17, it is apparent that the radiation patterns of an array depend strongly on the properties of the signal envelope $R_{bb}(i\Delta\tau)$, however from previous Section 7.4.1 it is apparent that the bandwidth of the signal plays a crucial role.

First, for OFDM systems, array factors can be used since the array actually transmits a large number of narrowband signals in parallel, rather than a short UWB impulse. However, the impact of the center frequency must be noted. In the language of the classical array factors, the phase shift between antenna elements is a variable of frequency. Thus, the array has different radiation properties for each subcarrier of the OFDM symbols. The practical implication is that MIMO beamforming has to be performed on a subcarrier basis.

In terms of IRs, it is noticed in Figure 7.11 that even for bandwidths of 500 MHz and 1 GHz, there is a little difference between the narrowband and UWB beam patterns. Consequently, [Sipal et al. 2012a] explored the upper bandwidth limit of array factor as a descriptor for IR. The work concluded that for impulse signals with bandwidth below B_{UL}, array factors can be used without a significant error. The bandwidth B_{UL} was empirically found to be [Sipal et al. 2012a]

$$B_{UL} = \frac{0.16}{d(N-1)}[\text{GHz}] \tag{7.18}$$

As observed in Equation 7.18, the upper limit in bandwidth is inversely proportional to the physical dimensions of the array. Thus, it can be shown that for practical arrays and bandwidths below 1 GHz, the classical array factors, evaluated at the signal's center frequency, can be used.

Grating lobe suppression is another interesting feature of sparse UWB arrays. Grating lobes appear in antenna arrays with element spacing larger than a half-wavelength. For spacing larger than a wavelength, the energy radiated in the direction of the grating lobes equals the energy radiated in the main lobe. The beamforming ability of the array is compromised because the grating lobes cannot be controlled independently of the main lobe. However, [Kaiser et al. 2009] reports that UWB arrays do not manifest grating lobes.

[Sipal et al. 2012c] studied this property in more detail and found that the transmitted pulse requires a relative bandwidth of more than 100% to effectively suppress grating lobes. Such a bandwidth is often used in theoretical works, but in current practical systems such as IEEE802.15.4 and IEEE802.15.6, the relative bandwidths are below 30% for the lowest center frequency in the US band. In fact, relative bandwidth of 100% is not currently allowed by the EIRP limits in Europe, Japan, and Korea.

To conclude the section on UWB arrays, it is noted that the case of UWB antenna arrays is different from the narrowband case. The UWB arrays therefore exhibit many interesting properties, but it is equally important to understand that these properties cannot be found in all UWB systems with bandwidths greater than 500 MHz. In other words, one should be aware of the limitations of statements relating to UWB arrays.

7.5 UWB MIMO

MIMO-based UWB OFDM can be considered similar to narrowband systems. This is the case especially for the high-speed UWB OFDM where multiple slower data streams are transmitted in parallel. In principle, the algorithms are the same as in narrowband OFDM systems because they are performed on the subcarrier basis, that is, the performance is optimized for multiple narrowband channels in parallel. In other words, many algorithms discussed in other chapters of this book can be directly applied to UWB OFDM systems. Rather than repeating information contained elsewhere in the book, some of the distinctive specifics associated with UWB MIMO systems are discussed here.

7.5.1 EIRP-Constrained Radiation

The first and foremost difference between UWB MIMO and IEEE802.11 MIMO is the fact that the transmitter for UWB is EIRP constrained, whereas in IEEE802.11 the transmitter is power constrained. Power-constrained systems cannot exceed the total radiated power but they may choose the spatial distribution of the radiation, that is, they can increase the radiation in a certain direction and reduce it in another (within certain limits). EIRP-constrained systems, however, can reduce the radiation in undesired direction but they cannot allocate more power to the desired direction [Vithanage et al. 2009].

This difference means that while UWB can benefit from transmit beamforming such that individual paths are summed constructively at the receiver, the algorithms for power-constrained MIMO systems cannot be used directly [Vithanage et al. 2009]. The optimum solution to the EIRP-constrained problem is a complex convex optimization not suitable for practical implementation [Vithanage et al. 2009].

UWB receivers are not limited by EIRP. Thus, a UWB SIMO system can use algorithms different from a UWB MISO system. Due to the limitations on the EIRP, a SIMO system can be expected to provide better performance enhancement than a MISO system.

7.5.2 Grating Lobes, Element Spacing, and Antenna Selection

Another challenge for UWB systems is antenna spacing for systems operating in the entire UWB band. The UWB band in the United States spans from 3.1 to 10.6 GHz. With spacing that optimizes performance at the lower end of the band, grating lobes appear at the upper end of the band which compromises the benefits of beamforming. With spacing that prevents grating lobes at the upper end of the spectrum, the antenna elements are too close at the lower end of the spectrum so that beamforming is effectively nonexistent and channel correlation is strong.

One approach to UWB MIMO not affected by grating lobes is antenna selection. This is a diversity scheme in which each subcarrier is transmitted by exactly one antenna. Through channel sounding, the system determines which antenna is more suitable for transmission of each subcarrier and the subcarriers are then assigned as required [Vithanage et al. 2009].

This approach has multiple advantages. First, it is inherently EIRP constrained [Vithanage et al. 2009]. Second, there are no issues with grating lobes [Vithanage et al. 2009]. Third, it is a spatial diversity scheme which improves performance by mitigating the impact of fading. Fading, as described in Section 7.2.5, is the main limiting factor for UWB OFDM systems. Due to fading, a nonnegligible number of subcarriers have very low or even negative SNR, even for mean signal energy well above the noise floor. These subcarriers contribute mostly to the overall BER [Sipal et al. 2012b]. Antenna selection can, despite its simplicity, address this issue and significantly improve the performance of an OFDM UWB system. This is illustrated in Figure 7.12.

Figure 7.12 shows the BER improvement a WiMedia OFDM system operating at 200 Mbps would achieve in a measured channel with a 4 × 1 antenna selection.

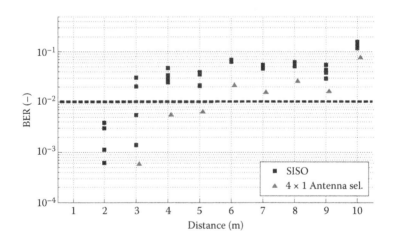

Figure 7.12 Gross BER performance of a WiMedia OFDM system at 200 Mbps for four SISO channels and a 4 × 1 antenna selection. The figure uses measured wireless channel in an office environment. The dashed line represents strength of the forward error correction in the WiMedia standard. (Reprinted from V. Sipal, Impact of the wireless channel on the performance of ultrawideband communication systems, PhD thesis, University of Oxford, Dec. 2012.)

Considering the highlighted strength of forward error correction, Figure 7.12 shows that increasing the number of antennas by a factor of 4 approximately doubles the range of the system from 3 m for a SISO system to 6 m with 4×1 antenna selection system.

Antenna selection is not suitable for IR systems because, due to the ability to resolve individual multipath components in the time domain, the energy received by all antennas is the same and spatial diversity typically cannot provide any significant performance boost.

7.5.3 Complexity of MIMO for OFDM Systems

Another challenging issue for OFDM-based UWB MIMO is implementation complexity. The optimum MIMO solution for UWB OFDM system (as well as other OFDM system) is performed in the frequency domain, that is, on the subcarrier basis. This means that a system with N antenna elements requires N OFDM chips.

The number of OFDM chips as well as the per-subcarrier optimization can be associated with a significant complexity/cost increase of the UWB OFDM system. Therefore, a cost–benefit analysis needs to be performed. The question of feasibility is not specific to UWB MIMO. However, the complexity/cost of an OFDM communication system is proportional to bandwidth [Sipal et al. 2012d]. As a result, a UWB OFDM chip is already more than 25-times more complex/costlier than the IEEE802.11a chip, and 65-times more complex than DVB-T chips. Thus, many MIMO techniques increase the cost of narrowband systems by a few cents, but they may not be feasible for UWB because they increase the cost on the order of a few dollars, which may be prohibitive for many consumer electronics applications.

7.6 UWB Localization

It has been suggested that we are about to enter an era of technological revolution triggered by the advent of machine-to-machine communications [Evans et al. 2012]. With the decrease of cost of wireless communication, billions of devices are expected to be equipped with the capability to communicate and share information and to be controlled wirelessly [Evans et al. 2012]. It is conceivable that for many of these applications, the localization of individual devices will play an important role. Therefore, much research effort has recently focused on localization in wireless sensor networks. UWB and UWB MIMO play a pivotal role in these activities because the properties of UWB signals and UWB wireless channels are suitable for these applications.

In this section, the main concepts of localization in wireless sensor networks will be introduced and it will be shown how localization can benefit from the properties of UWB and UWB MIMO systems.

There are three main localization principles considered in the literature which can all benefit from the properties of UWB systems. These are [Gezici et al. 2005]

■ Received signal strength (RSS) methods
■ Time-of-arrival (TOA) and time-different-of-arrival (TDOA)
■ Angle-of-arrival (AOA) methods

7.6.1 RSS Ranging

The objective of ranging is to determine the unknown distance between an anchor and a node. The position of the node can be uniquely determined by triangulation if the range to three anchors with known positions [Gezici et al. 2005]: Such systems based on RSS exploit the fact that the path gain, introduced in Section 7.2.1, depends on the range between the transmitter and the receiver (7.2). The range can be calculated from the RSS. The main systematic errors limiting the precision of the RSS method are fading, shadowing, and unknown path-gain properties (exact PG_0 and n in Equation 7.2 are unknown for a specific channel). As a result, RSS is typically used in conjunction with complex RSS fingerprinting [Malik and Allen 2006]. In other words, the RSS is measured for the entire area of interest. Later, measurements are compared to this database to determine the location.

UWB is typically not considered as a candidate technology for RSS ranging, but UWB and UWB MIMO can help to improve its precision. The first advantage of UWB systems compared to narrowband systems is the fact that wideband signals are robust to fading (see Section 7.2.5). In an indoor environment, the use of wideband signals with bandwidths exceeding 500 MHz effectively eliminates fading, removing one of the most significant sources of systematic errors in RSS systems.

If the RSS ranging system is employed in connection with MIMO, the precision further increases because beamforming can enable simplification of the complex multipath channel into a single-path directional channel [Malik and Molisch 2006]. This has two benefits. First, a single path channel does not suffer from fading. Second, the path gain in a single-path channel corresponds to free-space propagation, that is, PG_0 and n are known.

Despite the performance improvements UWB can provide to RSS systems, it is typically not considered because the precision of the other two localization methods is superior.

7.6.2 TOA Ranging

The principle of TOA-based ranging is the measurement of the propagation delay between the anchor and the node [Tarique et al. 2006a,b]. As with RSS methods, localization is performed when the range of the node to three anchors is known. It is immediately apparent that TOA methods are suitable only for LOS conditions as the delay in NLOS condition does not correspond to range. Complex algorithms have therefore been developed to identify NLOS conditions in order to reduce systematic error in localization [Gezici et al. 2005].

UWB systems are highly suitable for this type of localization, because the extremely short duration of UWB impulses enables very high precision in measurement of delay. The precision limit of TOA-based ranging arrival is given by the Cramer–Rao-lower-bound (CRLB), which is derived from the Fisher information matrix. Here, we present the CRLB for TOA ranging in free space. The minimum ranging error $\sqrt{\text{var}(d)}$ given by CRLB is given as follows [Gezici et al. 2005]:

$$\sqrt{\text{var}(d)} \geq \frac{c}{2\sqrt{2}\pi\sqrt{SNR}\beta} \tag{7.19}$$

where c is the speed of light; SNR is the signal-to-noise ratio; and $\beta = \sqrt{\int_{-\infty}^{\infty} f^2 |S(f)|^2 \, df \Big/ \int_{-\infty}^{\infty} |S(f)|^2 \, df}$ is the effective signal bandwidth of the signal $S(f)$.

Free-space ranging represents only a very simple case and practical UWB localization systems achieve precision above the CRLB. However, Equation 7.19 provides a good intuitive understanding of the main issues of ranging in practical systems.

Assuming that the ranging signal $S(f)$ is a bandpass signal, the first observation is that the precision of ranging increases with center frequency and bandwidth of $S(f)$. The second observation is that range d is not present in Equation 7.19.

Although both observations are correct, the impact of SNR in Equation 7.19 has to be considered. First, SNR decreases with range with an adverse effect on ranging precision. Second, the path-gain, that is, SNR, decreases with frequency. Third, due to the higher attenuation of higher frequency components, adding extra bandwidth increases β but also reduces SNR.

Since additional bandwidth also typically has adverse effect on noise figure of amplifiers [Sipal et al. 2012d], it is concluded here that while bandwidth enables precise ranging, for practical systems, a compromise between β (bandwidth and center frequency) and SNR must be considered. Also, even though CRLB offers high theoretical precision even for negative SNR (in dB), the establishment of communication for such a situation is highly challenging and for practical systems often require pulse bursts with hundreds of chips.

In practical systems, the ranging precision of TOA is further impaired by other factors. The first and most important factor is the multipath. Only the direct path carries the relevant information about range. Thus, not only do later multipath components in the channel act as noise and have adverse effect on the SNR, in some cases they may be stronger than the direct path making the detection of the direct path more challenging. These issues can, to some extent, be alleviated by the use of MIMO systems which employ spatial filtering to attenuate the later multipath components and enable improved detection of the direct path.

Another issue of TOA systems is clock synchronization between the transmitter and the receiver. Therefore, the practical ranging system in IEEE802.15.4a uses measurement of roundtrip delay. In principle, the transmitter initiates ranging and sends a message. This message is received by the node, which sends back confirmation of receipt followed by information about the delay incurred at the receiver due to processing of the ranging request. The initiator can then determine the round trip delay with high precision. For details about the scheme, the reader is referred to the IEEE802.15.4 standard (IEEE802.15.4).

An alternative solution to the synchronization problem is the use of the TDOA scheme [Gezici et al. 2005], where the node compares relative delays between signals received by the anchors. This scheme is used, for example, in the global positioning system.

7.6.3 AOA Localization

AOA localization aims to detect the direction of the incoming wave. If the direction of the node with respect to three anchors is known, the location can be determined. At first, it would appear that the precision of this localization scheme depends mainly on the physical dimensions of the array as these define the beam-width. However, the advantage of UWB is the fact that the wide bandwidth enables extremely short delays between the pulses received by adjacent antenna elements to be determined [Mallat et al. 2006].

In other words, the large bandwidth enables more precise detection of the AOA because the AOA problem is transformed into TDOA. As a result, the same challenges as in the case of the TOA/TDOA methods are faced by the AOA systems. In fact, the situation is more challenging as the AOA system must not match pulses originating from the direct path and indirect paths. Otherwise, fundamentally incorrect results are obtained.

7.6.4 UWB Localization: Summary

This section has introduced the fundamental concepts used in wireless localization, and explained the principles as well as their drawbacks and limitations of the methods. It is noted that despite the drawbacks the precision offered by UWB for node localization is significantly superior to the precision of any narrowband system. Thus, UWB localization remains one of the most interesting wireless technologies that can be seen as an enabler for many wireless sensor networks.

It is also noted that this section has barely scratched the surface of the topic of UWB localization. For instance, improvement in the precision of localization is achieved when cooperative localization is used, that is, the nodes also use and share the information about the relative position with other nodes to reduce the uncertainty of localization. Readers interested in the topic are referred to review the works of [Gezici et al. 2005] and [Zekavat et al. 2012].

7.7 Microwave Imaging

The use of UWB MIMO systems in microwave imaging is as exciting as its ability of precise localization. Microwave imaging for breast cancer detection has attracted significant attention from the research community for almost two decades and the research has moved from theoretical simulations to clinical trials with patients [Fear et al. 2003, Nikolova 2011].

The conventional techniques for breast cancer detection are mammography and MRI [Fear et al. 2003, Nikolova 2011]. The main issues of mammography are the use of ionizing radiation and patient discomfort due to breast compression. The disadvantage of MRI is the cost of the scan. Microwave imaging is of interest due to its potential to deliver a low-cost nonionizing method for breast cancer detection, but numerous obstacles have to be overcome before the potential of microwave imaging is fully achieved. In this section, we discuss some microwave imaging techniques and their limitations as well as the limitation of microwave imaging in general.

7.7.1 Techniques of Microwave Imaging

Most microwave imaging techniques rely on the contrast between the electromagnetic properties of malignant and healthy tissues. While the level of contrast is still subject to research especially due to differences found between *in vivo* and *ex vivo* samples [Nikolova 2011], there is a general agreement that malignant tissue has higher permittivity and conductivity than healthy tissue, for example, [Fear et al. 2003] and [Nikolova 2011]. Most microwave-imaging techniques aim to detect this contrast. The objective of most techniques is to achieve imaging with resolution below the diffraction limit of half-wavelength [Nikolova 2011].

The most promising microwave techniques that are capable of beating the diffraction limit and that have reached the stage of clinical trials are microwave tomography and UWB-pulsed radar [Fear et al. 2003, Klemm et al. 2009, Nikolova 2011].

The principle of *microwave tomography* is the following. A MIMO measurement is performed by antennas in the proximity of the breast, for instance one of the first practical setups from Dartmouth College consists of 32-channels and records data at multiple frequencies [Fear et al. 2003]. The measured channel matrix is then compared to a simulated channel matrix that is obtained using a breast model. The comparison is used in an iterative process to match the model parameters with the measured signals in order to obtain an accurate image of the breast [Fear et al. 2003]. This process is computationally challenging but it enables achieving precision on the order of a tenth of the wavelength [Fear et al. 2003, Nikolova 2011].

Pulsed radar is a time-domain technique using synthetic focusing. The technique relies on the assumption that reflections from significant scatterers will be present in all measured impulse responses [Klemm et al. 2009, Nikolova 2011]. In its simplest form, the technique can be explained using Figures 7.8 and 7.9. In Figures 7.8 and 7.9, the reflections in the channel impulse response originating from the same reflector can be matched. From the knowledge of the geometrical position of the antennas and the delays of the reflection footprints received by the antennas, the 3D angle of arrival of each ray can be determined. In microwave imaging, the same principle is applied considering more complex wave propagation in the heterogeneous medium of a breast.

Despite conceptually being a time-domain technique, the measurements are typically performed in the frequency domain using a VNA similar to the channel measurements described in Section 7.2.3. The precision of the method is given by the measurement bandwidth and the number of antennas. However, as will be discussed in Section 7.7.2, there are limits beyond which it is not feasible to increase the bandwidth or the number of antennas.

It is noted that, apart from the techniques mentioned above, there is a plethora of other microwave-imaging techniques that are being investigated, such as hybrid techniques. An example of a hybrid method is the thermoacoustic method [Nikolova 2011], which is based on the observation that various tissue types heat up differently (malignant tissue heats more rapidly due to higher conductivity). As a result of heating and cooling, the expansion and contraction of the tissue induces acoustic waves analyzed by ultrasound sensors.

7.7.2 Limitations of Microwave Imaging

The main limitations of microwave imaging are the impact of lossy tissue and antenna design [Fear et al. 2003, Nikolova 2011]. The first and perhaps the main challenge for microwave imaging is the fact that the tissues in human body are lossy as shown in Figure 7.13 presenting the electromagnetic properties of human tissue according to the Gabriel model [Gabriel 1996]. The wave coupled into the conductive medium is attenuated and has limited penetration depth. The skin depth for human tissue as a function of frequency is presented in Figure 7.14. For muscle and dry skin, the skin depth is less than 10 mm for frequencies above 2 GHz. Thus, the detection of signals reflected from small scatterers is challenging even for systems with a link budget as high as 120 dB [Fear et al. 2003].

This has a direct impact on the resolution and the application of these imaging systems. Practical systems typically do not use frequencies above 3 GHz [Klemm et al. 2009]. In the case of the pulsed radar, adding additional bandwidth above 3 GHz would actually reduce the SNR of the processed signal, in other words the resolution would not increase for increased signal bandwidth. In terms of

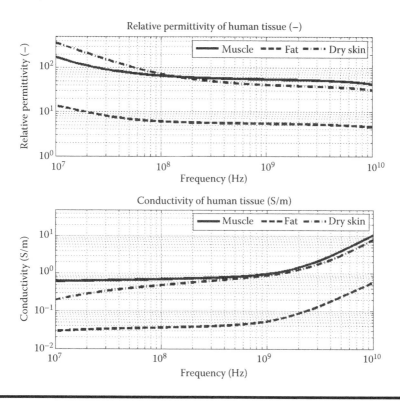

Figure 7.13 Electromagnetic properties of muscle, fat, and dry skin using tissue parameters according to [Gabriel 1996].

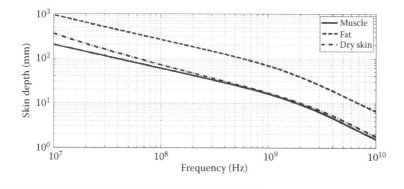

Figure 7.14 Skin depth as a function of frequency for muscle, fat, and dry skin.

applications, the losses in tissue mean that microwaves are not suitable for imaging of deep tissues and the applications are limited only to breast cancer imaging because breasts mainly consist of fatty tissue with relatively low conductivity (compared to other tissues in the human body) [Fear et al. 2003, Nikolova 2011].

In terms of antennas, there are two main challenges. As shown above, the electromagnetic properties of human tissue differ significantly from those of free space. Therefore, the antenna designs have to be matched in order for the signal to be radiated into the human body [Sipal et al. 2013a]. To increase the efficiency of radiation, a matching liquid is typically used. For breast placed directly on the antenna, the contact area between the antenna and the tissue depends on the patient, impacting the antenna efficiency. With the breast immersed in a matching liquid, this issue is resolved [Ruvio et al. 2010].

The second challenge is the physical size of the antennas and the coupling of adjacent antenna elements. It is noted that antennas designed to radiate into high permittivity tissue are significantly smaller than their free-space counterparts. However, the precision of the imaging depends strongly on the number of antennas employed. In typical proof of concept systems, two antennas are moved around the phantom by a positioner. For practical systems, the duration of these scans is not acceptable as the image would be distorted due to the movement of the patient. Thus, antenna arrays are typically used, but there is only a limited space around the breast and arrays with a large number of elements will be impacted by mutual coupling between antenna elements, which reduces the resolution capabilities.

There are numerous other challenges relating to microwave imaging. For instance, it is desirable to reduce the duration of the measurement to mitigate the impact of the patient's movement. This requires fast multichannel VNA. Another challenge is the design of low-loss fast-switching matrices so that a 4-channel VNA can service a 32-antenna imaging system [Fear et al. 2003]. Despite these challenges, microwave imaging remains a highly interesting research avenue. In the last 20 years, we have witnessed an enormous progress from theoretical simulations to clinical trials and it appears only a matter of time until microwave tomographic imaging based on UWB technology will find its way to the clinic.

7.8 Future Perspectives

The relevance of UWB technology for communications, localization, and medical imaging, among other applications, should not be underestimated. The main opportunity offered by UWB technology is the vast bandwidth. UWB bandwidth enables the emergence of many unique and interesting properties of the wireless channel and antenna arrays, including unprecedented resolution for localization and imaging systems.

The second opportunity is the possibility to reuse the system front-ends (amplifiers and antennas) that are already present in the handsets and base stations. In other words, from the system perspective, the integration of UWB in an existing system is easier than integration of a second front-end for mm-waves.

It has been 11 years since the U.S. Federal Communications Commission granted permission for unlicensed UWB radiation. During this time the understanding of the theoretical, practical, and economic challenges associated with the deployment of the technology has matured considerably.

Indeed, we are beginning to see the deployment of UWB-based technology in large-scale consumer applications. Not only can the IEEE802.11ac with 160 MHz bandwidth be considered almost a UWB system, but practical UWB chips for wireless communication and localization are entering the phase of mass production [DecaWave 2013].

With the massive performance boost offered by the synergy of UWB and MIMO technologies, it is reasonable to expect that UWB MIMO systems will enable advanced future wireless systems with wide-ranging applications.

References

Allen, B., M. Dohler, E. Okon, W.Q. Malik, A. Brown, and D.J. Edwards, Ultra wideband antennas and propagation for communications, *Radar and Imaging*, John Wiley & Sons, Chichester, UK, 2006.

Balanis, C. *Antenna Theory: Analysis and Design*, 3rd ed. Wiley-Blackwell, New Jersey, USA, May 2005.

de la Roche, G., A. A. Glazunov, and B. Allen, *LTE-Advanced and Next Generation Wireless Networks Channel Modelling and Propagation*, John Wiley & Sons, 1st edition, Chichester, UK, 2013.

Evans, P.C. and M. Annunziata, Industrial Internet: Pushing the boundaries of minds and machines, *GE Imagination at Work*, 26 November 2012.

Falciasecca, G. and B. Valotti, Guglielmo Marconi: The pioneer of wireless communications, *in Proceedings of the European Microwave Conference*, Rome, Italy, 2009, 544–546, Oct. 2009.

Fear, E.C., P.M. Meaney, and M.A. Stuchly, Microwaves for breast cancer detection? *Potentials, IEEE*, 22(1), 12,18, Feb/Mar 2003.

Gabriel, C. Compilation of the dielectric properties of body tissues at RF and microwave frequencies, 1996. Online: niremf.ifac.cnr.it/docs/dielectric/report.html

Gezici, S., Z. Tian, G.B. Giannakis, H. Kobayashi, A.F. Molisch, H.V. Poor, and Z. Sahinoglu, Localization via ultra-wideband radios: A look at positioning aspects for future sensor networks, *Signal Processing Magazine, IEEE*, 22(4), 70–84, July 2005.

Goldsmith, A., S. Jafar, I. Maric, and S. Srinivasa, Breaking spectrum gridlock with cognitive radios: An information theoretic perspective, *in Proceedings of the IEEE*, 97(5), 894–914, May 2009.

Heidari, G. *WiMedia UWB—Technology Choice for Wireless USB and Bluetooth*, John Wiley & Sons, Chichester, UK, 2008.

IEEE 802.11: Wireless LAN Medium Access Control (MAC) and Physical Layer (PHY) Specifications, 2012 revision, IEEE-SA. 5 April 2012.

IEEE 802.15.4: Low-rate wireless personal area networks (LR-WPANs) *IEEE-SA*, 16 June 2011.

IEEE 802.15.6: Wireless body area networks, *IEEE-SA*, 29 Feb. 2012.

Kaiser, T., F. Zheng, and E. Dimitrov, An overview of ultra-wide-band systems with MIMO, *Proceedings of the IEEE*, 97(2), 285–312, Feb. 2009.

Klemm, M., J.A. Leendertz, D. Gibbins, I.J. Craddock, A. Preece, and R. Benjamin, Microwave radar-based breast cancer detection: Imaging in inhomogeneous breast phantoms, *Antennas and Wireless Propagation Letters, IEEE*, 8, 1349–1352, 2009.

Malik, W.Q. and B. Allen, Wireless sensor positioning with ultrawideband fingerprinting, *European Conference on Antennas and Propagation* (EuCAP) 2006, Nice, France, pp.1–5, 6–10 Nov. 2006.

Malik, W.Q. and A.F. Molisch, Ultrawideband antenna arrays and directional propagation channels, *European Conference on Antennas and Propagation* (EuCAP) 2006, Nice, France, Nov. 2006.

Malik, W.Q., B. Allen, and D.J. Edwards, A simple adaptive beamformer for ultrawideband wireless systems, *in Proceedings of the 2006 IEEE International Conference on Ultra-Wideband*, Waltham, USA, 453–457, 24–27 Sept. 2006.

Malik, W.Q., B. Allen, and D.J. Edwards, Impact of bandwidth on small-scale fade depth, *GLOBECOM 2007, IEEE*, 3837–3841, 26–30 Nov. 2007.

Malik, W.Q., B. Allen, and D.J. Edwards, Bandwidth-dependent modelling of smallscale fade depth in wireless channels, *Microwaves, Antennas & Propagation, IET*, 2(6), 519–528, Sept. 2008.

Malik, W.Q., C.J. Stevens, and D.J. Edwards, Synthetic aperture analysis of multipath propagation in the ultra-wideband communications channel, *Signal Processing Advances in Wireless Communications, 2005 IEEE*, 375–379, 5–8 June 2005.

Malik, W.Q., C.J. Stevens, and D.J., Edwards, Ultrawideband antenna distortion compensation, *Transactions on Antennas and Propagation, IEEE*, 56(7), 1900–1907, July 2008.

Mallat, A. J., Louveaux, and L. Vandendorpe, UWB based positioning: Cramer Rao bound for angle of arrival and comparison with time of arrival, *in Proceedings of 2006 Symposium on Communications and Vehicular Technology*, 65–68, 23–23 Nov. 2006.

Mass production tape-out keeps Scensor on course for 2013 release, *DecaWave Newsletter*, Issue 9, Summer 2013.

Molisch, A.F. et al. IEEE 802.15.4a channel model—Final report, *IEEE, Technical Report*, 2004.

Nikolova, N.K. Microwave imaging for breast cancer, *Microwave Magazine, IEEE*, 12(7), 78,94, Dec. 2011.

Ohm, J.R. and H.D. Lueke, *Signalubertragung: Grundlagen der digitalen und analogen Nachrichtenubertragung*, 10th ed. Springer, Berlin, 2006.

Ruvio, G., R. Solimene, and M.J. Ammann, Evaluation of antenna types for RF breast cancer imaging using 2-layer planar tissue model, *in Proceedings of Microwave Conference (EuMC), 2010 European*, Paris, France, 212–215, 28–30 Sept. 2010.

Standard ECMA-368 High rate ultra wideband PHY and MAC standard, *ECMA*, 3rd ed. (December 2008)

Sipal, V. Impact of the wireless channel on the performance of ultrawideband communication systems, PhD thesis, University of Oxford, Dec. 2012.

Sipal, V., B. Allen, and D. Edwards, Effects of antenna impulse response on wideband wireless channel, *in Proc. of the Antennas and Propagation Conf. (LAPC)*, 2010 Loughborough, 129(132), 8–9 Nov. 2010.

Sipal, V., B. Allen, and B.D. Edwards, Descriptor choice for UWB antenna arrays, *in Proc. of the European Conference on Antennas and Propagation. EUCAP* 2012, Prague, Czech Republic, 603–606, 26–30 Mar. 2012a.

Sipal, V., D. Edwards, and B. Allen, On the range of Wimedia OFDM UWB wireless, Ultra-Wideband (ICUWB), 2012 *IEEE Conference*, Syracuse, USA, 421–425, 17–20 Sep. 2012b.

Sipal, V., D. Edwards, and B. Allen, Bandwidth requirement for suppression of grating lobes in ultrawideband antenna arrays, *in Proceedings of the 2012 IEEE International Conference on Ultra-Wideband (ICUWB)*, Syracuse, USA, 421–425, 17–20 Sept. 2012c.

Sipal, V., B. Allen, D. Edwards, and B. Honary, 20 Years of ultrawideband: Opportunities and challenges, *IET Communications*, 6(10), 1147–1162, July 3, 2012d.

Sipal, V., J.E. Browne, P. McEvoy, and M.J. Ammann, Evaluation of antenna suitability for the use in radiomyography, *in Proceedings of the Loughborough Antennas and Propagation Conference (LAPC)*, Loughborough, UK, 2013, Nov. 2013a.

Sipal, V., J. Gelabert, C. Stevens, B. Allen, and D. Edwards, Frequency-selective fading of ultrawideband wireless channels in confined environments, *Microwaves, Antennas & Propagation, IET*, 5(11), 1328–1335, August 19 2011.

Sipal, V., J. Gelabert, C. Stevens, B. Allen, and D. Edwards, Adaptive OFDM for wireless interconnect in confined enclosures, *Wireless Communications Letters, IEEE*, 2(5), 507–510, 2013b.

Tarique, Z., W.Q. Malik, and D.J. Edwards, Bandwidth requirements for accurate detection of direct path in multipath environment, *Electronics Letters*, 42(2), 100–102, Jan. 19, 2006a.

Tarique, Z., W.Q. Malik, and D.J. Edwards, Effect of bandwidth and antenna directivity on the range estimation accuracy in a multipath environment, in *Proc. of the Vehicular Technology Conf. (VTC)* 2006-Spring. *IEEE* 63rd, 6, 2887–2890, May 7–10, 2006b.

Vithanage, C., M. Sandell, J. Coon, and Y. Wang, Precoding in OFDM-based multi-antenna ultra-wideband systems, *Communications Magazine, IEEE*, 47(1), 41–47, Jan. 2009.

Zekavat, R. and R. Buehrer, Wireless localization using ultra-wideband signals, in *Handbook of Position Location: Theory, Practice and Advances*, Wiley-IEEE Press, New Jersey, USA, 1st edition, 2012.

Chapter 8

Frequency-Domain Packet Combining Techniques for UWB

Paulo Montezuma, Mário Marques da Silva, and Rui Dinis

Contents

8.1 Introduction

Currently, there is a big effort to standardize ultra wideband (UWB)-based communication systems for both military and civil applications (see [Siwiaik 2002] and the references therein). The reduced power spectral density (PSD) levels of the transmitted signals allow the coexistence of UWB signals with other narrowband systems without significant interference of the narrowband systems. Such reduced PSD results from the low transmit power and from the high bandwidth [Siwiak 2002] (UWB signal bandwidths may exceed 25% of the central frequency).

Initially, UWB systems were spread-spectrum based, with extremely high bandwidths and very high processing gains. This results in a system very effective against multipath propagation impairments, namely, fading and time-dispersion effects [Win and Scholtz 1998]. Spread-spectrum UWB-based systems can be implemented either in baseband or using a continuous wave. In the latter case, the carrier frequency should be of the order of the signal bandwidth.

Afterwards, impulse radio technique was widely adopted in UWB signals. This employs short pulses (typically with a duration below 1 ns), with zero mean and a small number of zero crossings, combined with time-hopping multiple access (TH-MA). These pulses are sent directly to the transmit antenna (i.e., a "baseband" transmission is employed), which results in a low implementation complexity. Nevertheless, since the achievable spectral efficiencies with impulse radio techniques are typically low, UWB systems employing continuous-wave techniques have received much attention [Ishiyama and Ohtsuki 2004]. Since these pulses are sent directly to the transmit antenna (i.e., a "baseband" transmission is employed), the implementation complexity at both the transmitter and the receiver can be relatively low.

Moreover, the despreading operations are usually simpler than for continuous-wave UWB options. However, since the achievable spectral efficiencies with impulse radio techniques are typically low, there is an increased interest on UWB systems employing continuous-wave techniques [Ishiyama and Ohtsuki 2004, Popescu et al. 2005]. Owing to the high signaling rates, the time-dispersion effects associated with multipath propagation can be severe in UWB systems. For this reason, continuous-wave UWB systems should employ block transmission techniques, with appropriate cyclic extensions and be combined with FDE techniques (frequency-domain equalization) [Gusmão et al. 2000, Falconer et al. 2002].

Orthogonal frequency division multiplexing (OFDM) schemes are the most popular modulations based on this concept, but the transmitted signals have high envelope fluctuations and a high peak-to-mean envelope power ratio (PMEPR), leading to amplification difficulties. For this reason, single-carrier with frequency-domain equalization (SC-FDE) schemes are particularly interesting for low-cost UWB terminals, especially when iterative receivers are employed [Benvenuto and Tomasin 2002, Dinis et al. 2003].

Block transmission techniques, with appropriate cyclic prefixes and employing FDE techniques, have been shown to be suitable for high data rate transmission over severely time-dispersive channels [Gusmão et al. 2000, Falconer et al. 2002]. Two possible alternatives based on this principle are OFDM and single carrier (SC) modulation using FDE (or SC-FDE). Owing to the lower envelope fluctuations of the transmitted signals (and, implicitly a lower PMEPR), SC-FDE schemes are especially interesting for the uplink transmission (i.e., the transmission from the mobile terminal to the base station) [Gusmão et al. 2000, Falconer et al. 2002], being considered for use in the upcoming LTE (long-term evolution) cellular system.

A promising iterative block–decision feedback equalization technique (IB-DFE) for SC-FDE was proposed in [Benvenuto and Tomasin 2002] and extended to other diversity [Marques da Silva and Dinis 2011] and spatial multiplexing scenarios [Vutecic and Yuan 2002, Gusmão et al. 2003]. These IB-DFE receivers can be regarded as iterative DFE receivers with the feedforward and the feedback operations implemented in the frequency domain. Since the feedback loop takes into account not just the hard decisions for each block but also the overall block reliability, error propagation is reduced. Consequently, IB-DFE techniques offer much better performance than noniterative methods [Benvenuto and Tomasin 2002, Dinis et al. 2003]. Within these IB-DFE receivers, the equalization and channel decoding procedures are performed separately (i.e., the feedback loop uses the equalizer outputs instead of the channel decoder outputs). However, it is known that higher performance gains can be achieved if these procedures are performed jointly. This can be done by employing turbo equalization schemes, where the equalization and decoding procedures are repeated in an iterative way [Marques da Silva and Dinis 2011], being essential in multiple-input multiple-output (MIMO) schemes with high-order modulations. For these reasons, SC-FDE schemes are particularly interesting for low-cost UWB terminals, especially when iterative receivers are employed.

It is well known that channel noise can lead to packet errors. This problem is particularly serious in wireless systems since fading and shadowing effects can lead to significant decrease in the received power and, consequently, significant packet error rates. The traditional approach to cope with an erroneous packet is to discard it and ask for their retransmission, which corresponds to employ conventional automatic repeat request (ARQ) techniques. The major problem with conventional ARQ schemes is that they can have very poor performance when the errors are due to unfavorable propagation conditions that remain for some time, since packet errors are likely to occur for several transmission attempts.

To cope with this type of scenarios, hybrid ARQ/forward error correction (FEC) strategies have been proposed [Hagenauer et al. 1988, Gusmão et al. 1999]. The basic idea behind these techniques is to retain the signal associated with an erroneous packet and to ask for additional redundancy. This is usually done by employing a basic error correction code that is very powerful and has low rate. This code is punctured, leading to a code with higher rate, but smaller error correction capabilities that is used in the first transmission attempt of the packet. If we have an error in the packet detection, the additional bits that were punctured are transmitted, instead of a conventional packet retransmission, increasing the error correction capabilities of the code. The major problem with these techniques is that we are limited by the performance of the basic code that was adopted. If we want a very powerful basic code, we need codes that are too complex and/or require very long blocks (e.g., turbo codes [Vucetic 2002]). The basic code needs to be designed for a worst-case scenario that is seldom used.

An alternative is to take advantage of the last received packets (regardless of the reason of the packet loss) and combine them until we end up receiving a packet successfully. In terms of the implementation complexity, the simplest way is to employ soft packet combining techniques that can be regarded as Hybrid-ARQ type II techniques based on repetition codes. They allow improved performance with the number of retransmission attempts. Hybrid-ARQ techniques are currently used in 3GPP's high-speed downlink packet access (HSDPA), high-speed uplink packet access (HSUPA), and LTE. In diversity combining techniques (DC), the usage of repetition codes with soft decision can be regarded as a low-complexity version of the conventional Hybrid-ARQ.

Moreover, soft packet combining techniques proposed in [Hagenauer et al. 1988] can be an alternative to conventional ARQ schemes. Within these techniques, packets associated with different transmission attempts are combined in a soft way, allowing improved performances [Hagenauer et al. 1988]. These techniques can be regarded as hybrid ARQ/FEC schemes based on repetition codes with soft decision, with low encoding/decoding complexity. Moreover, their performance is not bounded by the performance of the basic code, since it is not a conventional punctured code. A promising soft combining ARQ technique for SC-FDE was proposed in [Dinis et al. 2008], which was shown to be very effective to cope with fading effects.

However, when very severe fading conditions affect the transmission of several packets, the number of retransmissions can be very high, which compromises both spectral and energy efficiencies of the system. In the majority of dispersive environments, the antenna diversity is widely used to reduce the multipath fading and increase reliability. In addition, by using diversity, having multiple transmit/receive antennas should increase data throughput.

Diversity schemes can be applied either to the transmitter or to the receiver, the most common scheme being the use of several antennas in the receiver together with one combining method to enhance the quality of the received signal [Proakis and Salehi 2007]. However, in wireless mobile systems, constraints on size and power consumption of mobile devices can impose limitations to the implementation of this technique. However, from the base station's side, these limitations are less worrying because one of them can give service to hundreds of mobile subscribers.

Using multiple antennas can result in a smaller probability of error for the same throughput because of the diversity gain. A particularly elegant scheme for MIMO coding was developed by Alamouti [Alamouti 1998]. The MIMO Alamouti scheme is a transmit diversity scheme for two transmit antennas that does not require transmit channel knowledge. The main objective of space–time codes is to achieve the maximum possible diversity. Space–time codes provide a diversity gain equal to the product of the number of transmit and receive antennas. Also, the maximum throughput of the space–time codes is one symbol per channel use for any number of transmit antennas. The use of multiple antennas results in increasing the capacity of MIMO channels. Therefore, one may transmit at a higher throughput, compared to single-input single-output (SISO) channels, for a given probability of error.

The idea behind using a transmit diversity scheme is that maybe some of the redundant-sent signals can arrive in a better state to the receiver than others, and by exploiting them all together, the result should be better. The Alamouti scheme can also be easily generalized to two transmit antennas and *Nr* receive antennas, in a MIMO system. One of the biggest advantages is that the scheme requires no bandwidth increase because redundancy is applied in space and time across multiple antennas. It does not require higher transmit power either. These restrictions are the most important for wireless communication systems. The new scheme is able to improve error performance, data rate, or capacity of wireless systems without increasing bandwidth or transmit power. The smaller sensitivity to fading permits the system to use a higher-level modulation (a modulation that transmits more bits per symbol) to increase the bit rate or a smaller reuse factor to increase the capacity.

8.2 Single Carrier with Frequency-Domain Equalization

To explain the SC-FDE schemes, we start with a brief introduction where we consider an SC-based block $\{a_n; n = 0,1, \ldots, N-1\}$ transmission with N useful symbols per block resulting directly from a mapping operation of the original data into a selected signal constellation plus a suitable cyclic prefix (CP) longer than the overall delay spread of the channel. Let us consider block transmission schemes where the *l*th transmitted block has the form

$$s_l(t) = \sum_{n=-N_G}^{N-1} a_{n,l} h_T(t - nT_s), \tag{8.1}$$

with T_s denoting the symbol duration, N_G denoting the number of samples in the cyclic prefix, and $h_T(t)$ representing the adopted pulse-shaping filter. The signal $s_l(t)$ is transmitted over a time-dispersive channel. At the receiver, the received signal is sampled and after removing the samples associated with the CP results the following block in the time domain

$$\{y_{n,l}; n = 0,1,\ldots,N-1\} \tag{8.2}$$

The samples of the block are passed to the frequency domain by an N-point discrete Fourier transform (DFT), leading to the corresponding frequency-domain block samples $\{Y_{k,l}; k = 0,1,\dots,N-1\}$, where $Y_{k,l} = H_{k,l}A_{k,l} + N_{k,l}$. To minimize the effects of ISI (intersymbol interference) and channel noise, the equalized samples of the signal $\{\tilde{A}_{k,l}; k = 0,1,\dots,N-1\}$ are obtained based on the coefficients $\{F_k; k = 0,1, \dots, N-1\}$ that are optimized under the minimum mean square error (MMSE), which leads to the set of FDE coefficients

$$F_k = \frac{H_k^*}{\alpha + \left|H_k\right|^2}, \quad k = 1,2,\dots,N-1 \tag{8.3}$$

where α is the inverse of the signal-to-noise ratio (SNR) given by

$$\alpha = \frac{\sigma_N^2}{\sigma_s^2} \tag{8.4}$$

with $\alpha = E\left[\left|N_k\right|^2\right]/2$ and $\sigma_s^2 = E\left[\left|S_k\right|^2\right]/2$ denoting the variances of the real and imaginary parts of the channel noise $\{N_{k,l}; k = 0,1, \dots, N-1\}$ and the data samples $\{A_{k,l}; k = 0,1, \dots, N-1\}$, respectively.

In SC modulations, the data contents of a given block are transmitted in the time domain. Therefore, the equalized samples $\{\tilde{A}_{k,l}; k = 0,1,\dots,N-1\}$ are converted again into time domain by an inverse DFT (IDFT) operation leading to the block of time-domain-equalized samples $\{\tilde{a}_{n,l}; n = 0,1,\dots,N-1\}$. Then, these equalized samples are used in the decision process about the transmitted symbols or block.

In IB-DFE scheme (iterative block DFE), both the feedforward and feedback parts are implemented in the frequency domain. Thus, for a given *it*th iteration, the output samples are given by

$$\tilde{A}_k^{(it)} = F_k^{(it)}Y_k^{(it)} - B_k^{(it)}\hat{A}_k^{(it-1)} \tag{8.5}$$

where $\{F_k^{(it)}; k = 0,1,\dots,N-1\}$ and $\{B_k^{(it)}; k = 0,1,\dots,N-1\}$ denote the feedforward and feedback coefficients, respectively, and $\{A_{k,l}; k = 0,1, \dots, N-1\}$ is the DFT of the hard decision of the lth block $\{\breve{a}_{n,l}^{(it)}; n = 0,1,\dots,N-1\}$ for the ith iteration, associated with the transmitted time domain of the lth block $\{a_{n,l}; n = 0,1, \dots, N-1\}$. The feedforward and feedback coefficients are still chosen to maximize the signal-to-interference plus noise ratio (SINR), with the optimum coefficients given by

$$F_k^{(it)} = \frac{\mathrm{K}_F^{(it)} H_k^*}{\left[\alpha_k + (1-(\rho^{(it-1)})^2)\left|H_k\right|^2\right]}, \tag{8.6}$$

and

$$B_k^{(it)} = \rho^{(it-1)}\left(F_k^{(it)} H_k - 1\right), \tag{8.7}$$

respectively, where $\mathbf{K}_F^{(it)}$ is selected to assure that $1/N \sum_{it=0}^{N-1} F_k^{(it)} H_k = 1$ and $\rho^{(it)}$ is a measure of the reliability of the decisions used in the feedback loop. Since block estimates are used by the "feedback" filter (weighed by the log-likelihoods), the IB-DFE coefficients take into account the overall block reliability on the feedback loop, which leads to a small error propagation effect. In fact, IB-DFE has a turbo-like behavior and can be regarded as a low-complexity turbo equalizer [Tüchler and Hanenauer 2001].

8.3 Iterative Frequency-Domain Packet Combining Techniques Optimized for UWB

A promising soft combining ARQ technique for SC-FDE was proposed in [Dinis et al. 2008], which was shown to be very effective to cope with fading effects. We consider the uplink transmission in UWB wireless systems employing SC-FDE schemes. The basic receiver structure is depicted in Figure 8.1, where the DFT and IDFT blocks refer to implementations of discrete Fourier transform and its inverse. The time-domain block is associated with a given user (i.e., the corresponding packet) is $\{a_n; n = 0,1,\dots, N-1\}$ where a_n is selected from a given constellation and N is the fast Fourier transform (FFT) size. As usual, for SC-FDE block transmission techniques, a suitable cyclic prefix longer than the overall delay spread of the channel is added to each time-domain block. For the sake of simplicity, we assume a one-to-one correspondence between time-domain blocks and user packets.

If errors are detected in a received packet, we ask for its retransmission, but we store the signal associated with each transmission attempt. Although we could keep trying to transmit the packet until there were no errors, in practice, there is a maximum number of transmission attempts N_R. If we fail after N_R attempts, we need to change the transmission parameters (transmit power, carrier frequency, base station, etc.) since the channel is too bad. The packet associated with the rth transmission attempt of $\{a_n; n = 0,1,\dots,N-1\}$ is $\{a_n^{(r)}; n = 0,1,\dots, N-1\}$. The received signal associated with the rth transmission attempt is sampled and the cyclic prefix is removed, leading to the time-domain block $\{y_n^{(r)}; n = 0,1,\dots, N-1\}$. If the cyclic prefix is longer than the overall channel impulse response, then the corresponding frequency-domain block is $\{Y_k^{(r)}; k = 0,1,\dots, N-1\}$, where

$$Y_k^{(r)} = A_k^{(r)} H_k^{(r)} + N_k^{(r)}, \tag{8.8}$$

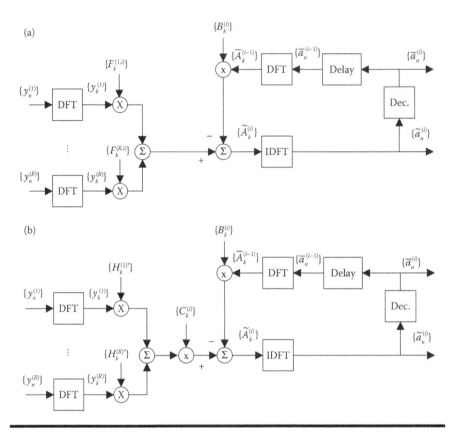

Figure 8.1 Receiver structure (a) and alternative format (b), since $F_k^{(r,i)} = H_k^{(r)} * C_k^{(i)}, (r = 1,2,...,R)$.

with $N_k^{(r)}$ denoting the channel noise. $\left\{ A_k^{(r)}; k = 0,1,...,N - 1 \right\}$ is the DFT of $\left\{ a_n^{(r)}; n = 0,1,...,N - 1 \right\}$ and $\left\{ H_k^{(r)}; k = 0,1,...,N - 1 \right\}$ is the overall channel frequency response for the rth transmission attempt.

8.4 Conventional Soft Packet Combining ARQ

Let us assume that we have R versions of the packet (i.e., there were R transmission attempts). Our receiver, depicted in Figure 8.1, consists of an iterative frequency-domain receiver where, for a given iteration i, the frequency-domain samples at the output are given by

$$\tilde{A}_k^{(i)} = \sum_{r=1}^{R} F_k^{(r,i)} Y_k^{(r)} - B_k^{(i)} \overline{A}_k^{(i-1)}, \tag{8.9}$$

where $\{F_k^{(r,i)}, k = 0,1,...,N-1\}(r = 1,2,...,R)$ and $\{B_k^{(i)}, k = 0,1,...,N-1\}$ are the feed-forward and the feedback coefficients, respectively. $\{\bar{A}_k^{(i-1)}, k = 0,1,...,N-1\}$ denotes the DFT of the average data estimates $\{\bar{A}_k^{(i-1)}, k = 0,1,...,N-1\} = \text{DFT}\{\bar{a}_n^{(i-1)}, k = 0,1,...,N-1\}$, where $\bar{a}_n$ denotes the average symbol values conditioned to the FDE output. For quadrature phase shift keying (QPSK) constellations, these average values are given by [Marques da Silva et al. 2010]

$$\bar{a}_n^{(i)} = \tanh\left(\frac{L_n^{I(i)}}{2}\right) + j\tanh\left(\frac{L_n^{Q(i)}}{2}\right),\qquad(8.10)$$

(without loss of generality, we assume that $|a_n|^2 = 2$, i.e., $a_n = \pm 1 \pm j$), with

$$L_n^{I(i)} = \frac{2}{\sigma_{eq}^2}\text{Re}\left\{\tilde{a}_n^{(i)}\right\}\qquad(8.11)$$

and

$$L_n^{Q(i)} = \frac{2}{\sigma_{eq}^2}\text{Im}\left\{\tilde{a}_n^{(i)}\right\}\qquad(8.12)$$

denoting the log-likelihood ratios (LLRs) of the "in-phase bit" and the "quadrature bit," associated with a_n, respectively, and $\{\tilde{a}_n^{(i)}; n = 0,1,...,N-1\} = \text{IDFT}\{\tilde{A}_k^{(i)}; k = 0,1,...,N-1\}$. The variance σ_{eq}^2 is given by

$$\sigma_{eq}^2 = \frac{1}{2}\text{E}\left[|a_n - \tilde{a}_n^{(i)}|^2\right] \approx \frac{1}{2N}\sum_{n=0}^{N-1}|\hat{a}_n^{(i)} - \tilde{a}_n^{(i)}|^2,\qquad(8.13)$$

where $\hat{a}_n^{(i)} = \pm 1 \pm j$ are the hard decisions associated with $\tilde{a}_n^{(i)}$. The simplest way to design the receiver is to consider only the channel noise, which corresponds to ignore the interference in its design. Under these conditions, the optimum feedforward coefficients, for a given iteration, can be written as

$$F_k^{(r,i)} = \frac{\breve{F}_k^{(r,i)}}{\gamma^{(i)}},\qquad(8.14)$$

with $\gamma^{(i)}$ and $\breve{F}_k^{(r,i)}$ as defined in [Dinis et al. 2008]. Moreover, the optimum feedback coefficients $B_k^{(i)}$ are also defined in [Dinis et al. 2008].

8.4.1 Packet Combining with Strong Interference Levels

The receiver can be designed taking into account the characteristics of the interference, which is added to the thermal channel noise.* This means that the FDE output is still given by Equation 8.9, with the feedforward coefficients given by Equation 8.14, but with

$$\breve{F}_k^{(r,1)} = \frac{H_k^{(r)*}}{\left[\alpha_k^{eq(r)} + (1 - (\rho^{(i-1)})^2) \sum_{r'=1}^{R} \left| H_k^{(r')} \right|^2 \right]}, \quad r = 1, 2, ..., R, \quad (8.15)$$

where

$$\alpha_k^{eq(r)} = \frac{E\left[\left| N_k^{(r)} \right|^2 \right] + E\left[\left| I_k^{(r)} \right|^2 \right]}{E\left[\left| A_k \right|^2 \right]}, \quad (8.16)$$

with $I_k^{(r)}$ denoting the interference at kth subcarrier and the rth retransmission attempt (i.e., $\alpha_k^{eq(r)}$ can be regarded as the inverse of the equivalent SINR). The correlation coefficient $\rho^{(i)}$ is given by

$$\rho^{(i)} = \frac{1}{2N} \sum_{n=0}^{N-1} \left(\left| \text{Re}\left\{ \overline{a}_n^{(i)} \right\} \right| + \left| \text{Im}\left\{ \overline{a}_n^{(i)} \right\} \right| \right). \quad (8.17)$$

The optimum feedback coefficients are the same as defined in [Gusmão et al. 2003].

8.5 Packet Combining with Diversity

The MIMO Alamouti code is a simple space–time block code (STBC), where the different replicas sent for exploiting diversity are generated by a space–time encoder, which encodes a single stream through space using all the transmit antennas and through time by sending each symbol at different times.

The Alamouti STBC scheme uses two transmit antennas and Nr receive antennas and can accomplish a maximum diversity order of $2Nr$ [Alamouti 1998]. Moreover, the Alamouti scheme has full rate (i.e., a rate of 1) since it transmits two symbols every two time intervals. Next, a description of the Alamouti scheme

---

* For the sake of simplicity, we will assume that the interference is Gaussian with zero mean and uncorrelated from subcarrier to subcarrier. The extension to other cases is straightforward.

is provided for both one and two receive antennas, followed by a generalization for *Nr* receive antennas. For an Alamouti STBC with two transmit antennas, the time-domain block to be transmitted by the *m*th antenna are $s_{n,i}$ and the encoding operation can be described in time domain as (see also Chapter 1, for a description of the STBC scheme)

$$
\begin{array}{cccc}
Time & Antenna\,1 & Antenna\,2 \\
2l-1 & \{s_{n,2l-1}\} & \{s_{n,2l}\} \\
2l & \{-s^*_{n,2l}\} & \{s^*_{n,2l-1}\},
\end{array}
\qquad (8.18)
$$

where we assume that the rows of each coding scheme represent a different time instant, while the columns represent the transmitted symbol through each different antenna, as shown in Figure 8.2. In this case, the first and second rows represent the transmission at the first and second time instant, respectively.

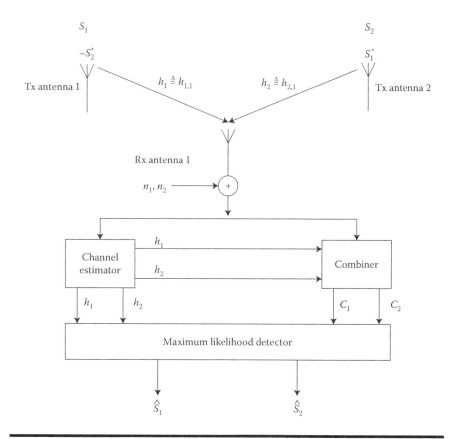

Figure 8.2 STBC with two transmit and one receive antenna.

Let us define $s_{n,2i} \triangleq s_1$ and $s_{n,2i+1} \triangleq s_2$. Considering the matrix–vector representation, the Alamouti scheme encoding operation becomes [Marques da Silva et al. 2012]

$$G_2 = \underbrace{\begin{bmatrix} s_1 & s_2 \\ -s_2^* & s_1^* \end{bmatrix}}_{\text{space}} \Bigg\} \text{time} \tag{8.19}$$

At a time $t = (2i - 1)T$, symbols s_1 and s_2 are transmitted from antenna 1 and antenna 2, respectively. Assuming a symbol duration of T, at time $t = 2iT$ are transmitted the symbols $-s_2^*$ and s_1^* from antenna 1 and antenna 2, respectively.

For one receive antenna, the received signals are

$$
\begin{aligned}
y_1 &= r(t) = h^{(1,1)} s_{n,2i} + h^{(2,1)} s_{n,2i+1} + n^{(1)} = h_1 s_{n,2i} + h_2 s_2 + n_1 \\
y_2 &= r(t + T) = -h^{(1,1)} s_{n,2i+1}^* + h^{(2,1)} s_{n,2i}^* + n_2^{(1)} = -h_1 s_2^* + h_2 s_1^* + n_2,
\end{aligned}
\tag{8.20}
$$

where $y_{n,i}^{(l)}$ denotes the received signal at antenna l (for one receive antenna, we have $l = 1$) at time iT, and $h_{n,i}^{(m,l)}$ denotes the channel response for the path between the mth transmit antenna and lth receive antenna. For the sake of simplicity, we assume $h_{n,i}^{(Ntr,Nr)} \triangleq h_{Ntr,Nr}$, where Ntr and Nr are the number of transmit and receive antennas, respectively. The subscript and superscript associated with the receive antenna are omitted when we have only one antenna at the receiver, so in this case, we may assume $h_{n,i}^{(1)} \triangleq h_1$ and $h_{n,i}^{(2)} \triangleq h_2$. Before being sent to the equalizer, the received signals are combined as follows:

$$
\begin{aligned}
c_1 &= h_1^* y_1 + h_2 y_2^* = h_1^* h_1 s_1 + h_1^* h_2 s_2 + h_1^* n_1 - h_2 h_1^* s_2 + h_2^* h_2 s_1 + h_2 n_2^* \\
c_2 &= h_2^* y_1 - h_1 y_2^* = h_2^* h_1 s_1 + h_2^* h_2 s_2 + h_2^* n_1 + h_1 h_1^* s_2 - h_1^* h_2 s_1 - h_1 n_2^*,
\end{aligned}
\tag{8.21}
$$

with the corresponding representation in frequency domain given by

$$
\begin{bmatrix} C_{k,2i} \\ C_{k,2i+1} \end{bmatrix} = \begin{bmatrix} H_{k,i}^{(1)*} & H_{k,i}^{(2)} \\ H_{k,i}^{(2)*} & -H_{k,i}^{(1)} \end{bmatrix} \begin{bmatrix} Y_{k,2i} \\ Y_{k,2i+1}^* \end{bmatrix},
\tag{8.22}
$$

where $\{H_k^{(m)}; k = 0,1,\dots,N - 1\} = \text{DFT}\,\{h_n^{(m)}; n = 0,1,\dots,N - 1\}$ represents the channel frequency response for the kth subcarrier and the mth transmit antenna, $N_{k,i} = \text{DFT}\,\{n_i\}$ is the noise term and $\{Y_{k,i}; k = 0,1,\dots,N - 1\} = \text{DFT}\,\{y_{n,i}; n = 0,1,\dots,N - 1\}$. Assuming for now a linear FDE for SC schemes, the Alamouti postprocessing becomes

$$
\tilde{A}_{k,2i+1} = \frac{\left[Y_{k,2i} H_{k,i}^{(1)*} + Y_{k,2i+1}^* H_{k,i}^{(2)*} \right]}{\alpha + \sum_{m=1}^{Ntr} \left| H_{k,i}^{(m)} \right|^2},
\tag{8.23}
$$

$$\tilde{A}_{k,2i} = \frac{\left[Y_{k,2i+1} H_{k,i}^{(1)*} + Y_{k,2i}^* H_{k,i}^{(2)} \right]}{\alpha + \sum_{m=1}^{N_{tr}} \left| H_{k,i}^{(m)} \right|^2}, \tag{8.24}$$

where $\{ A_{k,i}; k = 0,1,...,N-1 \} = \mathrm{DFT}\{ a_{n,i}; n = 0,1,...,N-1 \}$ and α can be regarded as the inverse of the equivalent SNR.

Let us consider now the case of two receive antennas, as depicted in Figure 8.3, in which the received signals are given by

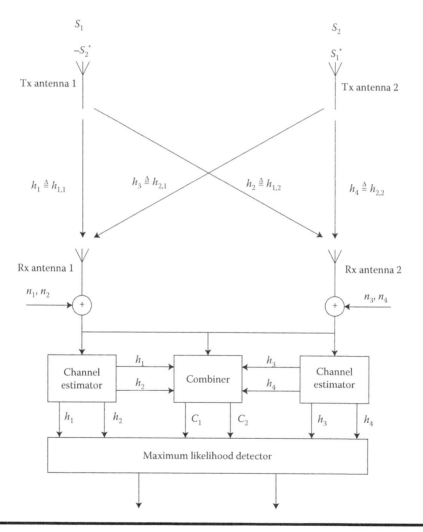

Figure 8.3   STBC for two transmit and two receive antennas.

$$
\begin{aligned}
y_1^{(1)} &= h_{1,1}s_1 + h_{2,1}s_2 + n_1^{(1)} \\
y_2^{(1)} &= -h_{1,1}s_2^* + h_{2,1}s_1^* + n_2^{(1)} \\
y_1^{(2)} &= h_{1,2}s_1 + h_{2,2}s_2 + n_1^{(2)} \\
y_2^{(2)} &= -h_{1,2}s_2^* + h_{2,2}s_1^* + n_2^{(2)},
\end{aligned}
\tag{8.25}
$$

where $y_1^{(l)}$ and $y_2^{(l)}$ denote the received signal by the lth antenna at time $2iT$ and $(2i+1)T$, respectively (the same notation applies to the noise terms $n_1^{(l)}$ and $n_1^{(l)}$). Assuming $y_1^{(1)} \triangleq y_1$, $y_1^{(2)} \triangleq y_2$, $y_2^{(1)} \triangleq y_3$, $y_2^{(2)} \triangleq y_4$, $h_{1,1} \triangleq h_1$, $h_{2,1} \triangleq h_2$, $h_{1,2} \triangleq h_3$, and $h_{2,2} \triangleq h_4$, we may write

$$
\begin{bmatrix}
y_1 \\ y_2 \\ y_3 \\ y_4
\end{bmatrix}
=
\begin{bmatrix}
s_1 & s_2 & 0 & 0 \\
-s_2^* & s_1^* & 0 & 0 \\
0 & 0 & s_1 & s_2 \\
0 & 0 & -s_2^* & s_1^*
\end{bmatrix}
\begin{bmatrix}
h_1 \\ h_2 \\ h_3 \\ h_4
\end{bmatrix}
+
\begin{bmatrix}
n_1 \\ n_2 \\ n_3 \\ n_4
\end{bmatrix}
\tag{8.26}
$$

with the resulting signals after the combiner

$$
\begin{aligned}
c_1 &= h_1^* y_1 + h_2 y_2^* + h_3^* y_3 + h_4 y_4^* \\
&= \left(|h_1|^2 + |h_2|^2 + |h_3|^2 + |h_4|^2 \right) s_1 + h_1^* n_1 + h_2 n_2^* + h_3^* n_3 + h_4 n_4^* \\
c_2 &= h_2^* y_1 - h_1 y_2^* + h_4^* r_3 - h_3 y_4^* \\
&= \left(|h_1|^2 + |h_2|^2 + |h_3|^2 + |h_4|^2 \right) s_2 - h_1^* n_2 + h_2^* n_1 - h_3^* n_4 + h_4^* n_3,
\end{aligned}
\tag{8.27}
$$

and in the matrix–vector representation results

$$
\begin{bmatrix} c_1 \\ c_2 \end{bmatrix}
=
\begin{bmatrix}
\sum_{l=1}^{4} |h_l|^2 & 0 \\
0 & \sum_{l=1}^{4} |h_l|^2
\end{bmatrix}
\cdot
\begin{bmatrix} s_1 \\ s_2 \end{bmatrix}
+
\begin{bmatrix}
h_1^* n_1 + h_2 n_2^* + h_3^* n_3 + h_4 n_4^* \\
- h_1^* n_2 + h_2^* n_1 - h_3^* n_4 + h_4^* n_3
\end{bmatrix}.
\tag{8.28}
$$

Now, for the conventional SC-FDE decoding, the postprocessing STBC for two transmit antennas and two receive antennas becomes (for the lth branch of receiver)

$$
\tilde{A}_{k,2i}^{(l)} = \frac{\sum_{l=1}^{2} Y_{k,2i+1}^{(l)} H_{k,i}^{(1,l)*} + Y_{k,2i}^{(l)*} H_{k,i}^{(2,l)}}{\alpha + \sum_{m=1}^{2} \sum_{l=1}^{2} \left| H_{k,i}^{(m,l)} \right|^2}
\tag{8.29}
$$

Table 8.1 Channel Definition

	Rx1	Rx2		Rx Nr
Tx1	$H_{k,2i}^{(1,1)}$	$H_{k,2i}^{(1,2)}$	$\cdots$	$H_{k,2i}^{(1,Nr)}$
Tx2	$H_{k,2i}^{(2,1)}$	$H_{k,2i}^{(2,2)}$	$\cdots$	$H_{k,2i}^{(2,Nr)}$

and

$$\tilde{A}_{k,2i+1}^{(l)} = \frac{\sum_{l=1}^{2} Y_{k,2i}^{(l)} H_{k,i}^{(2,l)*} - Y_{k,2i+1}^{(l)*} H_{k,i}^{(1,l)*}}{\alpha + \sum_{m=1}^{2} \sum_{l=1}^{2} \left| H_{k,i}^{(m,l)} \right|^2} \tag{8.30}$$

with $\tilde{A}_{k,2i+1} = \sum_{l=1}^{Ntr} \tilde{A}_{k,2i+1}^{(l)}$ and $\tilde{A}_{k,2i} = \sum_{l=1}^{Ntr} \tilde{A}_{k,2i}^{(l)}$.

As mentioned earlier, the Alamouti STBC can be used with two transmit antennas and one receive antenna accomplishing the full diversity of 2, and for two transmit and receive antennas the full diversity of 4. This is an important characteristic of Alamouti STBC as it reduces the effect of fading at mobile stations while only requiring extra antenna elements at the base station, where it is more economical than having multiple antennas at the receivers [Alamouti 1998].

However, if having more antennas at the receivers is not a problem, this scheme can be used with two transmit antennas and *Nr* receive antennas, while accomplishing a 2*Nr* full diversity.

The definition of the channels between transmit (*Tx*) and receive antennas (*Rx*) and the notation for the received signals in the antennas along symbol periods are those presented in Tables 8.1 and 8.2, respectively.

The transmission sequence is the same as in both cases explained earlier. Analyzing the received signals for the case of two transmit antennas and two receive antennas, it is possible to generalize for the case of *M* receive antennas. Therefore, similar to

Table 8.2 Received Signals Time Definition

	Rx1	Rx2		Rx Nr
$t + 2iT$	$y_{1,2i}$	$y_{2,2i}$	$\cdots$	y_{Nrr2i}
$t + (2i + 1)T$	$y_{1,2i+1}$	$y_{2,2i+1}$	$\cdots$	$y_{Nrr2i+1}$

the previous case, the received signals by each receiver are described by the set of equations:

$$
\begin{aligned}
y_1^{(1)} &= h_{1,1}s_1 + h_{2,1}s_2 + n_1^{(1)} \\
y_2^{(1)} &= -h_{1,1}s_2^* + h_{2,1}s_1^* + n_{2i+1}^{(1)} \\
y_1^{(2)} &= h_{1,2}s_1 + h_{2,2}s_2 + n_1^{(2)} \\
y_2^{(2)} &= -h_{1,2}s_2^* + h_{2,2}s_1^* + n_2^{(2)} \\
&\vdots \\
y_1^{(Nr)} &= h_{1,Nr}s_1 + h_{2,Nr}s_2 + n_2^{(Nr)} \\
y_2^{(Nr)} &= -h_{1,Nr}s_2^* + h_{2,Nr}s_1^* + n_2^{(Nr)},
\end{aligned}
\tag{8.31}
$$

which results in matrix–vector notation

$$
y = Hs + w \tag{8.32}
$$

with

$$
s = \begin{pmatrix} s_1 & -s_2^* \\ s_2 & s_1^* \end{pmatrix} = \begin{bmatrix} s_{[1]} & s_{[2]} \end{bmatrix}, \tag{8.33}
$$

$$
H = \begin{bmatrix} h_{1,1} & h_{2,1} \\ h_{1,2} & h_{2,2} \\ \vdots & \vdots \\ h_{1,Nr} & h_{2,Nr} \end{bmatrix} = \begin{bmatrix} h_{[1]} & h_{[2]} \end{bmatrix}, \tag{8.34}
$$

and

$$
w = \begin{bmatrix} w_{1,1} & w_{2,1} \\ w_{1,2} & w_{2,2} \\ \vdots & \vdots \\ w_{1,Nr} & w_{2,Nr} \end{bmatrix} = \begin{bmatrix} w_{[1]} & w_{[2]} \end{bmatrix}. \tag{8.35}
$$

It should be mentioned that the combined symbol still is a linear combination of the received signals and the channel coefficients. That fact makes very simple the receiver's design of the Alamouti scheme, independently of how many antennas are deployed in the receiver since we have

$$c_1 = h^*_{[1]} y_{[1]} + y^*_{[2]} h_{[2]}$$
$$c_2 = h^*_{[2]} y_{[1]} - y^*_{[2]} h_{[1]},$$

(8.36)

that is

$$
\begin{bmatrix} c_1 \\ c_2 \end{bmatrix} =
\begin{bmatrix}
\sum_{m=1}^{2}\sum_{l=1}^{Nr} |h_{m,l}|^2 & 0 \\
0 & \sum_{m=1}^{2}\sum_{l=1}^{Nr} |h_{m,l}|^2
\end{bmatrix}
\cdot
\begin{bmatrix} y_{[1]} \\ y_{[2]} \end{bmatrix}
+
\begin{bmatrix}
\sum_{l=1}^{Nr} h^*_{1,l} n_{1,l} + h_{2,l} n^*_{2,l} \\
\sum_{l=1}^{Nr} -h^*_{1,l} n^*_{2,l} + h^*_{2,l} n_{1,l}
\end{bmatrix}.
$$

(8.37)

This scheme can be applied to the structure of the receiver previously defined in Figure 8.1. The resulting structure, the IB-DFE for an Alamouti scheme 2XNr, is presented in Figure 8.4. The postprocessing to be adopted in the SC-FDE in the receiver at each retransmission for the branch associated with the *l*th receive antenna will be

$$
\tilde{A}^{(r,l)}_{k,2i} = \frac{\sum_{l=1}^{Nr} Y^{(r,l)}_{k,2i+1} H^{(r,1,l)^*}_{k,i} + Y^{(r,l)^*}_{k,2i} H^{(r,2,l)}_{k,i}}{\alpha + \sum_{m=1}^{Ntr}\sum_{l=1}^{Nr} \left| H^{(r,m,l)}_{k,i} \right|^2}
$$

(8.38)

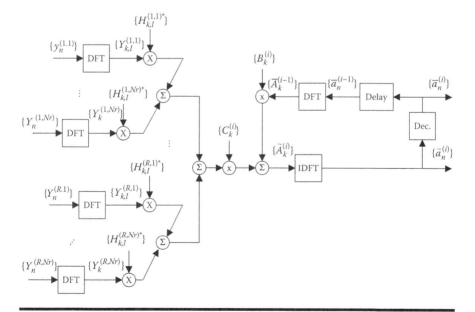

Figure 8.4 IB-DFE structure with *Nr* diversity order for *R* retransmissions.

and

$$\tilde{A}_{k,2i+1}^{(r,l)} = \frac{\sum_{l=1}^{Nr} Y_{k,2i}^{(r,l)} H_{k,i}^{(r,2,l)*} - Y_{k,2i+1}^{(r,l)*} H_{k,i}^{(r,1,l)*}}{\alpha + \sum_{m=1}^{Nr} \sum_{l=1}^{Ntr} \left| H_{k,i}^{(r,m,l)} \right|^2}. \tag{8.39}$$

Therefore, after R retransmissions, we will have $\tilde{A}_{k,2i+1} = \sum_{r=1}^{R} \tilde{A}_{k,2i+1}^{(r,l)}$ and $\tilde{A}_{k,2i} = \sum_{r=1}^{R} \tilde{A}_{k,2i}^{(r,l)}$. For each retransmission, the equivalent diversity coefficient $C_k^{(r)}$ is defined by

$$C_k^{(r)} = \frac{1}{\left[\alpha_k^{eq(r)} + (1 - (\rho^{(i-1)})^2) \sum_{r'=1}^{R} \sum_{m=1}^{Ntr} \sum_{l=1}^{Nr} \left| H_k^{(r',m,l)} \right|^2 \right]}, \quad r = 1,2,...,R. \tag{8.40}$$

The FDE output is still given by Equation 8.9, with the feedforward coefficients for each retransmission given by

$$F_k^{(r,l)} = H_k^{(r,l)*} C_k^{(r)}, \quad r = 1,2,...,R; \quad l = 1,2,..,Nr, \tag{8.41}$$

where

$$\alpha_k^{eq(r)} = \frac{E\left[\left| N_k^{(r)} \right|^2 \right] + E\left[\left| I_k^{(r)} \right|^2 \right]}{E\left[\left| A_k \right|^2 \right]}, \tag{8.42}$$

with $N_k^{(r)} = \sum_{l=1}^{Nr} N_k^{(r,l)}$ and $I_k^{(r)} = \sum_{l=1}^{Nr} I_k^{(r,l)}$.

8.5.1 Dealing with Fixed Channels

When strong interference remains over the several retransmission attempts, the performance can be rather poor unless we change our working band, as with frequency-hopping systems. Since changing the frequency leads to additional difficulties and delays, this is not practical in most UWB systems. In this case, we could assume that the frequency-domain block associated with the rth retransmission of a given packet $\left\{ A_k^{(r)}; k = 0,1,..., N-1 \right\}$ is an interleaved version of $\{A_k; k = 0,1,..., N-1\}$. Since this is formally equivalent to assume that $\left\{ H_k^{(r)}; k = 0,1,..., N-1 \right\}$ is an interleaved version of $\{H_k; k = 0,1,..., N-1\}$, the interference correlations for each frequency can be very small.

However, to avoid transmitting signals with very large envelope fluctuations, it is better to assume that $\{A_k^{(r)} = A_{k+\zeta_r}; k = 0,1,\ldots,N-1\}$, that is, corresponds to a cyclic-shifted version of $\{A_k; k = 0,1,\ldots,N-1\}$, with shift ζ_r. This means that the corresponding time-domain block is $\{a_n^{(r)} = a_n \exp(j2\pi\zeta_r n/N); n = 0,1,\ldots,N-1\}$, with a suitable ζ_r. It should be noted that this technique is formally equivalent to having $A_k^{(r)} = A_k$ and $H_k^{(r)}$ a cyclic-shifted version of $H_k^{(1)}$, with shift $-\zeta_r$. Moreover, we also have a cyclic shift in the interference, which means that when the interference characteristics remain constant, the mostly affected subcarriers will, in general, be different.* In general, the larger the ζ_r, the smaller the correlation between $H_k^{(r)}$ and $H_k^{(1)}$ and the smaller the correlation between $\alpha_k^{eq(r)}$ and $\alpha_k^{eq(1)}$, provided that $\zeta_r < N/2$ (since we consider cyclic shifts, $\zeta_r = N$ is equivalent to having $\zeta_r = 0$). In this chapter, we assume that the different ζ_r are the odd multiples of $N/2$, $N/4$, $N/4$, and so on, that is

$$\zeta_2 = N/2; \quad \zeta_3 = N/4; \quad \zeta_4 = 3N/2; \quad \zeta_5 = N/8;$$
$$\zeta_6 = 3N/8; \quad \zeta_7 = 5N/8; \quad \zeta_8 = 7N/8\ldots \tag{8.43}$$

This allows small correlation between different $H_k^{(r)}$ and $\alpha_k^{eq(r)}$, for each frequency (naturally, as r increases the correlations are higher). Moreover, envelope fluctuations associated with $\{a_n^{(r)}; n = 0,1,\ldots,N-1\}$ are not too different from the ones associated with $\{a_n; n = 0,1,\ldots,N-1\}$. One advantage of this approach relies on the fact that, for QPSK constellations, the constellation associated with $\{a_n^{(r)}; n = 0,1,\ldots,N-1\}$ is also a QPSK constellation for $r = 2, 3,$ and 4.

8.6 Performance Analysis

This section derives the analytical expression for the performance of ARQ schemes. It is assumed that the maximum number of transmission attempts for a given packet is N_R. This means that a packet is lost when it is not successfully decoded after N_R transmissions. We will assume that the probability of error detection (erroneous packet) after combining R packets is $P_F(R)$. For conventional ARQ schemes and a fixed channel, we may write

$$P_F(R) = [P_F(1)]^R. \tag{8.44}$$

The probability of detecting a packet successfully after N_R transmission attempts is

---

* Naturally, we are assuming that the interference only affects part of the spectrum. Otherwise, the "cyclic shift" would not have any impact on the performance.

$$P_S = 1 - P_F(N_R), \tag{8.45}$$

with the average number of transmitted packets given by

$$E[N_R] = \sum_{n=1}^{N_R-1} n(1 - P_F(n)) \prod_{m=1}^{n-1} P_F(m) + N_R \prod_{m=1}^{N_R-1} P_F(m). \tag{8.46}$$

The first term accounts for the packets that might require an additional transmission. The second term accounts for the packets that cannot have more retransmissions (regardless of being successfully detected or not). The average number in Equation 8.46 is closely related with the average packet delay.

8.6.1 Performance Results

In this section, a set of performance results are presented concerning the proposed ARQ techniques. The uplink of a continuous-wave UWB system employing SC-FDE schemes is considered using the proposed IB-DFE receiver. Each packet has $N = 256$ data QPSK symbols, corresponding to blocks with length 4 μs.

It is assumed to be a severely time-dispersive channel with rich multipath propagation and uncorrelated Rayleigh fading on the different paths.

Coded transmissions are considered, using the well-known 64-state, rate-1/2 convolutional code with generators $1 + D^2 + D^3 + D^5 + D^6$ and $1 + D + D^2 + D^3 + D^6$. The coded bits associated with a given FFT block are interleaved before being mapped into the QPSK constellation symbols and a suitable deinterleaver is employed before the channel decoder. Moreover, strong interference levels are considered, occupying 1/8 of the transmission band (i.e., we have very high interference levels). Also assumed are perfect synchronization and channel estimation conditions at the receiver.

As described previously, the proposed iterative receiver presents three different basic forms:

- Turbo receivers, corresponding to using the channel decoder output in the feedback
- IB-DFE receivers, corresponding to using uncoded data only in the iterative receiving chain
- Linear FDE, corresponding to the IB-DFE with a single iteration

With regard to the ability to mitigate the interferences, the proposed receiver presents two different configurations:

- Without knowledge of the interference characteristics (denoted "conventional FDE")
- With perfect knowledge of interference characteristics (denoted "ideal FDE")

With respect to the retransmissions, we have two different scenarios:

- Equal channels (EC): same channels for each packet retransmission
- Shifted channels (SC): same channels for each packet retransmission but different packet shifts for each retransmission

Figure 8.5 shows the average BER (bit error ratio) for a single transmission attempt, with or without interference. As can be seen, when the receiver does not know the interference characteristics, the performance can be far from the ones without interference. Notwithstanding, this difference is significantly reduced if the FDE receiver has knowledge about the interference characteristics, especially after four iterations.

Figures 8.6 and 8.7 show the packet error rate (PER) performance of the turbo FDE with four iterations for EC and SC cases, with strong interference levels. For the sake of comparison, we also include the PER for the linear FDE. In Figure 8.6, it is assumed that the receiver does not have knowledge about the PSD of the interference (i.e., we assume that $E[|I_k^{(r)}|^2] = 0$ in the computation of the feedforward coefficients), while in Figure 8.7 we assume that we use this knowledge in the

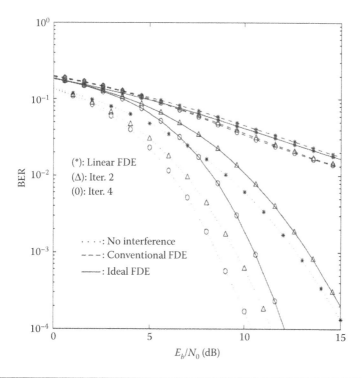

Figure 8.5 BER performance for a single packet transmission.

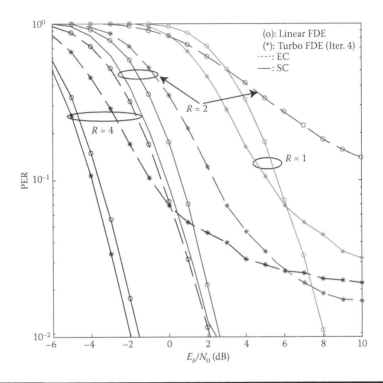

Figure 8.6 PER results for "conventional FDE" receiver.

computation of the feedforward coefficients (to avoid crowding these figures, we did not include the results with $R = 3$). Comparing both figures, the performance improvement achieved by the turbo FDE relating to the linear FDE becomes evident. Moreover, it is clear that a better performance tends to be achieved with the SC, relating to EC. Additionally, the PER performance is improved when the PSD of interference is taken into account by FDE coefficients (ideal receiver). In both cases, using the proposed ARQ technique, the performance improves significantly with R (transmissions attempts). From Figure 8.7, it is seen that, for two transmission attempts ($R = 2$) and EC, the performance tends to be worse than for $R = 1$. This results from the fact that combining packets using the same channel (EC) does not bring any added value (it even degrades, as we are combining multiple corrupted packets). Contrarily, when SC is employed, since the shift applied to the packets places the corrupted bits in different relative positions within the combined packets, the increase of the number of transmission attempts tends to achieve a performance improvement. Note that when an error is detected in the packet (using the punctured code), the additional bits that were punctured are requested for transmission.

The best overall performance is achieved with the turbo FDE associated with SC, for $R = 4$, using the "ideal receiver" receiver. Clearly, even when the turbo FDE

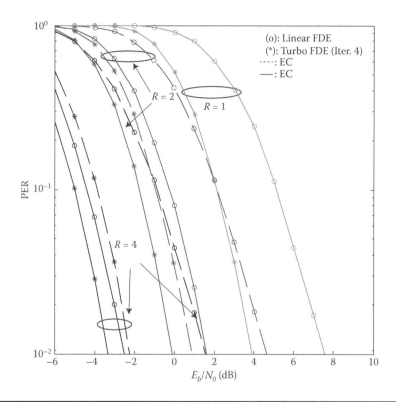

Figure 8.7 PER results for "ideal FDE" receiver.

is employed, the knowledge of the PSD of the interference is critical for the proposed receiver to achieve a good performance.

The performance results show that this technique is able to cope with strong interference levels as well as deep fadings, even for fixed channels. Moreover, the improvement of performance is obtained at the cost of very low complexity from both the transmitter and receiver.

Next, a set of figures are presented regarding a MIMO 2X2, that is, an Alamouti coding scheme at the transmitter and at the receiver with diversity of order 2. Only uncoded transmissions are considered, with the same conditions adopted for simulation proposed on previous cases. However, here we have a slight difference with respect to the retransmissions, since the studied scenarios are

- Uncorrelated channels (UC): uncorrelated channels for each packet retransmission
- Shifted channels (SC): same channels for each packet retransmission but different packet shifts for each retransmission

Figures 8.8 and 8.9 show the average BER behavior with transmission attempts in the presence of strong interference. It can be seen for each retransmission, the improvement associated with the successive iterations on the IB-DFE receiver, with a power gain of approximately 5 dB for four iterations in the first transmission attempt. Notwithstanding, when the number of combined packets increases with the number of retransmissions, the improvement due to iterations decreases for both channel types (as can be seen from Figures 8.8 and 8.9, for more than two retransmissions, there is no obvious advantage increasing the number of iterations from three to four). Again, for both cases, using the proposed ARQ technique, the performance improves significantly with R (transmissions attempts). As expected, the results for UC are better than those for SC, but the assumption of uncorrelated channels can be unrealistic in real conditions. From the comparison of these results with those from Figure 8.5, it is possible to conclude that the adopted MIMO technique assures a power gain near to 2.2 dB, even without the convolutional code adopted in the simulation scenario

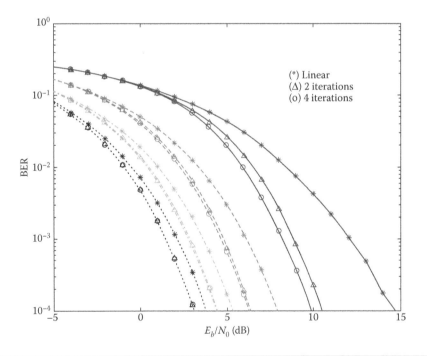

Figure 8.8 BER performance for an uncorrelated channel between retransmissions.

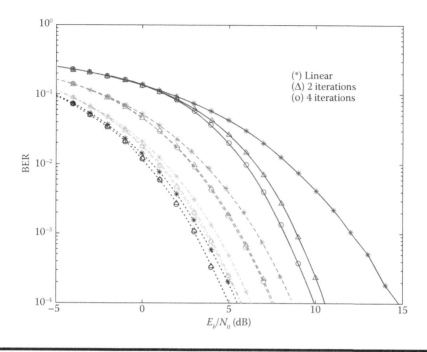

Figure 8.9   BER performance for a fixed channel between retransmissions.

of Figure 8.5. Obviously, if a coded transmission is adopted, we can expect higher power efficiency improvements.

Figures 8.10 and 8.11 show the PER performance behavior of the IB-DFE with different iterations for UC and SC cases, with strong interference levels. In both figures, it is assumed that the receiver has knowledge about the PSD of the interference, and uses this knowledge in the computation of the feedforward coefficients. Comparing both figures, it becomes clear about the performance improvement achieved by the IB-DFE relating to the linear FDE. Again, a better performance tends to be achieved with the UC relating to SC. Contrary to the EC case previously analyzed in Figures 8.6 and 8.7, the increase of the number of transmission attempts tends to achieve a performance improvement since, for both cases depicted in Figures 8.10 and 8.11, the proposed ARQ technique improves the performance significantly with R (transmissions attempts).

As expected, the best overall performance is achieved with UC, for $R = 4$. Clearly, when the IB-DFE is employed, the knowledge of the PSD of the interference is still critical for the receiver with diversity to achieve a good performance.

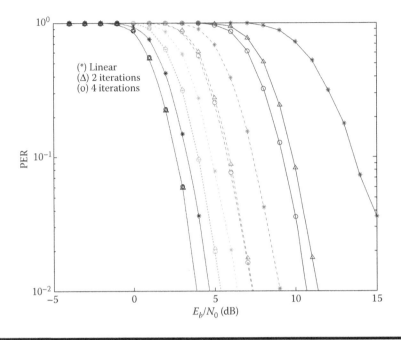

Figure 8.10   PER results for an uncorrelated channel between retransmissions.

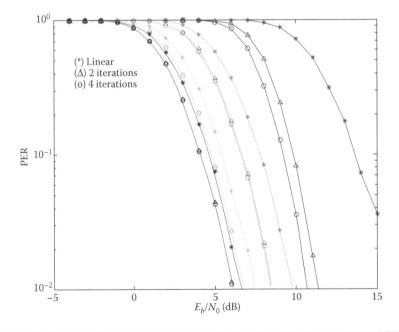

Figure 8.11   PER results for a fixed channel between retransmissions.

Acknowledgment

This work was supported by the FCT (Fundação para a Ciência e Tecnologia) via project PEst-OE/EEI/LA0008/2013.

References

Alamouti, S. A simple transmit diversity technique for wireless communications, *IEEE Journal on Selected Areas in Communications*, 16(8), 1451–1458, October 1998.

Benvenuto, N. and S. Tomasin, Block iterative DFE for single carrier modulation, *IEE Electronics Letters*, 39(19), 1144–1145, September 2002.

Dinis, R., P. Carvalho, and J. Martins, Soft combining ARQ techniques for wireless systems employing SC-FDE schemes, *IEEE ICCCN'08*, St. Thomas, US Virgin Islands, August 2008.

Dinis, R., A. Gusmão, and N. Esteves, On broadband block transmission over strongly frequency-selective fading channels, *Proc. Wireless 2003*, Calgary, Canada, July 2003.

Falconer, D., S. Ariyavisitakul, A. Benyamin-Seeyar, and B. Eidson, Frequency domain equalization for single-carrier broadband wireless systems, *IEEE Communications Magazine*, 4(4), 58–66, April 2002.

Gusmão, A., R. Dinis, and N. Esteves, Adaptive HARQ schemes using punctured RR codes for ATM-compatible broadband wireless communications, *IEEE VTC'99 (Fall)*, Amsterdam, Netherlands, September 1999.

Gusmão, A., R. Dinis, and N. Esteves, On frequency-domain equalization and diversity combining for broadband wireless communications, *IEEE Transactions on Communications*, 51(7), 1029–1033, July 2003.

Gusmão, A., R. Dinis, J. Conceicao, and N. Esteves, Comparison of two modulation choices for broadband wireless communications, *IEEE VTC'00 (Spring)*, Tokyo, Japan, May 2000.

Hagenauer, J. et al., Rate-compatible punctured convolutional codes (RCPC codes) and their application, *IEEE Transactions on Communications*, 36(4), 389–400, April 1988.

Ishiyama, Y. and T. Ohtsuki, Performance evaluation of UWBIR and DS-UWB with MMSE-frequency domain equalization (FDE), *IEEE GLOBECOM'04*, Dallas, TX, USA, November–December 2004.

Marques da Silva, M. and R. Dinis, Iterative frequency-domain detection and channel estimation for space-time block codes, *European Transactions on Telecommunications*, 22(7), 339–351, 2011.

Marques da Silva, M., A. Correia, R. Dinis, N. Souto, and J.C. Silva, *Transmission Techniques for Emergent Multicast and Broadcast Systems*, 1st edition, CRC Press Auerbach Publications, Boca Raton, USA, May 2010, ISBN: 9781439815939.

Marques da Silva, M., A. Correia, R. Dinis, N. Souto, and J. Silva, *Transmission Techniques for 4G Systems*, 1st edition, CRC Press Auerbach Publications, Boca Raton, USA, November 2012, ISBN: 9781466512337.

Popescu, D., P. Yaddanapudi, and R. Kondadasu, OFDM versus time-hopping in multiuser ultra wideband communication systems, *IEEE VTC'05 (Spring)*, Stockholm, Sweden, May 2005.

Proakis, J. and M. Salehi, *Digital Communications*, 5th edition, McGraw-Hill Science, New York, USA, 2007, ISBN-10: 0072957166.

Siwiak, K. Ultra-wide band radio: A new PAN and positioning technology, *IEEE Vehicular Technology Society News*, February 2002.

Tüchler, M. and J. Hanenauer, Linear time and frequency domain turbo equalization, *IEEE VTC'01 (Fall)*, pp. 2773–2777, Atlantic City, New Jersey, 2001.

Vutecic, B. and J. Yuan, *Turbo Codes: Principles and Applications*, Kluwer, Dordrecht, 2002.

Win, M. and R. Scholtz, On the robustness of ultra-wide bandwidth signals in dense multipath environments, *IEEE Communications Letters*, 2(2), 51–53, February 1998.

Chapter 9

Link and System Level Simulation for MIMO

Américo Correia and Mário Marques da Silva

Contents

9.1 Requirements and Scenarios

In this section, some important aspects related to long-term evolution (LTE), LTE-Advanced, and multiple-input and multiple-output (MIMO) requirements and scenarios are described. The baseline LTE radio access network (RAN) and evolved packet core (EPC) network were defined in 3GPP Release 8. This provided the world with a comprehensive and highly capable new cellular communications standard that, in November 2012, has been successfully launched in 113 commercial networks and in 51 countries. The main attributes that differentiate this new standard from previous standards are

- Single-channel peak data rates of up to 300 Mbps in the downlink and 75 Mbps in the uplink.
- Full integration of frequency division duplexing (FDD) and time division duplexing (TDD) access modes.
- Packet-based EPC network to eliminate cost and complexity associated with legacy circuit-switched networks.

Other key technologies introduced in Chapter 1 that enable the new capabilities include

- Adoption of orthogonal frequency-division multiple access (OFDMA) and single-carrier frequency-division multiple access (SC-FDMA) for the downlink and uplink air interfaces, respectively, to enable narrowband scheduling and efficient support of spatial multiplexing.
- Support for six channel bandwidths from 1.4 to 20 MHz to enable high data rates.
- Baseline support for spatial multiplexing (MIMO) of up to four layers on the downlink.
- Faster physical layer control mechanisms leading to lower latency.

Despite the substantial capabilities of LTE in Release 8, the 3GPP standard has continued to evolve through Releases 9, 10, 11, and 12.

Most LTE radio network designs prefer complying with the continuous networking principle. Continuous networking can improve cell edge user experience, reduce inter-radio access technologies (RAT) handover requirement, and facilitate evolution from 3G to LTE [3GPP 2010a]. LTE continuous networking needs considering continuous coverage of the reference signal received power (RSRP). This consists of the average of the power of all resource elements which carry cell-specific reference signals over the entire bandwidth. The reference signal received quality (RSRQ) measurement provides additional information when RSRP is not sufficient to make a reliable handover or cell reselection decision. RSRQ is the ratio between the RSRP and the received signal strength indicator (RSSI) and, depending on the measurement bandwidth, represents the number of resource blocks. RSSI is the total received wideband power including all interference and thermal noise. Best service cells, traffic channel for uplink and downlink, physical uplink shared channel (PUSCH), and physical downlink shared channel (PDSCH) are also required.

Special scenarios like femtocells may need LTE discontinuous coverage. Discontinuous coverage may dig out the advantages of high-order modulation and coding in good radio environment. Therefore, it improves cell throughput and cell center user experience.

The femtocell concept is not unique to LTE or LTE-Advanced [3GPP 2012a], but there was an opportunity for LTE to incorporate this technology from the start. From a radio deployment perspective, the femtocell operates over a small area within a larger cell. The radio channel can be the same as that of the macrocell (known as co-channel deployment) or on a dedicated channel. The femtocell concept is fundamentally different from relaying since the femtocell connection back into the core network is provided locally by an existing wired Internet connection rather than over the air back to the macrocell. Most femtocell deployments cover indoor areas, which helps provide isolation between the femtocell and the macrocell radio channels. There is also femtocell outside the coverage area of macrocell. This could be, for example, a way to provide local cellular coverage in rural areas where digital subscriber line exists but there is no cellular coverage provided by any operator.

It must be decided whether the femtocells are operated for closed subscriber group (CSG), user equipment (UE), or for open access. Other practical considerations such as pricing can be considered commercial issues, although in the co-channel CSG case, the probability that areas of dense femtocell deployment will block macrocells becomes an issue.

Several 3GPP studies have shown that increases in average data rates and 100 times greater capacity are possible with femtocells over what can be achieved from the macronetwork. However, femtocells do not provide the mobility of macrocellular systems, and differences exist in the use models of these systems.

The evolved multimedia broadcast and multicast service (E-MBMS) television service was specified at the physical layer in Release 8 but was not functionally complete until Release 9. The features in Release 9 provide a basic MBMS service carried over an MBMS single frequency network (MBSFN). One limitation of Release 9 definition was the lack of a feedback mechanism from the UEs that would inform the network if sufficient UEs were present in the target area to justify turning on the MBSFN locally. In Release 11, further MBMS enhancements for service continuity were specified including support on multiple frequencies, reception during radio resource control (RRC) idle and RRC connected states, and support to take UE positioning into account for further optimization of the received service.

E-MBMS is performed either in single cell or multicell mode. In single cell transmissions, E-MBMS traffic is mapped to the downlink shared channel (DL-SCH). In multicell mode, transmissions from cells are carefully synchronized to form an MBSFN [3GPP 2010b, Marques da Silva et al. 2010, 2012].

MBSFN is an elegant application of OFDM for cellular broadcast. The principle of operation is quite simple. Identical transmissions are broadcast from closely coordinated cells simultaneously on a common frequency [Marques da Silva et al. 2012]. Signals from adjacent cells arrive at the receiver and are dealt with in the same manner as multipath delayed signals. In this manner, UE can combine the energy from multiple transmitters with no additional receiver complexity.

If the UE is at a cell boundary, the relative delay between the two signals is quite small. However, if the UE is close to one base station and relatively distant from a second base station, the amount of delay between the two signals can be quite large. For this reason, MBSFN transmissions might be supported using 7.5 kHz subcarrier spacing (instead of 15 kHz) and a longer CP [3GPP 2010b], as depicted in Figure 9.1. MBSFN networks also use a common reference signal from all transmitters within the network to facilitate channel estimation. As a consequence of the MBSFN transmission scheme, UE can roam between cells with no handover procedure required. Signals from various cells will vary in strength and in relative delay, but in aggregate the received signal is still dealt with in the same manner as a conventional single channel OFDM transmission.

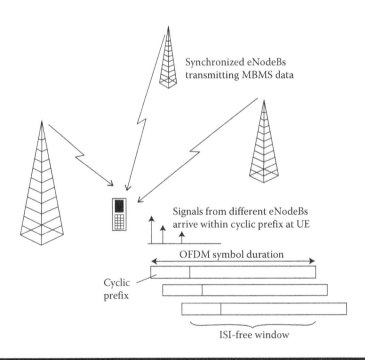

Figure 9.1 MBSFN transmission with long cyclic prefix to avoid intersymbol interference.

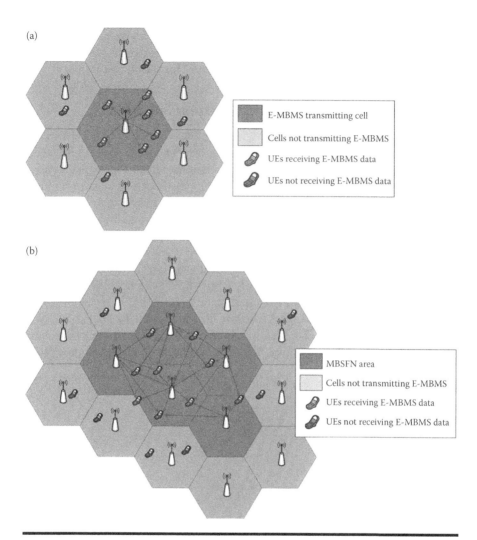

Figure 9.2 (a) SC-PMP transmission and (b) MBSFN transmission.

Figure 9.2 illustrates MBSFN transmission scheme including also an illustration of the single-cell point-to-multipoint (SC-PMP) transmission. The other scenario will be evaluated later in this chapter.

According to 3GPP specifications [3GPP 2007, 2009a,b,c], the deployment scenario where a dedicated carrier is used for broadcast only (MBSFN) has the following properties:

- E-UTRA MBMS is envisaged to achieve a cell edge spectrum efficiency of 1 bit/s/Hz, equivalent to the support of, at least, 16 mobile TV channels, at around 300 kbps per 5 MHz channel (in an urban or suburban environment).

■ In the deployment scenario, where a carrier is shared between broadcast and unicast traffic, the target performance at cell edge for broadcast traffic should be in line with the existing target performance for the unicast traffic.

9.1.1 Intercell Interference Coordination Schemes

Since the LTE system employs OFDMA in the downlink and SC-FDMA in the uplink, intracell interference is mitigated due to orthogonality between subcarriers [Marques da Silva 2012]. However, intercell interference remains as the main source of interference, especially at cell edges. Intercell interference occurs when multiple eNBs* are transmitting using the same frequency. Under these circumstances, the transmitted signals interfere with each other, collisions occur, and the UEs might not receive the packets correctly. Figure 9.3 depicts a situation where S1.1 is the signal sent from eNB1 to UE1, S2.1 is the signal sent from eNB2 to UE1, and S2.2 is the signal sent from eNB2 to UE2. Since UE1 is at the cell border, the signal S1.1 might suffer interference from S2.1. This occurs when there is no intercell interference coordination, such as in SC-PMP scenario. However, in MBSFN scenario, signals S1.1 and S2.1 are soft combined and therefore UE1 receives S1.1 + S2.1 correctly.

The ability of cells to coordinate their narrowband scheduling offers some potential for interference avoidance. Support for coordination of resource block

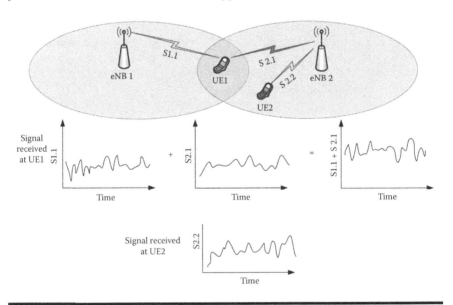

Figure 9.3   Example of potential intercell interference.

---

* Evolved Node-B (eNode-B) refers to the base station of LTE and 4G systems. This corresponds to the evolution of the node-B, considered by UMTS.

(RB) allocation between cells in the downlink was introduced in Release 8 with the inclusion of the relative narrowband transmit power (RNTP) indicator. This support feature is a bitmap that can be shared between base stations over the X2 interface. It represents those RB for which the base station intends to limit its output power to a configurable upper limit for some period of agreed-upon time. This allows schedulers to agree on how cell-edge RB will be used so that, for instance, cell-edge users who cause the most interference can be restricted to certain parts of the channel. This coordination could be implemented using a semistatic agreement for partial frequency reuse at the cell edge or might involve more dynamic scheduling based on real-time network loading. The next section covers partial frequency reuse schemes and hybrid reuse (HR), also referred to as soft reuse schemes.

9.1.2 Soft and Hybrid Reuse

To solve the limitations in terms of spectral efficiency of traditional frequency reuse schemes, HR, or soft fractional frequency reuse (SFFR), as it is sometimes mentioned in the literature, was proposed by several authors [Simonsson 2007, Sarperi et al. 2008, Zhou and Zein 2008, Fodor et al. 2009]. This consists of defining different zones within each cell where different reuse factors are applied. Figure 9.4 shows an example of HR 1 + 1/3 (reuse 1 + reuse 1/3), in which there is an area where all frequency

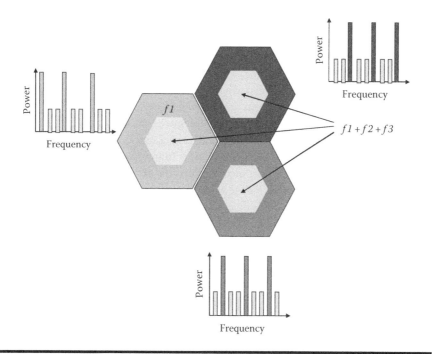

Figure 9.4   Example of HR with reuse 1 + reuse 1/3.

spectrums are used and another area at the border of the cells where only a fraction of the frequency is used. This way we can achieve maximum spectral efficiency in the center of each cell, and reduced intercell interference levels at the border of cells.

This technique is implemented using the same methodology used for normal fractional frequency reuse, but this time different levels of power P are applied to different frequencies, for each sector of every cell, as defined in the following.

$$P_{\max}(f,t) = P, \quad f \in f_n$$
$$P_{\max}(f,t) = p_{tx} < P, \quad f \notin f_n$$

(9.1)

The assignment of different power levels to different frequencies f_n is related to the characteristics of radio signals, more precisely the average path losses (L_{path}) that radio signals suffer over the air. This has direct impact on the total power received at the UE (P_{RxUE}). This way, at the center of the cell, all frequencies transmitted with power $P < p_{tx}$ will be received and, at the edge of cells, only frequencies transmitted with power $P > p_{tx}$ will be received and will interfere with each other.

To determine the power levels that each frequency should use, we first need to determine the location of the different reuse zones. This can be somewhat tricky, because it is hard for the UE to determine with precision its position, therefore making it hard to know when it should be using one or another reuse. To solve this, we can define certain levels or targets for carrier-to-interference ratio (C/I), and based on those target values the UE knows that it should apply for one type of reuse (e.g., if UE C/I is higher than the target C/I) or another (e.g., if UE C/I is below target C/I). However, the level of C/I received at UE is associated with a certain distance to the center of the cell (i.e., eNB site). This allows us to define distances of reuse (DR) that are directly related to a certain level of C/I and, in turn, these levels of C/I can be determined from the received power of pilot subcarriers at UEs.

Let us analyze the example depicted in Figure 9.5, in which HR 1 + 1/3 is considered, meaning that reuse 1 is used within a certain area (the small inner hexagons) and reuse 1/3 is used outside that area (outer region) for each cell sector. R is the cell radius and D is the distance between the center of cells using the same frequencies, and R_1 is the radius of the reuse 1 zone inside each cell, being D_1 the distance between reuse 1 zones inside each cell.

The carrier–interference ratio (C/I) in hexagonal cellular topologies is expressed as

$$\frac{c}{i} = \frac{1}{i_0}\left(\frac{D}{R}\right)^{\alpha}$$
$$\frac{C}{I} = 10\log_{10}\left(\frac{c}{i}\right) \text{dB}$$

(9.2)

where i_0 is the number of interfering cells interfering at distance D, and α is the average exponent of propagation path loss that can take values between 2 and 5

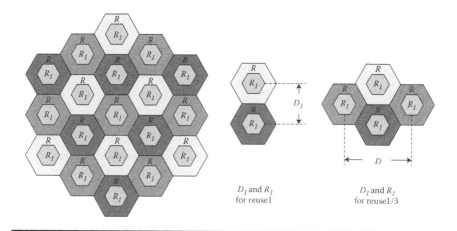

Figure 9.5   Hexagonal cellular topology, using HR 1 + 1/3 scheme.

(based on 3GPP specifications we use $\alpha = 2.2$). According to Figure 9.5, both reuse 1 and reuse 1/3 have $i_0 = 6$ interfering cells at distance D. If we want the DR to be 1/3 of R, for reuse 1 we have $R_1 = DR = R/3$ and $D_1 = 2R$. The C/I for users at DR is

$$\frac{c}{i} = \frac{1}{6}\left(\frac{2R}{\frac{R}{3}}\right)^{2.2} = 8.58$$

$$\frac{C}{I} = 10\log_{10}(8.58) = 9.34 \text{ dB} \tag{9.3}$$

We know that for a UE located at $DR = R/3$, the C/I target should be 9.34 dB, meaning that UEs with C/I equal or higher than this value will be using reuse 1, and UEs with C/I lower than that will be using reuse 1/3.

Next, we must find out what is the power that should be allocated to frequencies meant to be received within reuse 1, and the power that must be used by frequencies meant to be received on the entire cell (especially in the border of the cell). To do so, we need to first know the C/I level received at the edge of the cell; in our example the edge of the cell at the distance R from the center of cell. We have

$$\frac{c}{i} = \frac{1}{6}\left(\frac{3R}{R}\right)^{2.2} = 1.87$$

$$\frac{C}{I} = 10\log_{10}(1.87) = 2.71 \text{ dB} \tag{9.4}$$

The C/I for frequencies supposed to be received by UEs at the edge of the cell will be 2.71 dB. If we consider that for that C/I level the power is $P_{\max}$ 46 dBm = 40 W, the frequencies that are not supposed to be received must be transmitted with inferior power P_{tx}. This way we have

$$p_{tx} = \frac{40\,\text{W}}{\left(\dfrac{3R}{R}\right)^{2.2}} = \frac{40\,\text{W}}{11.21} = 3.57\,\text{W} \tag{9.5}$$

Looking at Equation 9.5, we conclude that, at the center of the cells, frequencies will be transmitted with $P_{max} = 3.57$ W and, at the border of the cells, frequencies will be transmitted using $P_{max} = 40$ W. The example described can be applied to other configurations such as HR 1 + 1/2 or HR 1/2 + 1/3. Furthermore, more than just two reuse zones can exist.

The following section describes another technique to improve the cell-edge spectral efficiency.

9.1.3 Coordinated Multipoint Transmission

The concept of cooperative MIMO was briefly introduced in Chapter 1, also named as network MIMO or CoMP [3GPP 2011]. The goal of CoMP is to improve the coverage of high data rates and cell-edge throughput, and also to increase system throughput.

The primary difference between standard MIMO and CoMP is that for the latter, the transmitters are not physically colocated. In the case of downlink CoMP, however, there is the possibility of linking the transmitters at baseband to enable sharing of payload data for the purposes of coordinated precoding. This sharing is not physically possible for the uplink, which limits the options for uplink CoMP. For the standard network topology in which the eNBs are physically distributed, the provision of a high capacity and low latency baseband link is challenging and would probably require augmentation of the X2 inter-eNB interface using fiber. However, a cost-effective solution for inter-eNB connectivity is offered by a network architecture in which the baseband and RF transceivers are located at a central site with distribution of the RF to the remote radio heads (RRHs) via fiber.

Four downlink deployment scenarios were defined for the feasibility study in Release 11 [3GPP 2011]:

■ CoMP scenario 1 is a homogeneous macronetwork (all cells have the same coverage area) with intrasite CoMP. This is the least complex form of CoMP and is limited to eNBs sharing the same site (see Figure 9.6).
■ CoMP scenario 2 is also a homogeneous network but with high Tx-power RRHs. This is an extension of scenario 1 in which the six sites adjacent to the central site are connected via fiber optic links to enable baseband cooperation across a wider area than is possible with scenario 1 (see Figure 9.7).

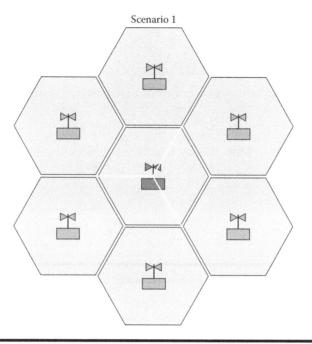

Figure 9.6   CoMP scenario 1 (considered in LTE Release 11).

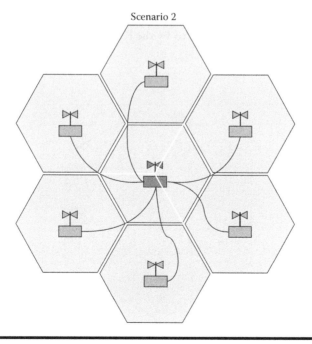

Figure 9.7   CoMP scenario 2 (considered in LTE Release 11).

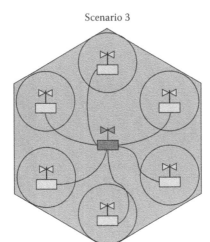

Figure 9.8   CoMP scenario 3 (considered in LTE Release 11).

■ CoMP scenario 3 is a heterogeneous network in which low power RRHs with limited coverage are located within the macrocell coverage area (see Figure 9.8).

■ CoMP scenario 4 is a heterogeneous network in which low power RRHs with limited coverage are located within the macrocell coverage area. The transmission/reception points created by the RRHs have the same cell identity as the macrocell (see Figure 9.9).

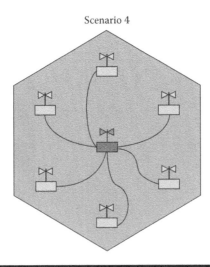

Figure 9.9   CoMP scenario 4 (considered in LTE Release 11).

Scenarios 3 and 4 are expected to be used in metropolitan areas where network deployment is dense and RRHs of different transmission power levels coexist.

The introduction of CoMP enables several new categories of network operation, which are defined for the downlink as follows [Lee et al. 2012, Rumney 2013, 3GPP 2011]:

■ Joint processing (JP): Data for a UE is available at more than one point in the CoMP cooperating set for the same time–frequency resource.
■ Joint transmission (JT): This is a form of spatial multiplexing that takes advantage of decorrelated transmissions from more than one point within the cooperating set. Data to a UE is simultaneously transmitted from multiple points to coherently improve the received signal quality or data throughput.
■ Dynamic point selection (DPS): The UE data is available at all points in the cooperating set but is only transmitted from one point based on dynamic selection in time and frequency. The DPS includes dynamic cell selection (DCS). DPS may be combined with JT, in which case multiple points can be selected for data transmission in the time–frequency resource.
■ Coordinated scheduling and beamforming (CS/CB): Data for a UE is only available at and transmitted from one point in the CoMP cooperating set but user scheduling and beamforming (BF) decisions are made across all points in the cooperating set. Semistatic point selection (SSPS) is used to make the transmission decisions. Dynamic or semistatic muting may be applied to both JP and CS/CB.

9.2 Evaluation Methodology

Radio networks resources are obviously finite and scarce (i.e., limited bandwidth for transmission). As the number of services that use the available bandwidth increases, it is essential to have techniques that optimize the usage and allocation of such limited resources. Scheduling consists precisely in performing the allocation of resources taking into consideration different factors. This allocation is typically performed in the time domain, meaning that scheduling algorithms divide the time domain in small pieces (time slots [TS]), and then allocate every one of those TS to different users.

With the introduction of LTE and OFDMA, the paradigm of scheduling has slightly changed since resources can be allocated not only in time domain but also in the frequency domain. This flexibility of allocation both in time and frequency will, in theory, enable OFDMA-based schedulers to allocate resources more efficiently, for example, allocating more time and bandwidth to users with more demanding needs.

With multiuser MIMO (MU-MIMO) specified for LTE-Advanced, the downlink of a MIMO broadcast channel requires both scheduling and BF. The well-known orthogonal random beamforming (ORBF) exploits properly the available MIMO spatial channel dimensions of the system.

A single simulation that takes into consideration both the link-level simulation (LLS) and system-level simulation (SLS) would be preferable. Nevertheless, that implies a significant increased degree of complexity, and huge simulation times. Therefore, two separate but interconnected approaches are the best solution.

The LLS objective is to model a radio link between a UE and an eNB, including channel estimation, interleaving, decoding, and so on, operating at symbol level frequency or time resolution. The output that is expected consists of one or more figures which show the relation between the minimum required signal-to-noise ratio (SNR) that the link should experience to achieve a given block error rate (BLER).

For the SLS, a 19-cell network topology with wrap-around will be used as the baseline network topology for all SLSs, with the following features [ITU-R 2008]:

1. The system is modeled as a network of 7 clusters. Each cluster has 19 hexagonal cells with 6 cells in the first tier and 12 cells in the second tier surrounding the central cell of each cluster. Each cell has 3 sectors. Frequency reuse is modeled by planning frequency allocations in different sectors in the network.

2. Users are dropped independently with uniform distribution throughout the system. Each mobile corresponds to an active user session that runs for the duration of the drop.

3. Mobiles are randomly assigned channel models. Depending on the simulation, these may be in support of a desired channel model mix, or separate statistical realizations of a single type of channel model.

4. Users are dropped according to the specified traffic mix.

5. For sectors belonging to the center cluster, sector assignment to a user is based on the received power at a user from all potential serving sectors. The sector with best path to the user, taking into account slow fading characteristics (path loss, shadowing, and antenna gains) is chosen as the serving sector.

6. Mobile stations are randomly dropped over the 57 sectors such that each sector has the required numbers of users. Although users may be in regions supporting handover, each user is assigned to only one sector for counting purposes. All sectors of the system shall continue accepting users until the desired fixed number of users per sector is achieved everywhere. Users dropped within 35 m of a sector antenna shall be redropped. User locations for six wrapping clusters are the same as the center cluster.

7. For simulations that do not involve handover performance evaluation, the location of each user remains unchanged during a drop, and the speed of a user is only used to determine the Doppler effect of fast fading. Additionally, the user is assumed to remain attached to the same eNB for the duration of the drop.

8. Fading signal and fading interference are computed from each mobile station into each sector and from each sector to each mobile for each simulation interval.

9. Packets are not blocked when they arrive into the system (i.e., queue depths are infinite). Users with a required traffic class shall be modeled according to the traffic models defined in this document. Start times for each traffic type for each user should be randomized, as specified in the traffic model being simulated.

10. Packets are scheduled with a packet scheduler using the required fairness metric. Channel quality feedback delay and protocol data unit (PDU) errors are modeled and packets are retransmitted as necessary. The hybrid automatic repeat request (HARQ) process is modeled by explicitly rescheduling a packet as part of the current packet call after a specified HARQ feedback delay period.

11. Simulation time is chosen to ensure convergence in user performance metrics. For a given drop, the simulation is run for this duration, and then the process is repeated with the users dropped at new random locations. A sufficient number of drops are simulated to ensure convergence in the system performance metrics.

12. Performance statistics are collected for users in all cells according to the output matrix requirements.

13. All 57 sectors in the system shall be dynamically simulated.

The system level simulators used in this chapter were based on the above 13 requirements (for this study only the downlink is considered). Another general description of an SLS is presented in Chapter 5 of [Marques da Silva et al. 2012].

9.2.1 Link-Level Simulations

For the LLSs, the urban macropropagation channel was used to simulate the point-to-point (PtP) and the point-to-multipoint (PMP) scenarios radio link. Since the number of subcarriers allocated to each user vary along the simulation, an extensive study had to be performed, testing all the possible number of subcarriers. Starting with one physical resource block, each one with 12 subcarriers, every time (12, 24, 36, 48 subcarriers, and so on) until reaching the maximum subcarrier value of 600 (corresponding to the total usage of the subcarriers available with a 10 MHz transmitting bandwidth).

The size of the transmitted block depends on the number of subcarriers used and the number of OFDM symbols per subframe, being expressed by

$$L_{TBcod} = N_{symb} \times N_{sc} \qquad (9.6)$$

where L_{TBcod} denotes the total size of the block transmitted in a single subframe including coding and cyclic redundancy check (CRC) bits, N_{symb} is the number of OFDM symbols per subframe and per subcarrier, and N_{sc} is the number of occupied subcarriers during one subframe.

The size of the effective data transmitted block L_{TB} (i.e., without the coding and correction bits) is inferior to L_{TBcod}, being expressed by

$$L_{TB} = L_{TBcod} \, R_c \qquad (9.7)$$

where R_c is the respective coding rate.

For the different combinations of environment (hierarchical modulation, coding rate, etc.), the LLS produced figures with the evolution of the BLER as a function of the link E_S/N_0. E_S/N_0 represents the ratio of the energy of each symbol (E_S) over the spectral noise density (N_0). For the PMP modes (SC-PMP and MBSFN scenarios), the target BLER to be achieved is 1% (BLER = 0.01). For PtP scenario, the target BLER is 10% (BLER = 0.1), that is, in every 10 transmitted packets only one of those is not correctly decoded at the receiver and needs to be retransmitted. The SNR values used in SLS are obtained using

$$SNR = \frac{E_S}{N_0} + 10 \log_{10} \left(\frac{R_b}{\log_2(M) \, B_w} \right) \qquad (9.8)$$

where E_S/N_0 is obtained from the figures produced by LLS, R_b is the specific bit-rate considered in the simulation run, M is the index of the hierarchical modulation used (e.g., for 16QAM, we have the modulation order $M = 16$), and B_w is the total bandwidth available for transmission (i.e., 10 MHz for all the cases). As defined in Equation 9.9, the R_b value is obtained by dividing the size of the block of bits being transmitted (before coding) over the duration in seconds that is needed to transmit the entire block of bits (typically this is the subframe duration or time transmission interval equal to 0.5 ms).

$$R_b = \frac{L_{TB}}{TTI} \qquad (9.9)$$

Some results from LLSs are presented in Figure 9.10 for one of the environments under consideration. The presented E_S/N_0 (and SNR values obtained from them) are required to receive the totality of the bits for each modulation and coding scheme (MCS). The target SNR values required to receive only a part of those bits (i.e., only the strong bits, or strong and average bits) are obviously lower than those.

Figure 9.10 plots the BLER performance of single-user MIMO (SU-MIMO) 2×2 with frequency diversity, as a function of E_S/N_0 values, for different frequency reuse R, two different streams transmitted by antennas A1 and A2 (in

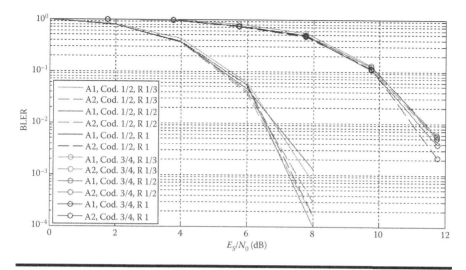

Figure 9.10 **BLER versus Es/No MIMO (2 × 2), QPSK, urban macrochannel.**

case of multiresolution, the BER of each spatial multiplexing* stream might be different) and coding rate Cod. The receiver is an iterative minimum mean square error (MMSE), as described in Chapter 2. Taking as reference the BLER = 0.01 (or BLER = 0.1 for PtP communications) then the corresponding E_S/N_0 values would provide the SNR values to correctly receive a block of N bits, being used as an input to run the system level simulator.

9.2.2 System-Level Simulations

The ITU-R IMT-Advanced MIMO channel model for SLS is a geometry-based stochastic model. It can also be called double-directional channel model. It does not explicitly specify the locations of the scatters, but rather the directions of the rays, like the well-known spatial channel model (SCM) [3GPP 2009a]. Geometry-based modeling of the radio channel enables separation of propagation parameters and antennas.

The channel parameters for individual snapshots are determined stochastically based on statistical distributions extracted from channel measurements. Antenna geometries and radiation patterns can be defined properly by the user of the model. Channel realizations are generated through the application of the geometrical principle by summing contributions of rays (plane waves) with specific small-scale (SS) parameters such as delay, power, angle-of-arrival (AoA) and angle-of-departure (AoD). Superposition results to correlation between antenna elements and temporal fading with geometry-dependent Doppler spectrum.

---

* A multilayer MIMO transmission is assumed (spatial multiplexing), as described in Chapter 1.

A number of rays constitute a cluster. In the terminology of this document, we equate the cluster with a propagation path diffused in space, either or both in delay and angle domains. Elements of the MIMO channel, for example, antenna arrays at both link ends and propagation paths, are illustrated in Figure 9.11. The generic MIMO channel model is applicable to all scenarios, for example, indoor, urban, and rural.

The time-variant impulse response matrix of the $U \times S$ MIMO channel is given by

$$\mathbf{H}(t;\tau) = \sum_{n=1}^{N} \mathbf{H}_n(t;\tau) \qquad (9.10)$$

where t is time, τ is delay, N is the number of paths, and n is path index.

The IMT-Advanced channel model for the evaluation of IMT-Advanced candidate technologies consists of a primary module and an extension module. Only the primary module is described, which is mandatory for IMT-Advanced evaluation.

The generic channel model is a stochastic model with two (or three) levels of randomness. First, large-scale (LS) parameters such as shadow fading, delay, and angular spreads are drawn randomly from tabulated distribution functions. Next, SS parameters such as delays, powers, and directions of arrival and departure are drawn randomly according to tabulated distribution functions and random LS parameters. At this stage, the geometric setup is fixed and the only free variables are the random initial phases of the scatters. By picking (randomly) different initial phases, an infinite number of different realizations of the model can be generated. When the initial phases are also fixed, there is no further randomness left.

Figure 9.12 shows the overview of the channel model creation. The first stage consists of two steps. First, the propagation scenario is selected. Then, the network

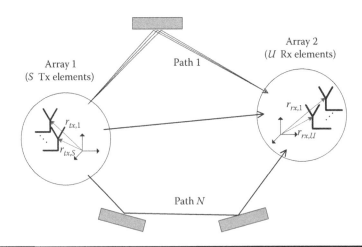

Figure 9.11 MIMO channel.

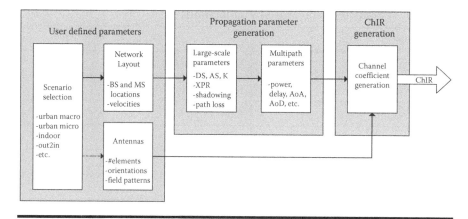

Figure 9.12   Channel model creation process.

layout and the antenna configuration are determined. In the second stage, LS and SS parameters are defined. In the third stage, channel impulse responses (ChIRs) are calculated.

The generic model is based on the drop concept. When using the generic model, the simulation of the system behavior is carried out as a sequence of drops, where a drop is defined as one simulation run over a certain time period. A drop (or snapshot or channel segment) is a simulation entity where the random properties of the channel remain constant except for the fast fading caused by the changing phases of the rays. The constant properties during a single drop are, for example, the powers, delays, and directions of the rays. In a simulation, the number and the length of drops have to be selected properly by the evaluation requirements and the deployed scenario. The generic model allows the user to simulate over several drops to get statistically representative results. Consecutive drops are independent.

Several different scenarios will be evaluated, some considering E-MBMS services in PtP mode and PMP modes. The scenario where PtP mode is evaluated will serve as reference result when comparing the results in PMP, which are expected to outperform the results of PtP mode.

The SU-MIMO, MU-MIMO, and finally CoMP scenarios will also be evaluated in the next section.

The most important parameters used for the different scenarios based on 3GPP recommendations [3GPP 2010c, 3GPP 2010d] can be observed from Table 9.1.

The average propagation loss with distance is dependent on the probability of line-of-sight (LOS). This is presented in Equation 9.11, where mobile users must be distant from the eNB antennas more than 18 m.

$$\Pr(LOS) = \frac{18}{d}(1 - e^{-d/63}) + e^{-d/63} \quad d > 18m \tag{9.11}$$

Table 9.1 Parameterization for SLS Simulations

Parameter	Values
Cell radius	500, 750, 1000, 2250 m
Schedulers	RR, Max C/I, FT, PF
Traffic model	CBR
Simulation time	500 (seg)
Subframe duration (TTI)	0.5 ms
Carrier frequency	2 GHz
Propagation model	3GPP urban macro
Distance attenuation (d = distance in meters)	$P_{NLOS} = -40.45 + 39.09 \log(d)$ (dB)
	$P_{LOS} = 34.04 + 22 \log(d)$ if $d < 360$ m (dB)
	$P_{LOS} = -11.02 + 40 \log(d)$ if $d > 360$ m (dB)
	$L = \Pr(LOS)P_{LOS} + (1-\Pr(LOS))P_{NLOS}$ (dB)
Channel model	Urban macro
Number of base station sites	19
Number sectors per base station site	3 sectors/site
User mobility	Random walk inside sector
UE antenna height	1.5 m
eNB antenna height	25 m
% of transmitted power by other cells	90
Modulations	QPSK, H-16QAM, H-64QAM
Coding rate	1/2, 3/4
Frequency reuse	1,1/3, HR: 1 + 1/3
Sites layout	19 eNBs
eNB base station power/sector	46 dBm or 40 W
RRH power/sector	34 dBm or 2.5 W
Number UEs per sector	Variable

9.3 Simulation Results

The problem of intercell interference is illustrated in Figure 9.13, which shows a cumulative distribution function plot of the geometry factor within a typical macrourban cell.

The geometry factor was the term used in universal mobile telecommunication system (UMTS) to indicate the ratio of the wanted signal relating to the interference plus noise. This corresponds to the signal-to-interference-plus-noise ratio (SINR). From the figure, it is viewed that 10% of the users experience better than 10.5 dB geometry factor but 50% of users experience worse than 1.5 dB. The exact shape of the curve varies significantly depending primarily on the frequency reuse factor followed by the cell size and cell loading. An isolated cell would exhibit a shift to the right, indicating that most users are experiencing very good signal conditions. A cell in an urban area with significant cochannel inter-cell interference would shift to the left. Penetration loss through buildings, as experienced when indoor coverage is provided from an external cell, would also cause a shift to the left. When the deployment in a particular area has resulted in a certain geometry factor distribution, the challenge becomes how to deal with the interference to improve cell-average and cell-edge performance.

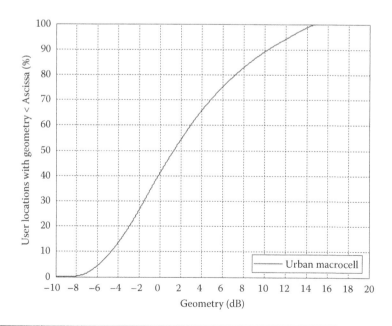

Figure 9.13   Geometry factor distribution in a typical macrourban cell with frequency reuse 1.

9.3.1 Results for PMP Scenarios

In 3G systems, the frequency reuse was optimized at one, and methods such as scrambling and spreading were used to minimize the impact of interference with resulting gains in average spectral efficiency. Later systems employed receive diversity, equalizer, transmit diversity, and limited spatial multiplexing (MIMO). For LTE-Advanced, the planned performance enhancement techniques will take further steps by using more advanced MIMO and BF, interference cancellation, fractional frequency reuse, and other advanced methods.

It is worth explaining how the cell-edge performance targets used by ITU-R and 3GPP were developed based on simulated geometry-factor distributions for the target deployment environments. Ten users were randomly distributed within each cell and the resulting geometry factor was calculated for each UE. This information was converted into a data throughput rate that was in turn used to plot a distribution of throughput. The process was repeated many times in a multidrop simulation to create a smooth throughput distribution. The cell-edge performance was then defined as the fifth percentile of the throughput distribution. It is not straightforward to take the cell-edge figure per user and multiply by 10 to predict the cell-edge average since the distribution is complicated by the type of scheduler used. The scheduler may have allocated more resources to the cell-edge users in a proportionally fair system.

The following six figures present several geometry distributions (SINR), illustrating several intercell interference cancellation (ICIC) schemes. The system simulated has 19 sites (eNBs), where each cell has 3 sectors, 57 sectors in total, and mobile users are distributed within the inner 7 cells but suffer intercell interference from all the eNBs. The cell radius is 2250 m.

In Figure 9.14, the spatial geometry distribution of the SC-PMP scenario with full frequency reuse of 1 is presented. As reuse 1 is considered, no ICIC is employed and as a result the SNR is low almost everywhere. The higher values of SNR occur closer to the sites aligned with the directions of the three sector antennas. The places with SNR$\leq$ −2 dB correspond to about 10% of the total area. There are no places with SNR$\geq$ 22 dB.

The simple form of reducing intercell interference is employing fractional frequency reuse. In Figure 9.15, the spatial geometry distribution of the SC-PMP scenario with fractional frequency reuse of 1/3 is presented. As a result, the SNR increases everywhere. As expected, the higher values of SNR continue to occur closer to the sites aligned with the directions of the three sector antennas. The places with SNR$\leq$ −2 dB have disappeared and were replaced by 0 dB $\leq$ SNR $\leq$ 2 dB corresponding to about 10% of the total area. The places with SNR $\geq$ 22 dB correspond to about 10% of the total area. Due to the significant co-channel intercell interference reduction, there is a shift to the right in terms of geometry curve presented in Figure 9.13.

In Figure 9.16, the spatial geometry distribution of the SC-PMP scenario with soft frequency reuse of 1 + 1/3 is presented. When compared with the case of simple

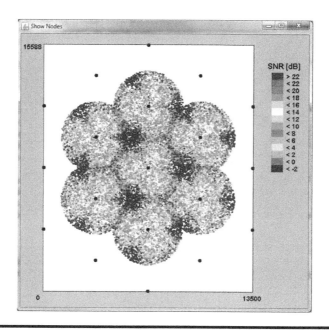

Figure 9.14 Spatial signal-to-interference-noise ratio of SC-PMP scenario with frequency reuse 1.

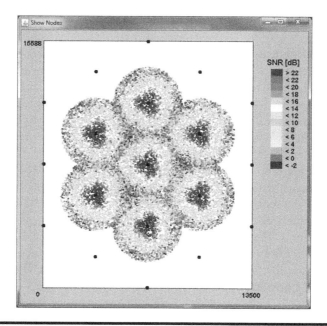

Figure 9.15 Spatial signal-to-interference-noise ratio of SC-PMP scenario with frequency reuse 1/3.

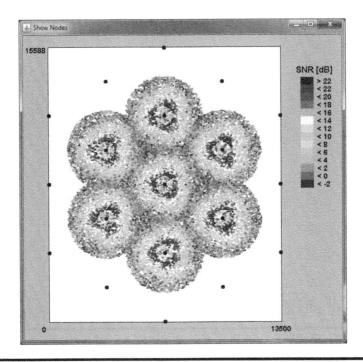

Figure 9.16   Spatial signal-to-interference-noise ratio of SC-PMP scenario with soft frequency reuse 1 + 1/3.

reuse 1, the SNR increases everywhere. The higher values of SNR continue to occur closer to the sites aligned with the directions of the three sector antennas. However, despite the higher available spectrum, when compared with the case of reuse 1/3, there is a decrease in SNR values close to eNBs due to the use of reuse 1 in those areas. The places with 0 dB ≤ SNR ≤ 2 dB correspond to about 10% of the total area, which is identical to simple reuse 1/3. However, the places with SNR ≥ 22 dB have decreased and correspond to about 3% of the total area.

In Figure 9.17, the spatial geometry distribution of the MBSFN scenario with full frequency reuse of 1 is presented. In spite of employing reuse 1, there is ICIC due to the considered MBSFN scenario. For reuse 1, MBSFN compared with SC-PMP presents spatial SNR higher due to coordination of the synchronized transmissions. The higher values of SNR continue to occur closer to the sites aligned with the directions of the three sector antennas. The places with values −2 dB ≤ SNR ≤ 0 dB correspond to about 5% of the total area. The places with values SNR ≥ 22 dB correspond also to about 5% of the total area.

To further reduce the intercell interference, we consider the fractional frequency reuse. In Figure 9.18, the spatial geometry distribution of the MBSFN scenario with fractional frequency reuse of 1/3 is presented. As a result, the SNR increases substantially everywhere. As expected, the higher values of SNR continue

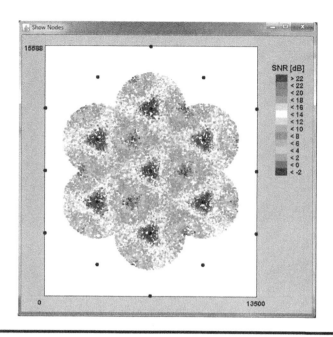

Figure 9.17 Spatial signal-to-interference-noise ratio of MBSFN scenario with frequency reuse 1.

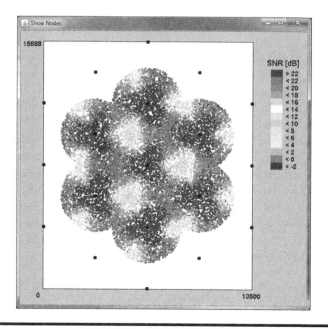

Figure 9.18 Spatial signal-to-interference-noise ratio of MBSFN scenario with frequency reuse 1/3.

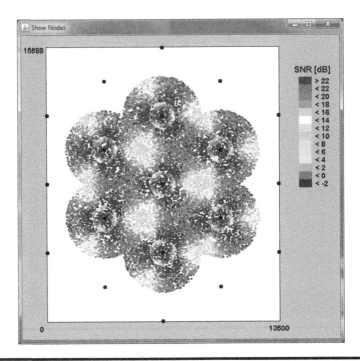

Figure 9.19 Spatial signal-to-interference-noise ratio of MBSFN scenario with soft frequency reuse 1 + 1/3.

to occur closer to the sites aligned with the directions of the three sector antennas. The places with −2 dB ≤ SNR ≤ 0 dB have disappeared and were replaced by 6 dB ≤ SNR ≤ 8 dB, corresponding to about 10% of the total area. The places with values SNR ≥ 22 dB have increased and correspond to about 20% of the total area.

In Figure 9.19, the spatial geometry distribution of the MBSFN scenario with soft frequency reuse of 1 + 1/3 is presented. When compared with the case of simple reuse 1, the SNR increases everywhere. The higher values of SNR continue to occur closer to the sites aligned with the directions of the three sector antennas. However, when compared with the case of reuse 1/3, there is a decrease in SNR values close to eNBs due to the use of reuse 1 in those areas. The places with 6 dB ≤ SNR ≤ 8 dB correspond to about 10% of the total area, which is identical to simple reuse 1/3. However, the places with values SNR ≥ 22 dB have decreased and correspond to about 15% of the total area.

9.3.2 Results for PtP Scenarios

In PtP scenario, every UE is served individually, and the link established by any given UE. If a user does not receive a packet properly, there is the option to retransmit the lost packet. Therefore, coverage in this type of system is assured. In this scenario, service delay or outage can be experienced (e.g., due to large waiting times when scheduling), outage being one of the aspects analyzed here. Another important aspect is the overall system capacity (i.e., how many users per cell can the system serve).

Since every UE is individually allocated with resources, and once these are finite, some sort of scheduling mechanism is necessary. Different scheduling mechanisms are tested, using different numbers of UEs in the system to better understand how every scheduling algorithm performs [Gomes 2010].

The following results only cover the traffic model, with constant bit-rate (CBR)* at 37,800 kbps. The CBR traffic model, as the name says, always generates the same amount of data, with exactly the same time intervals between consecutive data. This is the traffic model that comes closer to type of traffic generated in PMP transmissions. The CBR traffic model generates a packet of 37,800 bits every 1 ms (millisecond). This represents a traffic generator offering a load of 37,800 kbps per UE and a maximum spectral efficiency per user of 3.7 bps/Hz in a 10 MHz band with MIMO 4×4 [3GPP 2012b]. As we previously described, the exact shape of the geometry curve varies significantly depending on the frequency reuse factor followed by the cell size and cell loading. The influence of the type of employed scheduler, the cell size, and the cell loading will all be analyzed in the following sections.

9.3.2.1 Results for PtP with SU-MIMO

Figure 9.20 shows the cumulative distribution function of throughput (CDF(x)) as a function of the throughput, for SU-MIMO 4×4, with 10 channel quality indicators (CQIs) QPSK modulated, and five CQIs 16QAM modulated. The CDF(x) is the probability of the random variable% of UEs with throughput value less than or equal to x. Three different schedulers are analyzed, where the number of users per sector is 10, as recommended by ITU-R [ITU-R 2008] and by 3GPP [3GPP 2012a]. The CDF of throughput was chosen as the basis of comparison of different schemes in ITU-R and 3GPP. In this figure, the cell radius R is 750 m, the coding rate is 1/2 and the frequency reuse is one (high levels of intercell interference at cell borders). As can be seen, the results vary significantly depending on the scheduler. There are two "fair" schedulers, the simple round robin (RR) and the fair throughput (FT). In RR, users form a circular queue and the scheduler allocates equal timeslots for each and every user in the queue. The FT algorithm consists of serving users equally accordingly to the average throughput. Therefore, the FT scheduling aims at fairness in terms of user throughput (all users, no matter what their receiving conditions or position are inside cell will have the same average throughput). This is done by scheduling first users who have lowest average throughputs. The scheduler maximum carrier-interference (MCI), also referred to in the literature as "maximum SINR," or simply as "Max C/I," is a channel aware

---

* CBR stands for constant bit rate.

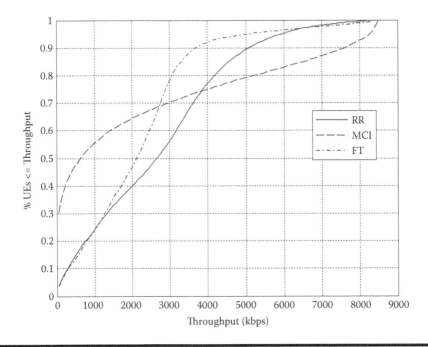

Figure 9.20   CDF of throughput versus throughput, SU-MIMO (4 × 4), different schedulers, Nu = 10.

scheduling algorithm where it is given more priority to users with good channel (users located closer to the base-station). The scheduler chooses the user *k* with maximum SINR at instant *t*. The measurement of SINR is performed via constant periodic CQI feedback done by every single user. The MCI is not fair.

Only RR and FT are able to transmit successfully to users located at cell borders. Remind that the cell-edge performance is defined as the fifth percentile of the CDF of the throughput. In this case, the cell-edge performance of RR and FT is 125 kbps. The average throughput (50% CDF) is 650 kbps for MCI, 2150 kbps for FT and 2650 kbps for RR. However, it can be viewed that 10% of users experience a better than 7600 kbps for MCI, 3700 kbps for FT, and 5000 kbps for RR.

The throughput distribution versus the geometry of SU-MIMO 4 × 4 is illustrated in Figure 9.21, for different schedulers, and corresponds to previous Figure 9.20. The geometry factor is the term used in UMTS to indicate the ratio of the wanted signal relating to the interference plus noise, and corresponds to the SINR. As expected the results depend on the scheduler. The curves corresponding to the two "fair" schedulers, RR and FT are less dependent of the parameter geometry. Indeed the FT algorithm aims at achieving a straight line along the geometry axe. Independent of the mobile position within the cell, each one gets the same average throughput. Due to the way it works, the RR scheduler presents the expected variation of the throughput with the geometry. Users located close to the base station,

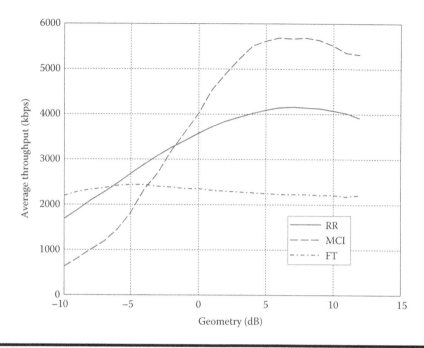

Figure 9.21 Throughput distribution versus geometry, SU-MIMO (4 × 4), different schedulers.

with high values of geometry (expressed in dB), achieve higher throughput than users at the border of the cell. The unfair scheduler MCI, giving more priority to users with good channel conditions, amplifies the difference of throughput versus the geometry. It has very small throughput for users at the cell edge and very high throughput for users located close to the base station.

The throughput distribution versus the geometry of SU-MIMO 4 × 4 is illustrated in Figure 9.22, for the RR scheduler, where the parameter cell radius is variable. As can be seen from Table 9.1 and Equation 9.11, the average propagation loss with the distance depends on the existence of LOS (or non-LOS [NLOS]). The probability of LOS decreases with the distance, and the probability of NLOS increases with the distance. In the model used for LOS at $d < 320$ m, there is an exponential power decay with 3.4 and for LOS at $d \geq 320$ m the power decay increases to 4. When we consider the cell with radius equal to $R = 500$ m, there is a predominance of places that are in LOS. In this case, there is a strong intercell interference, even for mobiles close to the base station. Increasing the cell radius to $R = 750$ m and to $R = 1000$ m decreases the probability of LOS and reduces the intercell interferences for users close to the base station increasing the throughput. This can be seen from Figure 9.22. However, for users located at the cell edge increasing the cell radius reduces the throughput due to decreasing levels of received power, while keeping the same geometry (the same signal to interference plus noise ratio).

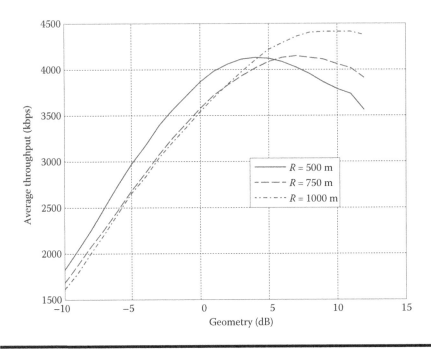

**Figure 9.22   Throughput distribution versus geometry, SU-MIMO (4 × 4), differ-
ent cell radius, Nu = 10.**

The CDF of throughput results for SU-MIMO 4 × 4, the RR scheduler, for different cell radius, Nu = 10, is illustrated in Figure 9.23, and corresponds to Figure
9.22. As expected, the case $R = 750$ m is the intermediate one. For $R = 1000$, there
is a reduction of the maximum throughput, and less than 10% of the users have
throughput higher than 4500 kbps. For $R = 500$ m, there are 10% of the users with
throughput higher than 5350 kbps. However, there is an average throughput (50%
of users) of 2650 kbps for $R = 1000$ m, while for $R = 500$ m the average value is
2500 kbps. For cell edge users, the throughput is also higher for $R = 1000$ m compared to $R = 500$ m (or $R = 750$ m).

The throughput distribution versus the geometry of SU-MIMO 4 × 4, where the
number of users per cell is Nu = 50, is illustrated in Figure 9.24, for different schedulers. As expected, there is a substantial throughput reduction per user when we increase
the number of users from 10 (Figure 9.22) to 50. For MCI, only 30% of the users
are served but 10% of these users get more than 2000 kbps. For RR and FT, 50% of
the users get 525 kbps which is around 10/50 of the throughput achieved when there
were 10 users. For FT, 10% of the users get more than 600 kbps and for RR there are
10% of users with more than 1000 kbps. In conclusion, there is a linearly proportional
decrease of the throughput for a linear increase of the number of users.

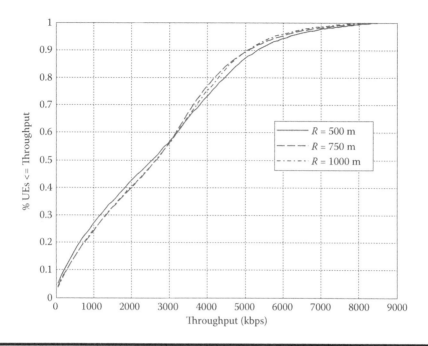

Figure 9.23 **CDF of throughput versus throughput, SU-MIMO (4 × 4), different cell radius, Nu = 10.**

The CDF of throughput results for SU-MIMO 2 × 2 with 10 CQIs QPSK modulated and five CQIs 16QAM modulated are illustrated in Figure 9.25. The same schedulers are analyzed, namely, RR, MCI, and FT. The number of users per sector is 10 and the cell radius R is 750 m, the coding rate is 1/2 for QPSK and 3/4 for 16QAM, the frequency reuse is one. This figure should be compared to Figure 9.22 with MIMO 4 × 4.

RR and FT are able to transmit successfully to users located at cell borders and MCI continues to not serve 30% of the users. The cell-edge performance (5% CDF) of RR and FT is 125 kbps. The average throughput (50% CDF) is 450 kbps for MCI, 1200 kbps for FT, and 1400 kbps for RR. However, 10% of users experience a throughput better than 7500 kbps for MCI, 1900 kbps for FT, and 3800 kbps for RR. When we compare the MIMO 2 × 2 results with MIMO 4 × 4, the cell edge throughput has not changed, but the average throughput is about half, independent of the scheduler. For the 90% CDF, the throughput results depend on the schedulers. For FT, this throughput is halved, for the RR scheduler is 76%, and for MCI is 96%. It should be noticed that the highest bit rate of the CQIs for MIMO 2 × 2 is almost the same as the CQIs for MIMO 4 × 4 because both employ 16QAM but the coding rates are different 3/4 and 1/2, respectively.

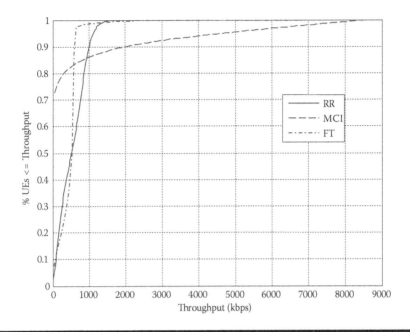

Figure 9.24   **CDF of throughput versus throughput, SU-MIMO (4 × 4), Nu = 50.**

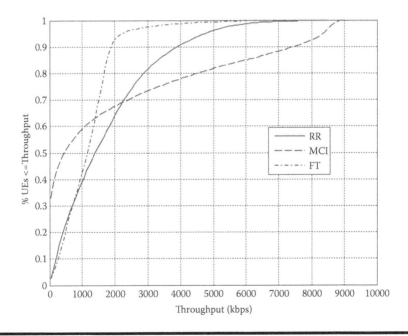

Figure 9.25   **CDF of throughput versus throughput, SU-MIMO (2 × 2), Nu = 10.**

9.3.2.2 Results for MU-MIMO

BF is a suboptimal strategy that can serve multiple users at a time, but with reduced complexity. In BF, each user stream is coded independently and multiplied by a BF weight vector for transmission through multiple antennas. Careful selection of weight vectors can reduce (or eliminate) mutual interference among different streams by taking advantage of spatial separation between users and thereby supporting multiple users simultaneously. This multiuser communication scheme is called space-division multiple access (SDMA).* In [Sharif and Hassibi 2005], the authors propose an orthogonal random beamforming (RBF) scheme. Another suboptimal BF strategy is zero-forcing beamforming (ZFBF) [Yoo and Goldsmith 2006], where the weight vectors are chosen to avoid interference among user streams. Both scheduling strategies essentially combine TDMA/OFDMA with SDMA.

In industry, SDMA with orthogonal BF, under the name "per user unitary and rate control" (PU2RC) [Samsung 2006], was proposed to the 3GPP-LTE standard. The main feature of PU2RC is limited feedback, where multiuser precoders or beamformers are selected from a codebook of multiple orthonormal bases. Based on limited feedback, PU2RC supports SDMA, scheduling, and adaptive modulation and coding. Because of its versatility and advanced features, PU2RC is one of the most promising solutions for high-speed downlink in 3GPP-LTE.

The cumulative distribution function of throughput results for MU-MIMO with $M = 4$ transmitting antennas, with the same adaptive MCSs of SU-MIMO are illustrated in Figure 9.26. Four different schedulers are analyzed, where the number of users per sector is 10. In current and next figures, the cell radius R is 750 m and perfect channel state information is assumed. All schedulers include beamformers selected from multiple orthonormal bases. There is a new scheduling algorithm named proportional fair (PF). Similar to Max C/I, PF is channel aware. In fact, we can look at PF as a less aggressive version of Max C/I scheduling algorithm. PF uses CQI feedback sent by users to determine the instantaneous possible data rate a user k can achieve at a given instant t, and also the average throughput a user k had until instant t. This way, users that have instantaneous throughputs higher than their average throughput are scheduled first. The unfairness of Max C/I scheduler, where only users with good SNR have resources allocated, is avoided [Ishizaki and Hwang 2009]. As expected, the results are varying depending on the scheduler type. The comparison with Figure 9.20 indicates that MCI increases the highest throughput achieved by 10% of users, better than 8000 kbps. However, there is a reduction of the highest throughput for FT (and RR), which was compensated by the increase of average throughput, specially the throughput for users at cell edge. The throughput performance curve of PF is following the MCI curve

---

* See Chapter 1.

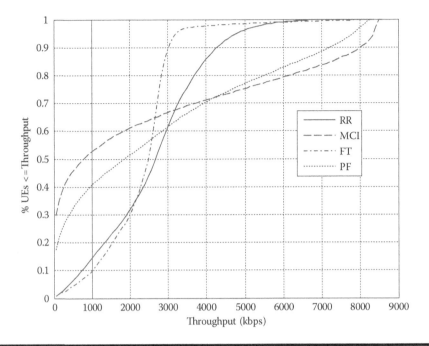

Figure 9.26 **CDF of throughput versus throughput, MU-MIMO (*M* = 4), different schedulers, Nu = 10.**

but with much more fairness. As a result, the average throughput of PF is more than the double of the MCI.

The throughput distribution versus the geometry of MU-MIMO, for *M* = 4 transmitting antennas, is illustrated in Figure 9.27, for different schedulers, and corresponds to Figure 9.26. As expected, the results depend on the scheduler and should be compared with Figure 9.21. An increase of throughput is observed for MCI and users with highest geometry. There is a reduction of throughput for RR and users with higher geometry. The performance of PF is between these two schedulers for all geometry values. The throughput of FT is the less dependent of the geometry parameter and presents an increase for the smaller geometry.

The throughput distribution versus the distance of MU-MIMO, *M* = 4 transmitting antennas, is illustrated in Figure 9.28, for different schedulers, and corresponds to previous Figure 9.27. As expected there is a strong correlation with the results presented in Figure 9.27. The correlation is due to the strong relation between geometry and location inside the cell of the users. Users at smaller distance of the base station are those with higher geometry. Inversely, users at the border of the cell (*R* = 750 m) are the ones with the smallest geometry. For MCI and users at *R* = 50 m, the throughput is higher than 8000 kbps, which should correspond to less than 10% of the users. For PF, RR, and FT algorithms, the

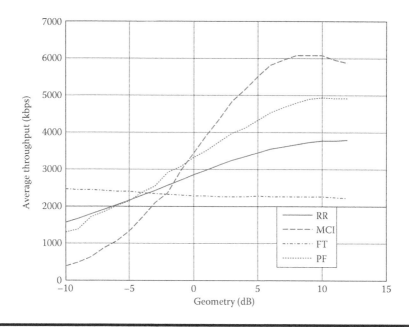

Figure 9.27   Throughput distribution versus geometry, MU-MIMO (*M* = 4), different schedulers, Nu = 10.

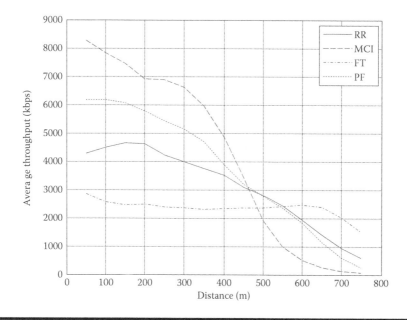

Figure 9.28   Throughput versus distance, MU-MIMO (*M* = 4), different schedulers, Nu = 10.

highest throughput is 6200 kbps, 4350 kbps, and 2900 kbps, respectively. At cell edge, there is an inversion and for MCI, PF, RR, and FT the throughput achieved is 90 kbps, 350 kbps, 650 kbps, and 1600 kbps, respectively.

The CDF of throughput results for MU-MIMO with $M = 4$ transmitting antennas and the same MCSs of SU-MIMO are illustrated in Figure 9.29, where the number of users per sector is 50. The same four different schedulers are analyzed. The comparison with Figure 9.24 indicates that MCI increases the highest throughput achieved by 10% of users, better than 2300 kbps. However, there is a reduction of the highest throughput for RR (and FT) which was compensated by the increase of the average throughput, specially the throughput of FT for users at cell edge. Again, the performance curve of PF is following the MCI curve but with much higher percentage of users served, in spite of small throughput. As a result, more than 55% of the users are scheduled by PF, while for MCI, only 28% are scheduled.

The throughput distribution versus the geometry of MU-MIMO, $M = 4$ transmitting antennas, for Nu = 50 users per sector is illustrated in Figure 9.30, and corresponds to CDF results of previous Figure 9.29. These results should be compared with Figure 9.27 for Nu = 10. For the FT algorithm with Nu = 50, there is an average throughput of 500 kbps that multiplied by 50/10 = 5 gives 2500 kbps which

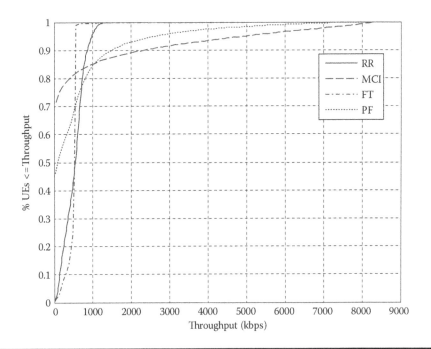

Figure 9.29 **CDF of throughput versus throughput, MU-MIMO ($M = 4$), different schedulers, Nu = 50.**

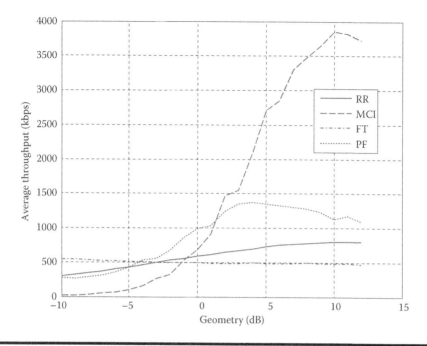

Figure 9.30 **Throughput distribution versus geometry, MU-MIMO (***M* **= 4), different schedulers, Nu = 50.**

is the average throughput when Nu = 10. There is no throughput gain (unitary gain) with this scheduler. For MCI with Nu = 50, in the region of higher geometry (interference-limited region), there is an average throughput of 3600 kbps that multiplied by 5 gives 1800 kbps which is three times the average throughput of 6000 kbps when Nu = 10. So there is a throughput gain of 3 for this scheduler. In the region of smaller geometry (interference plus noise-limited region), there is no throughput gain. In fact, a loss of throughput is observed. For PF in the interference-limited region, the throughput gain is 1.5. In the interference plus noise-limited region, the gain is unitary. With RR, there is a unitary gain in both regions. This is similar to FT.

To increase the spectral efficiency on average and/or at the cell borders, or when the number of users is increasing, the MU-MIMO with PF scheduling should be employed.

9.3.3 Results for CoMP with MU-MIMO

To allow a comparison with previous results, only CoMP scenario 2 is analyzed in this section. Data to a UE is simultaneously transmitted (joint transmission) from multiple points to coherently improve the received signal quality or data throughput. A combination of MU-MIMO with CoMP is employed. We take the simplification of assuming that the quality of channel state information is very good (perfect estimation).

The cumulative distribution function of throughput results for CoMP with MU-MIMO with $M = 4$ transmitting antennas per site, with the same adaptive MCSs of SU-MIMO are illustrated in Figure 9.31. The same four different schedulers are analyzed, where the number of users per sector is 10 and the cell radius R is 750 m. As expected, the results are varying depending on the scheduler. The comparison with Figure 9.26 indicates overall improvements independently of the schedulers. For MCI, the highest throughput achieved by 10% of users is around 8500 kbps. For PF, RR, and FT the corresponding values are 7750 kbps, 5900 kbps, and 4100 kbps, respectively. The average throughput has also increased for all the schedulers. There is an obvious increase of throughput of cell edge users for schedulers MCI and PF. There are also throughput increments at cell edge users for the FT and RR due to improvement of the received signal-to-noise ratio.

The throughput distribution versus the geometry of CoMP with MU-MIMO, $M = 4$ per site, for Nu = 10 users per sector is illustrated in Figure 9.32, and corresponds to CDF results of previous Figure 9.31. These results should be compared with Figure 9.27. There is an overall performance improvement independently of the schedulers. For the FT algorithm, there is an obvious increase of throughput of

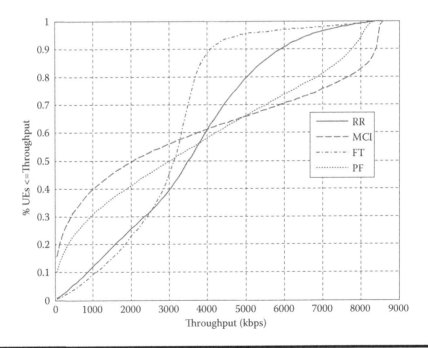

Figure 9.31   CDF of throughput versus throughput, CoMP with MU-MIMO, different schedulers, Nu = 10.

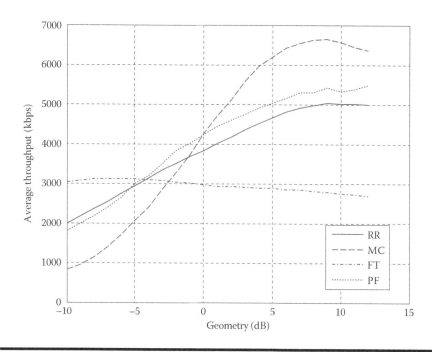

Figure 9.32 **Throughput distribution versus geometry, CoMP with MU-MIMO, different schedulers, Nu = 10.**

3000/2500 kbps = 1.2. For the other algorithms, it is not easy to compute the gain because we need to consider the performance improvement in all geometry values, but the same 20% increase seems to apply.

The CDF of throughput results for CoMP with MU-MIMO, and with $M = 4$ transmitting antennas per site, is illustrated in Figure 9.33, where the number of users per sector is 50. The comparison with Figure 9.29 indicates overall improvements independently of the schedulers. For MCI, PF, RR, and FT the highest throughput achieved by 10% of users is 3300 kbps, 2050 kbps, 1200 kbps, and 800 kbps, respectively. The performance curve of PF is following the MCI curve, but there is no improvement for users served with small throughput. In all, 55% of the users (the same percentage as without CoMP) are scheduled by PF while for MCI, 33% are now scheduled (5% increase).

The throughput distribution versus the geometry of CoMP with MU-MIMO, for Nu = 50 users per sector, is illustrated in Figure 9.34, and corresponds to CDF results of previous Figure 9.33. These results should be compared with Figure 9.32 for Nu = 10. For the FT algorithm, with Nu = 50, there is an average throughput of 700 kbps that multiplied by 50/10 = 5 gives 3500 kbps which is 500 kbps above the average throughput when Nu = 10. There is a throughput gain of 1.17 with this scheduler. For MCI with Nu = 50, in the region of higher geometry

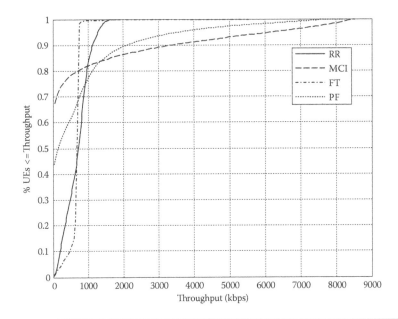

Figure 9.33 **CDF of throughput versus throughput, CoMP with MU-MIMO, different schedulers, Nu = 50.**

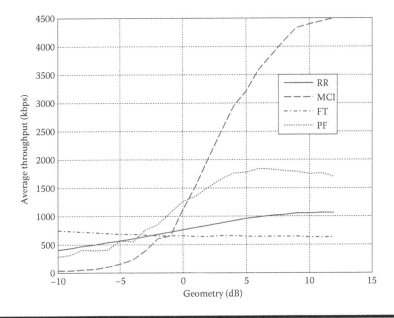

Figure 9.34 **Throughput distribution versus geometry, CoMP with MU-MIMO, different schedulers, Nu = 50.**

(interference-limited region), there is an average throughput of 4300 kbps that, multiplied by 5, gives 21,500 kbps, which is 3.3 times the average throughput of 6500 kbps when Nu = 10. The throughput gain is 3.3 for this scheduler. In the region of smaller geometry (interference plus noise-limited region), there is no throughput gain, but there is a loss. For PF in the interference-limited region, the throughput gain is 1.67. In the interference plus noise-limited region, the gain is unitary. With RR, there is a unitary gain in all regions. When these results are compared with Figure 9.30, for Nu = 50 but without CoMP, we get the following conclusions: for FT, the throughput gain is around 1.4 within both interference regions; for RR the gain is 1.25 within both interference regions; for MCI, in the interference-limited region, the gain is 1.15 and is unitary in the interference plus noise-limited region; for PF, in the interference-limited region, the gain is 1.4, being 1.10 in the interference plus noise-limited region. Due to the use of joint coordinated multipoint transmission, independently of the schedulers employed, throughput gains are observed. However, the gains achieved depend on the schedulers. The performance results indicate also that the criterion of proportional fairness is important when the number of users is increasing.

9.4 Conclusions

The main goal of this chapter was to describe a set of techniques that may be implemented in future LTE-Advanced networks, using PMP and PtP scenarios. These techniques use spatial multiplexing and introduce the concept of coordinated multipoint transmission.

The use of fractional reuse schemes helps to reduce intercell interference suffered by users, especially those at cell border. The disadvantage of this technique is that by partitioning the existing frequencies available, every cell will only have access to a part of the total available bandwidth, and this is unfair to UEs closer to the center of the cell (with good channel conditions) that could achieve higher throughputs if using the total bandwidth.

To increase the spectral efficiency on average and/or at the cell borders, or when the number of users is increasing, the MU-MIMO with fair scheduling should be employed. The use of joint coordinated multipoint transmission achieves additional throughput gains. However, the gains obtained depend on the schedulers employed. The performance results also indicate that the criterion of proportional fairness is important when the number of users is increasing.

Acknowledgments

This work was supported by Fundação para a Ciência e Tecnologia (FCT) via project PEst-OE/EEI/LA0008/2013.

References

3GPP, Improvements of the Multimedia Broadcast Multicast Service (MBMS) in UTRAN, TR 25.905 v7.2.0, December 2007.

3GPP, Requirements for Evolved UTRA (E-UTRA) and Evolved UTRAN (E-UTRAN), TR 25.913 v9.0.0, December 2009a.

3GPP, Feasibility study for evolved Universal Terrestrial Radio Access (UTRA) and Universal Terrestrial Radio Access Network (UTRAN), TR 25.912 V9.0.0, 2009b.

3GPP, Technical Specification Group Radio Access Network; MBMS synchronization protocol (SYNC), TS 25.446 v9.0.0, December 2009c.

3GPP, Further advancements for E-UTRA physical layer aspects TS 36.814, V9.0.0, 2010a.

3GPP, General aspects and principles for interfaces supporting Multimedia Broadcast Multicast Service (MBMS) within E-UTRAN, TS 36.440 v9.1.0, March 2010b

3GPP, Evolved Universal Terrestrial Radio Access (E-UTRA); Radio Frequency (RF) system scenarios, Technical Report TR 36.942 v9.0.1, April 2010c.

3GPP, Evolved Universal Terrestrial Radio Access (E-UTRA); User Equipment (UE) radio transmission and reception, TS 36.101 v9.3.0, 2010d.

3GPP, Coordinated multi-point operation for LTE physical layer aspects, TS 36.819, V11.1.0, 2011.

3GPP, Evolved Universal Terrestrial Radio Access (E-UTRA) and Evolved Universal Terrestrial Radio Access Network (E-UTRAN); Overall description, TS 36.300, V11.3.0, 2012a.

3GPP, Requirements for Further Advancements of E-UTRA (LTE-Advanced), TR 36.913 V11.0.0, 2012b.

Fodor, G., Koutsimanis, C., Rácz, A., Reider, N., Simonsson, A., and Muller, W., Intercell interference coordination in OFDMA networks and in the 3GPP long term evolution system. *Journal of Communications*, 4, 445–453, August 2009.

Gomes, P. S., Scheduling techniques to transmit multi-resolution in E-MBMS services of LTE-advanced, PhD thesis, ISCTE-IUL, September 2010.

Ishizaki, F., and Hwang, G., Throughput performance of quantized proportional fair scheduling with adaptive modulation and coding. In *Wireless Telecommunications Symposium*, Prague, April 2009, pp. 1–6.

ITU-R, Guidelines for evaluation of radio interface technologies for IMT-Advanced, M.2135, 2008.

Lee, J., Ng, B., and Mazzarese, D., Coordinated multipoint transmission and reception in LTE-advanced systems. *IEEE Communications Magazine*, 44–50, November 2012.

Marques da Silva, M., *Multimedia Communications and Networking*. CRC Press Auerbach Publications, 1st edition, ISBN: 9781439874844, Boca Raton, FL, March 2012.

Marques da Silva, M., Correia, A., Dinis, R., Souto, N., and Silva, J.C., *Transmission Techniques for Emergent Multicast and Broadcast Systems*. CRC Press Auerbach Publications, 1st edition, ISBN: 9781439815939, Boca Raton, FL, May 2010.

Marques da Silva, M., Correia, A., Dinis, R., Souto, N., and Silva, J.C., *Transmission Techniques for 4G Systems*. CRC Press Auerbach Publications, 1st edition, ISBN: 9781466512337, Boca Raton, FL, November 2012.

Rumney, M., *LTE and the Evolution to 4G Wireless: Design and Measurement Challenges*. Copyright Agilent Technologies, Inc. 2013, John Wiley & Sons, Ltd, United Kingdom.

Samsung Electronics, Downlink MIMO for EUTRA, in 3GPP TSG RAN WG1 # 44/R1-060335, Feb. 2006.

Sarperi, L., Hunukumbure, M., and Vadgama, S., Simulation study of Frequency Reuse in WiMAX Networks. *FUJITSU Scientific & Technical Journal (FSTJ)*, 44, 318–324, July 2008.

Sharif, M., and Hassibi, B., On the capacity of MIMO broadcast channels with partial side information. *IEEE Transactions on Information Theory*, 51, 506–522, Feb. 2005.

Simonsson, A., Frequency reuse and intercell interference co-ordination in E-UTRA. In *IEEE 65th Vehicular Technology Conference*, Dublin, April 2007, pp. 3091–3095.

Yoo, T. and Goldsmith, A., On the optimality of multiantenna broadcast scheduling using zero-forcing beamforming. *IEEE Journal on Selected Areas in Communications*, 24(3), 528–541, March 2006.

Zhou, Y., and Zein, N. Simulation study of fractional frequency reuse for mobile WiMAX. In *IEEE Vehicular Technology Conference*, Singapore, May 2008, pp. 2592–2595.

Chapter 10

Massive MIMO and Its Detection

Xiaoli Ma and Qi Zhou

Contents

10.1 Introduction and Motivation

MIMO techniques have gained considerable attention in modern wireless communications because of their high spectrum efficiencies and extended coverage. By transmitting data streams via multiple antennas and exploiting diversity gains offered by distinct channel propagations, MIMO techniques boost the data rate and enhance the data reception. Thanks to these benefits, MIMO systems have been widely adopted by modern wireless systems. For example, current long-term evolution (LTE) standard allows up to eight antennas equipped at the base station to transmit four streams simultaneously. To further harvest the benefits of MIMO systems, massive MIMO techniques have been proposed by installing a large number of antennas at base stations, possibly in the order of tens or hundreds. With the large number of antennas, massive MIMO offers several unique benefits for wireless communications:

■ Massive MIMO provides high orders of degrees of freedom of channels, and thus allows the base station to serve more terminals at the same time–frequency resources.
■ Massive MIMO can reap the benefits of rich propagation paths such that the fading effects can be averaged out, that is, diversity order goes to infinity.
■ When the number of antennas at the base station is much larger than at the terminals, massive MIMO enables simple precoding/detection with near-optimal performance.

To efficiently operate the communication systems with a large number of antennas, massive MIMO also imposes a number of challenges in hardware and signal processing, including

■ Fast and efficient detection algorithms given the massive channel matrices.
■ Stringent channel estimation issue, where the channel parameters grow linearly in terms of the number of antennas and the number of terminals.
■ Pilot contamination caused by the interference from the adjacent cells, which is becoming a bottleneck of multicell massive MIMO systems.
■ Low-cost/energy-efficient RF frontend.

This chapter introduces detection algorithms for Massive MIMO. We provide a brief introduction to massive MIMO and discuss its mutual information and precoding designs in Section 10.2. Then, the detection designs for massive MIMO are provided in Section 10.3. Hardware implementation concerns of massive MIMO detectors are stated in Section 10.4, and other issues of massive MIMO such as channel estimation and synchronization will be mentioned in Section 10.5.

10.2 System Model

Consider an uplink MIMO transmission model

$$\mathbf{y}^u = \mathbf{H}\mathbf{s}^u + \mathbf{w}^u, \tag{10.1}$$

where $\mathbf{y}^u$ is an $M \times 1$ received signal vector at the base station with M being the number of antennas at base station, $\mathbf{s}^u$ is an $N \times 1$ transmit signal vector with N being the number of antennas at the terminals, $\mathbf{H}$ is an $M \times N$ channel matrix, and $\mathbf{w}^u$ is the additive white Gaussian noise vector with zero mean and covariance matrix $N_0 \mathbf{I}_M$.

For the downlink MIMO transmission, we assume time-division duplex (TDD) transmission so that the downlink channel is the reciprocal of that of the uplink one [Rusek et al. 2013]. The corresponding downlink transmission model is

$$\mathbf{y}^d = \mathbf{H}^T \mathbf{s}^d + \mathbf{w}^d, \tag{10.2}$$

where superscript T denotes the matrix transpose, $\mathbf{y}^d$ is an $N \times 1$ received signal vector at the terminals, $\mathbf{s}^d$ is an $M \times 1$ transmitted signal at the base station, and $\mathbf{w}^d$ is the additive white Gaussian noise vector with zero mean and covariance matrix $N_0 \mathbf{I}_N$. The reason for using TDD here is to allow the base station to perform precoding which can obtain channel state information (CSI) with reciprocity. The CSI can also be acquired using the feedback from terminals, but it is expensive due to the large number of channel parameters of massive MIMO.

For massive MIMO, the number of antennas at both ends can be up to hundreds or even thousands. Since each terminal typically employs antennas less than 10 due to limited space, massive MIMO generally applies a large number of antennas at the base station, that is, $M \gg 0$. However, for multiuser MIMO (MU-MIMO), it is also possible that a group of synchronized terminals transmit simultaneously, which forms a large number of antennas at the mobile side such that $N \gg 0$. For example, for the LTE-A systems, each user may have four antennas. With more than 10 users, the total number of antennas becomes large if the base station needs to decode them simultaneously. Therefore, based on the number of antennas employed at the base station and the terminals, we categorize the following two general cases of massive MIMO.

- Case 1: $M \gg N$, which accounts for SU-MIMO or MU-MIMO when the number of serving terminal antennas is small.
- Case 2: $M \approx N$, which accounts for MU-MIMO when the total number of serving terminal antennas is close to the number of antennas at the base station.

These two setups have different impacts on MIMO transceiver designs. In general, the complexity of optimal symbol detection for MIMO grows exponentially in terms of the number of independent data streams for diversity and spatial multiplexing gains. However, when $M \gg N$, under the favorable channel propagation condition, the *channel orthogonalization* or *channel hardening* takes effect such that the columns of **H** become asymptotically orthogonal as in [Rusek et al. 2013]

$$\left(\frac{\mathbf{H}^H \mathbf{H}}{M} \right)_{M \gg N} \approx \mathbf{D}, \tag{10.3}$$

where the small-scale fading effect is averaged out and **D** is an $N \times N$ diagonal matrix with $D_{i,i}$ accounting for path loss and large-scale fading effects from the ith antenna to the base station (the superscript H in the expression is the Hermitian operator). The main benefit of channel orthogonality is that it allows efficient simple precoding and detection (e.g., zero-forcing detector) without losing "too much" performance (see [Ma and Zhang 2008a]). Another reason is that the diversity orders collected by optimal detector and linear detectors are M and $M - N + 1$, respectively. When $M \gg N$, $M - N + 1$ approaches M with negligible difference if M is large.

When $M \approx N$, the transceiver designs of massive MIMO systems are quite different because the linear detectors do not perform well and Equation 10.3 does not hold. Optimal detection such as sphere decoding algorithms (SDAs) is clearly unaffordable due to its high complexity (e.g., the search space for a 50×50 MIMO with 256-QAM (quadrature amplitude modulation) is $256^{50} \approx 2.5 \times 10^{120}$). Note that even for the first case, when N is 10 or 20, sphere decoding or maximum likelihood detector (MLD) does not have affordable complexity. Detection becomes one of the major obstacles to prevent applying massive MIMO in real systems. In Section 10.3, the recent development of low-complexity high-performance large MIMO detectors will be introduced.

10.2.1 Different Types of Massive MIMO

10.2.1.1 Single-User MIMO

An illustration of SU-MIMO systems is depicted in Figure 10.1. Owing to the physical limitation of terminals, the number of antennas N at the terminal is generally much less than M. Therefore, SU-MIMO systems fall into case 1 when a large number of antennas are equipped at the base station, and thus reap the benefits of channel orthogonalization if favorable channel propagation condition holds. However, the SU-MIMO channels could be highly correlated because of the compact distance of antennas at the terminal side and possible line-of-sight environment. From the power efficiency point of view, using a large antenna array to serve a single or a small number of users may not be wise. Hence, in this case, the gain of massive MIMO for SU-MIMO may be limited.

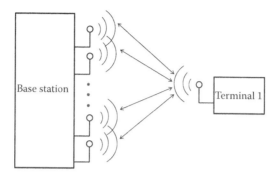

Figure 10.1 An illustration of SU-MIMO systems.

10.2.1.2 Multiuser MIMO

An illustration of MU-MIMO systems is depicted in Figure 10.2. When multiple terminals are allowed to access the same time–frequency resource, MU-MIMO provides higher system efficiency compared to SU-MIMO. Here, we consider single-cell MU-MIMO systems, where a base station is serving T terminals with each terminal being equipped with Q antennas (i.e., $N = TQ$). The transmission model of an uplink MU-MIMO system is

$$
\begin{aligned}
\mathbf{y}^u &= \sum_{t=1}^{T} \mathbf{H}_t \mathbf{s}_t^u + \mathbf{w}^u \\
&= \mathbf{H}\mathbf{s}^u + \mathbf{w}^u,
\end{aligned}
\tag{10.4}
$$

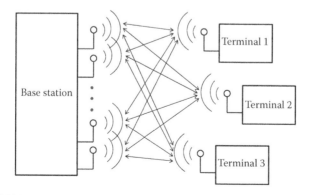

Figure 10.2 An illustration of MU-MIMO.

where $\mathbf{H}_t$ is an $M \times Q$ channel matrix from terminal t to the base station, $\mathbf{s}_t^u$ is a $Q \times 1$ transmitted signal from terminal t, $\mathbf{H} = [\mathbf{H}_1, \ldots, \mathbf{H}_t]$, and $\mathbf{s}^u = [(\mathbf{s}_1^u)^T, \ldots, (\mathbf{s}_T^u)^T]^T$.

When $T = 1$, the MU-MIMO reduces to SU-MIMO. When $K \geq 2$, the received signal of each terminal is interfered with those of the other terminals, and thus we could expect that the mutual information of each terminal for MU-MIMO is smaller than that for SU-MIMO given the same transmitted power at each terminal. However, when $M \gg N$, the channel orthogonalization kicks in such that the received signal of each terminal is almost orthogonal, that is, interference-free in the desired signal space under favorable channel propagation condition. In addition, since the terminals are autonomous, the favorable channel propagation condition is usually satisfied since the antennas at the terminals are almost uncorrelated and uncoupled [Gao et al. 2011, 2012]. This again shows that the massive MIMO is in favor of the MU-MIMO setup.

10.2.1.3 Distributed Massive MIMO

Distributed massive MIMO (Figure 10.3) can be treated as a special case of MU-MIMO to further provide higher system capacity by employing distributed-deployed antennas to transmit and receive signals. One form of distributed massive MIMO is to enable cooperation between the base stations in different cells that reduces the intercell interference. However, synchronization becomes a critical issue even for distributed antennas at the same base station. In some cases, the large

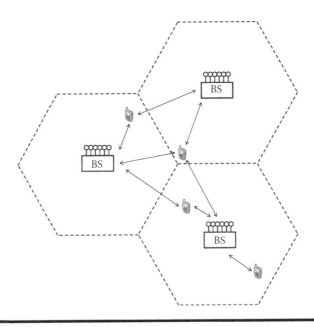

Figure 10.3 An illustration of distributed massive MIMO systems.

number of antennas at the base station can also be placed in different places (e.g., on tops of buildings) [Rusek et al. 2013]. In this case, synchronization is one issue, and the low-cost RF frontend may introduce more issues.

10.2.2 Mutual Information for Massive MIMO Uplink

Since SU-MIMO is a special case of MU-MIMO when $T = 1$, in this section, we focus on the mutual information of MU-MIMO systems. Unless stated otherwise, we assume that CSI is known at the receiver (CSIR), that is, the terminals for downlink transmissions, and the base station for uplink transmissions, and we impose power constraint at each terminal as

$$E\{\| s_t^u \|^2\} = E^u \quad \forall t = 1,\dots,T \tag{10.5}$$

and at the base station as

$$E\{\| s^d \|^2\} = E^d. \tag{10.6}$$

The signal-to-noise ratios (SNRs) for uplink and downlink are defined as $E^u/N_0 = \rho^u$ and $E^d/N_0 = \rho^d$, respectively.

To quantify the benefit of massive MIMO in information theory, we adopt achievable sum rate from [Rusek et al. 2013]. Achievable sum rate is defined as the mutual information between the received signal and transmitted signals. For the uplink MU-MIMO systems by assuming Gaussian-independent input $\mathbf{s}^u$, the sum rate is given as

$$C = I(\mathbf{y}^u;\mathbf{s}^u) = \log_2 \det\left(\mathbf{I}_M + \frac{E^u\mathbf{H}\mathbf{H}^H}{QN_0}\right). \tag{10.7}$$

When $M \gg N$, that is, uplink transmission with massive antennas at the base station and relatively small total number of antennas of all terminals, the asymptotic sum rate becomes

$$\begin{aligned}
C_{M\gg N}^u = I(\mathbf{y}^u;\mathbf{s}^u) &= \log_2 \det\left(\mathbf{I}_N + \frac{\rho^u\mathbf{H}^H\mathbf{H}}{Q}\right) \\
&\approx \log_2 \det\left(\mathbf{I}_N + \frac{\rho^u M\mathbf{D}}{Q}\right) \\
&= \sum_{t=1}^{T}\sum_{q=1}^{Q} \log_2\left(1 + \frac{\rho^u d_t M}{Q}\right),
\end{aligned} \tag{10.8}$$

where we employ channel orthogonalization in Equation 10.8, d_t is the path loss and large-scale fading effect from the tth terminal to the base station, and we assume that all antennas at the same terminal experience the same path loss and large-scale fading effect d_t, that is, $D_{(t-1)Q+q,(t-1)Q+q} = d_t, \forall q = 1,...,Q$.

From Equation 10.8, the gains of massive MIMO for uplink transmission are twofold: (i) the effective SNR at the receiver ($\rho^u d_t M/Q$) is linearly boosted as M increases and (ii) by rewriting Equation 10.8 as

$$\sum_{t=1}^{T}\sum_{q=1}^{Q} \log_2\left(1 + \frac{\rho^u d_t M}{Q}\right) = \sum_{t=1}^{T}(C_t^u)_{M \gg N}, \quad (10.9)$$

we could observe that the sum rate equals the rate summation of each terminal $(C_t^u)_{M \gg N}$, which is the rate of the terminal for SU-MIMO. Thus, with channel orthogonalization approximation, all terminals can simultaneously transmit the signals using the same time/frequency resource without interfering with each other, and as a result, the sum rate of massive MIMO for uplink MU-MIMO can be significantly increased.

10.2.3 Precoding Designs for Massive MIMO Downlink

For massive MIMO downlink transmissions, since each terminal cannot collectively process its received signal (because the system is underdetermined), MU-MIMO downlink usually precodes the transmit signals so that each terminal can decode their own signal separately (c.f. Chapter 3). The optimal precoding is the so-called dirty paper coding (DPC) [Caire and Shamai 2003, Vishwanath et al. 2003, Weingarten et al. 2006], which requires high complexity. To alleviate the complexity, some practical precoding schemes such as zero-forcing (ZF) precoding and matched filter (MF) precoding are found to be more efficient for massive MIMO systems.

The ZF precoding basically inverts the channel effects at the transmitter, where the transmitted signal is obtained as

$$\mathbf{s}^d = (\mathbf{H}^T)^\dagger \mathbf{d}^d, \quad (10.10)$$

where superscript T denotes the matrix transpose, $(\mathbf{H}^T)^\dagger = \mathbf{H}^*(\mathbf{H}^T \mathbf{H}^*)^{-1}$ is the Moore–Penrose pseudo inverse of $\mathbf{H}^T$ with superscript $*$ being the complex conjugate, and $\mathbf{d}^d$ are the downlink data symbols. When $M \approx N$, the performance gap of ZF precoding to the optimal one is significant. However, as the number of antennas at the base station grows, the performance of ZF precoding approaches the optimal and is almost the same if M is sufficiently greater than N [Rusek et al. 2013].

One observation of the pseudo inverse in Equation 10.10 is that $\mathbf{H}^*(\mathbf{H}^T \mathbf{H}^*)^{-1} \approx \mathbf{H}^* \mathbf{D}^{-1}$ under favorable channel propagation condition, and therefore, we can obtain an approximated ZF precoding, namely, MF precoding as

$$\mathbf{s}^d = \mathbf{H}^* \mathbf{D}^{-1} \mathbf{d}^d, \quad (10.11)$$

which does not require the computation of the inversion of matrix $\mathbf{H}^T \mathbf{H}^*$ and $\mathbf{D}^{-1}$ can be treated as a power normalization matrix. One drawback of MF precoding is that when M is not large enough, the approximation by favorable channel propagation condition may not be accurate, that is, $\mathbf{H}^T \mathbf{H}^*$ cannot be approximated as a diagonal matrix. As a consequence, the MF may experience error floor at high SNR regime if symbolwise detector is adopted at the receiver side.

10.3 Detection of MIMO Systems with Large Arrays

In this section, we introduce and compare the massive MIMO detection methods for uplink transmissions with QAM constellations. To simplify the notation, we drop the superscript of the received signal $\mathbf{y}^u$, transmitted signal $\mathbf{s}^u$, and noise $\mathbf{w}^u$ in this section. Given the model in Equation 10.1, the MLD is

$$\hat{\mathbf{s}} = \arg \min_{\tilde{\mathbf{s}} \in \mathcal{S}^N} \left\| \mathbf{y} - \mathbf{H}\tilde{\mathbf{s}} \right\|^2, \tag{10.12}$$

where $\mathcal{S}$ is the set of QAM constellation. In general, finding the solution of Equation 10.12 is a nondeterministic polynomial hard (NP-hard) problem (as described in Chapter 2), which suggests that no algorithm can guarantee to find the optimal solution efficiently, especially when N and/or the constellation size $|\mathcal{S}|$ is large. Since $\mathcal{S}$ is a finite set, one could resort to the exhaustive search to solve Equation 10.12, which suffers from exponential complexity with respect to problem size N. SDAs have been proposed to reduce the searching complexity, but the variance of the complexity is still high and the performance is degraded by fixing the complexity for large MIMO systems [Jaldén and Ottersten 2005, Rusek et al. 2013].

10.3.1 Linear Detectors

As seen in Chapter 2, linear detectors (LDs) are adopted in practical systems for their polynomial complexity. When $M \geq N$, the zero-forcing linear detector (ZF-LD) for the model in Equation 10.1 is given as

$$\hat{\mathbf{s}}^{ZF} = \mathcal{Q}(\mathbf{H}^\dagger \mathbf{x}) = \mathcal{Q}((\mathbf{H}^H \mathbf{H})^{-1} \mathbf{H}^H \mathbf{y}), \tag{10.13}$$

where $\mathbf{H}^\dagger = (\mathbf{H}^H \mathbf{H})^{-1} \mathbf{H}^H$ is the Moore–Penrose pseudo inverse of channel matrix $\mathbf{H}$ and $\mathcal{Q}(\cdot)$ is the symbolwise quantizer to the constellation set $\mathcal{S}$. ZF-LD is the most straightforward detector. However, it has inferior performance and does not exist when $M < N$.

The other widely used LD is minimum-mean-square-error linear detector (MMSE-LD), which aims at minimizing $E\{\|\mathbf{x} - \mathbf{Hs}\|^2\}$ as

$$\hat{\mathbf{s}}^{\text{MMSE}} = \mathcal{Q}\left(\left(\mathbf{H}^H\mathbf{H} + \frac{QN_0}{E^u}\mathbf{I}_N\right)^{-1}\mathbf{H}^H\mathbf{y}\right), \tag{10.14}$$

where $E\{\mathbf{ss}^H\} = E^u/Q\mathbf{I}_N$.

By exploiting some prior information of symbols, the MMSE-LD generally offers better error performance than ZF-LD. In addition, the MMSE-LD can have the same formula as in an extended model as

$$\bar{\mathbf{H}} = \begin{bmatrix} \mathbf{H} \\ \dfrac{QN_0}{E^u}\mathbf{I}_N \end{bmatrix}, \quad \bar{\mathbf{y}} = \begin{bmatrix} \mathbf{y} \\ \mathbf{0}_{N\times 1} \end{bmatrix}. \tag{10.15}$$

Regarding the performance and complexity of LDs for MIMO systems, although LDs (ZF-LD and MMSE-LD) have lower complexity than the MLD and SDAs, their error performance degrades for MIMO systems by only collecting diversity order $M - N + 1$ [Winters et al. 1994, Gore et al. 2002, Ma and Zhang 2008b]. However, in the case of massive MIMO systems with $M \gg N$, we could expect LDs to exhibit close error performance (similar diversity order) to MLD, since $M - N + 1$ is approaching M and the diversity order $M - N + 1$ looks the same as M in the error performance plot.

10.3.2 Successive Interference Cancelation and K-Best Detectors

Another class of polynomial-complexity detectors is hard-output decision feedback detectors, including successive interference cancelation (SIC) and *K*-best detectors. Since SIC can be treated as a special case of *K*-best, this section will mainly focus on *K*-best detectors.

A *K*-best detector first performs QR decomposition on the channel matrix as $\mathbf{H} = \mathbf{QR}$, where $\mathbf{Q}$ is an $M \times N$ orthonormal matrix and $\mathbf{R}$ is an $N \times N$ upper triangular matrix. Then, the problem in Equation 10.12 is reformulated as

$$\hat{\mathbf{s}} = \arg\min_{\tilde{\mathbf{s}} \in \mathcal{S}^N}\|\bar{\mathbf{y}} - \mathbf{R}\tilde{\mathbf{s}}\|^2, \tag{10.16}$$

where $\bar{\mathbf{y}} = \mathbf{Q}^H\mathbf{y}$.

Thanks to the upper triangular structure of $\mathbf{R}$, the *K*-best detector performs a layered search from the *N*th layer to the first layer to find suboptimal solutions to

Equation 10.16. To better illustrate layered search, we introduce some common terms used in the K-best detector:

■ *Partial candidate of the nth layer, $\mathbf{s}_i^{(n)}$*, which is defined as

$$\mathbf{s}_i^{(n)} = \left[s_{i,n}^{(n)},\dots,s_{i,N}^{(n)} \right]^T, \quad \forall n, s_{i,n}^{(n)} \in \mathcal{S} \tag{10.17}$$

■ *Cost associated with the partial candidate $\mathbf{s}_i^{(n)}$, $\mathrm{cost}_i^{(n)}$*, which is

$$\mathrm{cost}_i^{(n)} = \left\| \breve{\mathbf{y}} - \mathbf{R} \begin{bmatrix} \mathbf{0}_{(N-n)\times 1} \\ \mathbf{s}_i^{(n)} \end{bmatrix} \right\|^2 = \sum_{\ell=n}^{N} \left| \breve{y}_\ell - \sum_{k=\ell}^{N} R_{\ell,k} s_{i,k}^{(n)} \right|^2. \tag{10.18}$$

■ *Child of a partial candidate $\mathbf{s}_i^{(n)}$, $\mathbf{s}_j^{(n-1)}$*, which is a partial candidate of the $(n-1)$th layer and has the form $\mathbf{s}_j^{(n-1)} = [s_{j,n}^{(n-1)}, (\mathbf{s}_i^{(n)})^T]^T, s_{j,n}^{(n-1)} \in \mathcal{S}$.

Then, the K-best detector performs a breadth-first search from the Nth layer to the first layer. For each layer (e.g., the nth layer), the algorithm computes the K-best partial candidates $[\mathbf{s}_1^{(n)}, \mathbf{s}_2^{(n)},\dots,\mathbf{s}_K^{(n)}]$, that is, the K partial candidates with the minimum costs among all the possible children of the K partial candidates $[\mathbf{s}_1^{(n+1)}, \mathbf{s}_2^{(n+1)},\dots,\mathbf{s}_K^{(n+1)}]$ in the previous $(n+1)$th layer. The search procedure terminates once the first layer is reached, and the candidate with minimum cost in the first layer is denoted as the hard output of the K-best detector.

The complexity of the K-best detector is $\mathcal{O}(N^2 K + NK|\mathcal{S}|)$, and it is fixed when the number of candidates K and the constellation size $|\mathcal{S}|$ are constant, which is in favor of hardware complexity [Guo and Nilsson 2006]. In addition, when $K = 1$, the K-best detector reduces to the SIC detector. The performance of general K-best detector is hard to quantify. When $K = 1$, the SIC detector collects the same diversity as LDs (Ma and Zhang 2008a,b). One major issue of hard-output decision-feedback detectors is that the erroneous detected symbol could propagate to the successive layers and thus degrade the search performance. To combat this error propagation issue, one method is to apply sorted QR decomposition so that the detection starts from the stronger layers. Another way is to perform soft interference cancelation, which will be introduced in the following section.

10.3.3 Iterative Soft Interference Cancelation

To enhance the error performance, the iterative soft interference cancelation (ISIC) [Wang and Poor 1999, Liang et al. 2008] detectors incorporate soft information from the previous iteration to mitigate the interference.

Given the soft estimates of $\mathbf{s}^u = [\tilde{s}_{1,\ell-1}, \ldots, \tilde{s}_{N,\ell-1}]^T$ in the previous $(\ell-1)$th iteration, the ISIC detector reformulates the system model in Equation 10.1 to detect symbol s_k as

$$
\begin{aligned}
\mathbf{y}_{k,\ell} &= \mathbf{y} - \sum_{i \neq k} \mathbf{h}_i \tilde{s}_{i,\ell-1} \\
&= \mathbf{h}_k s_k + \sum_{i \neq k} \mathbf{h}_i (s_i - \tilde{s}_{i,\ell-1}) + \mathbf{w}.
\end{aligned} \tag{10.19}
$$

Then, the ISIC detector obtains an $M \times 1$ optimal MMSE filter vector $\mathbf{w}_{k,\ell}$ that minimizes the MSE between the filter output $\mathbf{w}_{k,\ell}^H \mathbf{y}_{k,\ell}$ and the information symbol s_k as

$$
\mathbf{w}_{k,\ell} = \left(\mathbf{h}_k \mathbf{h}_k^H + \sum_{i \neq k} d_{k,\ell} \mathbf{h}_i \mathbf{h}_i^H \right)^{-1} \mathbf{h}_k, \tag{10.20}
$$

where $d_{k,\ell}$ is the variance of the difference $s_k - \tilde{s}_{k,\ell-1}$ normalized by the variance of s_k (see [Wang and Poor 1999 and Liang et al. 2008] for details). The performance gain comes from the soft information, but there still exists a performance gap with the optimal detector, especially for high-order modulations.

10.3.4 Lattice-Reduction-Aided Linear Detectors

The performance gap between the MLD and LDs is mainly due to the nonorthogonality of the channel matrix $\mathbf{H}$ when $M \approx N$ (Ma and Zhang 2008a,b). As seen in Chapter 2, the motivation of lattice-reduction-aided (LRA) LDs is based on the fact that if channel matrix $\mathbf{H}$ is "close" to orthogonal, the decision region of LDs is also "close" to that of the MLD [Ma and Zhang 2008a,b, Ling 2011]. Hence, to improve the error performance of LDs, LR finds another "more orthogonal" basis $\tilde{\mathbf{H}}$ statistically that defines the same lattice as $\mathbf{H}$. As a result, LRA LDs yield error performance close to the MLD and have the same error performance as the MLD if the lattice-reduced basis is orthogonal. That is, the so-called reducing basis $\mathbf{H}$ is equivalent to finding a "more orthogonal" basis $\tilde{\mathbf{H}} = \mathbf{HT}$ [Agrell et al. 2002], where $\mathbf{T}$ is a unimodular matrix, such that all entries of $\mathbf{T}$ and $\mathbf{T}^{-1}$ are Gaussian integers, and the determinant of $\mathbf{T}$ is ± 1 or $\pm j$ (c.f. Chapter 2). Thus, the key of LRA detectors is to find the unimodular matrix T with low complexity while enabling better performance.

To find the unimodular matrix $\mathbf{T}$, there are several existing LR algorithms, including Minkowski reduction, Korkin–Zolotarev (KZ) reduction [Agrell et al. 2002], the LLL algorithm and its complex valued counterparts [Lenstra et al. 1982, Ma and Zhang 2008b, and Gan and Mow 2009], the Seysen's algorithm

(SA) [Seysen 1993, Zhang et al. 2010], and element-based lattice-reduction (ELR) algorithms [Zhou and Ma 2013a,b]. Among these algorithms, the LLL algorithms are well adopted because they allow worst-case polynomial complexity and yield a basis with certain orthogonality guaranteed, that is, the orthogonality deficiency is bounded. However, as shown in [Zhou and Ma 2013a], the LLL algorithms do not aim at minimizing the asymptotic error performance of LRA LDs and thus may exhibit unsatisfactory performance when N is large. To achieve better performance, ELR algorithms are proposed to minimize the asymptotic error performance and show considerable performance gain over LLL algorithms when N is large. In the following text, we provide the general framework for LRA detectors and the details of the LR algorithms can be referred to in the aforementioned papers.

The application of LRA to massive MIMO does not differ from the concept presented in Chapter 2: given the unimodular matrix **T** based on **H**, the model in Equation 10.1 can be rewritten as

$$\begin{aligned} \mathbf{y} &= \mathbf{HTT}^{-1}\mathbf{s} + \mathbf{w} \\ &= 2\mathbf{HTz} + (1 + j)\mathbf{H1}_{N\times1} + \mathbf{w}. \end{aligned} \tag{10.21}$$

The equivalent model becomes

$$\tilde{\mathbf{y}} = \tilde{\mathbf{H}}\mathbf{z} + \mathbf{w}, \tag{10.22}$$

where $\tilde{\mathbf{H}} = 2\mathbf{HT}$, $\tilde{\mathbf{s}} = 2\mathbf{T}\tilde{\mathbf{z}} + 1_{N\times1}(1 + j)$, $\mathbf{z} = \mathbf{T}^{-1}\mathbf{s}$ contains the information symbols in the lattice-reduced domain, and $\tilde{\mathbf{y}} = (\mathbf{y} - (1 + j)\mathbf{H1}_{N\times1})/2$. Since **T** is unimodular and the entries of **s** are drawn from QAM constellations, the entries of **z** are Gaussian integers in $\mathbb{Z}[j]$. Note that matrix **T** is obtained at the receiver and does not perform as a precoder.

Given the model in Equation 10.21, the LRA ZF-LD is given as

$$\hat{\mathbf{z}} = \mathcal{Q}((\tilde{\mathbf{H}}^{H}\tilde{\mathbf{H}})^{-1}\tilde{\mathbf{H}}^{H}\tilde{\mathbf{y}}) = \mathcal{Q}(\tilde{\mathbf{H}}^{\dagger}\tilde{\mathbf{y}}). \tag{10.23}$$

Then, the estimate of **s** is obtained by

$$\hat{\mathbf{s}} = \mathcal{Q}(2\mathbf{T}\hat{\mathbf{z}} + (1 + j)1_{N\times1}). \tag{10.24}$$

Similar to the LRA ZF-LD, LRA MMSE-LD can be derived by applying LR algorithms on the MMSE-extended model in Equation 10.15 to offer better performance than LRA ZF-LD. LR can also be applied to SIC, which further boosts the performance. For certain LR techniques (e.g., LLL), it has been proved that LRA detectors collect the same diversity as MLD does [Ma and Zhang 2008a].

10.3.5 Lattice-Reduction-Aided K-Best Algorithms

As the number of antennas grows, LR algorithms should also be combined with *K*-best detectors to boost the performance. Since the LRA detection assumes infinite lattice, LRA *K*-best detector relaxes the boundary constraints in Equation 10.12 to infinite lattice as

$$\hat{\mathbf{s}} = \arg\min_{\tilde{\mathbf{s}} \in \mathcal{U}^N} \left\| \mathbf{y} - \mathbf{H}\tilde{\mathbf{s}} \right\|^2, \tag{10.25}$$

where $\mathcal{U}$ is the unconstrained constellation set with the form $(2\mathbb{Z} + 1) + (2\mathbb{Z} + 1)j$.

By applying an LR algorithm to obtain a more "orthogonal" matrix $\tilde{\mathbf{H}} = \mathbf{HT}$, Equation 10.25 can be rewritten as

$$\hat{\mathbf{s}} = 2\mathbf{T} \arg\min_{\tilde{\mathbf{z}} \in \mathbb{Z}^N} \left\| \tilde{\mathbf{y}} - \tilde{\mathbf{H}}\tilde{\mathbf{z}} \right\|^2 + (1 + j)\mathbf{1}_{N \times 1}. \tag{10.26}$$

Similar to the *K*-best detector described in Section 10.3.2, by performing QR decomposition on $\tilde{\mathbf{H}} = \tilde{\mathbf{Q}}\tilde{\mathbf{R}}$, where $\tilde{\mathbf{Q}}$ is an $M \times N$ orthonormal matrix and $\tilde{\mathbf{R}}$ is an $N \times N$ upper triangular matrix, the problem in Equation 10.26 can be reformulated as

$$\hat{\mathbf{s}} = 2\mathbf{T} \arg\min_{\tilde{\mathbf{z}} \in \mathbb{Z}^N} \left\| \check{\mathbf{y}} - \tilde{\mathbf{R}}\tilde{\mathbf{z}} \right\|^2 + (1 + j)\mathbf{1}_{N \times 1}, \tag{10.27}$$

where $\check{\mathbf{y}} = \tilde{\mathbf{Q}}^H \tilde{\mathbf{y}}$. Therefore, LRA *K*-best detector performs a breadth-first search from the *N*th layer to the first layer, and for each layer, only *K*-best candidates survive and are served as the parents of the next layer.

Compared to the conventional *K*-best detector, the main difference of the LRA *K*-best is that the information symbol in the LR domain **z** is unbounded while **s** is constrained in a finite QAM constellation set. This results in a challenging problem to the LRA *K*-best algorithm, that is, how to efficiently find *K*-best candidates among all infinite children of *K* parents for each layer. This problem can be addressed by using the Schnorr–Euchner (SE) strategy [Shabany and Gulak 2008] and priority queue [Zhou and Ma 2012, Wen et al. 2013] such that the overall complexity of LRA *K*-best detector (not including the LR algorithm) is on the order of $\mathcal{O}(N^2 K + NK \log_2(K))$. This method has great potential on hardware realization with controllable performance and complexity trade-offs.

10.3.6 Local Neighborhood Search Methods

Local neighborhood search methods [Datta et al. 2010, Srinidhi et al. 2011] start with an initial symbol vector, and iteratively improve the current symbol vector by moving to the best symbol vector in the neighborhood of the current vector. The best symbol vector means the symbol vector with the minimum cost in the neighborhood $\mathcal{N}(\tilde{\mathbf{s}}_k)$ of current symbol vector $\tilde{\mathbf{s}}_k$

$$\tilde{\mathbf{s}}_{k+1} = \arg \min_{\breve{\mathbf{s}} \in \mathcal{N}(\tilde{\mathbf{s}}_k)} \left\| \mathbf{y} - \mathbf{H}\breve{\mathbf{s}} \right\|^2. \tag{10.28}$$

To allow simple enumeration of all neighbors with low complexity, a vector $\breve{\mathbf{s}}$ is defined as a neighbor of the symbol vector $\tilde{\mathbf{s}}$ if the vector $\breve{\mathbf{s}}$ has only one symbol that is different from the corresponding symbol of $\tilde{\mathbf{s}}$, that is, $\exists i, \breve{s}_i \neq \tilde{s}_i$ and $\forall i \neq \ell, \breve{s}_\ell = \tilde{s}_\ell$ while the different one $\breve{s}_i$ is the neighbor of $\tilde{s}_i$ in the constellation set S (see [Datta et al. 2010, Srinidhi et al. 2011] for details). After the search is terminated, the symbol vector with the minimum cost among the vectors visited during the search becomes the estimate of $\mathbf{s}$.

One issue of local neighborhood search methods is that the search process can be easily trapped in a local minimum. To address this issue, *tabu search* [Datta et al. 2010] is proposed to allow a worse move (i.e., the best symbol vector in the neighborhood has higher cost than the current one) to jump out of the local minima. However, to prevent cycle search paths, tabu search maintains a tabu table of past moves and prohibits some duplicated visits by looking up the table. To further improve the performance of tabu search methods, layered tabu search algorithm is proposed in [Srinidhi et al. 2011].

One drawback of local neighborhood search methods is that their complexity has a large variance with respect to channel, noise, and/or symbol realizations, and thus may pose great challenges in hardware implementation.

10.3.7 Performance and Complexity Comparisons

In this section, we show the performance and complexity of various detectors for massive MIMO via Monte-Carlo simulations. We consider an uplink MU-MIMO transmission with one base station equipped with $M = 32$ antennas and serving $T = M$ terminals such that $Q = 1$ and $N = M$. The entries of channel matrix $\mathbf{H}$ are modeled as independent and identically distributed (i.i.d.) complex Gaussian variables with zero mean and unit variance. We consider the following detectors for comparisons: (i) MMMSE-LD; (ii) dual ELR-shortest-longest-basis-aided MMSE sorted-variance SIC (D-ELR-SLB-aided MMSE SV-SIC) detector; (iii) LTS detector; (iv) MMSE-ISIC detector with five iterations; (v) K-best detector with sorted QR decomposition (SQRD) preprocessing [Wübben et al. 2003]; and (vi) LLL-aided MMSE K-best detector. The performance of interference-free (IF), that is, $K = 1$, is also considered as a benchmark.

Figures 10.4 and 10.5 demonstrate the error performance of various detectors with different SNR ρ^us. From the figures, we have the following observations: (i) MMSE-LD shows inferior performance compared to the rest of the detectors and IF case, and its diversity order is just 1. (ii) D-ELR-aided MMSE-SV-SIC

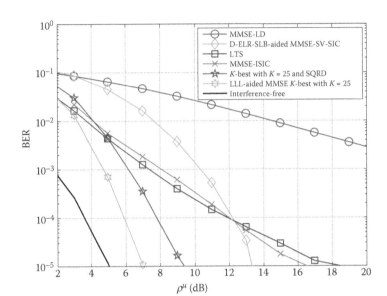

Figure 10.4 **Performance comparisons of various detectors for MIMO systems** $M = N = T = 32$, $Q = 1$, **16-QAM, and different** ρ^u**s.**

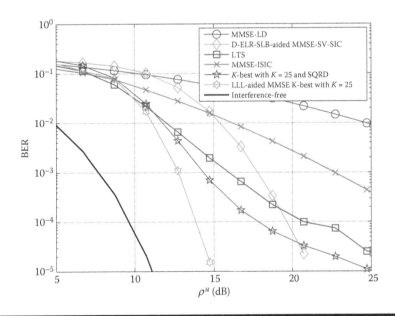

Figure 10.5 **Performance comparisons of various detectors for MIMO systems** $M = N = I = 32$, $Q = 1$, **64-QAM, and different** ρ^u**s.**

improves the performance of MMSE-LD, especially for high SNR and outperforms MMSE-ISIC and LTS when ρ^u is large (e.g., $\rho^u > 19$ dB for 64-QAM). (iii) LTS and MMSE-ISIC exhibit almost the same performance for 16-QAM, while for 64-QAM, LTS shows superior performance to MMSE-ISIC. However, both of them lose some diversity orders as SNR increases. (iv) K-best detector with $K = 25$ and SQRD preprocessing generally yields better performance than LTS and MMSE-ISIC for medium-to-high SNR, and its gain over LTS at BER = 10^{-4} is about 4 dB for 16-QAM. However, it still cannot achieve the same diversity order as MLD at high SNR, as demonstrated by the case with 64-QAM. (v) By exploiting the LR algorithm, the LLL-aided MMSE K-best detector with $K = 25$ achieves the best performance among all detectors for medium-to-high SNR (e.g., $\rho^u \geq 11$ for 64-QAM). The performance gains of the LLL-aided MMSE K-best detector over K-best at BER = 10^{-4} are about 2 dB and 4 dB for 16-QAM and 64-QAM, respectively. In addition, the gaps of the LLL-aided MMSE K-best to IF case are about 2 dB and 4 dB at BER = 10^{-5}, respectively.

Figure 10.6 illustrates the number of arithmetic operations of various detectors. Several remarks about the complexity of the detectors can be drawn: (i) Although the instantaneous complexity of the D-ELR-SLB and LLL algorithms depends on the channel realizations, the worst-case complexity (99.9% upper percentile) of the D-ELR-SLB-aided MMSE-SV-SIC and LLL-aided MMSE K-best detectors is the closest to their average complexity, respectively. (ii) The complexity of the

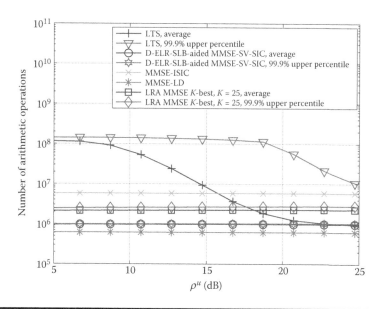

Figure 10.6 Number of arithmetic operations of various MIMO detectors with $M = N = T = 32$, $Q = 1$, 64-QAM, and different ρ^us for MU-MIMO systems.

D-ELR-SLB-aided MMSE-SV-SIC detector is only slightly higher than that of the low-complexity MMSE-LD. The reason for similar levels of complexity of the D-ELR-SLB-aided MMSE-SV-SIC detector is that, while the complexity of the D-ELR-SLB detector constitutes only a small portion of the overall detection complexity, the main computational complexity consists of the preprocessing operations (i.e., matrix inversion of the Gram matrix and QR decomposition), which requires the same polynomial order complexity of the MMSE-LD. (iii) The number of arithmetic operations of LLL-aided MMSE K-best detector is about twice higher than that of the D-ELR-SLB-aided MMSE-SV-SIC detector. The price for the extra complexity mainly comes from the K-best algorithm. For illustration purpose, we do not plot the complexity of the K-best algorithm without LR, which is almost the same as that of LLL-aided MMSE K-best detector. (iv) The MMSE-ISIC detector requires higher complexity than the LLL-aided MMSE K-best detector. LTS requires highest average complexity when ρ^u is small ($\rho^u \leq 15$), but its average complexity decreases as ρ^u increases. However, when ρ^u is large, the worst-case complexity of LTS, which makes a critical contribution to the error performance of the LTS [Zhou and Ma 2013a], becomes much higher than the average one because of the neighborhood search nature of the algorithm.

Figure 10.7 illustrates the performance of various detectors for MIMO systems with $M = 48$, $N = 32$, $Q = 1$, and 64-QAM. Compared to the results with $M = N = 32$ in Figure 10.5, all detectors obtain considerable performance gain by equipping additional antennas at the base station. For example, the MMSE-LD

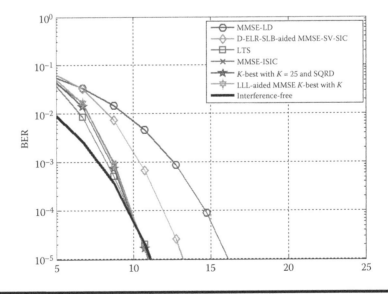

Figure 10.7 Performance comparisons of various detectors for MIMO systems with *M* = 48, *N* = 32, *Q* = 1, 64-QAM, and different *ρ*^us.

exhibits about 4 dB loss compared to IF case at BER = 10^{-2} (in contrast, the loss is 20 dB for $M = N = 32$). The D-ELR-SLB-aided MMSE-SV-SIC detector has about 2.5 dB gain over the MMSE-LD at BER = 10^{-5}. The LTS, MMSE-ISIC, K-best, and LLL-aided MMSE K-best detectors have almost the same performance and reach the performance of IF at BER = 10^{-5}. In addition, $K = 5$ candidates are sufficient for both K-best and LLL-aided to have near-optimal performance.

10.3.8 Detection for Other Channels

In the previous sections, we focussed on detector designs for uncoded systems with i.i.d. channels. However, for massive MIMO systems, the antennas may not be spaced enough and thus the channels may be correlated. At the same time, in practical systems, error control codes have to be considered. As one may observe from the previous sections, the LRA detectors have great potential on hardware realization for massive MIMO. In the literature, LRA detectors have been extended to other channel conditions. In [Zhang and Ma 2010], LRA detectors with soft output are proposed, which show low complexity and high performance for coded systems. In [Gestner et al. 2012], correlated channels are considered and again LRA detectors show great performance. In [Zhang and Ma 2009] and [Zhou and Ma 2013c], LRA precoders are developed when the CSI is available at the transmitter. Although these results have shown that LRA detectors are strong candidates for MIMO systems, there are still many open issues for massive MIMO. For example, when the channels are correlated, how are the performance and complexity of the detectors going to change?

10.4 Hardware Implementation of Massive MIMO Detectors

To realize the detectors in hardware, one faces several fundamental challenges [Zhang et al. 2009]. One of these challenges is that hardware realizations necessarily have only finite numerical resolution. Furthermore, in practical systems, floating-point representations are abandoned for the improved efficiency afforded by fixed-point operations. Fixed-point signal-processing solutions have generally been chosen due to advantages in speed, power consumption, cost, and suitability for portable applications. However, so far, the existing theoretical analysis of detectors is derived within the real or complex field. There are only a limited number of tractable equalizer algorithms that have been adequately characterized, implemented, and measured as to the performance for fixed-point applications (see [Zhang and Ma 2008, Gestner et al. 2011]).

In the literature, people have prototyped different types of detectors. However, the major discrepancy here is that those who developed the algorithms do not have enough knowledge of hardware on VLSI-integrated circuits, and those who try to prototype the detection algorithms do not know much about the math behind it. The applicable and practical detectors should be implemented with low complexity in hardware while keeping the performance as high as possible. Note that blindly reducing the complexity of an algorithm without considering the effects on numerical precision, parallelism, and memory requirements can result in a poor hardware realization and similarly, blind application of hardware optimizations can result in an unexpected increase in algorithm complexity and a reduction in system performance. By carefully considering both perspectives, one will realize reduced-complexity LR algorithms that take into consideration the finite hardware resources and processing constraints of real-time communication systems.

10.5 Other Issues of Massive MIMO Systems

Massive MIMO has shown great potential benefit for future wireless communications. It also reveals "massive" challenges that deserve further research.

10.5.1 Channel State Information

The benefit of massive MIMO highly relies on the perfect CSI at the receiver and/or transmitter. Channel models have to capture the new behavior/phenomenon of the radio channel so that the performance assessment can be more realistic. For example, due to the large number of antennas, the channels may be correlated. How to quantify the correlation deserves more efforts on measurements and analysis. At the same time, owing to the large number of antennas and possible time variation and frequency selectivity of the channels, channel estimation (either training based or semiblind) becomes challenging—large number of parameters in three dimensions. Furthermore, how the channel estimation error affects the detection and also the precoder design is an interesting topic for research.

10.5.2 Synchronization

Massive MIMO requires hundreds of RF chains, ADC/DAC, and power amplifiers connecting to the antennas. Synchronization is critical in this case to guarantee the performance of massive MIMO systems. For example, if the antennas are distributed at different places, the timing offsets due to propagation delays and processing delays of RF chains will degrade the performance of the detectors since the aggregated streams are not aligned. Together with the phase noise, frequency offsets, and DC offset, the performance is severely affected by the synchronization errors.

10.5.3 Hardware Issues

In practice, RF chains introduce DC offset, phase noise, I/Q imbalance, power amplifier (PA) nonlinearity, and other nonideal issues. Since massive MIMO systems require large numbers of components, to enable economic designs, usually low-cost devices are adopted. In this case, the PA nonlinearity may redirect the beams and introduce the interference to other cells. The nonideal characteristics of the hardware will also reduce the efficiency of massive MIMO. To support the high data rate transmissions, the connection from the RF frontend to baseband (e.g., in the common public radio interface) is also important. Massive MIMO makes it more challenging and has higher demand on the RF designs.

References

Agrell, E., T. Eriksson, A. Vardy, and K. Zeger, Closest point search in lattices, *IEEE Trans. Inf. Theory*, 48(8), 2201–2214, August 2002.

Caire, G. and S. Shamai, On the achievable throughput of a multiantenna Gaussian broadcast channel, *IEEE Trans. Inf. Theory*, 49(7), 1691–1706, July 2003.

Datta, T., N. Srinidhi, A. Chockalingam, and B.S. Rajan, Random-restart reactive tabu search algorithm for detection in large-MIMO systems, *IEEE Commun. Lett.*, 14(12), 1107–1109, December 2010.

Gan, Y.H., C. Ling, and W.H. Mow, Complex lattice reduction algorithm for low-complexity full-diversity MIMO detection, *IEEE Trans. Signal Process.*, 57(7), 2701–2710, July 2009.

Gao, X., O. Edfors, F. Rusek, and F. Tufvesson, Linear pre-coding performance in measured very-large MIMO channels, in *Proc. IEEE Veh. Tech. Conf. (VTC Fall)*, San Francisco, September 5–8, 2011, pp. 1–5.

Gao, X., F. Tufvesson, O. Edfors, and F. Rusek, Measured propagation characteristics for very-large MIMO at 2.6 GHz, in *Proc. IEEE 46th Annual Asiomar Conf. Signals Systems Computers (ASILOMAR)*, California, November 4–7, 2012, pp. 295–299.

Gestner, B., X. Ma, and D.V. Anderson, Incremental lattice reduction: Motivation, theory, and practical implementation, *IEEE Trans. Wireless Commun.*, 11(1), 188–198, January 2012.

Gestner, B., W. Zhang, X. Ma, and D. V. Anderson, Lattice reduction for MIMO detection: From theoretical analysis to hardware realization, *IEEE Trans. Circuits Systems I Regular Papers*, 58(4), 813–826, April 2011.

Gore, D.A., R. Heath Jr, and A.J. Paulraj, On performance of the zero forcing receiver in presence of transmit correlation, in *Proc. IEEE Int. Symp. Info. Theory*, Lausanne, Switzerland, June 30–July 5, 2002, p. 159.

Guo, Z. and P. Nilsson, Algorithm and implementation of the K-best sphere decoding for MIMO detection, *IEEE J. Sel. Areas Commun.*, 24(3), 491–503, March 2006.

Jaldén, J. and B. Ottersten, On the complexity of sphere decoding in digital communications, *IEEE Trans. Signal Process.*, 53(4), 1474–1484, April 2005.

Lenstra, A.K., H.W. Lenstra, and L. Lovász, Factoring polynomials with rational coefficients, *Math. Annalen*, 261(4), 515–534, 1982.

Liang, Y.-C., E.Y. Cheu, L. Bai, and G. Pan, On the relationship between MMSE-SIC and bi-GDFE receivers for large multiple-input multiple-output channels, *IEEE Trans. Signal Process.*, 56(8), 3627–3637, August 2008.

Ling, C. On the proximity factors of lattice reduction-aided decoding, *IEEE Trans. Signal Process.*, 59(6), 2795–2808, June 2011.

Ma, X. and W. Zhang, Fundamental limits of linear equalizers: Diversity, capacity and complexity, *IEEE Trans. Inf. Theory*, 54(8), 3442–3456, August 2008a.

Ma, X. and W. Zhang, Performance analysis for MIMO systems with lattice-reduction aided linear equalization, *IEEE Trans. Commun.*, 56(2), 309–318, February 2008b.

Rusek, F., D. Persson, B. Lau, E. Larsson, T. Marzetta, O. Edfors, and F. Tufvesson, Scaling up MIMO: Opportunities and challenges with very large arrays, *IEEE Signal Process. Mag.*, 30(1), 40–60, January 2013.

Seysen, M. Simultaneous reduction of a lattice basis and its reciprocal basis, *Combinatorica*, 13(3), 363–376, September 1993.

Shabany, M. and P. Glenn Gulak, The application of lattice-reduction to the K-best algorithm for near-optimal MIMO detection, in *Proc. IEEE Int. Symp. Circuits Systems (ISCAS)*, Washington, USA, May 18–21, 2008, pp. 316–319.

Srinidhi, N., T. Datta, A. Chockalingam, and B.S. Rajan, Layered tabu search algorithm for large-MIMO detection and a lower bound on ML performance, *IEEE Trans. Commun.*, 59(11), 2955–2963, July 2011

Vishwanath, S., N. Jindal, and A. Goldsmith, Duality, achievable rates, and sum-rate capacity of Gaussian MIMO broadcast channels, *IEEE Trans. Inf. Theory*, 49(10), 2658–2668, October 2003.

Wang, X. and H.V. Poor, Iterative (turbo) soft interference cancellation and decoding for coded CDMA, *IEEE Trans. Commun.*, 47(7), 1046–1061, July 1999.

Wen, Q., Q. Zhou, C. Zhao, and X. Ma, Fixed-point realization of lattice-reduction aided MIMO receivers with complex K-best algorithm, in *Proc. Int. Conf. Acoustic Speech Signal Proc. (ICASSP)*, Vancouver, Canada, May 26–29, 2013.

Weingarten, H., Y. Steinberg, and S. Shamai, The capacity region of the Gaussian multiple-input multiple-output broadcast channel, *IEEE Trans. Inf. Theory*, 52(9), 3936–3964, September 2006.

Winters, J.H., J. Salz, and R.D. Gitlin, The impact of antenna diversity on the capacity of wireless communication systems, *IEEE Trans. Commun.*, 42(234), 1740–1751, February/March/April 1994.

Wübben, D., R. Böhnke, V. Kühn, and K.D. Kammeyer, MMSE extension of V-BLAST based on sorted QR decomposition, in *Proc. Veh. Tech. Conf. (VTC)*, Orlando, FL, October 6–9, 2003, pp. 508–512.

Zhang, W. and X. Ma, Quantifying diversity for wireless systems with finite-bit representation, in *Proc. IEEE Int. Conf. Acoustics Speech Signal Process.*, Las Vegas, NV, March 30–April 4, 2008, pp. 2841–2844.

Zhang, W. and X. Ma, A novel lattice reduction aided linear precoding scheme, in *Proc. 43rd Conf. Information Sci. Syst.*, Johns Hopkins University, Baltimore, MD, March 18–20, 2009, pp. 518–523.

Zhang, W. and X. Ma, Low-complexity soft-output decoding with lattice-reduction-aided detectors, *IEEE Trans. Commun.*, 58(9), 2621–2629, September 2010.

Zhang, W., X. Ma, B. Gestner, and D.V. Anderson, Designing low-complexity equalizers for wireless systems, *IEEE Commun. Mag.*, 47(1), 56–64, January 2009.

Zhang, W., X. Ma, and A. Swami, Designing low-complexity detectors based on Seysen's algorithm, *IEEE Trans. Wireless Commun.*, 9(10), 3301–3311, October 2010.

Zhou, Q. and X. Ma, An improved LR-aided K-best algorithm for MIMO detection, in *Proc. IEEE Int. Conf. Wireless Commun. Signal Process. (WCSP)*, Huangshan, China, September 2012, pp. 1–5.

Zhou, Q. and X. Ma, Element-based lattice reduction algorithms for large MIMO detection, *IEEE J. Sel. Areas Commun.*, 31(2), 274–286, February 2013a.

Zhou, Q. and X. Ma, Improved element-based lattice reduction algorithms for wireless communications, *IEEE Trans. Wireless Commun.*, 2013, 12(6), 2806–2816, June 2013b.

Zhou, Q. and X. Ma, Joint transceiver designs using lattice reduction algorithms, in *Proc. China Summit Int. Conf. Signal Info. Process. (ChinaSIP)*, Beijing, China, July 6–10, 2013c.

Chapter 11

MIMO Two-Way Relay Channel with Superposition Coding

Ioannis Krikidis and John S. Thompson

Contents

11.1 Two-Way Relay Channels

The two-way relay channel is an important information theoretic network structure [Shannon 1961], and it is becoming a central concept in the design of the physical layer of future communication systems. It consists of two users who are unable to communicate directly, hence they establish communication via a shared relay node. This scenario characterizes networks with central controllers as well as ad hoc networks with limited relaying resources. In two-way relay channels, the two terminals exchange messages with each other via the relay, and the objective of a two-way relay protocol design is to maximize the spectral efficiency of the two communication links that are formed between the two links.

Recently, there is a lot of interest in the design of efficient cooperative protocols for two-way relay channels. The proposed schemes can be divided into two main categories based on the number of the required time phases: (a) in the two-phase protocol, called multiple access broadcast (MABC) protocol [Kim et al. 2008], both users simultaneously transmit their data to the relay during the first phase and then the relay transmits during the second phase, while (b) in the three-phase protocol the two users sequentially transmit to the relay followed by a transmission from the relay [Saleh et al. 2009]. Furthermore, each protocol category can be combined with any relaying strategy resulting in numerous two-way protocols with different complexities and performance. In [Kim et al. 2011], the authors highlight the most significant two-way relaying protocols and characterize their achievable rate region from an information theoretic point of view. An interesting MABC two-way scheme that employs superposition coding (SPC) at the relay node has been proposed in [Hammerstrom et al. 2007], where the authors analyze its capacity performance for a general multiple-input multiple-output (MIMO) two-way relay channel with different channel side information (CSI) requirements. The use of SPC as a broadcast approach for two-way relay configurations has been reported in several studies for different contexts [e.g., Chen and Yener 2010; Chen et al. 2010; Oechtering and Boche 2008]; in addition a combination of SPC with network coding appropriate for asymmetric two-way relay topologies is discussed in [Park and Oh 2009].

Although there is a lot of work on the design of efficient two-way relay protocols, the majority of them assume perfect channel estimation, which is not always a realistic assumption. The impact of imperfect channel estimation on the achievable system performance is a classical problem in the literature. In [Medard 2000], the authors characterized the Shannon capacity of a conventional single-input single-output (SISO) network under imperfect channel estimation. A related power allocation strategy that maximizes the achieved capacity of SISO with imperfect channel estimation has been proposed in [Klein and Gallager 2001]. The impact of an imperfect channel estimation on the capacity performance of a MIMO network as well as a related waterfilling power allocation (WF-PA) strategy have been proposed in [Yoo and Goldsmith 2006]. However, the effects of channel estimation error on the reception reliability of a decode-and-forward two-way relay channel with physical layer network coding have been reported in [Ding and Leung 2011] in terms of error probability. In addition, different channel-estimation techniques for a two-way relay channel are discussed in [Jiang et al. 2010 and references therein], while the achieved sum rate for a two-way relay channel with AF and imperfect channel estimation is investigated in [Panah and Heath 2010] and [Jia and Vosoughi 2011]. However, the impact of imperfect channel estimation on the achievable performance (rate) of a MIMO two-way relay protocol with SPC from an information theoretic standpoint is still an open problem in the literature.

This chapter analyses the effects of imperfect channel estimation on the achievable rate region of a MIMO MABC-SPC two-way relay protocol. By extending the work presented in [Hammerstrom et al. 2007] and [Yoo and Goldsmith 2006], we characterize the information theoretic performance of the MIMO MABC-SPC protocol under imperfect channel estimation for two main CSI assumptions: (a) without CSI at the users where a symmetric power allocation between the two data flows is used and (b) with an imperfect CSI at both users where a WF-PA is employed. Another issue that is discussed through this chapter is the impact of the SPC power split on the achievable system performance. We show that an appropriate power split between the two data flows can maximize the achievable performance by simultaneously supporting user fairness. In addition, a power split that maximizes the achievable sum rate is presented, and a theoretical framework that calculates the optimal SPC power split between the two data flows for both optimization targets (fairness, sum rate) is proposed. It is shown that the optimal SPC allocation is common for both CSI assumptions and independent of the instantaneous channels; a result that makes the proposed scheme suitable for applications with critical complexity constraints.

The remaining part of the chapter is organized as follows. In Section 11.2, we present the system model and we describe the considered MIMO MABC-SPC protocol as well as its related achievable rate region. In Section 11.3, we present the two CSI assumptions and we introduce an SPC power split under a user fairness constraint as well as for sum-rate maximization. Numerical results are shown and discussed in Section 11.4, followed by concluding remarks in Section 11.5.

Note that in this chapter $\log(\cdot)$ denotes the logarithm of base 2.

11.2 Communication Model

We assume a three-node MIMO two-way relay channel consisting of two users A and B and a shared relay node R. All nodes are equipped with $M > 1$ antennas and both users can establish communication using an MABC-SPC cooperative protocol. Figure 11.1 depicts the system model and the two phases of the cooperative protocol. We consider flat fading spatially uncorrelated Rayleigh MIMO channels where $\mathbf{H}_{Ri} \in \mathcal{C}^{M \times M}$ with $i \in \{A, B\}$ denotes the channel matrix for the $i \rightarrow R$ link. The entries of the channel matrices $\mathbf{H}_{Ri}$ are independently and identically distributed (i.i.d.) zero-mean circularly symmetric complex Gaussian (ZMCSCG) random variables with unit variance (i.e., $\text{rank}(\mathbf{H}_{Ri}) = M$). In addition, the channel matrices remain constant for the whole transmission (during the two phases of the adopted cooperative protocol) and change to an independent realization for the next transmission. We assume that the channel matrices are subject to a channel estimation error and therefore are *imperfectly* known at the receivers with an MMSE estimation error $\mathbf{E}_i \triangleq \mathbf{H}_{Ri} - \hat{\mathbf{H}}_{Ri}$ for the $i \rightarrow R$ link, where the entries of

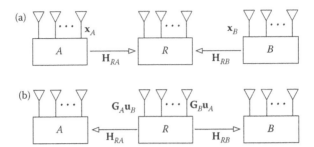

Figure 11.1 The system model: (a) the first phase of the protocol, (b) the second phase of the protocol. (Adapted from Krikidis I. and J. S. Thompson, *J. Netw. Comput. Appl.*, 35(1), 510–516, January 2012.)

$\mathbf{E}_i$ are ZMCSCG with variance $\sigma_{e_i}^2$ and the entries of $\hat{\mathbf{H}}_{Ri}$ are also i.i.d. ZMCSCG with variance $1 - \sigma_{e_i}^2$ [Yoo and Goldsmith 2006]. The two phases of the MABC-SPC cooperative protocol are described as follows:

Phase 1: In the first phase of the protocol, both users transmit their messages to the common relay by forming a conventional MIMO multiple-access channel. At the relay node the received signal can be expressed as

$$\mathbf{r}_R = \mathbf{H}_{RA}\mathbf{x}_A + \mathbf{H}_{RB}\mathbf{x}_B + \mathbf{n}_R, \tag{11.1}$$

where $\mathbf{x}_i \in \mathcal{C}^{M \times 1}$ denotes the transmitted message for the ith user, $\mathbf{n}_R \sim \mathcal{CN}(0, \sigma_n^2 \mathbf{I}_M)$ represents a noise vector having ZMCSCG entries of variances σ_n^2 and both users transmit subject to a power constraint trace$(\mathbf{P}_i) \leq P$ where $\mathbf{P}_i \triangleq \mathbb{E}[\mathbf{x}_i \mathbf{x}_i^H]$ denotes the input covariance matrix. Due to the considered channel estimation error at the relay node, the instantaneous rate region is the closure of the convex hull of the set points $(\mathcal{R}_{A_1}, \mathcal{R}_{B_1})$ satisfying [Tse 2005, Section 10.1.2]:

$$\mathcal{R}_{A_1} \leq I_{A_1} \triangleq \log \det\left(\mathbf{I}_M + \frac{\hat{\mathbf{H}}_{RA} \hat{\mathbf{H}}_{RA}^H}{\sigma_n^2 + (\sigma_{\varepsilon_A}^2 + \sigma_{\varepsilon_B}^2)P} \right), \tag{11.2}$$

$$\mathcal{R}_{B_1} \leq I_{B_1} \triangleq \log \det\left(\mathbf{I}_M + \frac{\hat{\mathbf{H}}_{RB} \hat{\mathbf{H}}_{RB}^H}{\sigma_n^2 + (\sigma_{\varepsilon_A}^2 + \sigma_{\varepsilon_B}^2)P} \right), \tag{11.3}$$

$$\mathcal{R}_{A_1} + \mathcal{R}_{B_1} \leq I_{\text{sum}} \triangleq \log \det\left(\mathbf{I}_M + \frac{\hat{\mathbf{H}}_{RA} \hat{\mathbf{H}}_{RA}^H + \hat{\mathbf{H}}_{RB} \hat{\mathbf{H}}_{RB}^H}{\sigma_n^2 + (\sigma_{\varepsilon_A}^2 + \sigma_{\varepsilon_B}^2)P} \right), \tag{11.4}$$

where the above expressions consist of a generalization of the analysis presented in [Yoo and Goldsmith 2006, Eq. (7)].

Phase 2: In the second phase of the protocol, the relay node re-encodes the users' messages using the same or a different codebook with the first phase of the protocol and broadcasts them to both users via an SPC scheme [Tse 2005, Section 6.2.2; Cover and Thomas 2006, Section 15.6]. With SPC, the relay node superposes the sources' messages in the modulation domain [Larsson and Vojcic 2005] (e.g., linear combination of the two signals) with an appropriate power split and broadcasts the resulting signal without further processing. It is worth noting that SPC is an efficient approach in order to boost the broadcast performance and achieve capacity.

Given that each user knows its own transmitted message (self-interference), it suppresses its contribution from the received signal subject to the imperfect channel estimate and decodes the message of the opposite user. More specifically, the SPC scheme is defined as

$$\mathbf{u}_R = \mathbf{G}_B \mathbf{u}_A + \mathbf{G}_A \mathbf{u}_B, \tag{11.5}$$

where $\mathbf{u}_i \in \mathcal{C}^{M \times 1}$ and $\mathbf{G}_i \in \mathcal{C}^{M \times M}$ denote respectively the re-encoded message and a precoding matrix (unitary matrix), respectively, for the ith user. The received signals are expressed as

$$\mathbf{r}_A = \underbrace{\mathbf{H}_{RA}\mathbf{G}_B\mathbf{u}_A}_{\text{self-interference}} + \mathbf{H}_{RA}\mathbf{G}_A\mathbf{u}_B + \mathbf{v}_A, \tag{11.6}$$

$$\mathbf{r}_B = \mathbf{H}_{RB}\mathbf{G}_B\mathbf{u}_A + \underbrace{\mathbf{H}_{RB}\mathbf{G}_A\mathbf{u}_B}_{\text{self-interference}} + \mathbf{v}_B, \tag{11.7}$$

where $\mathbf{r}_i \in \mathcal{C}^{M \times 1}$ represents the receiver signal at ith user and $\mathbf{v}_i \sim \mathcal{CN}(0, \sigma_v^2 \mathbf{I}_M)$ represents a noise vector having ZMCSCG entries of variances σ_v^2. The relay node transmits with a total power P which is split between the two data flows with $\text{trace}(\mathbf{G}_B \mathbf{P}_A' \mathbf{G}_B^H) \leq (1 - \alpha)P$ and $\text{trace}(\mathbf{G}_A \mathbf{P}_B' \mathbf{G}_A^H) \leq \alpha P$, where $\mathbf{P}_i' \triangleq \mathbb{E}[\mathbf{u}_i \mathbf{u}_i^H]$ and $\alpha \in [0\ 1]$ denote the power split factor. Due to imperfect channel estimation, the self-interference terms cannot be perfectly removed from the received signals and affect the rate region computation related to the second phase of the protocol. By using the analysis presented in [Yoo and Goldsmith 2006, Eq. (7)] and taking into account the power split that characterizes the SPC design (for user A the imperfect self-interference suppression corresponds to $\mathbb{E}[\mathbf{E}_A \mathbf{G}_B \mathbf{u}_A] = \sigma_{\epsilon_A}^2 \alpha P$, which is added to the channel estimation error of the main-link $\sigma_{\epsilon_A}^2 (1 - \alpha)P$ [Yoo and Goldsmith 2006] resulting in a total degradation of $\sigma_{\epsilon_A}^2 P$), we can prove that the rates of the second phase of the protocol satisfy the constraints:

$$\mathcal{R}_{A_2} \leq I_{A_2}(\alpha) \triangleq \log \det \left(\mathbf{I}_M + \frac{\hat{\mathbf{H}}_{RA}\mathbf{G}_A\mathbf{P}_B'\mathbf{G}_A^H\hat{\mathbf{H}}_{RA}^H}{\sigma_v^2 + \sigma_{\epsilon_A}^2 P} \right), \tag{11.8}$$

$$\mathcal{R}_{B_2} \leq I_{B_2}(\alpha) \triangleq \log\det\left(\mathbf{I}_M + \frac{\hat{\mathbf{H}}_{RB}\mathbf{G}_B\mathbf{P}'_A\mathbf{G}_B^H\hat{\mathbf{H}}_{RB}^H}{\sigma_v^2 + \sigma_{\varepsilon_B}^2 P}\right). \tag{11.9}$$

By intersecting the above instantaneous rate regions related to the two phases of the protocol and by averaging over many independent fades of the channels [Tse 2005], we define the overall achievable (average) rate region of the system as

$$\mathcal{R}_A \leq \frac{1}{2}\mathbb{E}\left[\min\left[I_{A_1}, I_{B_2}(\alpha)\right]\right], \tag{11.10}$$

$$\mathcal{R}_B \leq \frac{1}{2}\mathbb{E}\left[\min\left[I_{A_2}(\alpha), I_{B_1}\right]\right] \tag{11.11}$$

$$\mathcal{R}_A + \mathcal{R}_B \leq \frac{1}{2}\mathbb{E}\left[I_{\text{sum}}\right], \tag{11.12}$$

where the factor 1/2 represents the rate loss caused by the two phases of the cooperative protocol.

11.3 SPC and Power Split

In this section, we deal with the impact of the CSI on the SPC design and investigate a power-split technique that (a) maximizes the achievable rate region under a user-fairness constraint and (b) maximizes the achievable sum rate.

11.3.1 Conventional SPC without CSI at the Users

The conventional SPC (C-SPC) design does not apply a precoding processing to the bidirectional data at the relay node and therefore corresponds to $\mathbf{G}_i = \mathbf{I}_M$ for $i \in \{A, B\}$. In this case, the power allocated to each data flow is symmetrically distributed among the spatial direction of the MIMO channels. If we use the single-value decomposition (SVD) of the channel matrices $\hat{\mathbf{H}}_{RA}$ and $\hat{\mathbf{H}}_{RB}$, the (instantaneous) maximum user rates during the second phase of the protocol can be written as

$$I_{A_2}^{(\text{C-SPC})}(\alpha) \triangleq \sum_{m=1}^M \log\left(1 + \frac{\alpha P}{M}\frac{\lambda_m^2}{\sigma_v^2 + \sigma_{\varepsilon_A}^2 P}\right), \tag{11.13}$$

$$I_{B_2}^{(\text{C-SPC})}(\alpha) \triangleq \sum_{m=1}^M \log\left(1 + \frac{(1-\alpha)P}{M}\frac{\mu_m^2}{\sigma_v^2 + \sigma_{\varepsilon_B}^2 P}\right), \tag{11.14}$$

where λ_m and μ_m (with $m = 1,...,M$) denote the mth singular values for the channel matrices $\hat{\mathbf{H}}_{RA}$ and $\hat{\mathbf{H}}_{RB}$, respectively.

11.3.2 SPC with CSI at the Users

This SPC technique (called WF-SPC) assumes that imperfect channel knowledge is available to all the nodes of the network and elaborates a WF-PA in each bidirectional link. More specifically, we assume that the relay node has a global knowledge of the estimated channels $\hat{\mathbf{H}}_{RA}$ and $\hat{\mathbf{H}}_{RB}$ (this estimation is performed during the first phase of the protocol) while each user has a local channel knowledge ($\hat{\mathbf{H}}_{Ri}$ is available at ith user). This CSI assumption enables a power allocation (PA) optimization for each bidirectional link and therefore the power assigned to the ith data flow is distributed among the spatial directions (eigenmodes) of the $\hat{\mathbf{H}}_{Ri}$ channel according to the principles of the WF-PA. We note that in the WF-SPC scheme the precoding matrix $\mathbf{G}_i$ is given by the right-hand matrix of the SVD of the $\hat{\mathbf{H}}_{Ri}$ while its left-hand matrix defines the postprocessing matrix applied at the ith user [Hammerstrom et al. 2007; Tse 2005]. By using a similar analysis with the C-SPC scheme, the instantaneous user rates achieved during the second phase of the cooperative protocol with WF-SPC are given as

$$I_{A_2}^{(\text{WF-SPC})}(\alpha) \triangleq \sum_{m=1}^{M} \log\left(1 + \frac{P_m^{(A)}}{\sigma_v^2 + \sigma_{\varepsilon_A}^2 P} \lambda_m^2\right), \tag{11.15}$$

$$I_{B_2}^{(\text{WF-SPC})}(\alpha) \triangleq \sum_{m=1}^{M} \log\left(1 + \frac{P_m^{(B)}}{\sigma_v^2 + \sigma_{\varepsilon_B}^2 P} \mu_m^2\right), \tag{11.16}$$

where $P_m^{(A)}, P_m^{(B)}$, denote the WF power allocated to mth eigenmode of the $R \to A$ and $R \to B$ links, respectively, which are equal to

$$P_m^{(A)} = \left(v_A - \frac{\sigma_v^2 + \sigma_{\varepsilon_A}^2 P}{\lambda_m^2}\right)^+, \tag{11.17}$$

$$P_m^{(B)} = \left(v_B - \frac{\sigma_v^2 + \sigma_{\varepsilon_B}^2 P}{\mu_m^2}\right)^+, \tag{11.18}$$

where $[x]^+ \triangleq \max[0,x]$ and the constant parameters v_A, v_B are chosen to satisfy the total power constraints $\sum_{m=0}^{M} P_m^{(A)} = \alpha P$ and $\sum_{m=0}^{M} P_m^{(B)} = (1 - \alpha)P$, respectively.

11.3.3 Fairness and SC Power Split

As it can be seen from the instantaneous user rates for both the C-SPC and the WF-SPC schemes, the power split between the two data flows is a critical parameter for the overall performance of the system. In this subsection, we wish to identify an SPC power-split technique that jointly maximizes the achievable rate region of the system and supports data-rate fairness between the two users. To focus our analysis on the second phase of the protocol, which refers to the SPC power split, we assume that $I_{A_1} > I_{B_2}(\alpha)$ and $I_{B_1} > I_{A_2}(\alpha)$ for all the channel realizations which indicates that the single-user rate constraints in Equations 11.10 and 11.11 are dominated by the second-phase user rates. In this case, a power split factor α^* that maximizes the minimum instantaneous single-user rate jointly achieves a maximization of the rate region and ensures data-rate fairness between the two users. More specifically, the power split factor that ensures user fairness is defined as

$$\alpha^* = \arg\max_{\alpha \in [0\,1]} \min \left[I_{A_2}(\alpha), I_{B_2}(\alpha) \right], \tag{11.19}$$

where the solution of the above optimization problem requires that both single-user instantaneous rates become equal. Therefore, the optimal power split can be found by solving the following equation in respect to the parameter α^*

$$I_{A_2}(\alpha^*) = I_{B_2}(\alpha^*). \tag{11.20}$$

It is obvious that the above solution requires Equation 11.20 to be solved for the optimal power split for each channel realization, which will increase complexity and is undesirable for real-time applications.

However, as the adopted performance metric is the average achievable data rate, we can approximate the solution of the optimization problem by replacing the instantaneous rates with their average values. More specifically, the SPC power split based on the average rates can be found by equating the average single-user rates and solving the resulting equation in respect to the parameter $\bar{\alpha}^*$

$$\mathbb{E}\left[I_{A_2}(\bar{\alpha}^*) \right] = \mathbb{E}\left[I_{B_2}(\bar{\alpha}^*) \right]. \tag{11.21}$$

For the C-SPC policy, we can solve Equation 11.21 as follows:

$$\mathbb{E}\left[\log \det \left(\mathbf{I}_M + \frac{\bar{\alpha}^* P \, \text{diag}(\hat{\mathbf{H}}_{RA} \hat{\mathbf{H}}_{RA}^H)}{M(\sigma_v^2 + \sigma_{\epsilon_A}^2 P)} \right) \right]$$

$$= \mathbb{E}\left[\log \det \left(\mathbf{I}_M + \frac{(1 - \bar{\alpha}^*) P \, \text{diag}(\mathbf{H}_{RB} \mathbf{H}_{RB}^H)}{M(\sigma_v^2 + \sigma_{\epsilon_B}^2 P)} \right) \right]$$

$$\Rightarrow \mathbb{E}\left[\det\left(\mathbf{I}_M + \frac{\bar{\alpha}^* P \operatorname{diag}(\hat{\mathbf{H}}_{RA}\hat{\mathbf{H}}_{RA}^H)}{M(\sigma_v^2 + \sigma_{\epsilon_A}^2 P)}\right)\right]$$

$$= \mathbb{E}\left[\det\left(\mathbf{I}_M + \frac{(1-\bar{\alpha}^*) P \operatorname{diag}(\hat{\mathbf{H}}_{RB}\hat{\mathbf{H}}_{RB}^H)}{M(\sigma_v^2 + \sigma_{\epsilon_B}^2 P)}\right)\right]$$

$$\Rightarrow \left(1 + \frac{\bar{\alpha}^* P(1-\sigma_{\epsilon_A}^2)}{M(\sigma_v^2 + \sigma_{\epsilon_A}^2 P)}\right)^M = \left(1 + \frac{(1-\bar{\alpha}^*) P(1-\sigma_{\epsilon_B}^2)}{M(\sigma_v^2 + \sigma_{\epsilon_B}^2 P)}\right)^M$$

$$\Rightarrow \bar{\alpha}^* = \frac{(1-\sigma_{\epsilon_B}^2)(\sigma_v^2 + \sigma_{\epsilon_A}^2 P)}{(1-\sigma_{\epsilon_A}^2)(\sigma_v^2 + \sigma_{\epsilon_B}^2 P) + (1-\sigma_{\epsilon_B}^2)(\sigma_v^2 + \sigma_{\epsilon_A}^2 P)}, \qquad (11.22)$$

where $\bar{\alpha}^*$ denotes the optimal power split which can be shown to be equal to $\bar{\alpha}^* = \mathbb{E}[\alpha^*]$ and the operator $\operatorname{diag}(\mathbf{X})$ gives $\mathbf{X}_{i,j} = 0$ if $i \neq j$ (matrix diagonalization). This proposed power split does not require an instantaneous adjustment of the allocated power and requires less complexity than adjusting the power for each channel realization. The simulation results to be presented in Section 11.4 will show that this simplification does not significantly affect the (average) achievable rate region.

Although the above optimal power split given by Equation 11.22 refers to the C-SPC scheme, it is a good approximation for the WF-SPC scheme. The simulations in Section 11.4 show that the above optimal value can be applied to both C-SPC and WF-SPC schemes in order to identify the optimal power split from a user-fairness standpoint. This main observation reveals that the optimal power split can be calculated for the C-SPC technique, whose corresponding equation is simpler than the WF-SPC scheme, and can be used for both SPC schemes.

11.3.4 Optimization

The question now is to find the SPC power split that maximizes the achievable (average) sum rate. Based on the rate region of the SPC scheme given in Equations 11.10 through 11.12, the achievable sum rate is written as [Hammerstrom et al. 2007]

$$\mathcal{R}_A + \mathcal{R}_B \leq \frac{1}{2}\mathbb{E}\left[\min\left[\min[I_{A_1}, I_{B_2}(\alpha)] + \min[I_{A_2}(\alpha), I_{B_1}], I_{\text{sum}}\right]\right], \qquad (11.23)$$

Following the approach of Section 11.3.3, by assuming that $I_{A_1} > I_{B_2}(\alpha)$ and $I_{B_1} > I_{A_2}(\alpha)$, Equation 11.23 can be simplified to

$$\mathcal{R}_A + \mathcal{R}_B \leq \frac{1}{2}\mathbb{E}\left[\min\left[I_{B_2}(\alpha) + I_{A_2}(\alpha), I_{\text{sum}}\right]\right], \qquad (11.24)$$

where the parameter α (power split) affects only the term $I_{B_2}(\alpha) + I_{A_2}(\alpha)$; given that the optimization criterion is the average sum rate, the optimal power split for the second phase of the SPC protocols is given by

$$\overline{\alpha}^* = \arg\max_{\alpha \in [0\,1]} \mathbb{E}\left[\left[I_{B_2}(\alpha) + I_{A_2}(\alpha)\right]\right]. \tag{11.25}$$

The above optimization problem can be solved numerically and the simulations presented in the next section show that the optimal solution is common for both the C-SPC and WF-SPC schemes. It is worth noting that for scenarios where $\mathbb{E}[I_{A_2}(\overline{\alpha}^*)] + \mathbb{E}[I_{B_2}(\overline{\alpha}^*)] > \mathbb{E}[I_{\text{sum}}]$, the achievable sum rate is independent on the power split; I_{sum} dominates the sum rate.

11.4 Numerical Results

Monte Carlo simulations were carried out in order to validate the performance of the proposed schemes. The simulation system follows the system model described in Section 11.2, and the performance metric considered is the achievable average user rate; the simulation parameters are defined in the following discussion for each simulation example and summarized in Table 11.1. It is worth noting that the main contribution of this chapter is to characterize the achievable rate of a MIMO MABC-SPC protocol with imperfect channel estimation for different CSI assumptions and study the optimal power allocation for different optimization targets. The simulation results are in line with this main purpose of the chapter

Table 11.1 Simulation Examples and Corresponding Parameters

Simulation example 1	$\sigma_n^2 = 0.01, \sigma_v^2 = 1, \sigma_{\varepsilon_A}^2 = 0.01, \sigma_{\varepsilon_B}^2 = 0.1,$ $P = 5\,\text{dB}, M = 2,4,6,8$
Simulation example 2	$\sigma_n^2 = 0.01, \sigma_v^2 = 1, \sigma_{\varepsilon_A}^2 = 0.01, \sigma_{\varepsilon_B}^2 = 0.1,$ $P = 30\,\text{dB}, M = 2,4,6,8$
Simulation example 3	$\sigma_n^2 = 0.01, \sigma_v^2 = 1, \sigma_{\varepsilon_B}^2 = 0.1, P = 5,30\,\text{dB},$ $M = 2,4$
Simulation example 4	$\sigma_n^2 = 0.01, \sigma_v^2 = 1, \sigma_{\varepsilon_A}^2 = 0.01, \sigma_{\varepsilon_B}^2 = 0.1,$ $P = 0,5,10,30\,\text{dB}, M = 2,4$

and therefore a comparison of the considered protocol (MIMO MABC-SPC) with other MIMO–MABC approaches is beyond the scope of this work.

Simulation 1: In Figures 11.2 and 11.3, we plot the max–min achievable (average) single-user rate, which is half the sum-rate expression for both users shown in Equation 11.24. This is plotted versus the SPC power split factor for a simulation setup with $\sigma_n^2 = 0.01$, $\sigma_v^2 = 1$, $\sigma_{\epsilon_A}^2 = 0.1$, $\sigma_{\epsilon_B}^2 = 0.1$, $P = 5$ dB and $M = 2, 4, 6, 8$ antennas. We note that these simulation parameters correspond to a low SNR scenario and ensure that the second phase of the cooperative protocol determines the maximum single-user rates (e.g., $\sigma_n^2 > \sigma_v^2 \Rightarrow I_{A_1} > I_{B_2}$ and $I_{B_1} > I_{A_2}$). This simulation scenario is not a symmetric one since $\sigma_{\epsilon_B} > \sigma_{\epsilon_A}^2$, which means that selecting $\alpha = 0.5$ to split the relay transmit power equally between the links is not optimal in this case. More specifically, Figure 11.2 deals with the C-SPC scheme and plots the max–min achievable user rate for different values of α while the maximum average achievable rate for both A and B users as well as the max–min achievable user rate corresponding to an instantaneous power split adjustment α^* are used as reference curves.

As can be seen the power split that maximizes the max–min user performance is equal to $\alpha = 0.42$; a power split value that corresponds to equal maximum average

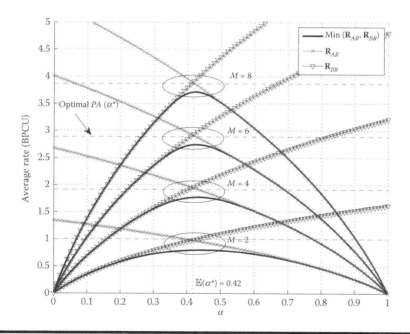

Figure 11.2 Achievable (average) user rates versus α for C-SPC; $\sigma_n^2 = 0.01$, $\sigma_v^2 = 1$, $\sigma_{\epsilon_A}^2 = 0.01$, $\sigma_{\epsilon_B}^2 = 0.1$, $M = 2,4,6,8$ antennas and $P = 5$ dB; the dotted line denotes the achievable user rates for an instantaneous SPC power split (α^*). (Adapted from Krikidis I. and J. S. Thompson, *J. Netw. Comput. Appl.*, 35(1), 510–516, January 2012.)

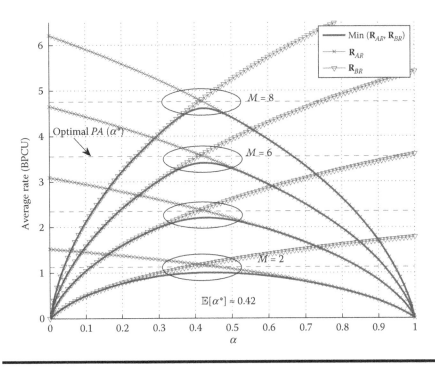

Figure 11.3 Achievable (average) user rates versus α for WF-SPC; $\sigma_n^2 = 0.01, \sigma_v^2 = 1, \sigma_{\epsilon_A}^2 = 0.01, \sigma_{\epsilon_B}^2 = 0.1, M = 2,4,6,8$ antennas and $P = 5$ dB; the dotted line denotes the achievable user rates for an instantaneous SPC power split (α^*). (Adapted from Krikidis I. and J. S. Thompson, *J. Netw. Comput. Appl.*, 35(1), 510–516, January 2012.)

rates for both users. Therefore, by equating the maximum average user rates, we can calculate the optimal power split value; an observation that validates the efficiency of the proposed power-split technique, given in Equation 11.21. Furthermore, the performance of the instantaneous power-split technique (corresponding to α^*) validates the fact that the fixed α value, obtained from the solution of Equation 11.21, is equal to $\mathbb{E}[\alpha^*]$. In addition, we can see that the proposed fixed power-split technique efficiently approximates the optimal max–min performance with a performance loss almost equal to 0.2 bits per channel use (BPCU). However, Figure 11.3 deals with the WF-SPC scheme and shows the impact of the power split on the max–min achievable single-user rate. The first important observation is that the WF-SPC scheme outperforms the C-SPC scheme for the considered low SNR scenario. This observation is in line with conventional MIMO configurations where the WF is beneficial for low SNRs [Tse 2005]. As for the performance of the proposed SPC power-split techniques, we can see that the main observations follow the previously reported conclusions.

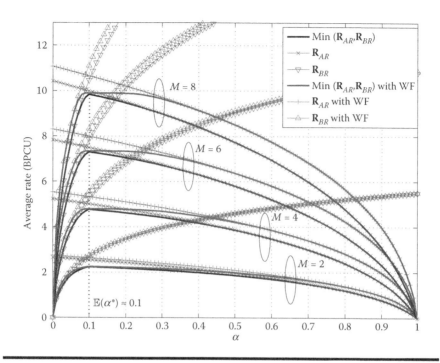

Figure 11.4 **Achievable (average) user rates versus** α **for both the C-SPC and WF-SPC;** $\sigma_n^2 = 0.01, \sigma_v^2 = 1, \sigma_{\epsilon_A}^2 = 0.01, \sigma_{\epsilon_B}^2 = 0.1,$ $M = 2, 4, 6, 8$ **antennas and** $P = 30$ **dB. (Adapted from Krikidis I. and J. S. Thompson,** *J. Netw. Comput. Appl.,* **35(1), 510–516, January 2012.)**

Another significant observation is that the optimal fixed power split value for the WF-SPC scheme is equal with this one used for the C-SPC scheme ($\alpha \approx 0.42$). This result motivates the use of the C-SPC scheme for the computation of the optimal α in order to further reduce the computational complexity. Furthermore, it is worth noting that due to the asymmetric channel estimation error for the $R \to A$, $R \to B$ links, the optimal SPC power split is not symmetric ($\alpha = 0.5$) and assigns more power to the link with the higher channel estimation error ($R \to B$).

Simulation 2: Figure 11.4 follows the parameter settings used in Figures 11.2 and 11.3 but with a fundamental change; $P = 30$ dB which refers to a high SNR scenario. As can be seen from the curves corresponding to the C-SPC and the WF-SPC schemes, both SPC schemes achieve a similar performance. This observation is in line with conventional point-to-point MIMO systems where WF is useful only for the low SNR regime while it converges to a symmetric PA for high SNRs. As for the power-split techniques, we can see that both the instantaneous and average power allocations yield virtually the same data-rate performance.

Simulation 3: In addition, it can be seen that for $P = 30$ dB the optimal fixed power split corresponds to $\alpha \approx 0.1$ and therefore the 90% of the available power is allocated to the $R \rightarrow B$ link that suffers from the higher channel estimation error. It is worth noting that for the low SNR regime the optimal power split factor is closer to the symmetric power allocation as the Gaussian noise dominates the system performance degradation. In addition, in Figure 11.5, we plot the optimal fixed SPC power split ($\mathbb{E}[\alpha^*]$) for both the C-SPC and the WF-SPC schemes versus different channel estimation errors of the $R \rightarrow A$ link for $P = 5$, 30 dB, $\sigma_n^2 = 0.01, \sigma_v^2 = 1, M = 4$ antennas and $\sigma_{\epsilon_B}^2 = 0.1$. The curves show that both SPC schemes correspond to the same optimal power split allocation while more power is allocated to the link with the higher channel estimation error. We note that the optimal SPC power split is more sensitive to high SNR regime and therefore the interval value as well as the variability of $\mathbb{E}[\alpha^*]$ are higher for high SNRs than low SNRs. Furthermore, Figure 11.5 validates the analytical derivation of the optimal power split given by Equation 11.22; as can be seen in the curves corresponding to Equation 11.22 that match the simulation results.

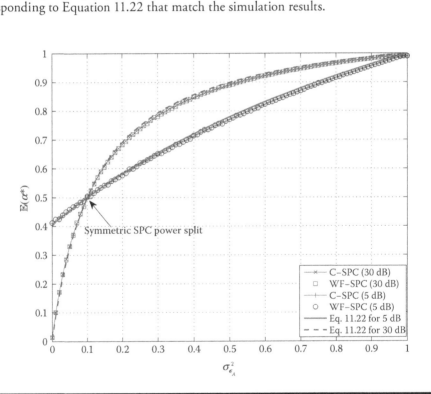

Figure 11.5   The optimal fixed SPC power split versus the channel estimation error of the $R \rightarrow A$ link for both the C-SPC and the WF-SPC; $\sigma_n^2 = 0.01$, $\sigma_v^2 = 1$, $\sigma_{\epsilon_B}^2 = 0.1$, $M = 4$ antennas and $P = 5$, 30 dB. (Adapted from Krikidis I. and J. S. Thompson, *J. Netw. Comput. Appl.*, 35(1), 510–516, January 2012.)

Simulation 4: Figure 11.6 shows the impact of the SPC power split on the achievable (average) sum rate for both the C-SPC and the WF-SPC schemes. More specifically, Figure 11.6 plots the achievable sum rate versus α (power split) for a simulation setup with $\sigma_n^2 = 0.01, \sigma_v^2 = 1, \sigma_{\epsilon_A}^2 = 0.01, \sigma_{\epsilon_B}^2 = 0.1$, $M = 2, 4$ antennas and (a) $P = 0$ dB, (b) $P = 5$ dB, (c) $P = 10$ dB, and (d) $P = 30$ dB; the sum rate $1/2\mathbb{E}[I_{\text{sum}}]$ in Equation 11.12 is also depicted as a reference curve. As can be seen, for the low and intermediate SNR ($P = 0, 5, 10$ dB) the $1/2\mathbb{E}[I_{A_2}(\alpha) + I_{B_2}(\alpha)]$ dominates the achievable sum rate for all α and therefore the SPC power split becomes a critical issue. We can see that the power split that maximizes the achievable sum rate is equal to $\overline{\alpha}^* = 0.6$ for $P = 0$ dB, $\overline{\alpha}^* = 0.56$ for $P = 5$ dB, $\overline{\alpha}^* = 0.55$ for $P = 0$ dB; these values are common for both the C-SPC and WF-SPC schemes. On the other hand, for high SNRs ($P = 30$ dB), although $1/2\mathbb{E}[I_{\text{sum}}]$ dominates the achievable sum rate, the maximization of the $1/2\mathbb{E}[I_{A_2}(\alpha) + I_{B_2}(\alpha)]$ guarantees the

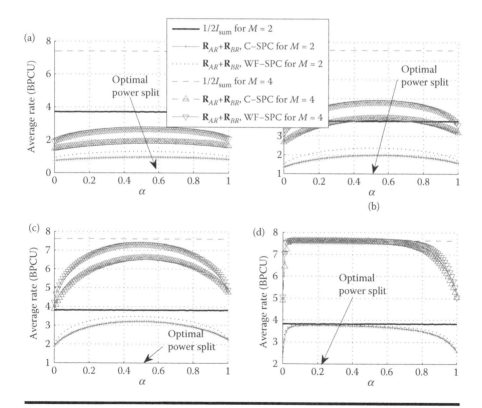

Figure 11.6 Achievable (average) sum rate versus α for both the C-SPC and the WF-SPC ($0.5I_{\text{sum}}$ and) $0.5(I_{A_2}(\sigma) + I_{B_2}(\alpha))$; $\sigma_n^2 = 0.01, \sigma_v^2 = 1, \sigma_{\epsilon_A}^2 = 0.01, \sigma_{\epsilon_B}^2 = 0.1$, $M = 2, 4$ antennas and (a) $P = -0$ dB, (b) $P = -5$ dB, (c) $P = -10$ dB, and (d) $P = 30$ dB. (Adapted from Krikidis I. and J. S. Thompson, *J. Netw. Comput. Appl.*, 35(1), 510–516, January 2012.)

maximization of the achievable sum rate (i.e., $\overline{\alpha}^* = 0.9$ results in a significant loss on the achievable sum-rate performance). As for the C-SPC and WF-SPC schemes, we can see that the WF-SPC scheme outperforms C-SPC at low SNRs, an observation that is in line with our previous remarks.

11.5 Chapter Summary

This chapter investigated the achievable rate region as well as the power-split process for a MIMO bidirectional network with superposition coding relaying and imperfect channel estimation. A power-allocation technique that incorporates the average channel statistics and maximizes the achievable rates under a user-fairness constraint has been described, and it was shown that an optimal power allocation strategy allocates more power to the link with the higher channel estimation error while it is more sensitive at high SNRs. A power split that maximizes the achievable sum rate has also been investigated. It was seen that, in contrast to the fairness target, the power allocation becomes less sensitive to estimation errors and thus the optimal power allocation is close to the symmetric power allocation. Additionally, it was demonstrated that the SPC power allocation is independent of the CSI and therefore both C-SPC and WF-SPC schemes are optimized with the same power split.

References

Chen, M. and A. Yener, Power allocation for F/TDMA multiuser two- way relay networks, *IEEE Trans. Wireless Commun.*, 9, 546–551, February 2010.

Chen, L., L. Cao, X. Zhang, and D. Yang, Integrating network coding and superposition coding in extended two-way relay networks, *in Proc. of the IEEE Glob. Commun. Conf.*, Miami, USA, December 2010, pp. 1481–1486.

Cover, T. M. and J. A. Thomas, *Elements of Information Theory*, John Wiley & Sons, Chischester, UK, 2006.

Ding, Z. and K. K. Leung, Impact of imperfect channel state information on bi-directional communications with relay selection, *IEEE Trans. Sign. Proc.*, 59, 5657–5662, Nov. 2011.

Hammerstrom, I., M. Kuhn, C. Esli, J. Zhao, A. Wittneben, and G. Bauch, MIMO two-way relaying with transmit CSI at the relay, *in Proc. of the IEEE Workshop on Sign. Proc. Adv. Wireless Commun.*, Helsinki, Finland, June 2007, pp. 1–5.

Jia, Y. and A. Vosoughi, Sum-rate maximization of two-way amplify-and-forward relay networks with imperfect channel state information, *in Proc. of the IEEE Inter. Conf. Acoustics Speech Sign. Proc.*, Prague, Czech Republic, May 2011, pp. 2808–2811.

Jiang, B., F. Gao, X. Gao, and A. Nallanathan, Channel estimation and training design for two-way relay networks with power allocation, *IEEE Trans. Wireless Commun.*, 9, 2022–2032, June 2010.

Kim, S., J. P. Mitran, and V. Tarokh, Performance bounds for bidirectional coded cooperation protocols, *IEEE Trans. Inform. Theory*, 54, 5235–5241, November 2008.

Kim, S., J. N. Devroye, P. Mitran, and V. Tarokh, Achievable rate regions and performance comparison of half duplex bi-directional relaying protocols, *IEEE Trans. Inf. Theory*, 57, 6405–6418, Oct. 2011.

Klein, T. E. and R. G. Gallager, Power control for the additive white Gaussian noise channel under channel estimation errors, *in Proc. of the IEEE Inter. Symp. Inf. Theory*, June 2001, p. 304.

Krikidis, I. and J. S. Thompson, MIMO two-way relay channel with superposition coding and imperfect channel estimation, *J. Netw. Comput. Appl.*, 35(1), 510–516, January 2012.

Larsson, E. and B. Vojcic, Cooperative transmit diversity based on superposition modulation, *IEEE Commun. Lett.*, 9, 778–780, September 2005.

Medard, M. The effect upon channel capacity in wireless communications of perfect and imperfect knowledge of the channel, *IEEE Trans. Inf. Theory*, 46(3), 933–946, May 2000.

Oechtering, T. J. and H. Boche, Bidirectional regenerative half-duplex relaying using relay selection, *IEEE Trans. Wireless Commun.*, 7, 1879–1888, May 2008.

Park, M. and S. K. Oh, A hybrid network-superposition coding for asymmetrical two-way relay channels, *in Proc. of the IEEE Vehicular Tech. Conf.*, Anchorage, Alaska, September 2009, pp. 1–5.

Panah, A. and R. W. Heath, Sum-rate of MIMO two-way relaying with imperfect CSI, *in Proc. of the IEEE Inter. Conf. Acoustics Speech Sign. Proc.*, Dallas, TX, March 2010, pp. 3418–3421.

Shannon, C. E. Two-way communications channels, in *4th Berkeley Symp. Math. Stat. Prob.*, Chicago, IL, June 1961, pp. 611–644.

Saleh, A. B., C. Hausl, and R. Kotter, Outage behavior of bidirectional half-duplex relaying schemes, *in Proc. of the IEEE Inform. Theory Workshop*, Taormina, Italy, October 2009, pp. 50–54.

Tse, D. *Fundamentals of Wireless Communication*, Cambridge University Press, Cambridge, UK, 2005.

Yoo, T. and A. Goldsmith, Capacity and power allocation for fading MIMO channels with channel estimation error, *IEEE Trans. Inf. Theory*, 52, 2203–2214, May 2006.

Chapter 12

Physical-Layer Network Coding with Multiple-Antenna Relays

Dirk Wübben, Meng Wu, and Armin Dekorsy

Contents

12.1 Introduction to Bidirectional Communication/ PLNC

In wireless communication, the concept of cooperation is a very promising aspect in realizing enhanced system performance, reliability, and coverage extension. For example, relaying nodes have been introduced to support communication between terminals such that the source transmits its message to the relay, which then forwards a processed message to the destination. As full-duplex relays, that is, relays transmitting and receiving at the same time and frequency, are difficult to implement, it is common to restrict to the application of half-duplex transceivers causing a substantial loss in spectral efficiency. This drawback has to be compensated by the performance improvements achieved by cooperative communication techniques.

In order to compensate this drawback, such a one-way relaying system can be expanded in a bidirectional manner based on the concept of network coding by allowing the intermediate node to perform operations on the incoming data [Ahlswede et al. 2000]. In this two-way relaying system setup, both terminals (also denoted as sources) intend to exchange information with each other with the help of a relay. The direct approach is to let the sources send their messages successively to the relay, and the relay transmits a network-coded signal containing, for example, the bit-level exclusive-or (XOR) of both received messages back to the sources in a third time slot by exploiting the broadcast nature of wireless channels [Fragouli et al. 2006, Katti et al. 2006]. Since both sources are aware of what they have transmitted previously, they can use this *a priori* information to estimate the message of the other source without ambiguity. As a result, doubled amount of data can be maximally transmitted in three time slots compared to one-way relaying requiring two time slots. Furthermore, the number of required time slots can be reduced to two by allowing both sources to transmit simultaneously to the relay. As the information of the sources are then combined during the transmission to the relay, such a scheme is termed physical-layer network coding (PLNC) [Zhang et al. 2006], which maps the receive signal at the relay directly to the network-coded signal. Additionally, an essentially identical concept termed denoise-and-forward (DNF) was proposed in [Popovski and Yomo 2006] and [Koike et al. 2009]. In these literatures, the impact of channel coding is not considered. Extensions to joint consideration of channel decoding and PLNC are referred to [Zhan and He 2008]. More generalized approaches herein requiring modified decoders are introduced in [Zhang and Liew 2009] for repeat accumulate codes and in [Wübben and Lang 2010, Wübben 2010, and Pfletschinger 2011] for low-density parity-check (LDPC) codes with a modified sum-product algorithm (SPA). For convolutional codes jointly considered with PLNC, an extended Trellis diagram is applied [To and Choi 2010]. Furthermore, precoding strategies are investigated in [Schmidt et al. 2013a,b] to combat multipath fading for PLNC.

The above-mentioned literatures are focused on single-antenna relay nodes. When multiple antennas are available at the relay, the increased spatial degrees of freedom basically lead to a multiuser MIMO configuration for the two-way relaying system using PLNC. Therefore, common MIMO processing techniques can be adopted in this context and thus provide a variety of design flexibility. In [Xu et al. 2010], the uplink transmissions from both sources to the relay are treated as a conventional multiuser MIMO channel, where different multilayer MIMO detection techniques are applied. After this separate detection of the source messages, a network-coded signal is generated for broadcasting. The joint consideration of MIMO detection and PLNC is investigated in [Zhang and Liew 2010] and [Chung et al. 2012] for linear detection and in [Zhang et al. 2012] for nonlinear detection. However, the impact of channel coding is ignored in these works. In this chapter, the existing schemes for PLNC in bidirectional communications as well as the MIMO-related extensions are studied and compared in detail with the emphasis on multiple-antenna relays and channel coding.

In bidirectional relaying systems, two source nodes A and B intend to exchange information with each other supported by a relaying node R. The communication consists of two phases as shown in Figure 12.1. In the multiple-access (MA) phase, both sources transmit their packets of the same length to the relay simultaneously, resulting in a superimposed receive signal $\mathbf{y}_R$ at relay R. Upon reception, R estimates a network-coded message $\mathbf{x}_R$ from the receive signal and broadcasts it back to the sources in the broadcast (BC) phase. Here, both sources A and B are assumed to be equipped with a single antenna, whereas the relay is equipped with $J \geq 1$ antennas. It is noted that the direct link between A and B is not available due to the half-duplex constraint.

The binary information words of the same length K for source A and B are denoted as $\mathbf{b}_A$ and $\mathbf{b}_B$, respectively. These two information words are encoded by the same linear channel code Γ with code rate $R_C = K/N$ into binary *source codewords* $\mathbf{c}_A = \Gamma(\mathbf{b}_A)$ and $\mathbf{c}_B = \Gamma(\mathbf{b}_B)$ of length N at the sources. Applying a mapper $\mathcal{M}$ with an M-ary modulation alphabet, the *source codewords* are mapped to the symbol vectors $\mathbf{x}_A = \mathcal{M}\{\mathbf{c}_A\}$ and $\mathbf{x}_B = \mathcal{M}\{\mathbf{c}_B\}$ of length $L = N/m$ with $m = \log_2 M$. Both symbol vectors are transmitted to the relay simultaneously in the MA phase. The symbol vector $\mathbf{x}_A = [x_A(1) \quad x_A(2) \quad \cdots \quad x_A(L)]$ of source A consists of L symbols

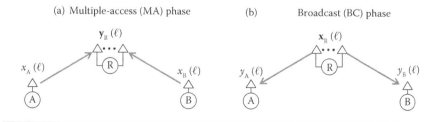

Figure 12.1   A two-way relaying network where two sources A and B exchange information with each other via the relay R. The communication consists of an MA phase and a BC phase.

$x_A(\ell)$, $\ell = 1, 2, \ldots, L$, where the symbol $x_A(\ell) = \mathcal{M}\{c_A(\ell)\}$ depends on m code bits that are collected to the code bit tuple $c_A(\ell) = [c_{A,1}(\ell) \quad c_{A,2}(\ell) \quad \cdots \quad c_{A,m}(\ell)]$. Here, m denotes the number of bits contained in one modulated symbol $x_A(\ell)$ with index $\upsilon = 1, \ldots, m$.

The ℓth receive signal $y_{R,j}(\ell)$ on the jth antenna at relay R, $j = 1,2,\ldots, J$, is given by

$$y_{R,j}(\ell) = h_{A,j}(\ell)x_A(\ell) + h_{B,j}(\ell)x_B(\ell) + n_{R,j}(\ell). \tag{12.1}$$

When flat fading channels are considered, $h_{A,j}(\ell)$ and $h_{B,j}(\ell)$ represent the channel coefficients in time domain. In case of frequency selective fading channels using orthogonal frequency division multiplexing (OFDM), the channel coefficients are defined in frequency domain for the ℓth subcarrier. The additive white Gaussian noise (AWGN) $n_{R,j}(\ell)$ is circularly symmetric complex, which has zero mean and variance σ_n^2.

With the definition of the 2×1 transmit symbol vector $\mathbf{x}(\ell) = [x_A(\ell) \quad x_B(\ell)]^T$, the corresponding $J \times 1$ receive signal vector $\mathbf{y}_R(\ell) = [y_{R,1}(\ell) \quad y_{R,2}(\ell) \quad \cdots \quad y_{R,J}(\ell)]^T$ at relay R is given by

$$\mathbf{y}_R(\ell) = \mathbf{H}(\ell)\mathbf{x}(\ell) + \mathbf{n}_R(\ell), \tag{12.2}$$

where $\mathbf{H}(\ell) = [\mathbf{h}_1(\ell) \quad \mathbf{h}_2(\ell) \quad \cdots \quad \mathbf{h}_J(\ell)]^T$ denotes the $J \times 2$ MIMO channel matrix and $\mathbf{n}_R(\ell)$ denotes the noise vector at R. Note that $\mathbf{h}_j(\ell) = [h_{A,j}(\ell) \quad h_{B,j}(\ell)]^T$ holds. In order to represent the complete receive signals at the relay for $\ell = 1,2, \ldots, L$, we additionally define the receive signal matrix $\mathbf{Y}_R = [\mathbf{y}_R(1) \quad \mathbf{y}_R(2) \quad \cdots \quad \mathbf{y}_R(L)]$ of dimension $J \times L$ that collects the L receive signal vectors.

Upon receiving the superimposed signal, the relay performs an estimation with respect to the XOR of the *source codewords* $\mathbf{c}_{A \oplus B} = \mathbf{c}_A \oplus \mathbf{c}_B$ based on $\mathbf{Y}_R$. This estimated *relay codeword* $\mathbf{c}_R = \hat{\mathbf{c}}_{A \oplus B}$ is further mapped to the symbol vector $\mathbf{x}_R = \mathcal{M}\{\mathbf{c}_R\}$, which is broadcasted toward both sources in the BC phase. In case of multiple antennas at the relay, diversity-exploiting schemes can be applied, for example, orthogonal space-time block code (OSTBC) [Alamouti 1998, Tarokh et al. 1999], space-time trellis code (STTC) [Tarokh et al. 1998], linear dispersion code (LDC) [Hassibi and Hochwald 2002], and so on. For example, assuming that OSTBC is applied in the BC phase while ignoring the data rate loss due to orthogonal code designs for $J > 2$, the equivalent system equation on the ℓth element from relay R to source A is given by [Tarokh et al. 1999]

$$y_A(\ell) = \sqrt{\frac{1}{J}\sum_{j=1}^{J}\left|h'_{A,j}(\ell)\right|^2}\, x_R(\ell) + n_A(\ell), \tag{12.3}$$

where $y_A(\ell)$ represents the receive signal after OSTBC detection. Here, $h'_{A,j}(\ell)$ denotes the channel coefficient for the link from the jth antenna at relay R to source A. As can be observed in Equation 12.3, a diversity gain of order J can be achieved. After reception and OSTBC detection in the BC phase, both source nodes A and B estimate the relay codeword $\hat{\mathbf{c}}_{R,A}$ and $\hat{\mathbf{c}}_{R,B}$ from the receive signals $\mathbf{y}_A = [y_A(1) \quad y_A(2) \quad \cdots \quad y_A(L)]$ and $\mathbf{y}_B = [y_B(1) \quad y_B(2) \quad \cdots \quad y_B(L)]$, respectively. Based on the fact that each source knows what it has transmitted in the MA phase as *a priori* information, the information from the counterpart can be obtained by simply performing XOR operation again at the sources between the estimated relay codeword and this *a priori* knowledge. Specifically, the estimated codeword from source B at source A is achieved by $\hat{\mathbf{c}}_B = \hat{\mathbf{c}}_{R,A} \oplus \mathbf{c}_A$. The estimate $\hat{\mathbf{c}}_A = \hat{\mathbf{c}}_{R,B} \oplus \mathbf{c}_B$ is obtained similarly at source B.

The above decode-forward (DF)-based approach for PLNC requires decoding at relay R to estimate the relay codeword. However, when decoding errors occur at R, that is, $\hat{\mathbf{c}}_{A \oplus B} \neq \mathbf{c}_{A \oplus B}$, these errors will then propagate to the sources in the BC phase, which severely degrades the end-to-end performance of the bidirectional communication. Therefore, the overall system performance is highly dependent on the decoding results at the relay. To this end, we focus on the crucial MA phase in the sequel, where different detection and decoding schemes at the single-antenna or multiple-antenna relay are presented and compared with respect to the error rate performance of the estimated relay codeword.

12.2 APP-Based Detection Schemes

12.2.1 Calculation of APPs

In this section, several *a posteriori* probability (APP)-based detection and decoding schemes are presented to estimate the relay codeword $\mathbf{c}_R$ from the receive signal $\mathbf{Y}_R$ at the relay. We may either estimate the source messages explicitly or directly estimate the relay message. In order to simplify the description, we will first introduce some basic relations between the occurring signals and their corresponding probabilities. Note that the element index ℓ is omitted in the sequel for the sake of simplicity unless otherwise stated.

Since source A and B transmit simultaneously to relay R in the MA phase, each receive signal $y_{R,j}$ at the jth antenna defined in Equation 12.1 is determined by both source messages $x_A = \mathcal{M}\{c_A\}$ and $x_B = \mathcal{M}\{c_B\}$. This leads to M^2 different noise-free receive signals at R

$$s_{AB,j} = h_{A,j}x_A + h_{B,j}x_B, \tag{12.4}$$

where each of these hypothesis is defined by the bit tuples $c_A = [c_{A,1} \quad c_{A,2} \quad \cdots \quad c_{A,m}]$ and $c_B = [c_{B,1} \quad c_{B,2} \quad \cdots \quad c_{B,m}]$. As a short-hand notation, we define the combined bit tuple $c_{AB} = [c_A \quad c_B]$ and introduce the polynomial description with indeterminate D as

$$c_{AB} = [c_{A,1} \quad c_{A,2} \quad \cdots \quad c_{A,m} \quad c_{B,1} \quad c_{B,2} \quad \cdots \quad c_{B,m}]$$
$$= c_{A,1} + \cdots + c_{A,m}D^{m-1} + c_{B,1}D^m + \cdots + c_{B,m}D^{2m-1}. \tag{12.5}$$

Thus, c_{AB} belongs to a Galois field $\mathcal{C}_{AB} = \mathbb{F}_{M^2}$ and $c_{AB} = \mathcal{C}_{AB}(i)$ represents the jth event in $\mathbb{F}_{M^2}$, $0 \le i \le M^2 - 1$. Based on the instantaneous channel knowledge $h_{A,j}$ and $h_{B,j}$, the relations between these notations are shown in Tables 12.1 and 12.2 for BPSK and QPSK modulation as examples, respectively. For the later derivations, we also include the element-wise XOR combination of c_A and c_B given by

$$c_{A \oplus B} = c_A \oplus c_B = [c_{A,1} \oplus c_{B,1} \quad c_{A,2} \oplus c_{B,2} \quad \cdots \quad c_{A,m} \oplus c_{B,m}]. \tag{12.6}$$

It is noted that $j = \sqrt{-1}$ in the tables denotes the imaginary unit. Additionally, graphical examples are presented in Figure 12.2 to illustrate the complex constellation set for the noise-free receive signal $s_{AB,j}$ on the jth antenna at the relay for one fading channel realization. As can be observed, BPSK and QPSK lead to 4 and 16 hypotheses, which correspond to Tables 12.1 and 12.2, respectively. Note that some hypotheses can be completely superimposed in certain channel conditions,

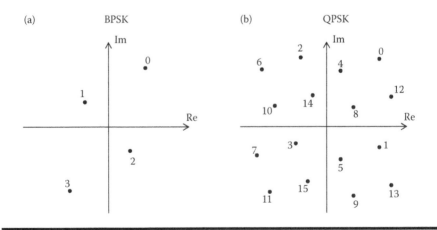

Figure 12.2 Graphical example for the noise-free receive signal constellation set on the jth antenna at the relay for one fading channel realization, (a) for BPSK with $h_{A,j} = 0.8 + 0.6j$, $h_{B,j} = 0.3 + j$, $i = 0,\ldots,3$ and (b) for QPSK with $h_{A,j} = j$, $h_{B,j} = -0.2 + 0.5j$, $i = 0,\ldots,15$.

Table 12.1 Mapping Rules of Code Bit Tuple (c_A, c_B), Transmit Signals (x_A, x_B), and Noise-Free Receive Signal $s_{AB,j}$ on the jth Antenna at Relay R for BPSK Modulation

i	c_A	c_B	$c_{A \oplus B}$	c_{AB}	x_A	x_B	$s_{AB,j}$
0	0	0	0	0	1	1	$h_{A,j} + h_{B,j}$
1	1	0	1	1	−1	1	$-h_{A,j} + h_{B,j}$
2	0	1	1	D	1	−1	$h_{A,j} - h_{B,j}$
3	1	1	0	$1 + D$	−1	−1	$-h_{A,j} - h_{B,j}$

for example, only three hypotheses for BPSK and nine hypotheses for QPSK are present in AWGN channels, as shown in Figure 12.9.

Since all transmitted code symbols are assumed to be equally likely, the *a priori* probabilities for $c_{AB} = \mathcal{C}_{AB}(i)$ are given by

$$\Pr\{c_{AB} = \mathcal{C}_{AB}(i)\} = \frac{1}{M^2}. \tag{12.7}$$

Furthermore, the probability density for $\mathbf{y}_R$ conditioned on the code bit combination $c_{AB} \in \mathcal{C}_{AB}$ with deterministic channel gains can be deduced from AWGN channels as

$$p\{\mathbf{y}_R \mid c_{AB} = \mathcal{C}_{AB}(i)\} = \frac{1}{\left(\pi\sigma_n^2\right)^K} \exp\left\{-\frac{1}{\sigma_n^2} \| \mathbf{y}_R - \mathbf{H}\mathcal{X}(i) \|^2\right\}. \tag{12.8}$$

Here, $\mathcal{X}(i) = \mathcal{M}\{c_{AB}\} = [\mathcal{M}\{c_A\} \quad \mathcal{M}\{c_B\}]$ denotes the ith modulated symbol tuple from the sources given that $c_{AB} = \mathcal{C}_{AB}(i)$ is transmitted. Applying Bayes' rule, the *a posteriori* probability P_i that $c_{AB} = \mathcal{C}_{AB}(i)$ was transmitted, given the current receive signal vector $\mathbf{y}_R$, is

$$P_i = \Pr\{c_{AB} = \mathcal{C}_{AB}(i) \mid \mathbf{y}_R\}$$
$$= p\{\mathbf{y}_R \mid c_{AB} = \mathcal{C}_{AB}(i)\} \frac{\Pr\{c_{AB} = \mathcal{C}_{AB}(i)\}}{\Pr\{\mathbf{y}_R\}}$$
$$= p\{\mathbf{y}_R \mid c_{AB} = \mathcal{C}_{AB}(i)\}\alpha. \tag{12.9}$$

The constant $\alpha = \frac{\Pr\{c_{AB} = \mathcal{C}_{AB}(i)\}}{\Pr\{\mathbf{y}_R\}} = \frac{1}{M^2 \Pr\{\mathbf{y}_R\}}$ can be calculated using the completeness condition $\sum_i P_i = 1$ to normalize the probabilities P_i.

Table 12.2 Mapping Rules of Code Bit Tuple (c_A, c_B), Transmit Signals (x_A, x_B), and Noise-Free Receive Signal $s_{AB,j}$ on the jth Antenna at Relay R for QPSK Modulation

i	c_A	c_B	$c_{A \oplus B}$	c_{AB}	x_A	x_B	$s_{AB,j}$
0	00	00	00	0	1	1	$h_{A,j} + h_{B,j}$
1	10	00	10	1	$\jmath$	1	$\jmath h_{A,j} + h_{B,j}$
2	01	00	01	D	-1	1	$-h_{-A,j} + h_{B,j}$
3	11	00	11	$1 + D$	$-\jmath$	1	$-\jmath h_{A,j} + h_{B,j}$
4	00	10	10	D^2	1	$\jmath$	$h_{A,j} + \jmath h_{B,j}$
5	10	10	00	$1 + D^2$	$\jmath$	$\jmath$	$\jmath h_{A,j} + \jmath h_{B,j}$
6	01	10	11	$D + D^2$	-1	$\jmath$	$-h_{A,j} + \jmath h_{B,j}$
7	11	10	01	$1 + D + D^2$	$-$	$\jmath$	$-\jmath h_{A,j} + \jmath h_{B,j}$
8	00	01	01	D^3	1	-1	$h_{A,j} - h_{B,j}$
9	10	01	11	$1 + D^3$	$\jmath$	-1	$\jmath h_{A,j} - h_{B,j}$
10	01	01	00	$D + D^3$	-1	-1	$-h_{A,j} - h_{B,j}$
11	11	01	10	$1 + D + D^3$	$-\jmath$	-1	$-\jmath h_{A,j} - h_{B,j}$
12	00	11	11	$D^2 + D^3$	1	$-\jmath$	$h_{A,j} - \jmath h_{B,j}$
13	10	11	01	$1 + D^2 + D^3$	$\jmath$	$-\jmath$	$\jmath h_{A,j} - \jmath h_{B,j}$
14	01	11	10	$D + D^2 + D^3$	-1	$-\jmath$	$-h_{A,j} - \jmath h_{B,j}$
15	11	11	00	$1 + D + D^2 + D^3$	$-\jmath$	$-\jmath$	$-\jmath h_{A,j} - \jmath h_{B,j}$

The APPs determine the probabilities of different transmit code bit combinations c_{AB} based on the receive signal $\mathbf{y}_R$. These APPs can be used to calculate log-likelihood ratios (LLRs) for the individual code bits c_A and c_B, or the XORed code bit $c_{A \oplus B}$. In the sequel, the LLRs are obtained from the APPs that facilitate different detection and decoding schemes, which are termed APP-based schemes.

12.2.2 Separated Channel Decoding

The estimation of the source information at the relay can be interpreted as a traditional MA problem, which targets at estimating the individual messages c_A and c_B explicitly by separated channel decoding (SCD). Based on the APPs defined in Equation 12.9, the probability for $c_{A,\upsilon}$, $\upsilon = 1, \ldots, m$, given the receive signal vector y_R, can be calculated as

$$\Pr\{c_{A,\upsilon} = \xi \mid y_R\} = \sum_{i \in \Omega^{\xi}_{A,\upsilon}} P_i,$$
(12.10)

where $\Omega^{\xi}_{A,\upsilon}$ denotes the set of indices with code bit $c_{A,\upsilon}$ equal to $\xi \in \{0,1\}$ according to Tables 12.1 and 12.2. For example, $\Omega^0_{A,1} = \{0, 2\}$ collects all events i with $c_{A,1} = 0$ for BPSK. Correspondingly, the LLR $L_{A,\upsilon}$ for the code bit $c_{A,\upsilon}$ can be formulated as

$$L_{A,\upsilon} = \ln\left(\frac{\Pr\{c_{A,\upsilon} = 0 \mid y_R\}}{\Pr\{c_{A,\upsilon} = 1 \mid y_R\}}\right),$$
(12.11)

which is fed to a soft-input channel decoder for each code bit of the codeword. In this way, the estimated $\hat{c}_A$ for the source codeword transmitted by source A is achieved. In a similar manner, LLR calculation and channel decoding are performed with respect to the codeword from source B, resulting in the estimate $\hat{c}_B$.

Graphical Illustration

In Figure 12.3, graphical examples are given that illustrate the set $\Omega^{\xi}_{A,\upsilon}$ and $\Omega^{\xi}_{B,\upsilon}$ used to calculate LLRs in Equation 12.11 for BPSK. Here, the event sets are defined as $\Omega^0_{A,1} = \{0, 2\}$ and $\Omega^1_{A,1} = \{1, 3\}$ for the calculation of $L_{A,1}$ by

$$L_{A,1} = \ln\left(\frac{P_0 + P_2}{P_1 + P_3}\right),$$
(12.12)

which correspond to the solid lines and dashed lines in the figure, respectively. Similarly, $\Omega^0_{B,1} = \{0, 1\}$ and $\Omega^1_{B,1} = \{2, 3\}$ hold for calculating $L_{B,1}$ leading to

$$L_{B,1} = \ln\left(\frac{P_0 + P_1}{P_2 + P_3}\right).$$
(12.13)

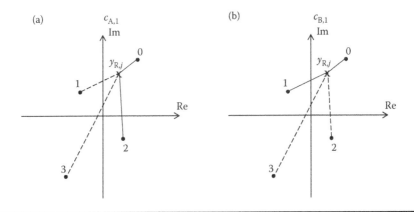

Figure 12.3 **Graphical example of the LLR calculation for SCD with BPSK. The receive signal $y_{R,j}$ on the jth antenna at R is represented by "x." Solid lines and dashed lines correspond to entries $s_{AB,j}$ with code bit equal to 0 and 1, respectively, (a) for code bit $c_{A,1}$ and (b) for code bit $c_{B,1}$, $i = 0, \ldots, 3$.**

It is noted that for QPSK, the graphical illustration for LLR calculations is focused on the first code bit in one modulated symbol, that is, $c_{A,1}$ and $c_{B,1}$, as visualized in Figure 12.4. Specifically, $\Omega_{A,1}^0 = \{0, 2, 4, 6, 8, 10, 12, 14\}$ and $\Omega_{A,1}^1 = \{1, 3, 5, 7, 9, 11, 13, 15\}$ are used to calculate the LLR $L_{A,1}$, whereas $L_{B,1}$ is calculated using $\Omega_{B,1}^0 = \{0, 1, 2, 3, 8, 9, 10, 11\}$ and $\Omega_{B,1}^1 = \{4, 5, 6, 7, 12, 13, 14, 15\}$.

Furthermore, it can be observed in Figure 12.4 for QPSK that SCD will lead to relatively larger LLR for $c_{A,1}$ as the sets $\Omega_{A,1}^0$ and $\Omega_{A,1}^1$ are spatially separated for this

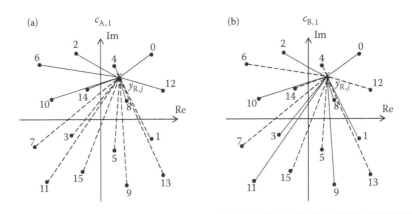

Figure 12.4 **Graphical example of the LLR calculation for SCD with QPSK. The receive signal $y_{R,j}$ on the jth antenna at R is represented by "x." Solid lines and dashed lines correspond to entries $s_{AB,j}$ with code bit equal to 0 and 1, respectively, (a) for code bit $c_{A,1}$ and (b) for code bit $c_{B,1}$, $i = 0, \ldots, 15$.**

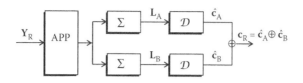

Figure 12.5 Block diagram for parallel separated channel decoding (P-SCD), which estimates c_A and c_B individually and subsequently applies network coding $c_R = \hat{c}_A \oplus \hat{c}_B$.

channel realization. In contrast, $\Omega_{B,1}^0$ and $\Omega_{B,1}^1$ are not well distinguished and thus result in smaller LLRs. A closer look on the LLR distributions taking into account the fading channel characteristic is presented in the upcoming section.

Finally, XOR-based network coding is applied to combine the decoding output vectors and generate the relay codeword as $c_R = \hat{c}_A \oplus \hat{c}_B$, which is then modulated and broadcasted back to the sources in the BC phase.

Note that in the above approach, the source codewords c_A and c_B are decoded in parallel. Therefore, such a separated decoding scheme is termed P-SCD with the block diagram shown in Figure 12.5. Alternatively, the decoding result of the channel with larger signal-to-noise ratio (SNR) can be subtracted from the receive signal and a common decoding for the second codeword with respect to this interference-reduced signal is performed. Such a successive decoding scheme is termed S-SCD.

12.2.3 Joint Channel Decoding and Physical-Layer Network Coding

For PLNC, the functionality of the relay node is to generate a network-coded message c_R from the receive signal $\mathbf{Y}_R$. In order to build such a relay codeword, it is not necessary that the relay has to estimate the source codewords c_A and c_B explicitly as performed by SCD presented in the previous section. Alternatively, the relay codeword $c_R = c_{A\oplus B}$ can be directly estimated from $\mathbf{Y}_R$ without knowing the individual messages c_A and c_B, which is termed joint channel decoding and physical-layer network coding (JCNC) [Zhan and He 2008, Zhang and Liew 2009].

Based on the assumption that both sources apply the same linear channel code, the modulo-2 sum $c_{A\oplus B} = c_A \oplus c_B$ is also a valid codeword of the code Γ. In this case, the APPs for each XOR code bit $c_{A\oplus B,v} = 0$ and $c_{A\oplus B,v} = 1$ can be determined as

$$\Pr\{c_{A\oplus B,v} = \xi \mid \mathbf{y}_R\} = \sum_{i \in \Psi_v^\xi} P_i, \tag{12.14}$$

using the APPs P_i in Equation 12.9. Here, Ψ_v^ξ defines the set of indices with the XOR code bit $c_{A\oplus B,v} = c_{A,v} \oplus c_{B,v}$ equal to $\xi \in \{0,1\}$ For example, the set $\Psi_1^0 = \{0,3\}$ indicates all events i with $c_{A\oplus B,v} = 0$ for BPSK. Thereafter, the corresponding LLR

$$L_{A\oplus B,v} = \ln\left(\frac{\Pr\{c_{A\oplus B,v} = 0 \mid y_R\}}{\Pr\{c_{A\oplus B,v} = 1 \mid y_R\}}\right) \tag{12.15}$$

is fed to the channel decoder, which produces the relay codeword $c_R = \hat{c}_{A\oplus B}$ directly.

Graphical Illustration

The LLR calculation based on the set Ψ_v^ξ is illustrated graphically in Figure 12.6 for both BPSK and QPSK modulation according to the occurrence of the XOR code bit $c_{A\oplus B}$ in Tables 12.1 and 12.2. The event sets $\Psi_1^0 = \{0,3\}$ and $\Psi_1^1 = \{1,2\}$ are used to calculate the LLR $L_{A\oplus B,1}$ for BPSK by

$$L_{A\oplus B,1} = \ln\left(\frac{P_0 + P_3}{P_1 + P_2}\right), \tag{12.16}$$

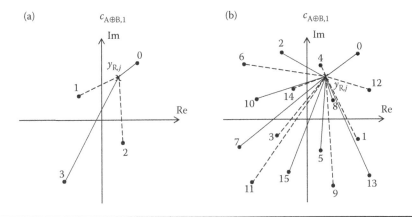

Figure 12.6 Graphical example of the LLR calculation for JCNC. The receive signal $y_{R,j}$ on the *j*th antenna at R is represented by "x." Solid lines and dashed lines correspond to entries $s_{AB,j}$ with the first bit in the XOR code bit tuple $c_{A\oplus B,1}$ equal to 0 and 1, respectively, (a) for BPSK with $i = 0, \ldots, 3$ and (b) for QPSK with $i = 0, \ldots, 15$.

which is only different from the LLR calculations (Equations 12.12 and 12.13) for SCD in summation of the APPs. Contrarily, for QPSK, the LLR of the first bit in the XOR code bit tuple is calculated using $\Psi_1^0 = \{0, 2, 5, 7, 8, 10, 13, 15\}$ and $\Psi_1^1 = \{1, 3, 4, 6, 9, 11, 12, 14\}$. Note that for this channel realization, JCNC will produce relatively smaller LLRs for $c_{A\oplus B}$ since the sets Ψ_1^0 and Ψ_1^1 are not well separated in space. This will also be elaborated and compared to SCD subsequently by considering the distribution of the LLRs.

Finally, the estimated relay codeword $\mathbf{c}_R$ is modulated and broadcasted to both sources in the BC phase. Compared to SCD with two decoding chains, JCNC requires channel decoding only once per MA channel use and, thus, computational efforts are reduced. The block diagram for JCNC is depicted in Figure 12.7.

LLR Distributions

The LLRs (Equations 12.11 and 12.15) calculated using the APPs are fed to a channel decoder for SCD and JCNC, respectively. It is noted that larger amplitude of LLRs provides more reliability and thus is able to benefit more from channel decoding. Therefore, the distributions of the LLR amplitudes of the individual code bit c_A for P-SCD and XOR code bit $c_{A\oplus B}$ for JCNC are investigated as follows. As an example, the cumulative distribution function (CDF) of the absolute LLRs (denoted as |LLR|) at the relay is shown in Figure 12.8 for both AWGN channels and fading channels. QPSK modulation is adopted at both source nodes A and B.

As can be observed in the figure, approximately half of the LLRs for P-SCD are extremely small (close to 0) in AWGN channels. This is because some constellation points in the noise-free receive signal plain are completely superimposed and therefore cannot be distinguished by SCD at all due to this ambiguity. These superimposed points can be observed in Figure 12.9 for BPSK and QPSK modulation. When BPSK is applied at the sources, the transmit symbol tuple $(x_A, x_B) = (1, -1)$ and $(x_A, x_B) = (-1, 1)$ lead to the same hypothesis $s_{AB,j} = 0$ at each antenna. For QPSK, 16 different transmit symbol tuples (x_A, x_B) lead to only 9 points in the complex plain. For example, the events $i = 3, 6, 9, 12$ all result in $s_{AB,j} = 0$, which degrades the performance of SCD significantly. However, these hypotheses correspond to the same XOR code bit tuple $c_{A\oplus B} = [1 \quad 1]$, all of which belong to the set Ψ_1^1 or Ψ_2^1. Therefore, JCNC is immune to this ambiguity and produces larger LLRs compared to SCD as shown Figure 12.8a.

Figure 12.7 Block diagram for joint channel decoding and physical-layer network coding (JCNC), which estimates $\mathbf{c}_R = \hat{\mathbf{c}}_{A\oplus B}$ directly from the receive signal $\mathbf{Y}_R$.

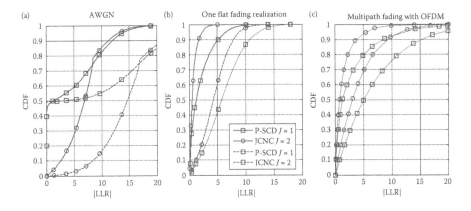

Figure 12.8 |LLR| distributions (CDF) of code bits before decoding for P-SCD and JCNC in (a) AWGN channels, (b) fading channels with one channel realization $h_{A,j} = j$ and $h_{B,j} = -0.2 + 0.5j$ and (c) multipath fading channels using OFDM averaged over different channel realizations, SNR = 5 dB and QPSK modulation.

When fading channels are considered, we investigate the LLR distribution with respect to one flat fading realization used in Figure 12.2, where the channel coefficients lead to distributed noise-free constellation points $s_{AB,j}$ at the relay. In this case, SCD benefits from this spatial separation and thus produces larger LLRs compared to JCNC, as visualized in Figure 12.8b. For practical concerns, frequency-selective fading channels in combination with OFDM are considered, resulting in different constellation maps over the subcarriers. It can be observed in Figure 12.8c that SCD still achieves larger LLR amplitudes compared to JCNC by averaging

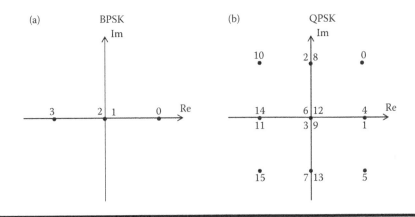

Figure 12.9 Graphical example for the noise-free receive signal constellation set on each antenna at the relay in AWGN channels, (a) for BPSK with $i = 0, \ldots, 3$ and (b) for QPSK with $i = 0, \ldots, 15$.

over different channel realizations. This scenario will also be considered later on in Section 12.4 for performance evaluations. Furthermore, multiple antennas at the relay lead to improved CDF curves while the comparison between SCD and JCNC remains unchanged.

12.2.4 Generalized Joint Channel Decoding and Physical-Layer Network Coding

The separated decoding of both source codewords $\mathbf{c}_A$ and $\mathbf{c}_B$ as well as the joint decoding of the XOR codeword $\mathbf{c}_{A\oplus B}$ from the receive signal $\mathbf{Y}_R$ are operated in binary bit level. Alternatively, the decoding at the relay can be performed with respect to the nonbinary field $\mathbb{F}_{M^2}$ such that the M^2-ary code combination $\mathbf{c}_{AB}$ is estimated from $\mathbf{Y}_R$. Subsequently, PLNC is executed that maps the estimated $\hat{\mathbf{c}}_{AB}$ to the relay codeword $\mathbf{c}_R$. This approach is able to fully exploit the coding gain of both channel codes at the sources and is termed generalized joint channel decoding and physical-layer network coding (G-JCNC) (Wübben and Lang 2010, Wübben 2010].

In order to examine the validity of the G-JCNC scheme operated in nonbinary field, the code structure for $\mathbf{c}_{AB}$ applied to the MA phase is observed. Here, the encoding process is assumed to be slightly different compared to the system defined in Section 12.1, where m identical encoders are applied to the information bit streams $\mathbf{b}_{A,1}$ to $\mathbf{b}_{A,m}$, that is, the encoding process, for example, at source A reads $\mathbf{c}_{A,\upsilon} = [c_{A,\upsilon}(1) \quad c_{A,\upsilon}(2) \quad \cdots \quad c_{A,\upsilon}(L)] = \Gamma(\mathbf{b}_{A,\upsilon})$. In this way, the code bits $c_{A,\upsilon}(\ell)$ with $\upsilon = 1, \ldots, m$ contained in one modulated symbol $x_A(\ell)$ have similar connections in the code structure of the applied channel code Γ. The encoding at source B is performed likewise. Basically, each code bit of a linear channel code consists of the modulo-2 sum of some information bits, for example, the code bit $c_{A,\upsilon}(\ell)$ in the codeword is achieved by the sum of the uth and the wth information bits, that is

$$c_{A,\upsilon}(\ell) = b_{A,\upsilon}(u) \oplus b_{A,\upsilon}(w). \tag{12.17}$$

Since both sources employ the same channel code, the ℓth element in the code bit polynomial can be written as

$$
\begin{aligned}
c_{AB}(\ell) &= [c_{A,1}(\ell) \quad \cdots \quad c_{A,m}(\ell) \quad c_{B,1}(\ell) \quad \cdots \quad c_{B,m}(\ell)] \\
&= [b_{A,1}(u) \oplus b_{A,1}(w) \quad \cdots \quad b_{A,m}(u) \oplus b_{A,m}(w) \quad b_{B,1}(u) \oplus b_{B,1}(w) \quad \cdots \\
&\quad b_{B,m}(u) \oplus b_{B,m}(w)] \\
&= b_{A,1}(u) + \cdots + b_{A,m}(u)D^{m-1} + b_{B,1}(u)D^m + \cdots + b_{B,m}(u)D^{2m-1} \\
&\quad \oplus b_{A,1}(w) + \cdots + b_{A,m}(w)D^{m-1} + b_{B,1}(w)D^m + \cdots + b_{B,m}(w)D^{2m-1} \\
&= b_{AB}(u) \oplus b_{AB}(w).
\end{aligned}
$$

$$\tag{12.18}$$

It can be observed that the code bit polynomial $c_{AB}(\ell)$ is given by the modulo-2 sum of the information bit polynomials $b_{AB}(u) = [b_{A,1}(u) \; \cdots \; b_{A,m}(u) \; b_{B,1}(u) \; \cdots \; b_{B,m}(u)]$ and $b_{AB}(u) = [b_{A,1}(w) \; \cdots \; b_{A,m}(w) \; b_{B,1}(w) \; \cdots \; b_{B,m}(w)]$. Similarly, a code bit equal to the modulo-2 sum of more than 2 information bits can be calculated by nesting the operation in Equation 12.18, which still belongs to $\mathbb{F}_{M^2}$. Correspondingly, channel decoding is performed at the relay with respect to the codeword $\mathbf{c}_{AB}$.

In [Wübben and Lang 2010] and [Wübben 2010], a modified SPA at the relay is applied considering LDPC codes at the sources for the G-JCNC scheme. The modified SPA is based on the investigations in [Kschischang et al. 2001] and [Davey and Mackay 1998]. As shown in Figure 12.10 for the block diagram, instead of the LLRs, the APPs are directly fed to the channel decoder in $\mathbb{F}_{M^2}$ and updated iteratively within the modified SPA. As a result, the APP vector $\mathbf{p} = [p_0 \; p_1 \; \cdots \; p_{M^2-1}]$ with $p_i = \Pr\{c_{AB} = \mathcal{C}_{AB}(i) \mid \mathbf{y}_R\}$ is generated from the decoding algorithm. Finally, PLNC is performed that maps the updated APPs $\mathbf{p}$ to the binary code bit $c_R(\ell)$ in the relay codeword. Exemplarily, the mapping rules are given by

$$c_R(\ell) = \hat{c}_{A\oplus B}(\ell) = \begin{cases} 1 & \text{if } \arg\max_i p_i = \{1,2\} \\ 0 & \text{else} \end{cases} \tag{12.19}$$

and

$$c_R(\ell) = \hat{c}_{A\oplus B}(\ell) = \begin{cases} 00 & \text{if } \arg\max_i p_i = \{0,5,10,15\} \\ 10 & \text{if } \arg\max_i p_i = \{1,4,11,14\} \\ 01 & \text{if } \arg\max_i p_i = \{2,7,8,13\} \\ 11 & \text{if } \arg\max_i p_i = \{3,6,9,12\} \end{cases} \tag{12.20}$$

for BPSK and for QPSK, respectively.

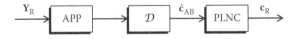

Figure 12.10 Block diagram for generalized joint channel decoding and physical-layer network coding (G-JCNC). The estimation $\hat{c}_{AB}$ is mapped to the corresponding network-coded signal after a channel decoding in $\mathbb{F}_{M^2}$.

Compared to SCD and JCNC, G-JCNC takes into account both channel codes from the sources and therefore is able to fully exploit the available information about the superimposed receive signal $\mathbf{Y}_R$ at the relay. This leads to improved system performance, as can be observed in Section 12.4 later on. Since the decoding process is performed in $\mathbb{F}_{M^2}$ instead of in binary field for SCD and JCNC, higher computational complexity is required for G-JCNC.

12.3 MIMO Detection Techniques

The APP-based detection and decoding schemes presented in the previous section can be applied to systems with an arbitrary number of antennas J at the relay. Specifically, introducing multiple antennas at relay R only influences the APP calculations defined in Equation 12.9, whereas the decoding schemes based on the APPs remain unchanged. However, when equipping $J > 1$ antennas at R, the MA phase corresponds to a $2 \times J$ MIMO channel, which allows the application of MIMO processing techniques. In the sequel, several common multilayer MIMO detection techniques are applied to two-way relaying networks with two time slots, which estimate $\mathbf{c}_A$ and $\mathbf{c}_B$ explicitly from the received signal $\mathbf{Y}_R$. Network coding is then applied to generate the relay codeword $\mathbf{c}_R = \hat{\mathbf{c}}_A \oplus \hat{\mathbf{c}}_B$ for broadcasting. Although common MIMO detection techniques also fall into the category "separated channel decoding" presented in Section 12.2.2, since $\mathbf{c}_A$ and $\mathbf{c}_B$ are individually estimated, the term "SCD" still indicates the APP-based scheme exclusively. Note that the MIMO detection techniques are applied independently to PLNC in this chapter. Alternative schemes with joint consideration of MIMO detection and PLNC are referred to [Zhang and Liew 2010, Chung et al. 2012, and Zhang et al. 2012].

12.3.1 Linear Equalization

For linear equalization (LE) schemes, the receive signal vector $\mathbf{y}_R$ is filtered by a $2 \times J$ matrix $\mathbf{G}$, resulting in the filtered signal vector $\tilde{\mathbf{x}}_R = [\tilde{x}_A \ \tilde{x}_B]^T = \mathbf{G}\mathbf{y}_R$. The filter matrix $\mathbf{G}$ for linear detectors employing the zero-forcing (ZF) or minimum mean squared error (MMSE) criteria are given as

$$\mathbf{G} = \begin{cases} \mathbf{H}^+ = (\mathbf{H}^H\mathbf{H})^{-1}\mathbf{H}^H & \text{ZF} \\ \underline{\mathbf{H}}^+ = (\underline{\mathbf{H}}^H\underline{\mathbf{H}})^{-1}\underline{\mathbf{H}}^H & \text{MMSE} \end{cases}. \tag{12.21}$$

Here, the operator $(\cdot)^+$ denotes the pseudo-inverse of a matrix. Note that MMSE corresponds to ZF for the extended system, which has an extended channel matrix $\underline{\mathbf{H}}$ of dimension $(J + 2) \times 2$

$$\underline{\mathbf{H}} = \begin{bmatrix} \mathbf{H} \\ \sigma_n \mathbf{I}_2 \end{bmatrix}. \tag{12.22}$$

Furthermore, the receive signal vector also has to be extended for MMSE to fit the dimension as

$$\underline{\mathbf{y}}_R = \begin{bmatrix} \mathbf{y}_R \\ \mathbf{0}_{2\times1} \end{bmatrix}. \tag{12.23}$$

The corresponding estimation errors of different layers are determined by the diagonal elements of the error covariance matrix $\boldsymbol{\Phi}$, which is given by

$$\boldsymbol{\Phi} = \begin{cases} \sigma_n^2(\mathbf{H}^H\mathbf{H})^{-1} & \text{ZF} \\ \sigma_n^2(\underline{\mathbf{H}}^H\underline{\mathbf{H}})^{-1} & \text{MMSE} \end{cases}. \tag{12.24}$$

Applying the filtering matrix to the receive signal vector, the filtered signal vector $\tilde{\mathbf{x}}_R = \mathbf{H}^+\mathbf{y}_R$ for ZF or $\tilde{\mathbf{x}}_R = \underline{\mathbf{H}}^+\underline{\mathbf{y}}_R$ for MMSE of dimension 2×1 contains the estimated messages for x_A and x_B in each layer, respectively. When ZF is performed, the cross-layer interference is removed completely with greater noise amplification. For MMSE, a compromise between the interference and the amplified noise is achieved. Subsequently, the filtered signal $\tilde{\mathbf{x}}_R$ is demodulated and decoded for both layers to estimate the source codewords $\hat{\mathbf{c}}_A$ and $\hat{\mathbf{c}}_B$, separately. This further leads to the relay codeword $\mathbf{c}_R = \hat{\mathbf{c}}_A \oplus \hat{\mathbf{c}}_B$ for broadcasting.

12.3.2 Successive Interference Cancellation

The linear estimation schemes detect different layers in parallel. Alternatively, the layers can be detected successively, where the cross-layer interference resulting from the layer already detected in the layer to be detected is reconstructed and eliminated. Such a nonlinear detection technique is termed successive interference cancellation (SIC). In this section, we emphasize on SIC based on QR decomposition (QRD) of the channel matrix [Wübben et al. 2003].

Employing QR matrix decomposition, the $2 \times J$ channel matrix $\mathbf{H}$ in the MA phase can be decomposed into $\mathbf{H} = \mathbf{QR}$, where $\mathbf{Q}$ denotes a $J \times 2$ matrix with orthonormal columns and $\mathbf{R}$ denotes a 2×2 upper triangular matrix. The matrix $\mathbf{Q}^H$ is used as the filter matrix, yielding $\tilde{\mathbf{x}}_R = \mathbf{Q}^H\mathbf{y}_R$. Owing to the upper triangular structure of $\mathbf{R}$, the second layer of the filtered signal vector $\tilde{\mathbf{x}}_R$ is free of interference. This estimated symbol is decoded and then used to reconstruct the interference contributing to the first layer. After subtracting this cross-layer interference, common decoding is performed to the first layer. In this way, the source codewords $\mathbf{c}_A$ and $\mathbf{c}_B$ are estimated successively, which are further network coded to generate the relay codeword $\mathbf{c}_R$ broadcasted in the BC phase. Note that the above SIC approach with QRD is based on the ZF criterion as the interference is completely removed in the second layer. When the MMSE criterion is considered, QRD is applied to the extended channel matrix as $\underline{\mathbf{H}} = \underline{\mathbf{Q}}\underline{\mathbf{R}}$ while the subsequent operations are performed according to $\underline{\mathbf{Q}}$ and $\underline{\mathbf{R}}$ instead.

Since erroneous decisions of the layer detected first will propagate to the second detection step, the performance of SIC can be improved by first detecting the layer with higher reliability. Note that for OFDM systems with channel coding applied to each OFDM symbol, the same detection order is required on all subcarriers. For such an ordered SIC (OSIC) detection scheme, we take the ordering criteria based on SINR optimization from [Wübben and Kammeyer 2006] for performance evaluations later on. With this concern, the layer with the smaller averaged estimation error $\bar{\phi}_\kappa$ over all subcarriers of one OFDM symbol is detected first. Here, $\bar{\phi}_\kappa$ is defined as

$$\bar{\phi}_\kappa = \frac{1}{L}\sum_{\ell=1}^{L}[\mathbf{\Phi}(\ell)]_{\kappa,\kappa}, \tag{12.25}$$

with $\mathbf{\Phi}(\ell)$ denoting the error covariance matrix on the ℓth subcarrier defined in Equation 12.24. The operator $[\cdot]_{\kappa,\kappa}$ indicates the κth diagonal term of a squared matrix with $\kappa = 1,2$.

12.4 Performance Analysis

For performance evaluations, a two-phase two-way relaying system is considered where the relay is located in the middle of the two sources and the three nodes are on a line. In the link-level simulations, all links are assumed to be multipath Rayleigh block fading with $N_\mathrm{H} = 5$ normalized equal power taps in time domain. In order to ease the equalization efforts for such frequency selective channels, OFDM is applied for all transmissions, where each OFDM symbol is individually encoded by an optimized LDPC code and contains $L = 1024$ subcarriers with QPSK modulation. The optimized degree distributions of the LDPC code are achieved from [Urbanke 2010] and the progressive edge growth (PEG) algorithm [Hu et al. 2005] is performed to construct the parity-check matrix. This code with both a medium code rate ($R_\mathrm{C} = 0.5$) and a high code rate ($R_\mathrm{C} = 0.875$) is considered. Additionally, a maximum number of 100 iterations are employed for both binary and nonbinary channel decoding. The iterative decoding process is terminated when the parity-check condition is fulfilled or the maximum number of iterations is reached. For performance evaluations, the dependency of the frame error rate (FER) for the relay codeword on SNR $= 1/\sigma_n^2$ assuming normalized transmit power at both sources is presented for the detection and decoding schemes investigated in Sections 12.2 and 12.3.

12.4.1 Single-Antenna Relay

For the single-antenna relay scenario, the FER performance of the relay codeword for the APP-based detection and decoding schemes presented in Section 12.2 is shown in Figure 12.11. It can be observed that G-JCNC achieves the best performance due to full exploitation of both channel codes from the sources. When an LDPC code of medium rate $R_C = 0.5$ is applied, P-SCD performs approximately 1 dB better than JCNC. This is because P-SCD produces for the individual code bits LLRs with larger amplitude compared to JCNC for the XOR code bits, as demonstrated by the LLR distributions in Figure 12.8c. Therefore, P-SCD is more capable of exploiting the coding gain from the frequency selective fading channel environment compared to JCNC. Furthermore, S-SCD improves the performance by approximately 2 dB in contrast to P-SCD and is still 1 dB worse than G-JCNC. For a high code rate scenario, for example, $R_C = 0.875$, JCNC approaches P-SCD with a very slight performance degradation as shown in Figure 12.11b, whereas G-JCNC still achieves tremendous performance gain.

12.4.2 Multiple-Antenna Relay

When the relay employs multiple antennas, the common multilayer MIMO detection techniques can be applied in addition to the APP-based schemes to estimate the source messages separately, which is followed by network coding as $c_R = \hat{c}_A \oplus \hat{c}_B$. In Figure 12.12, the FER performance at the relay is presented for several common multilayer MIMO detection techniques demonstrated in Section 12.3 with $J = 2$ antennas at the relay. It is shown that MMSE outperforms ZF and SIC outperforms LE in general. It is also well known and verified in the figure that the MMSE-OSIC scheme with ordering criterion (12.25) achieves significant performance gain in

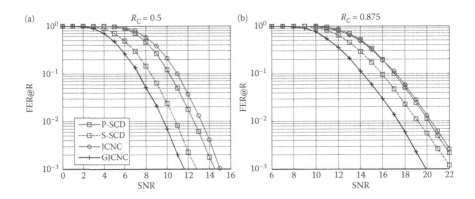

Figure 12.11 FER performance of the estimated relay codeword for APP-based schemes. The relay is equipped with $J = 1$ antenna. The code rate is $R_C = 0.5$ in (a) and $R_C = 0.875$ in (b).

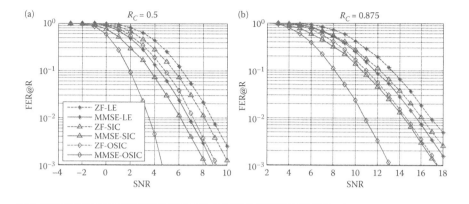

Figure 12.12 **FER performance of the estimated relay codeword for common multilayer MIMO detection techniques. The relay is equipped with $J = 2$ antennas. The code rate is $R_C = 0.5$ in (a) and $R_C = 0.875$ in (b).**

contrast to the other schemes for both the medium and high code rate scenarios, which is employed for comparison with the APP-based schemes subsequently.

In Figure 12.13, the FER performance of the APP-based schemes is compared to MMSE-OSIC when the relay is equipped with $J = 2$ antennas. In contrast to Figure 12.11, the performance loss of P-SCD to G-JCNC is getting smaller with increasing number of antennas at the relay. Furthermore, MMSE-OSIC approaches G-JCNC for the medium code rate case with $R_C = 0.5$ but requires less computational complexity. However, the performance degrades significantly for the high

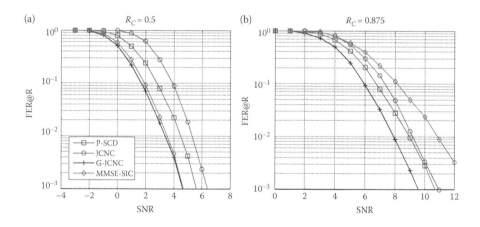

Figure 12.13 **FER performance of the estimated relay codeword for APP-based schemes and MMSE-OSIC. The relay is equipped with $J = 2$ antennas. The code rate is $R_C = 0.5$ in (a) and $R_C = 0.875$ in (b).**

rate scenario, which indicates that the code rate results in dramatic influence on the system performance for different schemes.

Acknowledgment

This work was supported in part by the German Research Foundation (DFG) under grant Wu 499/8-1 within the priority program "Communication in Interference Limited Networks (COIN)," SPP 1397.

References

Ahlswede, R., Cai, N., Li, S.-Y.R., and Yeung, R.W., Network information flow, *IEEE Transactions on Information Theory*, 46(4), 1024–1016, July 2000.

Alamouti, S.M., A simple transmit diversity technique for wireless communications, *IEEE Journal on Selected Areas in Communications*, 16(8), 1451–1458, October 1998.

Chung, H.-H., Kuo, S.-H., and Lin, M.-C., A physical-layer network coding scheme based on linear MIMO detection, *IEEE 75th Vehicular Technology Conference (VTC'12— Spring)*, Yokohama, Japan, May 2012.

Davey, M.C. and Mackay, D., Lowe density parity check codes over GF(q), *IEEE Communications Letters*, 2(6), 165–167, June 1998.

Fragouli, C., Boudec, J.Y., and Widmer, J., Network coding: An instant primer, *ACM SIGCOMM Computer Communication Review*, 36(1), 63–68, March 2006.

Hassibi B. and Hochwald, B.M., High-rate codes that are linear in space and time, *IEEE Transactions on Information Theory*, 48(7), 1804–1824, July 2002.

Hu, X.-Y., Eleftheriou, E., and Arnold, D.M., Regular and irregular progressive edge-growth tanner graphs, *IEEE Transactions on Information Theory*, 51(1), 386–398, January 2005.

Katti, S.R., Rahul, H., Hu, W., Katabi, D., Medard, M., and Crowcroft, J., XORs in the air: Practical wireless network coding, *Conference on Applications, Technologies, Architectures, and Protocols for Computer Communications (ACM SIGCOMM)*, Pisa, Italy, September 2006.

Koike-Akino, T., Popovski, P., and Tarokh, V., Optimized constellations for two-way wireless relaying with physical network coding, *IEEE Journal on Selected Areas in Communications*, 27(5), 788–796, October 2009.

Kschischang, F.R., Frey, B.J., and Loeliger, H.A., Factor graphs and the sum-product algorithm, *IEEE Transactions on Information Theory*, 47(2), 498–519, February 2001.

Pfletschinger, S., A practical physical-layer network coding scheme for the uplink of the two-way relay channel, *45th Asilomar Conference on Signals, Systems and Computers (ASILOMAR'101)*, Monterey, California, USA, November 2011.

Popovski, P. and Yomo, H., The antiPackets can increase the achievable throughput of a wireless multiHop network, *IEEE International Conference on Communications (ICC'06)*, Istanbul, Turkey, June 2006.

Schmidt, A., Gerstacker, W., and Schober, R., Maximum SNR transmit filtering for decision-feedback equalization in physical layer network coding, *International Conference on Systems, Communications and Coding (SCC'13)*, Munich, Germany, January 2013a.

Schmidt, A., Gerstacker, W., and Schober, R., Optimal Tomlinson–Harashima precoding for physical layer network coding, *IEEE Wireless Communications and Networking Conference (WCNC'13) Workshop on New Advances for Physical Layer Network Coding*, Shanghai, China, April 2013b.

Tarokh, V., Jafarkhani, H., and Calderbank, A.R., Space-time block codes from orthogonal designs, *IEEE Transactions on Information Theory*, 45(5), 1456–1467, July 1999.

Tarokh, V., Seshadri, N., and Calderbank, A.R., Space-time codes for high data rate wireless communication: Performance criterion and code construction, *IEEE Transactions on Information Theory*, 44(2), 744–765, March 1998.

To, D. and Choi, J., Convolutional codes in two-way relay networks with physical-layer network coding, *IEEE Transactions on Wireless Communications*, 9(9), 2724–2729, September 2010.

Urbanke, R., Degree distribution optimizer for LDPC code ensembles, http://ipgdemos.epfl. ch/ldpcopt/, 2010.

Wübben, D., Joint channel decoding and physical-layer network coding in two-way QPSK relay systems by a generalized sum-product algorithm, *7th International Symposium on Wireless Communication Systems (ISWCS'10)*, York, UK, September 2010.

Wübben, D., Böhnke, R., Kühn, V., and Kammeyer, K.-D., MMSE extension of V-BLAST based on sorted QR decomposition, *IEEE 58th Vehicular Technology Conference (VTC'03—Fall)*, Orlando, Florida, USA, October 2003.

Wübben, D. and Kammeyer, K.-D., Low complexity successive interference cancellation for per-antenna-coded MIMO-ODFM schemes by applying Parallel-SQRD, *IEEE 63th Vehicular Technology Conference (VTC'06—Spring)*, Melbourne, Australia, May 2006.

Wübben, D. and Lang, Y., Generalized sum-product algorithm for joint channel decoding and physical-layer network coding in two-way relay systems, *IEEE Global Telecommunications Conference (GLOBECOM'10)*, Miami, Florida, USA, December 2010.

Xu, D., Bai, Z., Waadt, A., Bruck, G.H., and Jung, P., Combining MIMO with network coding: A viable means to provide multiplexing and diversity in wireless relay networks, *IEEE International Conference on Communications (ICC'10)*, Cape Town, South Africa, May 2010.

Zhan, A. and He, C., Joint design of channel coding and physical network coding for wireless network, *International Conference on Neural Networks and Signal Processing*, Zhejiang, China, June 2008.

Zhang, S. and Liew, S.C., Channel coding and decoding in a relay system operated with physical-layer network coding, *IEEE Journal on Selected Areas in Communications*, 27(5), 788–796, October 2009.

Zhang, S. and Liew, S.C., Physical-layer network coding with multiple antennas, *IEEE Wireless Communications and Networking Conference (WCNC'10)*, Sydney, Australia, April 2010.

Zhang, S., Liew, S.C., and Lam, P., Hot topic: Physical layer network coding, *International Conference on Mobile Computing and Networking (MobiCom'06)*, Los Angeles, CA, USA, September 2006.

Zhang, S., Nie, C., Lu, L., Zhang, S., and Qian, G., MIMO physical-layer network coding based on V-BLAST detection, *2012 International Conference on Wireless Communications & Signal Processing (WCSP'12)*, Huangshan, China, October 2012.

Index